D0821639

Plasma Etching

An Introduction

Plasma–Materials Interactions

A Series Edited by

A list of titles in this series appears at the end of this volume.

Plasma Etching

An Introduction

Edited by

Dennis M. Manos
Plasma Physics Laboratory
Princeton University
Princeton, New Jersey

Daniel L. Flamm
AT & T Bell Laboratories
Murray Hill, New Jersey

ACADEMIC PRESS, INC.
Harcourt Brace Jovanovich, Publishers
Boston San Diego New York
London Sydney Tokyo Toronto

This book is printed on acid-free paper. ♾

ACADEMIC PRESS, INC.
1250 Sixth Avenue, San Diego, CA 92101-4311

United Kingdom Edition published by
ACADEMIC PRESS LIMITED
24–28 Oval Road, London NW1 7DX

Library of Congress Cataloging-in-Publication Data

Plasma etching.
(Plasma: materials interactions)
Bibliography: p.
Includes index.
1. Plasma etching. I. Manos, Dennis M. II. Flamm, Daniel L. III. Series: Plasma.
TA2020.P5 1988 621.044 87-37419
ISBN 0-12-469370-9

Alkaline paper

Printed in the United States of America
92 93 94 95 96 EB 9 8 7 6 5 4 3

Contents

Contributors

Numbers in parentheses refer to the pages on which the authors' contributions begin.

SAMUEL A. COHEN (185), *Plasma Physics Laboratory, Princeton University, Princeton, New Jersey 08544*

H. F. DYLLA (259), *Plasma Physics Laboratory, Princeton University, Princeton, New Jersey 08544*

DANIEL L. FLAMM (1, 91), *University of California at Berkeley, Berkeley, California 94720 and AT&T Bell Laboratories, 600 Mountain Ave., Murray Hill, NJ 07974-2070*

JAMES M. E. HARPER (391), *IBM Thomas J. Watson Research Center, Yorktown Heights, New York 10598*

G. K. HERB (1, 425), *AT&T Bell Laboratories, Allentown, Pennsylvania 18103*

DENNIS M. MANOS (259), *Plasma Physics Laboratory, Princeton University, Princeton, New Jersey 08544*

ALAN R. REINBERG (339), *Perkin-Elmer Corporation, Norwalk, Connecticut 06859*

Preface

The material contained in this book grew out of a series of lectures that comprised a short course in plasma chemistry, which we have taught in Princeton, Washington, Orlando, and Palo Alto over the last five years. This book is designed to serve a broad audience with widely divergent educational backgrounds. We hope that those who are new to the field of dry processing will find the tutorial style useful. Those who are familiar with some aspects of dry processing should find the in-depth treatments interesting and helpful.

Most of our students wanted a text or reference book that covered the course material but, unfortunately, nothing appropriate was available. This new Academic Press series, "Plasma-Materials Interactions," became a logical focus for such a project and stimulated this volume. In *Plasma Etching: An Introduction* experts and teachers cover the subject in seven chapters.

In the first chapter, a broad overview of the entire field of plasma etching is given. Here, virtually all of the material in the subsequent chapters is briefly introduced, related to material in other chapters, and placed in perspective for practitioners in the field. The terminology of the field, fundamental definitions, and concepts of plasma physics and chemistry are given in a simple, non-mathematical manner so that they will seem familiar in the more detailed treatments of the following chapters. Although any chapter can stand alone, each chapter does tend to draw on material presented in previous chapters, and we strongly recommend that the novice begin with a reading of at least the first chapter before moving on.

The next two chapters in the book are designed to give the reader a thorough presentation of the fundamentals of plasma physics and chemistry. Chapter 2 conveys the important fundamental chemistry involved in the etching of silicon-based materials, including resists. The effects of pressure, frequency, and temperature on the chemical processes are discussed here. Chapter 3 provides a thorough grounding in the plasma physics of reactors, including those that contain magnetic fields. The reader will leave these chapters familiar with the arcane nomenclature of plasma processing and comfortable with the concepts that this nomenclature sometimes fails to convey.

Chapter 4 presents a detailed treatment of several of the most important types of diagnostics used in process plasma research and development and in the production of ULSI devices. These include established diagnostics, such as electrostatic probes and emission spectroscopy, as well as the newer techniques of microwave interferometry and laser-induced fluorescence. The chapter serves as a bridge between the theory developed in the first two chapters and the practical applications that follow in chapters 5 through 7.

Chapter 5 reviews etching equipment geometries and operation. It begins with an analysis of the relative merits and drawbacks of various etch configurations and leads to a set of detailed recipes for etching a wide variety of materials. A discussion of methods for line profile control is also presented. Chapter 6 discusses the design, construction, and operation of broad-beam ion sources in the fabrication of silicon devices. Methods for combining etching with growth or deposition are presented. Chapter 7 presents an analysis of the various safety concerns associated with the production line in semiconductor fabrication. Techniques for proper handling of gases—from the inlet to the exhaust stack—are reviewed in detail.

Throughout the book we attempt to use examples that have the broadest significance to the topic of discussion. In this way, we hope the book will remain current as the specifics of production practice change at an accelerated pace owing to the ferocious competition that characterizes the industry.

We would like to acknowledge the constructive comments and encouragement of our colleagues and the many students who have participated in the course over the years. We hope you like the result.

Dennis M. Manos

Daniel L. Flamm

1 Plasma Etching Technology—An Overview

Daniel L. Flamm[a] and G. Kenneth Herb[b]
AT&T Bell Laboratories
[a]*Murray Hill, New Jersey*
[b]*Allentown, Pennsylvania*

ISBN 0-12-469370-9

I. Introduction

The rapid development of plasma etching technology was stimulated by its application to the manufacture of microelectronic devices. Since the early 1970s, when plasma etching was first widely adapted to device manufacturing, a new recognition and understanding of plasma chemistry has emerged. Today, state-of-the-art integrated circuit manufacture depends on the mass replication of tightly controlled, micron-sized features in a variety of materials. Plasma etching has become central to this process because it is the *only* current technology that can do this job efficiently and with high yield. Conversely, most basic research in plasma etching has been done using electronic materials and with the aid of microfabrication. To understand this interplay, we briefly turn to the evolution of microelectronic fabrication.

A. DEVELOPMENT OF MICROELECTRONICS

Bardeen, Brattain and Shockley could not anticipate that the transistor they invented in 1947 [1, 2] would influence virtually every aspect of human life. The point-contact transistor, shown in Fig. 1, was rapidly followed by Shockley's theory of the p-n junction and the junction transistor [3]. Two years later, Shockley, M. Sparks and G. K. Teal verified Shockley's theoretical predictions and produced the first junction-transistor amplifier [4]. This work formed the basis for all the transistor technology that was to follow and spurred the development of integrated circuits invented by J. Kilby and R. Noyce [5].

The use of SiO_2 masks for controlling silicon doping by impurities was invented by C. J. Forsch and L. Derick in 1955 [6]. Oxide masking evolved as a key step in the manufacture of silicon integrated circuits. An early integrated circuit (1961 design of Fairchild Camera) is shown in Fig. 2. These first circuits typically had 25–40 micrometer feature dimensions and,

FIGURE 1. The first point contact transistor made at Bell Laboratories in 1947 (magnified 2X).

as the photograph in Fig. 2 clearly shows, the defect density [7] was high. Better processes with lower defect densities were essential to the unprecedented advance of this technology over the next 25 years.

Figure 3 shows how integrated circuits have evolved since then and suggests what the future may offer. Gordon Moore observed that the number of individual components per circuit *and* the component density of IC designs doubled every year [8]. This yearly doubling rate held true from the first IC in 1960 at Texas Instruments until just recently, when the pace slowed somewhat. The Moore projection would put a 4,000,000 component device in existence in 1983, and this functionality finally appeared as the 4 megabit dynamic memory design announced in 1985 [9, 10]. This 25-year

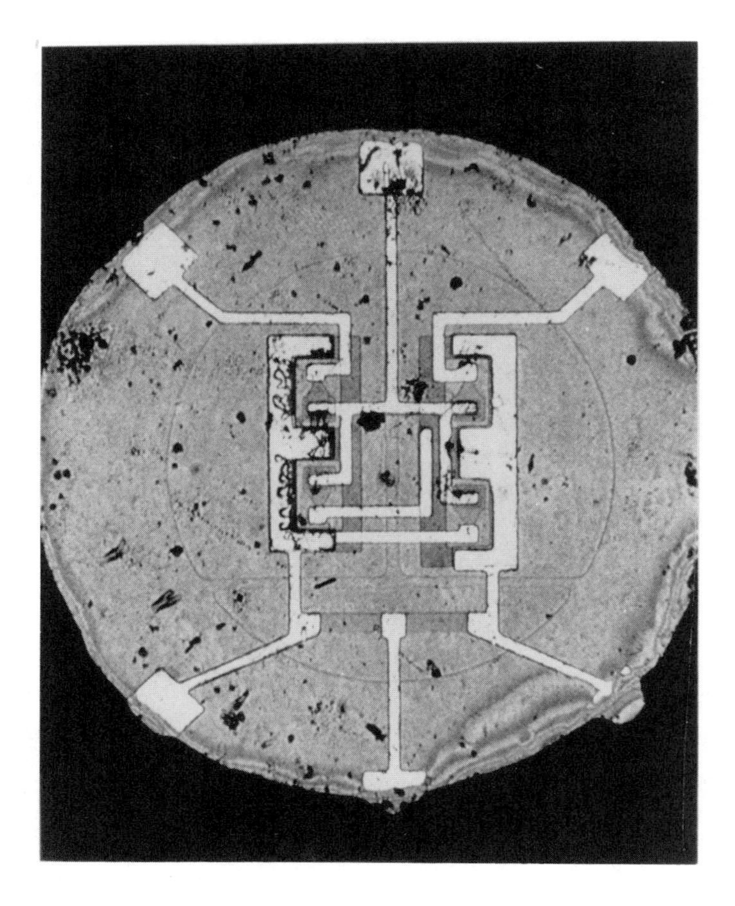

FIGURE 2. The first Fairchild Camera integrated circuit, made in 1961.

record of rapid innovation in the microelectronic industry is unmatched in history.

B. CRITICAL DIMENSION DECREASE

The cost of a unit memory cell decreases with increasing circuit density. Figure 4 shows that unit memory cost dropped a factor of 33 over the past decade. This dramatic cost reduction provides a strong incentive for industry to continue developing memory technology. Figure 5 traces the history

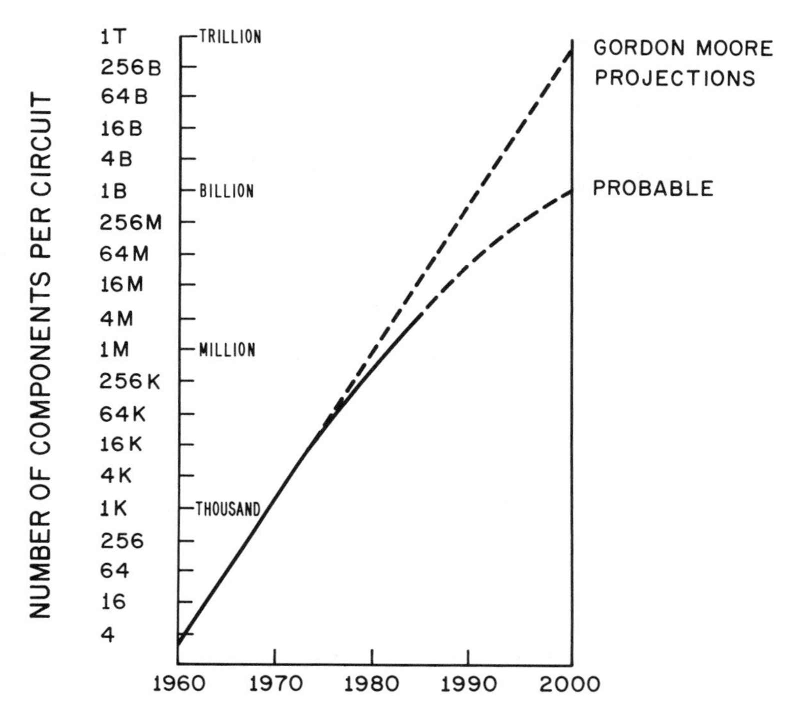

FIGURE 3. Gordon Moore observed that the number of components per integrated circuit doubled each year. Historical experience with component density from 1960 to 1987 is plotted and extrapolated to the year 2000. Even with the decrease below Moore's yearly doubling, the number of components per circuit could reach a billion elements by the year 2000.

of the critical feature linesize with the technological advances [11]. The two fundamental types of random access memory, dynamic (DRAM) and static (SRAM), are often said to be the technologies that drive IC design and processing. The more common dimensions used in most other ICs are larger than those of memory, often by a factor of 2–4. The design rule size [12] has shrunk from 10.0 μm to 0.75 μm in the last decade, and the smallest dimensions are expected to decrease to 0.5 μm in 1990 for the 16M DRAM. As the design dimension decreased, device complexity has progressed beyond large-scale integration (LSI, 10^3–10^5 devices/circuit), through very large scale integration (VLSI, 10^5–10^6 devices/circuit) to ultra-large-scale integration (ULSI, $> 10^6$ devices/circuit).

Table 1 traces the evolutionary changes in memory production for the period 1980–1985. The average feature size decreased from 5.0 μm to

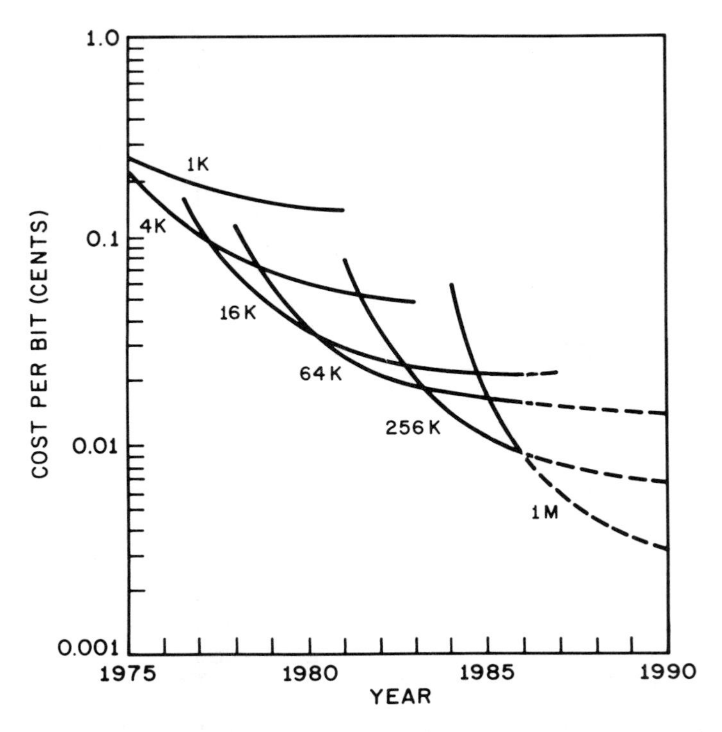

FIGURE 4. As succeeding generations of dynamic memories were manufactured, the cost per bit decreased markedly. The average yearly price decrease over the past ten years has been approximately 0.008 cents per bit per year.

2.5 μm, while the minimum feature dimension in leading edge designs dropped from 2.0 μm to 0.75 μm. The complexity and functionality of circuit design has grown along with decreasing feature size. This complexity introduced additional masking steps in device production. At the same time alignment tolerance between the mask levels has decreased from one micron to one-fourth that value (0.25 μm edge-to-edge).

To take full advantage of the cost decrease from increased integration capabilities, larger wafers were introduced to get more circuits per wafer. There is now a trend to 6- and even 8-inch diameter wafers (150 mm, 200 mm). Table 1 shows the time period during which 4-inch wafers were replaced by 5-inch material. Note that an increase in wafer size usually coincides with the opening of a new fabrication facility, since converting an operational clean room from one size to another is excessively disruptive.

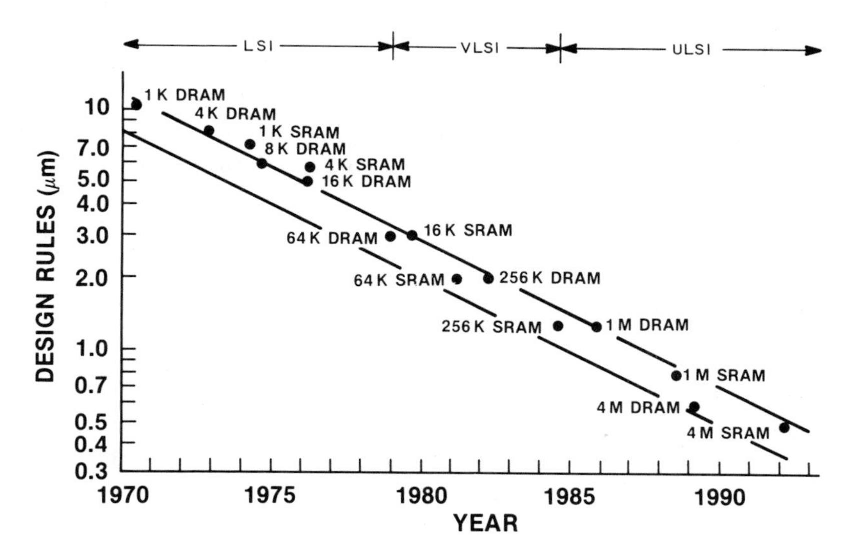

FIGURE 5. The minimum feature dimension in both dynamic and static memory designs has shown a linear reduction rate of slightly less than 0.5 micrometer per year. Somewhat arbitrary divisions between LSI, VLSI and ULSI are 3.0 μm and 1.0 μm features sizes, respectively (after Tobey [11]).

Table 1 Evolutionary changes in device design parameters during the six-year period, 1980–1985. The feature size, alignment tolerance, dielectric thickness and junction depth decreased, while the wafer size, number of masking steps and chip size all increased. Manufacturability of these designs required decreasing defect densities as the chip size grew.

CHARACTERISTIC	1980	1981	1982	1983	1984	1985
FEATURE SIZE (LOOSEST - MICRONS)	5.0	4.0	3.5	3.0	3.0	2.5
FEATURE SIZE (TIGHTEST - MICRONS)	2.0	2.0	1.5	1.5	1.0	0.7
ALIGNMENT (TIGHTEST OVERALL FIT - MICRONS)	1.0	1.0	0.7	0.5	0.3	0.2
WAFER SIZE (% 100 mm / % 125 mm)	99	97	95	90	80	60
MASKING STEPS (MINIMUM)	5 + 1	5 + 1	6 + 1	6 + 1	6 + 1	6 + 1
MASKING STEPS (MAXIMUM)	10 + 1	10 + 1	11 + 1	11 + 1	12 + 1	12 + 1
DIELECTRIC THICKNESS (MINIMUM - Å)	400	400	350	300	200	150
JUNCTION DEPTH (MINIMUM - MICRONS)	0.4	0.4	0.3	0.3	0.2	0.2
DEFECT / cm^2 (MINIMUM)	3	2	1.5	1.5	1	1
DIE SIZE (MAXIMUM - MILS SQUARE)	300	350	400	430	460	500

The chip size also increased, from 0.5 cm^2 to more than 1.5 cm^2, as each generation incorporated smaller dimensions and more functionality. Vertical dimensions have also decreased. The "critical" dielectric gate film thicknesses thinned to 150 Å from 400 Å during this period, and the junction depth of the active element structure has shrunk from 4000 Å to 2000 Å, or less.

C. PROCESS DEFECT DENSITY

To make these advanced designs possible, the fabrication process must provide an improved "zero defect" capability. The key process parameter that measures this capability is the *fatal defect density*, or D_o. With 1980 technology, a D_o of 3 defects per cm^2 gave about 70% circuit yield for 16K DRAM's. Yield scales with circuit area according to [13]

$$Y = Y_o \frac{1 - e^{-D_o A}}{D_o A}. \tag{1}$$

This relationship is plotted as a function of the chip linear dimension (chips are usually square) in Fig. 6. As shown in Fig. 6 the 1980 memory chip area was 0.25 cm^2 (5000 μm on a side). To maintain the same 70% yield in 1985 when the 256K DRAM chip area increased by a factor of 5 to 1.5 cm^2, the process defect density had to decline fivefold, to a D_o of 0.6. This was an extraordinary accomplishment since the number of circuit levels increased from 10 to 12. The above D_o is the *total* defect density for the *entire* process sequence; hence, the average density at each circuit level could not exceed 1/n of the listed value, which means the 12 level device had less than 0.05 *fatal* defects per cm^2 per step!

D. PATTERN TRANSFER

Table 2 shows essential process steps in an early (1974) MOS (metal on silicon) process and the equipment used. There were only two plasma etch patterning steps and three wet etching steps (resist strip and simple oxide operations not included). The motivation to change the three wet pattern steps to dry (plasma) processing was strong and still is a focus in many manufacturing facilities.

A pattern is commonly formed by either the *additive* or *subtractive* method, as illustrated in Fig. 7. In both methods the objective is to produce

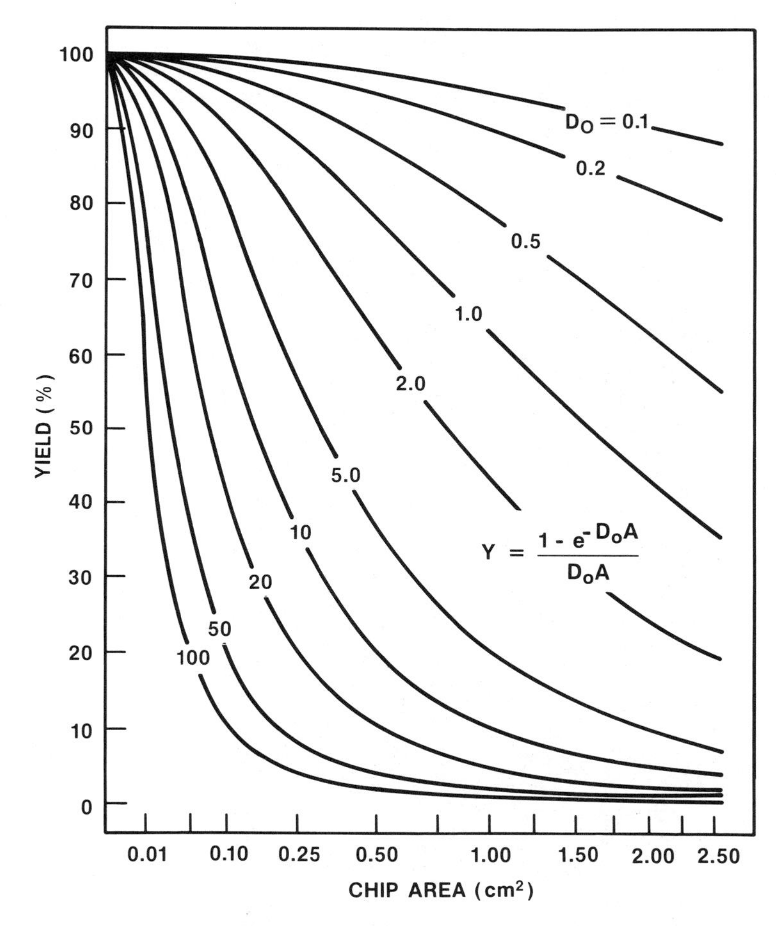

FIGURE 6. The equation in the figure is plotted to show how wafer yield varies with chip area (with defect density as the process variable). As the chip size increases, the process must provide a lower D_0 technology to maintain wafer yield (after Murphy) [13].

a patterned thin film on a substrate. In the subtractive process this film is first deposited as a uniform layer by chemical vapor deposition (CVD) or sputtering. To pattern this film, a mask is formed next, usually by photolithography. A light-sensitive organic polymer "resist" film is "spun on" the wafer from a viscous solution of the polymer. It is dried and then exposed to light through a mask containing the device pattern appropriate for that particular level of device fabrication. Depending on the kind of resist (positive or negative), exposed areas of the polymer are rendered

Table 2 Steps in a simplified MOS process are shown with the type of equipment used for various deposition, masking and etching steps. Wet chemical etching is being replaced by dry plasma processes, where possible.

PROCESS STEP	EQUIPMENT USED: FURNACE HIPOX	FURNACE	FURNACE LPCVD	ION IMPLANT	PHOTORESIST	DRY PLASMA	WET CHEMISTRY	SPUTTERING
GROW THIN OXIDE		X						
DEPOSIT NITRIDE			X					
CONDITION NITRIDE		X						
MASK #1-DEFINE FIELD					X			
PATTERN NITRIDE						X		
IMPLANT FIELD-BORON				X				
REMOVE RESIST							X	
OXIDIZE FIELD	X							
REMOVE NITRIDE						X		
REMOVE OXIDE							X	
MASK #2-DEPLETION DEVICES					X			
IMPLANT ARSENIC				X				
REMOVE RESIST							X	
GROW GATE OXIDE		X						
MASK #3-BURIED CONTACTS					X			
ETCH OXIDE							X	
DEPOSIT POLYSILICON			X					
MASK #4-POLY/GATE					X			
ETCH POLY						X		
ETCH OXIDE							X	
OXIDIZE POLY		X						
IMPLANT ARSENIC SOURCE/DRAIN				X				
DEPOSIT SILOX			X					
MASK #5-CONTACTS					X			
ETCH CONTACT HOLES							X	
REFLOW/DIFFUSE POCL.		X						
ETCH TO CLEAR CONTACTS							X	
DEPOSIT ALUMINUM								X
MASK #6-METAL					X			
ETCH METAL							X	
ANNEAL (H_2)		X						
DEPOSIT GLASSIVATION			X					
MASK #7-PADS					X			
ETCH PADS							X	
SUMMARY TOTALS	1	6	4	3	7	3	9	1

soluble (or insoluble) to a developer-solvent. The developer dissolves unwanted areas creating the final mask pattern.

The resist pattern now becomes a stencil that will protect the underlying film from attack by the etching process, which is the next step in the pattern transfer operation. The etchant can either be a wet chemical agent (acids like HF or H_3PO_4) or a gaseous plasma (dry etchant) using a feed such as CF_4/O_2 or Cl_2. After etching is complete, the mask is stripped off using a strong solvent, or by etching in a second plasma.

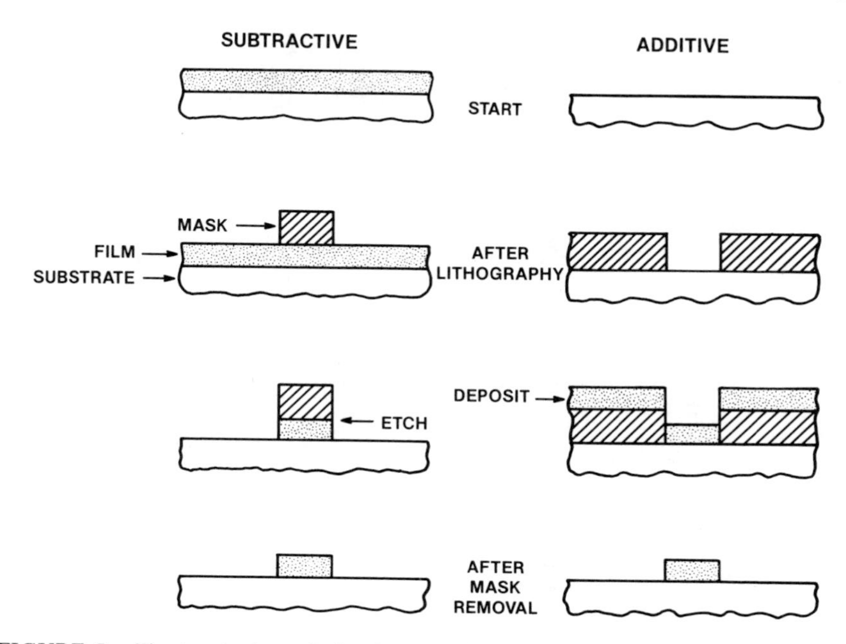

FIGURE 7. The two basic methods of pattern transfer are: subtractive, where the planar film is removed from areas not protected by the mask; and additive, where the film is deposited over the mask pattern and the mask with undesired film deposits is later removed.

In the additive process, the pattern is formed at the same time the film is deposited. A resist pattern-mask is first formed on the base substrate and afterwards, a film material is deposited over the mask and onto uncovered substrate areas, thus creating the desired pattern. When deposition is complete, the resist is swelled by a strong solvent and dissolves away, lifting off unwanted deposits from the masked areas of the pattern.

The subtractive process dominates production pattern transfer. The additive method is mainly used for masking laboratory prototype devices. Plasma etching can be applied to steps in both processes.

E. ISOTROPIC AND ANISOTROPIC ETCHING

Differences in etching mechanisms have an immediate effect on the profiles of features in both "wet" chemical and "dry" plasma chemical etching. *Purely chemical* etching usually has no preferential direction. This leads to isotropic circular profiles, which undercut a mask [14] (see Fig. 8). As

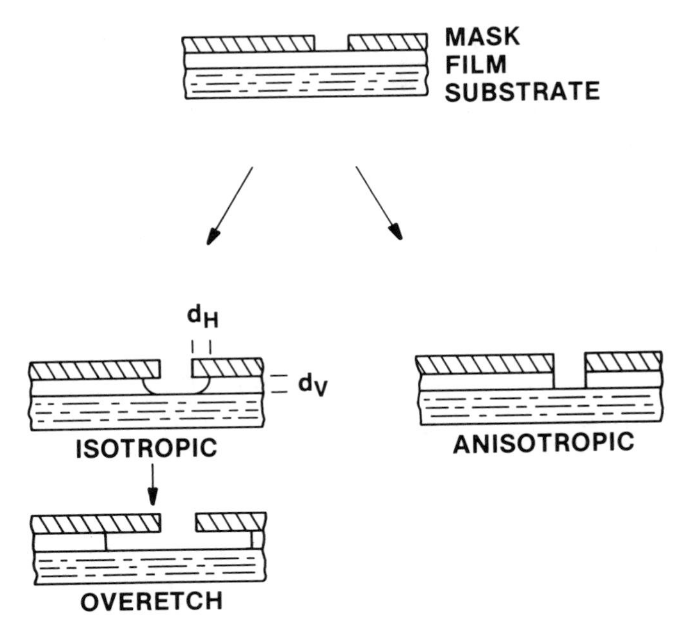

FIGURE 8. Replicating the pattern from a mask stencil into a film depends on the ability to control horizontal etching. Left: Isotropic etching; the horizontal and vertical etch rates are equal, creating an undercut, d_H, in the film beneath the mask, equal to the film thickness, d_V. If etching is continued, this lateral attack will continue and the etched film edge profile will become almost vertical. Right: Anisotropic etching; the horizontal etch component is very small, resulting in faithful pattern transfer.

shown, the thickness etched (d_V) equals the undercut (d_H) until etching reaches the film-substrate boundary. Overetching past the point where the underlying substrate is just exposed increases the undercut and radius of the undercut cross section. At large degrees of overetching the walls are essentially vertical. In practice, overetching is necessary because of wafer surface topography, nonuniform film thicknesses and variations in etch rate in the reactor (see Section IV.C.2). Moreover, in isotropic plasma etching it is difficult to control the degree of overetching beyond the endpoint because *the rate of undercut may accelerate at the end point* where (unfortunately) control is needed most. This loading effect is discussed further in Chapter 2.

Ideal *ion enhanced* plasma etching, on the other hand, produces an anisotropic profile (see Fig. 8). When the mean free path of ions in a plasma is long compared to feature depth, electrical fields (the sheath field) make ions strike horizontal surfaces almost exclusively at normal incidence. This ion bombardment preferentially accelerates the chemical reaction in the vertical direction so that vertical sidewalls are formed along the edges of masked features at right angles to the substrate. In practice, tapered profiles can also be produced (see Sections IV.C.1, IV.C.2) because of ion scattering and superimposed isotropic chemical etching effects.

It is important to add that the formation of vertical walls on etched features does not necessarily mean that etching was ion enhanced. Vertical profiles are formed in isotropic chemical etching during severe undercutting (as noted before), and can also be produced when there is preferential etching along specific crystal planes. Such crystallographic attack is more common in wet chemical etching [15], but it also occurs in plasmas, for instance when III-V compound semiconductors are etched by gaseous halogen radicals. Dry crystallographic etching can be used to form vertical or tapered walls with only a small undercut (see Chapter 2). While undercutting is usually obvious when the mask is compared to the feature, in practice the edges of a mask may be tapered rather than vertical, and mask erosion (e.g., by sputtering, or chemical attack) leads to a more complex profile. If the mask erodes at an appropriate rate relative to the undercutting, it is possible to attain features in which the edge of the mask aligns with the walls of etched features, even though the etching is isotropic.

F. THE TRANSITION FROM WET TO PLASMA ETCHING

Plasma etching was explored as a cheaper alternative to wet solvent resist stripping [16] for integrated circuit manufacture in the late 1960s and early 1970s. By the early 1970s, CF_4/O_2 plasma etching was widely adopted for patterning silicon nitride passivation of its selectivity over resist masks and underlying metalization. The wet chemical alternatives were complicated and indirect. Around the same time, oxygen plasma resist stripping, or ashing in oxygen plasmas, was finally integrated into many manufacturing lines. Production plasma etch processes were next developed for polysilicon and a host of other materials. These initial dry plasma-assisted etching processes were purely chemical and isotropic. At first, their advantages seemed to lie in unique processing sequences, substitution of safe nontoxic gases such as O_2 and CF_4 for corrosive liquids, easily discharged waste products and simple automation. By the late 1970s, however, it was widely recognized that plasma etching offered the possibility of a vertical etch rate that greatly exceeded the horizontal rate (e.g., anisotropic etching).

Because of the isotropic undercut, wet etching a 1 μm thick film through a 1 μm mask opening at best yields a cross-sectional profile that measures

3 μm at the top and 1 μm on the bottom. Thus with increasing circuit density and narrower linewidth requirements, anisotropic etching became *necessary* to pattern features smaller than 2–3 microns. As designs advanced, wet etch operations were converted to dry anisotropic ones for control of critical feature dimensions. Higher circuit density required low defect densities and precise process control, but while dry processing offered anisotropy, unfortunately it also introduced new sources of variability such as the loading effect (Chapter 2), and new types of defects caused by ion bombardment and other plasma phenomena. Hence, there was an urgent requirement to improve understanding of plasma chemistry and physics.

II. What Is a Plasma?

The simplest plasma reactors may consist of opposed parallel plate electrodes in a chamber that can be maintained at low pressure, typically ranging from 0.01 to 1 Torr (1.33–133 Pa) (Fig. 9). When a high frequency voltage is applied between the electrodes, current flows forming a *plasma*, which emits a characteristic glow. Reactive radicals are generated by this *electrical discharge*. Semiconductor wafers or other substrate material on the electrode surfaces are exposed to reactive neutral and charged species. Some of these species combine with the substrate and form volatile products that evaporate, etching the substrate (Fig. 10). Hence in some sense the plasma acts as a "chemical factory," which supplies reagents that remove a film selectively, such that the mask and substrate are not attacked.

Before going into more detail about reactors and the etching process, we will briefly describe some essential basic characteristics of the plasma environment. To the physicist, a plasma is an ionized gas [17] with equal numbers of free positive and negative charges. The free charge is produced by the passage of electric current through the discharge. For most plasmas of interest for etching, the extent of ionization is very small. Typically there is only one charged particle per 100,000 to 1,000,000 neutral atoms and molecules (Fig. 11). The positive charge is mostly in the form of singly ionized neutrals, (i.e., atoms, radicals or molecules) from which a single electron has been stripped (removed). The majority of negatively charged particles are usually free electrons; although in very electronegative gases such as chlorine, negative ions can be more abundant.

In any event, electrons are the main current-carriers because they are light and mobile. Because electrons are much lighter than the neutrals they

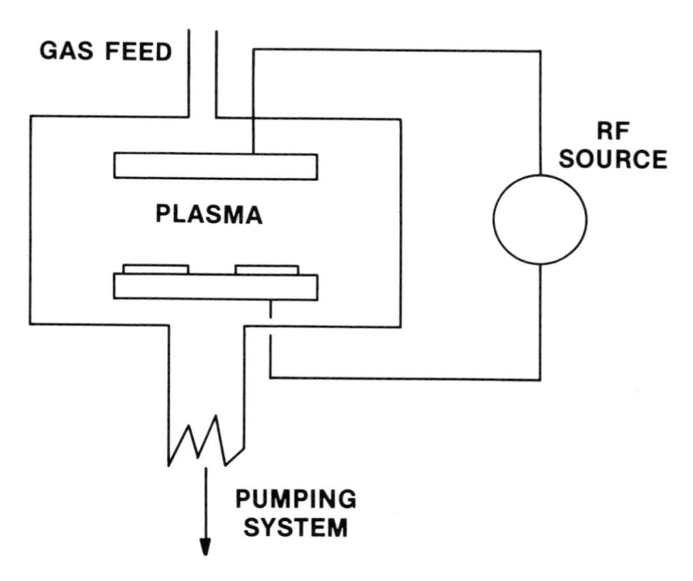

FIGURE 9. A plasma reactor consists of a vacuum enclosure with at least two electrically separate electrodes. One of these must be isolated, while the second electrode can either be a similar structure, or can be the interior surface of the vacuum chamber itself. The system has provision for continuously introducing a feed gas, and a port for pumping. A source of radiofrequency power coupled to the electrodes creates a plasma. Radicals and ions formed in the plasma give controlled etching or deposition of a wafer which is supported on one of the electrodes.

collide with (generally by a factor of $\sim 5 \times 10^{-4}$ to 5×10^{-6}), energy transfer from electrons to gas molecules is inefficient and electrons can attain a high average energy, often many electron volts (equivalent to tens of thousands of degrees above the gas temperature). The elevated electron temperature permits electron-molecule collisions to excite high temperature type reactions, which form free radicals, in a low temperature neutral gas. Generating the same reactive species without a plasma would require temperatures in the $\sim 10^3$–10^4°K range that would incinerate organic photoresist and melt many inorganic films. The coexistence of a warm gas and high temperature active species distinguishes the plasma reactor from conventional thermal processing.

Although formally there are equal numbers of positive and negative charged particles in a plasma, diffusion of charge to the walls and recombination on boundary surfaces tends to deplete charge in the adjacent gas

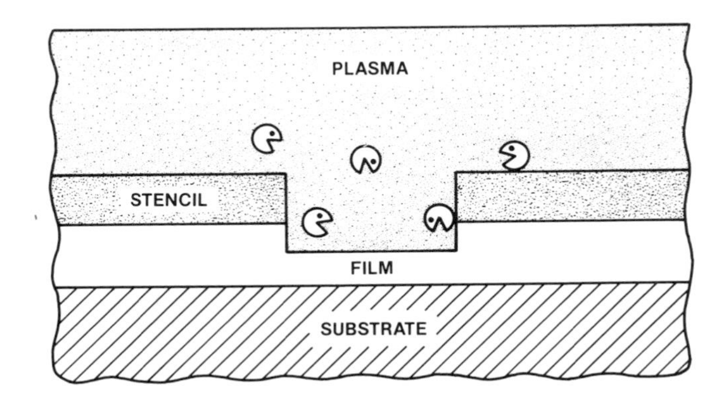

FIGURE 10. We may visualize the plasma etching process as being similar to an attack by trained "pacmen." The pacmen cannot penetrate the stencil material, and consume only film that is exposed. Well trained pacmen might eat the film in a vertical direction (anisotropy) and find the substrate material unpalatable (selectivity).

- A plasma is an ionized gas with equal numbers of positive and negative charges

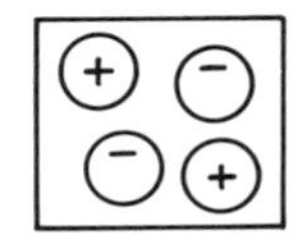

- For most plasmas of interest here, the extent of ionization is small, typically only 1 charged particle per 1,000,000 neutral atoms and molecules

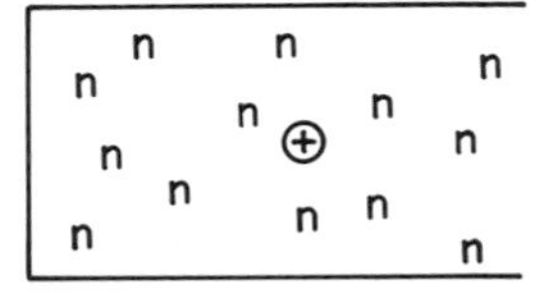

- The negative particles are predominantly electrons. Energy transfer from electron in a plasma is inefficient. Because of this, the electrons have high energy (many eV) permitting high temperature type reactions (which make free radicals) in a low temperature neutral gas

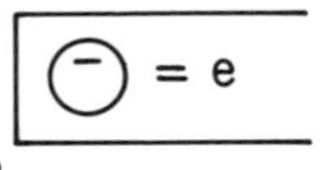

FIGURE 11. Some basic characteristics of plasmas used for microelectronic processing.

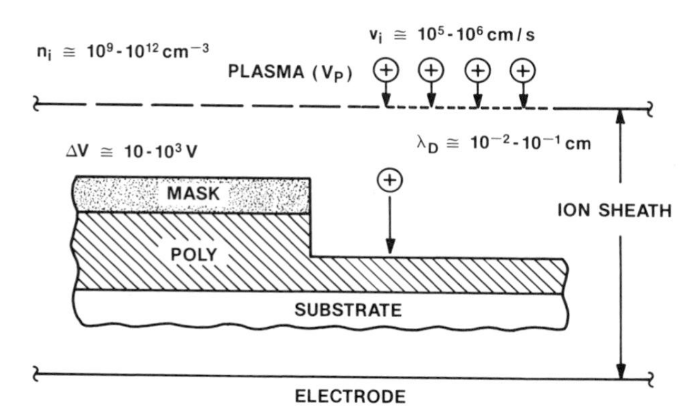

FIGURE 12. Since the "hot" electrons are more mobile than ions, they diffuse to the walls faster and charge them negatively, leaving the plasma center (only a few Debye lengths, λ_D, from the walls) positive. The potential drop V_P across these sheaths is generally 10–1000 V. The thin sheaths shield the low field glow zone from the applied RF field. Ion motion is random in the central glow but when a positive ion drifts to the sheath boundary (dotted line), the perpendicular electric field accelerates it toward the wall and wafer surfaces. Typical parameters are shown.

phase, forming a thin boundary layer or *sheath*. Because electrons are light and have high energy, they diffuse fastest, leaving an excess of positive charge and a plasma potential that is positive relative to the walls (Fig. 12). Since charged particles are most abundant in the central glow of the plasma, it is a fairly good conductor and most of the potential drop appears across the sheath. Voltage across the sheath ranges from a few volts to thousands of volts, depending on other parameters. Positive ions are accelerated through the sheath and strike the walls at near-normal incidence. As shown in Fig. 12, the sheath thickness is on the order of a *Debye length* (discussed in Chapter 2) and features smaller than this characteristic length can be patterned directionally, because the vertical feature walls largely escape the effects of the perpendicular-going ion bombardment.

In the bulk of the plasma (glow), the plasma density is normally in the range of 10^9–10^{12} cm^{-3}. This range is set by fundamental limits. That is, when the number density of electrons is below $\sim 10^9$ cm^{-3}, the electrostatic force is weak enough for charge to separate over a fairly large distance and neutrality is no longer maintained. This is usually unstable as an operating region because higher density will result in a lower charged particle loss rate. Plasma densities above $\sim 10^{12}$ cm^{-3} correspond to high currents and significant gas heating. High temperature can be harmful to substrate material and leads to plasma instability and nonuniformity.

A. WHAT BASIC MECHANISMS CAUSE ETCHING?

Plasma etching can proceed by physical sputtering or chemical reaction and ion-assisted mechanisms. During sputtering, substrate material is removed by purely physical processes; energetic ions crossing the sheath transfer large amounts of energy and momentum to the substrate. This causes surface material to be ejected. At low pressure where the mean free path is long, the ejected sputtered material can cross the reactor vessel and reach opposing walls. Since this sputter etching is purely physical and requires high energy, it is the least selective mechanism. Sputtering also suffers from the disadvantages of low rate, surface facetting and trenching, and electrical damage to the substrate from ion bombardment and implantation (see Section VII).

Chemical and ion/chemical mechanisms differ from sputtering in that the surface material is first converted into high vapor pressure products to facilitate removal. Mass loss is then evaporative, so that etching can be done at higher pressure where any sputtered material is scattered back onto the surface by gas collisions. Volatility is essential for chemical and ion enhanced plasma etching processes.

All known basic plasma etching processes can be grouped into four categories as shown schematically in Fig. 13: 1) sputtering, 2) chemical gasification, 3) energetic ion-enhanced chemistry and 4) inhibitor ion-enhanced chemistry. We have already mentioned sputtering. In sputtering, ions accelerated across the sheath potential bombard a surface with high energy. The sudden energy impulse can immediately eject surface atoms outward, or by a billiard ball-like collision cascade can even stimulate the ejection of subsurface species. If there is to be net material removal, however, molecules sputtered from the surface must not return. This requires a low gas pressure or, equivalently, a mean free path that is comparable to the vessel dimensions. If the mean free path is too short, collisions in the gas phase will reflect and redeposit the sputtered species.

Sputtering requires plasma conditions where ion energies are high. These conditions exist in low pressure plasmas (< 50 mTorr) where the mean free paths are long as well. As a mechanical process, sputtering lacks selectivity. It is sensitive to the magnitude of bonding forces and structure of a surface, rather than its chemical nature and quite different materials can sputter at similar rates. In a way this is symptomatic of using ion bombardment with energy far higher than the surface binding energy.

In chemical etching, gas phase species merely react with a surface according to elementary chemistry. Fluorine atom etching of silicon is a good example of this mechanism. The key, and really the only requirement for this kind of process, is that a *volatile* reaction product be formed. In

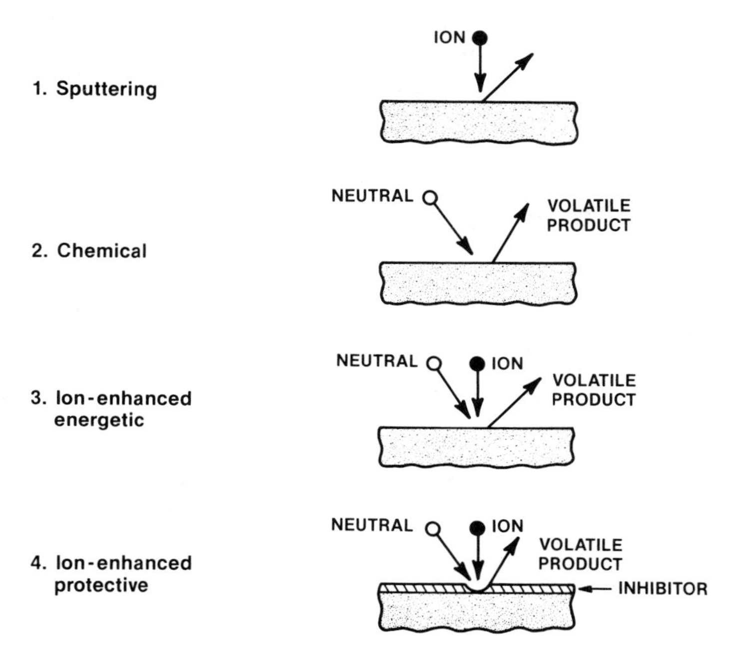

FIGURE 13. Four basic methods of plasma etching are: 1) Sputtering—the ion energy mechanically ejects substrate material, 2) Chemical—thermalized neutral radicals chemically combine with substrate material forming volatile products, 3) Ion-enhanced Energetic—neutral radicals alone reaching the surface cannot form a volatile product; energetic ions alter the substrate or product layer so chemical reactions can gasify the material; since the ion flux mainly bombards surfaces oriented perpendicular to the electric field, feature sidewalls receive minimal ion flux and the reaction is directional (anisotropic), 4) Ion-enhanced Protective—an inhibitor, or recombinant, film coats the surface forming a protective barrier which excludes the neutral etchant. Moderate ion flux disrupts this protective film (on surfaces that are at right angles to the flux) and allows chemical etching to proceed anisotropically, in the vertical direction. The inhibitor film on the sidewalls protects them from attack.

silicon/F-atom etching, spontaneous reactions between F-atoms and the substrate form SiF_4, a gas. The only purpose of the plasma, in chemical etching, is to make the reactive etchant species, F-atoms in this example. The etchant species are formed through collisions between energetic free electrons and gas molecules, which stimulate dissociation and reaction of the feed gas (i.e., plasma feeds such as F_2, NF_3 and CF_4/O_2 all make F atoms).

Chemical etching is the most selective kind of process because it is inherently sensitive to differences in bonds and the chemical consistency of

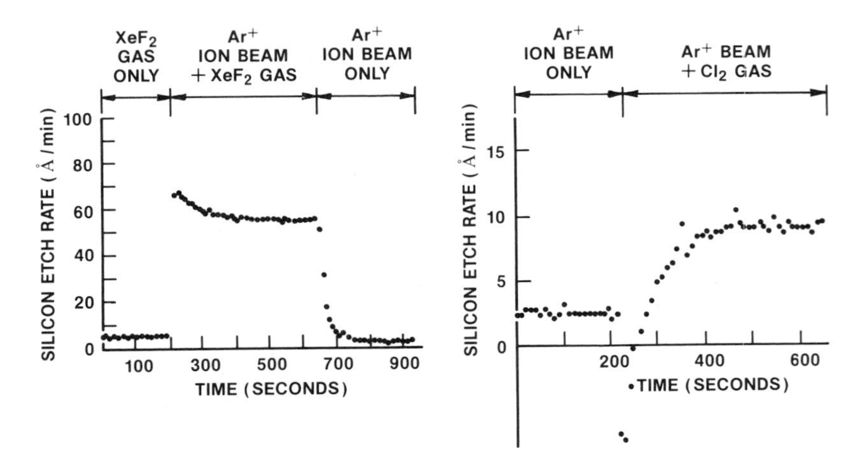

FIGURE 14. Laboratory demonstration of ion-enhanced etch reactions. Left: the synergistic reaction of the XeF_2 (chemical etch only) with Ar^+ ion bombardment (sputter etch only) cannot be explained by simple addition of these separate rates. Right: Ar^+-enhanced etching of silicon by chlorine also shows a similar synergistic etching effect (after Ref. 20).

a substrate. However, the process is usually *isotropic* or nondirectional, [18] which is sometimes a disadvantage. With isotropic etching, both vertical and horizontal material removal proceed at the same rate, making it impossible to form the fine lines (less than about 3 microns in the usual ~ 1 μ-thick films).

There are directional etching mechanisms that are stimulated by the perpendicular-going ion flux: *energetic* ion and *inhibitor* induced anisotropy. In *energetic* ion-enhanced etching, impinging ions damage the surface, which increases its reactivity. For example, an undoped single crystal silicon surface is not etched by Cl_2 or Cl atoms at room temperature [19]. When the surface is simultaneously exposed to a high energy ion flux, however, the result is a rapid reaction that forms silicon chlorides and removes material much faster than the physical sputtering rate, as shown in Fig. 14.

We mean the word *damage* in a general sense. For some systems, damage may be the introduction of defects or dislocations into the lattice. In other cases, it may be interpreted to be the formation of dangling bonds on the surface. Sometimes a surface compound can be partly dissociated. Whatever the microscopic details (which undoubtedly can vary greatly from one surface/etchant system to another) the generic mechanism is one in which ions impart energy to the surface, which serves to modify it and to render the impact zone and its environment more reactive.

The second class, *inhibitor* ion-enhanced etching, requires two conceptually different species: etchants and inhibitors. The substrates and etchants in this mechanism would react spontaneously and etch isotropically, if it were not for the inhibitor species. The inhibitors form a very thin film on (vertical) surfaces that see little or no ion bombardment. The film acts as a barrier to etchant and prevents attack of the feature sidewalls, thereby making the process anisotropic.

III. Processes in a Plasma

In this section we introduce basic plasma processes and some of their effects on etching. We first examine collisional processes that transfer energy from electrons to neutrals and show why the electron temperature in a discharge is high. Then basic formation and loss processes for ions and electrons are considered. The difference in the mobility of positive and negative charge causes a sheath to form at the boundaries of a discharge, which accelerates the ions responsible for anisotropic etching. From an equivalent circuit representing the sheath, it is shown that a DC bias potential is produced in an unsymmetrical reactor. Nomenclature and common misconceptions associated with these phenomena are discussed, followed by a discussion of some basic parameters, and important general steps, which model plasma etching.

A. ELECTRON ENERGY

In a system in thermal equilibrium, kinetic and internal energy are uniformly distributed among all the species and degrees of freedom. As a result, the energy in any internal or translational state can be expressed in terms of a single parameter, the temperature T. On the other hand, plasmas are dynamic systems with large power fluxes and diverse loss processes, where many species are not in equilibrium with each other. Still, the exchange of kinetic and rotational energy among atoms and molecules is often fast compared to power external input, so heat transfer is efficient and the translational energy (temperatures) of these species can be nearly in equilibrium with each other. Only charged particles are accelerated by the

RF electric field applied to a plasma, so they are the agent through which external energy is supplied. This energy is distributed to neutral species by collisions; in plasma etching, it is a fair approximation to treat most neutral species as having a common temperature, or average energy. It is customary to speak of a temperature for each class of species (e.g., T_i for species i). Temperature can be defined in terms of the average energy, ϵ, of the particles:

$$\epsilon_i = \frac{3}{2} k T_i, \tag{2}$$

where k is the Boltzman constant, 1.38×10^{-23} J°K^{-1} or 8.62×10^{-5} eV°K^{-1}, so temperature and average energy amount to the same thing. In a plasma most of the radio frequency (RF) power is input by way of the electrons because (as will be shown next) they are much more mobile and easily accelerated than ions. Thus the power input to electrons, $\langle n_e E v_e \rangle$, dominates $\langle n_i E v_i \rangle$ where n_e, n_i are the number densities of electrons and ions respectively, v_e and v_i are the respective velocities, and E is the electric field.

When particles collide without dissociating or absorbing energy into internal states (as with two billiard balls), the collision is said to be *elastic*. The energy transferred from higher-kinetic-energy particles to lower-energy particles in these collisions depends on the distance of closest approach, with the average energy transfer per collision being [21]

$$\epsilon_{coll} = \frac{2mM}{(m+M)^2} \Delta\epsilon \approx \frac{2m}{M} \Delta\epsilon, \tag{3}$$

where $\Delta\epsilon$ is the initial energy difference. Since neutrals and ions have comparable mass ($m/M \approx 1$), this expression predicts that energy exchange among these particles is efficient. In addition, fast-moving ions can be quickly converted to fast-moving neutrals by a process called charge exchange, which tends to reduce further the difference between neutral and ion energies in the discharge.

The mass of electrons, on the other hand, is small compared with ion and neutral particle masses. For example, in collisions between electrons (m_e), and argon atoms (M_A) the ratio m_e/M_A is only 1.3×10^{-5}. Thus collisions between electrons and neutrals provide comparatively poor energy transfer.

We see that thermal conduction in a gas is high compared with energy transfer from electrons to neutrals. Therefore, energy input to the gas can pass to the vessel walls across a small temperature gradient. There is also rapid temperature equilibration between ions and neutrals. Hence, the ion temperature is usually close to the neutral gas and only moderately warmer than the surrounding walls. Because of this fact, low pressure glow discharges are called cold plasmas.

B. TYPES OF COLLISIONS

In the last section we introduced elastic collisions in which particles exchange momentum and kinetic energy without affecting their internal states. *Inelastic* collisions are a second broad category of particle-particle interactions. When inelastic collisions take place, the interacting species are typically put into excited states. Whether energy exchange in electron molecule interactions is predominantly elastic or inelastic depends somewhat on the kind of gas being discharged. Figure 15 compares the situation in a noble gas where the first excited levels are above the average electron energy to a molecular gas where there are many excited states at low energy. When the average electron energy is too small to cause excitation, most electron molecule collisions will be elastic, as in helium where the first excited state (^3S) is at 19.8 eV. Molecular gases have lower energy vibration, excitation and dissociation channels open so that a larger fraction of

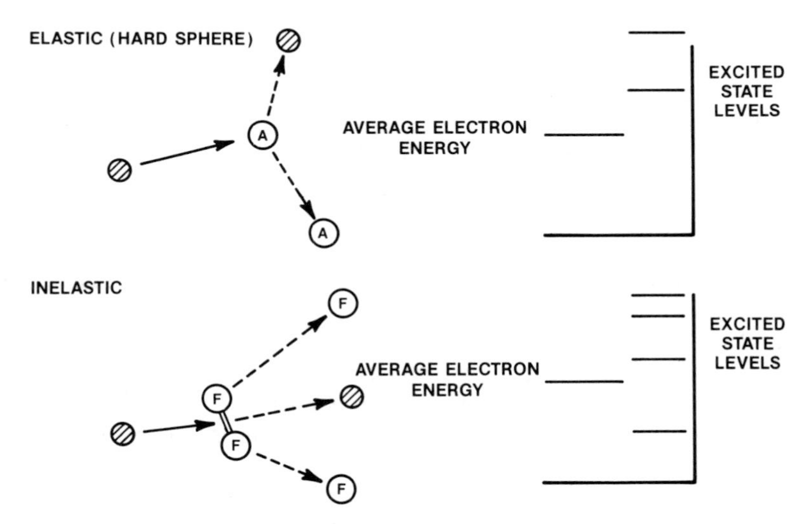

FIGURE 15. Energy exchange in collisions between electrons and molecules in inert gas plasmas is mainly through elastic collisions—the energy required to reach excited levels is greater than the average electron energy. By contrast, in a typical molecular gas there are lower lying levels and these inelastic collisions transfer much more energy.

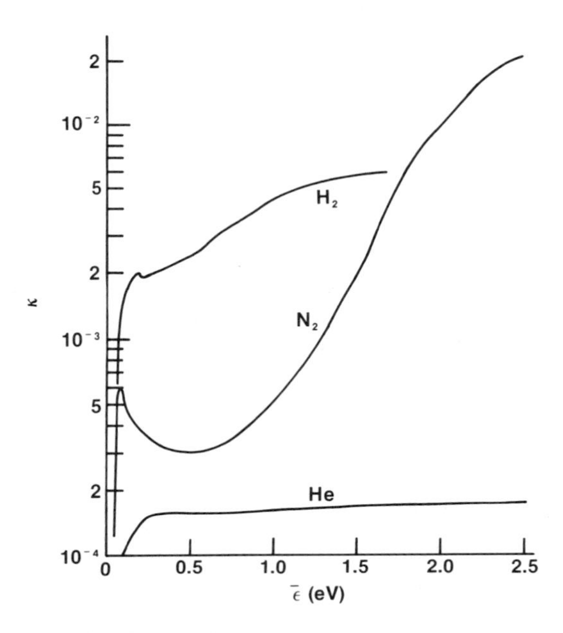

FIGURE 16. Mean fractional energy loss per collision (κ) versus mean energy for He, H_2 and N_2 (from Hunter and Christophorou [51]).

collisions are inelastic. Of course the number and probability of collisions going through available channels will influence the average electron energy, but this simplified point of view reflects the important trends.

Electrons can lose a much larger fraction of their kinetic energy through inelastic excitation and dissociative collisions with molecules and atoms. Still, even in molecular gases, the average fraction of energy lost per collision is usually low. Figure 16 shows the average fraction of energy lost per collision in He, N_2 and H_2 (including elastic collisions) as a function of the average electron energy. Note, the fractional energy transferred by elastic collisions is *not* a function of electron energy.

In experiments based on drift velocity or microwave measurements [22] the frequency of electron molecule collisions is measured by the average time it takes for the direction of electron momentum to be randomized. This parameter is known as the electron collision frequency, ν, and for plasmas we define it to be the rate at which an electron collides with gas molecules. A second kind of frequency ν_u is defined as the rate at which the energy of an electron is lost through collisions. This number is quite different from ν. For example, if all collisions were elastic, ν_u would be given by:

$$\nu_u = \frac{2m_e}{M}\nu, \tag{4}$$

in accordance with Eq. (3). The parameter κ has often been introduced for the average fraction of energy lost per collision (e.g., $\kappa = 2m_e/M$ for elastic collisions).

C. WHY IS $T_e \gg T_g$?

At this point we present a simple model that shows some important qualitative features of the relationship between the electron and neutral gas temperatures. As an electron moves over a mean free path λ, it is acted on by a force eE (where E is the electric field), and gains an energy $eE\lambda$. After transversing the mean free path, on the average, there is a collision with a gas molecule and a fraction κ of the energy difference $(\epsilon_e - \epsilon_g)$ is transferred, where ϵ_e and ϵ_g are the mean electron and gas energies respectively. Equating these terms, we obtain the energy balance

$$eE\lambda = \kappa(\epsilon_e - \epsilon_g), \tag{5}$$

and then solve for $(\epsilon_e - \epsilon_g)$,

$$\epsilon_e - \epsilon_g = \frac{eE\lambda}{\kappa} \approx \epsilon_e = \frac{E\lambda}{\kappa}\ \text{eV}. \tag{6}$$

We see that a small κ means high energy electrons. For typical values of these quantities electron energies are on the order of 10 V ($\epsilon_e - \epsilon_g \approx 10$ eV, $E \sim 10$ V/cm, $\kappa \lesssim 10^{-3}$, $\lambda \sim 10^{-3}$ cm, which is equivalent to an electron temperature more than 116,000°K higher than the gas temperature. Since $\epsilon_g \ll \epsilon_e$, the ϵ_g is often dropped, as shown in Eq. (6). This simplified model points out another important parameter, E/n. The mean free path, λ, is inversely proportional to gas density ($\lambda \propto 1/n$), so

$$\epsilon_e \approx \frac{E\lambda}{\kappa} \propto \frac{E}{n} \propto \frac{E}{p_o}, \tag{7}$$

where p_o is the pressure (equivalent to gas density) at a standard temperature (usually 25°C or 0°C). More sophisticated analyses [23] and experimental data show the relationship between electron energy and E/n is more complex than Eq. (7). But they confirm the important result that electron energy ϵ_e, depends mainly on E/n.

D. PRODUCTION OF SPECIES

Electron energy, channelled into inelastic electron-neutral collisions, maintains the supply of ions and radical species which are continuously lost by reaction and recombination. Simple examples of the most important types of electron-neutral collisions for plasma chemistry are illustrated by an oxygen plasma discharge:

a) ion and electron formation

$$e + O_2 \rightarrow O_2^+ + 2e, \tag{8a}$$

b) atom and radical formation

$$e + O_2 \rightarrow O + O, \text{ and} \tag{8b}$$

c) generation of heat and light

$$e + O_2 \rightarrow O_2^*, \tag{8c}$$

$$O_2^* \rightarrow h\nu, \tag{8d}$$

$$e + O \rightarrow O^*, \text{ and} \tag{8e}$$

$$O^* \rightarrow h\nu, \tag{8f}$$

where O_2^* and O^* are excited states of O_2 and O. Oxygen plasmas are widely used in integrated circuit manufacture for stripping resist (isotropically), for anisotropic patterning of intermediate resist layers (trilevel process), and to a limited extent for plasma enhanced anodization (oxide growth). In these systems, Reaction (8a) forms the charge carriers that sustain the plasma, Reaction (8b) is responsible for producing atomic oxygen, the active chemical species, and electronic excitation (Reactions 8c, 8e) stimulate characteristic plasma induced optical emission (Eqs. (8d), (8f)), which is used for process control and endpoint detection (see Section VI).

In the plasma, each formation step balances various loss processes. The equilibrium between formation and loss determines the steady concentrations of species in a discharge. For the charged species in a plasma, these formation and loss processes may be grouped into a few categories as follows.

1. Ionization and Detachment

Ionization reactions are the main source of ions and electrons. The general form of these reactions is

$$e + M \rightarrow A^{+} + 2e(+B), \tag{9}$$

where M is either a molecule (AB) or an atom. If the molecule dissociates in this process (yielding the neutral fragment B), it is called dissociative ionization. When the species M is a negative ion, the process is called "detachment," since the negatively charged electron is said to be attached when a negative ion is formed in the first place. The detachment process,

$$e + A^{-} \rightarrow A + 2e, \tag{10}$$

is no less an ionization than Reaction (9) and similarly creates a free electron. Less frequently, another process, Penning ionization, can make an important contribution. In Penning ionization, an excited metastable state is formed by electron impact,

$$e + C \rightarrow C^{*} + e, \tag{11}$$

and the excitation energy of the metastable is enough to ionize a second species via

$$C^{*} + M \rightarrow C + M^{+} + e, \tag{12a}$$

or (where again $M = AB$)

$$C^{*} + AB \rightarrow C + A^{+} + B + e, \tag{12b}$$

or

$$C^{*} + M \rightarrow CM^{+} + e. \tag{12c}$$

Penning ionization has been found to be significant in mixtures where C is a rare gas with a metastable state excitation energy (e.g., Ne) that is just above the ionization energy for M (near-resonance, as in Ne-Ar mixtures).

2. Recombination, Detachment and Diffusion

A series of loss processes balance the formation steps outlined before. Some of the most important charge loss mechanisms are electron-ion recombination,

$$e + M^{+} \rightarrow A + B, \tag{13}$$

attachment

$$e + M \rightarrow A^{-} + B, \tag{14}$$

and diffusion of ions and electrons to the walls of the reaction vessel. These reactions take place in a variety of ways, depending on the species involved. For example, while the dissociative *recombination* Eq. (13) would be the most rapid ion-electron recombination process in an oxygen discharge,

$$e + O_2^+ \rightarrow O + O, \tag{15}$$

in a pure argon discharge only simple electron-ion recombination is possible

$$e + A^+ \rightarrow A. \tag{16}$$

In highly exothermic reactions, reaction channels that form two or more product fragments with comparable mass are generally favored, because they make it easier (there are more configurations) to conserve both energy and momentum. Plasma production and loss processes are covered further in Chapters 2 and 4.

E. SHEATH REGION

1. Sheath Potentials and Bias

The voltage applied to sustain the plasma is usually concentrated across thin sheaths close to the boundaries of the plasma. The reason is really simple. The plasma cannot sustain itself until the charge is greater than a certain critical density (usually around 10^9–10^{10} cm^{-3}); diffusion is too rapid otherwise (see Chapter 4). At this charge density the plasma is a comparatively good conductor, which means that it cannot sustain much of a field without high current flow. This is a consequence of Ohm's Law. However, since charged species rapidly recombine on walls at the plasma boundaries, charge density there drops nearly to zero. In the adjacent region electrons and ions move to the walls by diffusion, leaving an electrical boundary-layer zone, the sheath. Charge density is low in the sheath region, so there can be high fields here.

Kinetic theory[24] shows that *random* thermal motion of a gas results in a flux,

$$\bar{j} = \bar{n}\frac{\bar{v}}{4} = \frac{\bar{n}}{4}\sqrt{\frac{8kT}{\pi m_i}} \tag{17}$$

of particles to the wall, where m_i is the mass of gas molecules. Since there is an intense electric field *in the sheath*, the charged particle motion there is not "random thermal" and derivations leading to Eq. (17) do not apply.

Nevertheless, in the main body of the plasma, *adjacent to the sheath*, the electric force may be neglected so that the number of ions that *enter* the sheath are given by Eq. (17). Hence by a mass balance (e.g., ions that enter the sheath are accelerated to the wall) it is apparent that Eq. (17), evaluated at the glow-plasma boundary, does give the ion current in the sheath.

There are important differences between the ion and electron fluxes, despite the fact that the same equations apply. First, we notice that the random thermal flux of electrons is far greater than that of ions or neutrals,

$$j_e \gg j_i, j_n \tag{18}$$

because $T_e \gg T_g, T_i, m_e \ll m_g, m_i$ and the velocity in Eq. (17) is proportional to $\sqrt{T_e/m_e}$ as shown in Eq. (2). Thus, electrons are much more mobile than ions and hence the plasma can, as we will see, serve as a rectifier. Second, since electrons carry negative charge, the excess random thermal electron flux over the ion flux (Eq. (18)) makes insulating walls charge negatively. This negative charge accumulates on the walls until the repulsive electric sheath field is intense enough to reduce electron current to the level of the ion current flux (Fig. 17). The wall is then said to be at *floating potential*, and at this point, the current of electrons entering the boundary layer no longer corresponds to random motion described by Eq. (17).

These phenomena usually force the average wall potential to be negative relative to the higher density plasma in the (more nearly equipotential) glow region. The potential in the glow is termed the plasma potential. Since electrons have an average energy kT_e/e, this is, roughly, the magnitude of the floating potential. It is proper to think of insulating walls being charged to floating potential by the kinetic energy of electrons in the plasma.

At conductive boundaries, such as the parallel plate electrodes that sustain the plasma, the situation is different. When a voltage is applied between conductors in contact with the plasma, a large electron current will move toward the momentary positive electrode in response to the field. Electrons lost to this positive conducting wall are continuously replenished from the higher charge density in the glow region of the plasma, and there is a relatively low potential drop. Ions are far less mobile (c.f. Eqs. (2), (17)) and their motion is negligible by comparison. Near the momentary negative electrode, the applied potential forces electrons away, leaving behind an ion-rich sheath zone with net positive charge. These ions reach and bombard the electrode with a current set by the random thermal ion current entering the sheath (Eq. (17)). The high resistance electron poor negative ion sheath drops most of the applied potential, and determines the maximum energy with which ions cross the sheath and bombard the negative electrode.

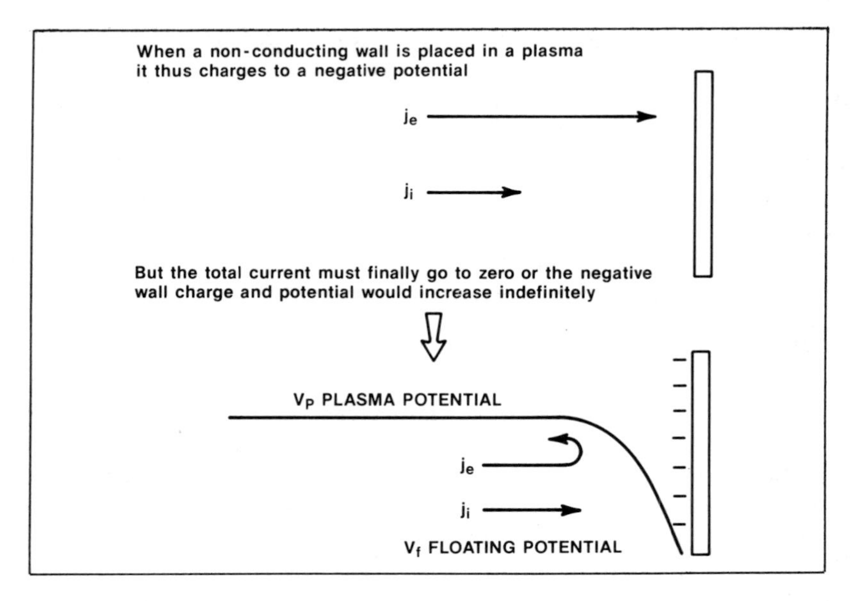

When a non-conducting wall is placed in a plasma
it thus charges to a negative potential
je
ji
But the total current must finally go to zero or the negative
wall charge and potential would increase indefinitely
Vp PLASMA POTENTIAL
je
ji
Vf FLOATING POTENTIAL

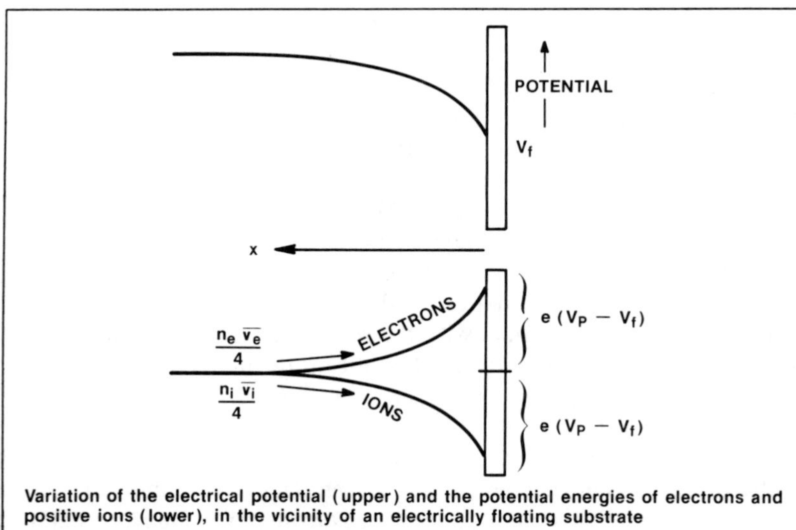

Variation of the electrical potential (upper) and the potential energies of electrons and positive ions (lower), in the vicinity of an electrically floating substrate

FIGURE 17. An insulating wall placed in a plasma charges to a negative potential because the random electron flux is larger than the ion flux (upper diagram). But the steady current must finally fall to zero or the negative wall potential would increase indefinitely. As the wall charges, the electron flux is retarded. The wall becomes increasingly negative until electron and ion fluxes are equal. At this point the wall has reached a floating potential, V_f, which is negative relative to the plasma potential, V_P.

2. *The Single Probe Curve*

A useful and somewhat complementary view of these effects comes from considering the current-voltage characteristic, Fig. 18, of a small wire or plane surface inserted into the plasma glow. This I-V curve contains information about the density and energy of the plasma and was first used as a diagnostic by I. Langmuir [25] in the early part of this century. The bias is usually applied relative to the RF common (grounded) electrode as a reference potential. When the probe is highly negative (left-hand-side of the figure), it collects all positive ions entering the plasma sheath while the electrons are repelled. This is the "saturation" ion current,

$$i = \frac{n_+ v_+ A}{4}. \tag{19}$$

As the probe is made more positive, the *net* current flow starts to decrease (that is, the signed value of current increases toward zero) because the

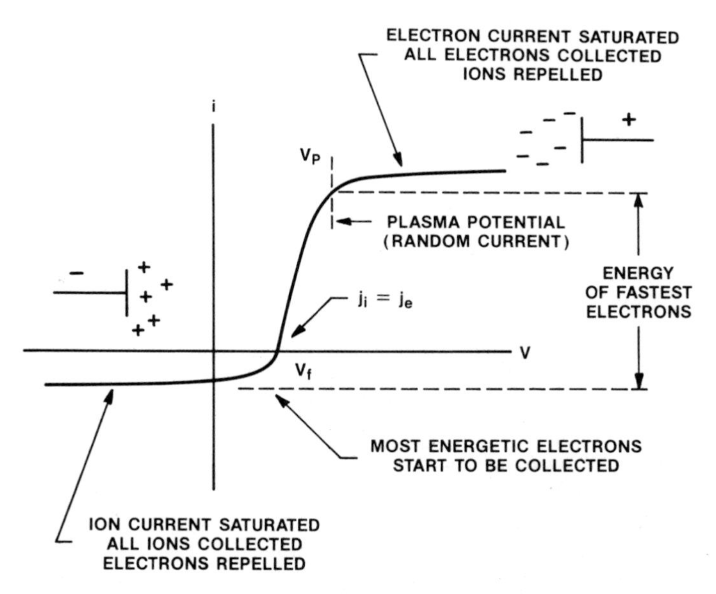

FIGURE 18. The I-V characteristics of a probe introduced into a plasma (*positive* current flow is from the probe to the plasma). At highly negative bias, positive ions are collected and electrons are repelled. As the potential is made more positive, electron current increases. At floating potential V_f the electron and ion fluxes are equal so *net* current is zero. Plasma potential, V_P, is the point where random ion and electron currents are unaffected by the bias and reach the probe. Finally, ions are repelled and the electron current is saturated.

highest energy electrons (at the top of the electron energy distribution function) overcome the applied bias and reach the probe. When the potential reaches V_f, there is no net current to the probe. At V_f, electrons are repelled to the point that their current is just equal to the ion current. This is precisely the negative potential reached by an insulating wall, which was mentioned before. Finally, as the probe is made more positive, electrons with still lower energies can reach the probe surface and the current increases further. At the point marked V_p, the probe is at plasma potential, without relative bias, and the fluxes of ions and electrons correspond to the random currents given by Eq. (17). The energy of the fastest electrons corresponds to the difference between plasma potential and the point where almost all electrons are repelled (marked in the figure). When the probe is made more positive than plasma potential, ions are repelled so the net (signed) current increases further by an amount equal to the ion saturation current. Note that the saturation asymptotes are "soft," not sharp, because the number of electrons or ions above the "maximum" energies (distribution functions) declines exponentially rather than suddenly. Secondary effects, such as electron emission induced by high energy ion bombardment of the probe also play a role in the gradual increase of current with voltage on the asymptotes.

3. *Potentials and DC Bias at the Electrode Sheaths*

In the last section, it was pointed out that most of the applied potential appears across the plasma sheaths; however, the analyses treated uniform steady fields and voltages in contrast to the RF alternating current generally used for device processing. The high mobility of electrons means that most of the applied potential appears across the momentary negative sheath, which is denuded of electrons. Hence, conductivity in the negative sheath is low, so at high frequency ($\gtrsim$ 5 MHz) continuity through the sheath (in the negative half cycle) is provided by displacement current rather than movement of free charge. This requires a high potential across the electrode-to-glow (sheath) capacitance. In the momentary positive sheath, on the other hand, mobile electrons readily conduct current to the metal electrode surface across a small potential fall.

This asymmetric electrode-plasma current-voltage characteristic approximates a leaky diode in parallel with the sheath capacitance (Fig. 19). In most cases, the RF source, which supplies power, has an unbalanced feed (e.g., one lead is connected to chassis common or ground) and there is a "blocking" capacitor in series with the high potential connection (in trade jargon, the electrode connected to the high potential is called the driven

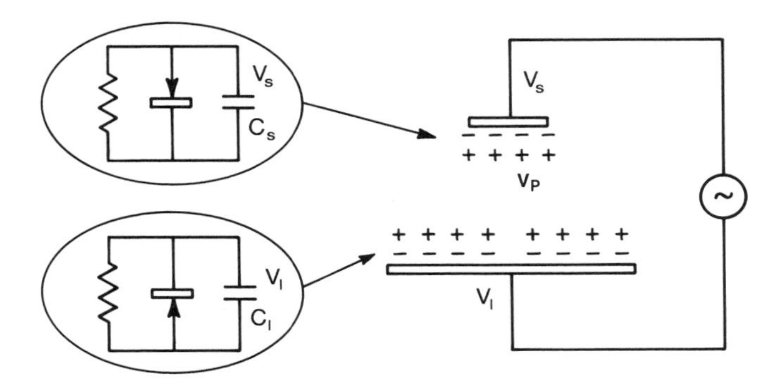

FIGURE 19. Equivalent circuit for reactor electrodes of unequal area. The sheaths can each be represented by a leaky diode and capacitor in parallel.

electrode). The size of the blocking capacitor has no effect on the bias or plasma when it is very large compared to the effective capacitances of the sheaths. These sheath equivalent capacitances are proportional to electrode area. Therefore when both electrodes are of equal size, $C_s = C_l$, equal voltages, 180° out of phase with each other, will be developed across the cathodic sheaths (adjacent to the negative momentary electrodes). The voltage across these equal area sheaths will be the sum of 1) a rectified DC component equal to 0.5 times the peak RF voltage and 2) the applied AC radiofrequency voltage.

When one of the electrodes is smaller than the other, a larger voltage is developed across the smaller (C_s) sheath capacitance. Again, if the blocking capacitor is large, a simplified circuit analysis suggests the voltage across the smaller electrode sheath modelled in Fig. 19 is given by

$$V_{s,\text{sheath}} = \frac{V_0}{1 + C_s/C_1}[\sin \omega t - 1]. \tag{20}$$

Although this model is oversimplified, it shows correctly, that with highly asymmetric electrode geometries, $C_s \ll C_1$, the peak applied RF potential, V_0 is rectified and the total voltage appearing between the smaller electrode and the plasma is the sum of a negative DC component, added to a RF potential determined by the capacitive voltage diffusion. The relationship between these various potentials is shown in Fig. 20. The voltage drop across the sheath at the larger electrode is smaller, because the capacitance scales with area and a higher capacitance results in a smaller potential fall.

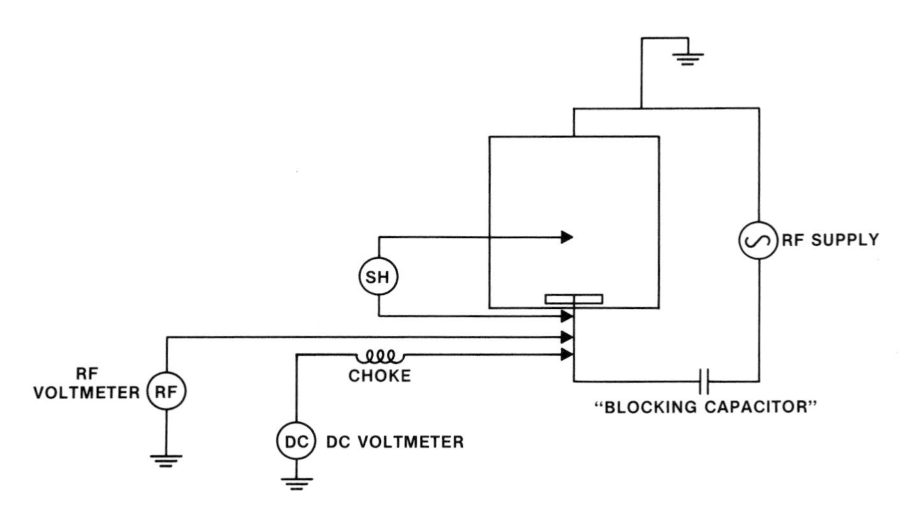

FIGURE 20. Definitions of measured DC bias (DC voltmeter), RF voltage and sheath potentials in a plasma reactor. RF and DC voltages appear between the smaller "driven" reactor electrode and chassis common, or "ground." The RF voltage (RF) is commonly measured with an RF voltmeter or oscilloscope, while the DC "bias" component of the waveform (DC) or can be separated from the RF component with a choke. The sheath voltage, which accelerates ions onto the surface of a wafer (SH) can be a complex function of these quantities because both DC and RF potentials can be developed across the "blocking capacitor" and the sheath capacitances.

The larger RF voltage drop causes intensified ion bombardment at the smaller electrode.

There is another minor effect, which is not represented in Figs. 19 and 20. Electrons can cross potentials as large as their average energy, usually a few volts. This energy acts to offset the plasma potential a small amount, which is sometimes represented by small equivalent voltage sources, $V_{fp} \approx kT/e$ placed in series with each of the sheath diodes in Fig. 19.

Although the ion bombardment energy is controlled by the voltage drop across the plasma sheath, this quantity is not easily measured. In practice process engineers usually monitor the DC potential of the small electrode instead (relative to ground). This negative voltage is called the "DC bias." Empirically, DC bias is known to increase with the ratio of electrode areas, ($[A_1/A_s]$), and with the sheath potential and ion energy at the smaller electrode. However, these relationships have not been analyzed soundly and the DC bias may be as much a function of reactor geometry, the blocking capacitor value, and matching network parameters as of ion bombardment energy. Therefore, while the DC bias is a useful control variable for monitoring or characterizing processes on a given piece of equipment, it should not be used as a basic parameter for transferring processes between different reactors.

In low frequency plasmas (50 kHz $\lesssim f \lesssim$ 1 MHz), the sheath capacitors have a high impedance, so resistances control the potential distribution next to the electrodes. The oversimplified representation of ion and electron conduction through the sheaths by *simple* linear resistors (as shown in Fig. 19), however, is probably not good enough for estimating the potentials in low frequency discharges where highly non-linear voltage current relationships are involved. These sheath potentials and voltage-current characteristics need further study.

4. Effects of Pressure and Frequency on Sheath Potentials

As pressure is lowered below ~ 0.05–0.1 Torr, the sheath thickness and voltage across the sheaths begin to increase from tens of volts to 100s of volts or more. Correspondingly, in this regime, the plasma potential goes up and the ion-substrate bombardment energy rises sharply with decreasing pressure. These effects are a consequence of longer mean free paths and reduced collision rates between electrons and molecules. Electron energy and potentials increase in order to raise the probability of ionization and sustain the plasmas despite lower collision rates.

The change in ion energy tends to shift material removal processes as shown in Fig. 21. At higher pressures chemical etching by radicals is favored when the chemistry provides suitable reactions. Then, at lower pressure, ion energy increases, the density of reactive gas phase neutrals decreases, and energy-driven damage mechanisms tend to be more important. Finally, at very high ion energy and low reactant pressure, physical sputtering will dominate.

Low RF excitation frequency has a similar influence. When gas pressure is in the 0.1–1.0 Torr range and frequency is lowered from around 5 MHz to $\lesssim$ 1 MHz, once again sheath potentials increase dramatically and favor energy-driven ion assisted etching. In this instance the potential increase is attributed to a change in the plasma sustaining mechanism, which is discussed further in Chapter 2. For substrate-gas combinations that show slow chemical etch rates at high pressure, low frequency can be used to speed etching and induce anisotropy, as illustrated in Fig. 22.

Consequently, we see that frequency and pressure are, to an extent, interchangable variables. Either low frequency or low pressure can be used to increase sheath potentials. The important differences are that with low

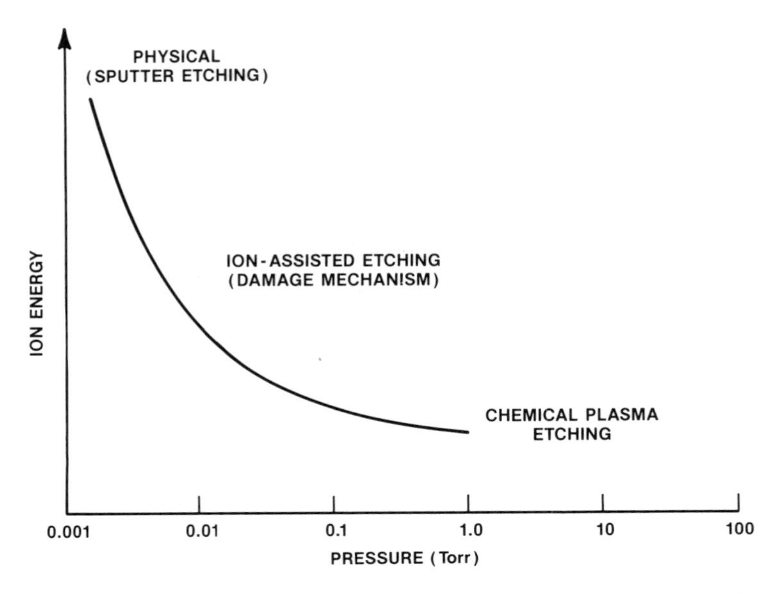

FIGURE 21. Ion bombardment energy decreases and the flux of neutral radicals increases with pressure. Accordingly, the mechanism of plasma etching that is favored changes.

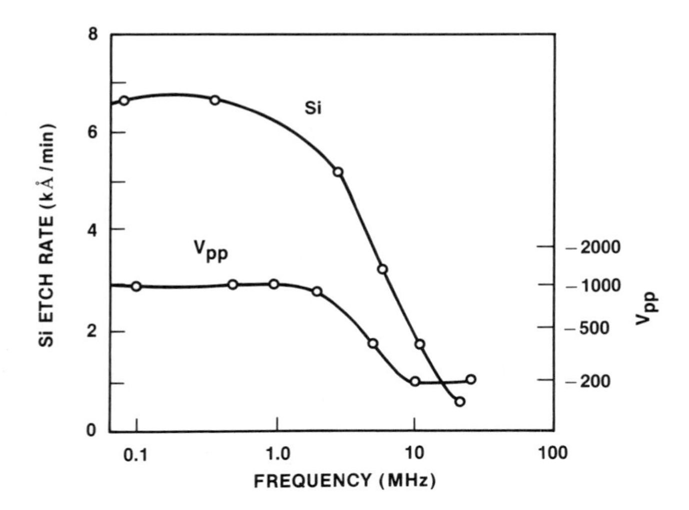

FIGURE 22. The peak-peak voltage and silicon etch rate in a chlorine plasma are plotted against the power frequency. The applied voltage V_{PP} and sheath potential increase with decreasing frequency and provide a more energetic ion flux, which gives higher etch rates.

pressure, the concentration of neutrals decreases relative to charged species (evidently the minimum plasma density is constrained by the ambipolar limit and the maximum plasma density is limited by charge and energy loss processes) and at low pressure there is a correspondingly longer mean free path. A long mean free path facilitates the ejection of sputtered material, while large neutral concentrations tend to favor chemical-surface gasification.

5. *Plasma Etching Jargon*

Process engineers and scientists have used a confusing variety of terms to refer to processing at various etching pressures and sheath potentials. Since these terms are common in the literature, a brief commentary is appropriate at this point.

Plasma etching is the practice of etching or gasifying substrate material in a plasma. Unfortunately, sometimes the phrase has been colloquially used to refer to etching at higher pressure, (e.g., above ~ 0.1 Torr) where ion energies are often low. There is an innuendo that etching in this range is done by neutrals, while etching at lower pressure is done by ions. These concepts and usage are *wrong*, since (1) ions are rarely the etchant, neutrals are responsible for almost *all* reactive etching at pressures above about 0.001–0.005 Torr, and (2) literally, etching is done in a plasma over the entire range.

Reactive ion etching (RIE) is another misused phrase. By definition this term should be used for etching by ions that react with and remove substrate material (e.g., "hungry ions"). As we will see, with this meaning any reactive ion etching is rare, unless beam ion sources are used at lower pressure (see Chapter 6). Unfortunately, the term RIE is commonly used to refer to plasma etching at lower pressures, below ~ 0.1 Torr. A similar phrase, *Reactive sputter* etching (sometimes RSE) was apparently made up to avoid using the words reactive ion etching because RIE was once associated with processes from a major computer company. Taken literally, RSE seems to imply reactive etching with sputtering effects, which may indeed be an accurate description of some plasma etching processes.

Sputter etching refers to etching brought about by the physical effects of ion impact. To avoid confusion, this term should be used only for etching by physical material ejection.

F. GAS PARAMETERS AND MATERIAL FLUXES

Table 3 presents expressions showing how important gas kinetic quantities vary with temperature, pressure and molecular weight. Several important

Table 3 Gas number density, molecular velocity, wall flux and mean free path for molecules of gas 1 moving through gas 2, as a function of temperature (T), normalized pressure ($p_o = 273p/T$), molecular weight (M), and molecular radii (r_1, r_2). Working formulae are given as a function of normalized pressure p_o (in Torr corrected to 293°K: $p_o = p \times 293/T$), M and d_o (in Å), with tabulated values for some gases at 1 Torr (data from Ref. 26).

	Number Density cm^{-3}	Molecular Velocity cm/s	Wall Flux cm^{-2}/s	Mean Free Path (λ) cm
Formula	$3.3 \times 10^{16}\, p_o$	$\frac{8kT}{\pi M}$	$\frac{nv_o}{4}$	$\frac{1}{\pi n_2 (d_2 + d_1)^2 (1 + \frac{m_1}{m_2})^{1/2}}$
Pure Gas, ($m_1 = m_2$, $d_2 = d_2$, T=293 K)				
Pressure Dependence	$3.3 \times 10^{16} p_o$	$\frac{2.5 \times 10^5}{M}$	$\frac{3.26 \times 10^{21} p_o}{\sqrt{M}}$	$\frac{0.0171}{r_1^2 p_o}$
298 K, 1 Torr				
He (M=4, r_o = 1.09Å)	3.3×10^{16}	12350	1.6×10^{21}	0.014
O_2 (M=32, r_o = 1.81Å)	3.3×10^{16}	43700	0.58×10^{21}	0.0052
NH_3 (M=17, r_o = 2.22Å)	3.3×10^{16}	59800	0.79×10^{21}	0.0034

dependencies should be noticed. Gas density and molecular flux to a surface are proportional to pressure at constant temperature. The mean free path λ of species is inversely proportional to density, pressure, and collision cross sections (square of the molecular radius sum). Mean free paths and molecular collision diameters for common gases at 1 Torr and 298°K are on the order of 0.5 mM, as shown in Table 3. Since λ is inversely proportional to pressure, this means that at 1 mTorr the mean free path will be a sizable fraction of reactor dimensions (centimeters).

A simple comparison of fluxes, shown in Fig. 23, indicates that neutral radicals, not ions, must be the agent responsible for substrate material removal. That is why reactive ion etching is an inappropriate synonym for low pressure etching. The role of ions, when they participate in plasma etching, is more physical than chemical—they enhance the etch rate and lend directionality. At usual etch rates, though, they are not the agent that combines with substrate material to form gaseous products. Within the

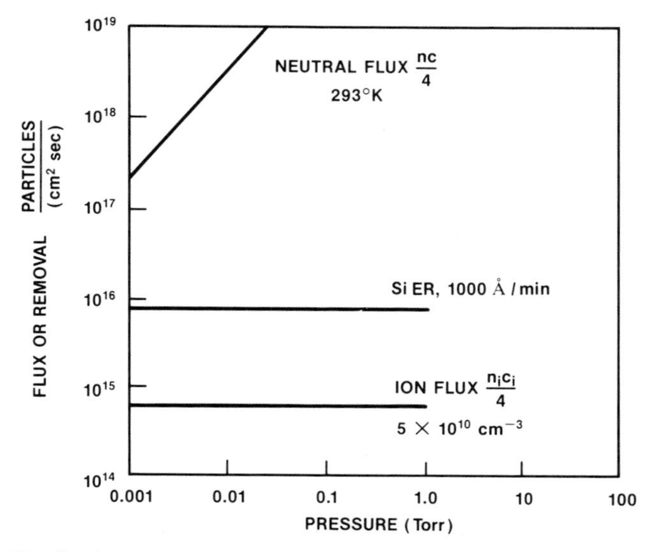

FIGURE 23. Product, neutral and ion fluxes versus pressure for a modest 1000 Å/min. etch rate. Even at the lowest pressure shown, the neutral flux could support far higher etch rates while the ion flux is more than 10 times smaller than the surface atom removal rate.

usual range of plasma densities, 10^{10}–10^{11} cm^{-3}, the flux of ions to the wall is probably no greater than $\approx 10^{15}$ cm^{-2} s^{-1}. By contrast, a moderately low etch rate, 1000 Å/min, corresponds to almost 10^{16} Si-atoms removed per cm^2-sec. Therefore, even if every ion reaching a substrate surface reacted and removed a surface atom, they would be far too few to account for this etch rate. Actually, 1000 Å/min is relatively low as etch rates go, and a 10^{15} cm^{-2} s^{-1} ion flux may be uncharacteristically intense. The flux of neutral particles, on the other hand, is more than enough to sustain the etch rate, even at pressures as low as 1 mTorr. And, it is common for etchant radicals to comprise 5–50 percent of the neutral population [27, 28, 29].

G. SEQUENTIAL STEPS IN ETCHING

The etching process can be viewed as a series of steps illustrated by the energy-driven ion assisted etching of *undoped* silicon by chlorine:

(I) ion and electron formation (which balances loss of charged species and is part of a more complex reaction set)

$$e + \begin{cases} Cl \\ Cl_2 \end{cases} \rightarrow \begin{cases} Cl^+ \\ Cl_2^+ \end{cases} + 2e, \tag{21}$$

(II) etchant formation

$$e + Cl_2 \rightarrow 2Cl + e, \tag{22}$$

(III) adsorption of etchant on the substrate

$$\begin{cases} Cl \\ Cl_2 \end{cases} \rightarrow Si_{surf} - nCl, \tag{23}$$

(IV) reaction to form product

$$Si{-}nCl \xrightarrow{(ions)} SiCl_{x(ads)}, \tag{24}$$

and, finally

(V) product desorption

$$SiCl_{x(ads)} \rightarrow SiCl_{x(gas)}. \tag{25}$$

For completeness, surface diffusion of adsorbed reactants or reactive intermediates should also be considered since some evidence for such a step has been observed. Chemical etching proceeds similarly, except that Step IV (Eq. (24)) would be replaced by one that does not require ion bombardment,

$$Si{-}nCl \rightarrow SiCl_{x(ads)}. \tag{26}$$

Any one of these can potentially be the slowest step, limiting the etching rate. The purely chemical reaction between undoped silicon and Cl atoms or Cl_2 is slow at ordinary temperatures. However, heavily doped silicon and heavily doped polysilicon react rapidly with Cl atoms, so that anisotropic ion-assisted etching of doped polysilicon (a conductor material in integrated circuits) must be done with an inhibitor chemistry. Note that the relative significance of ions in Step IV [Eq. (24) versus Eq. (26)] can vary with the chemical system. For example, in F-atom etching of silicon by CF_4/O_2 plasmas, the reaction analogous to Eq. (26) is rapid, so ion-assisted etching (Eq. (24)) makes a negligible contribution.

H. VOLATILITY AND EVAPORATION

We already pointed out that a key distinction between plasma etching and sputtering is that the products of plasma etching are volatile. There are many examples of plasma etching in which product desorption is rate limiting. We describe the temperature dependence of evaporation and sublimation rates and present two examples of volatility limited plasma etching.

The evaporation rates of a material (A) of molecular weight M_a is proportional to its vapor pressure (30), p_a,

$$\mu_A = \alpha \left[\frac{M_a}{2\pi RT} \right]^{-1/2} p_A. \tag{27}$$

Experiments show the material-dependent efficiency factor α, is usually between 0.1 and 1.0. The vapor pressure p_A in turn depends on temperature (T), according to the Clausius Clapeyron equation, (31)

$$p_A = C_A e^{-\Delta H/RT}, \tag{28}$$

where ΔH is the latent heat of vaporization. Equations (27) and (28) can be combined to obtain:

$$\mu_A = \alpha C_a \left[\frac{M_a}{2\pi RT} \right]^{-1/2} e^{-\Delta H/RT}. \tag{29}$$

The exponential term dominates the temperature dependence. Thus, when volatility is rate limiting, the slope of material removal rate, μ_A, plotted against 1/RT, is ΔH, the latent heat of evaporation.

Experiments show that plasma etching of InP at ~ 0.1–1 Torr is rate-limited by $InCl_3$ sublimation [32]. The etch rate has an activation energy (slope of rate versus 1/RT) of 35 kCal/mole, in good agreement with the latent heat of $InCl_3$ sublimation. At 250°C Eq. (27) predicts a maximum evaporation rate of nearly 45 μm/min. for this system (assuming $\alpha = 1$), whereas the measured etch rate of about 10 μm/min, corresponds to $\alpha \approx 0.2$—a reasonable value for the evaporation efficiency.

Aluminum etching presents a case where an involatile product is thought to seriously limit the etching rate. Aluminum is usually etched in chlorine-containing plasmas where the principal etch product is $AlCl_3$. Since the chloride has a vapor pressure just under 100 Torr at 150°C [33], its removal should present no problem at slightly elevated temperatures. Copper is often added to aluminum metallization to suppress electromigration, however, and the copper chlorides are far less volatile. Figure 24 shows recent data on the rate of Cu_3Cl_3 evaporation at low temperatures [34]. A surface temperature around 150°C is required for reasonable rates of Al removal with 5% copper.

This brief discussion ignores other factors and concerns in Al-Cu etching. For instance, small amounts of surface copper may be sputtered, especially if there is high ion energy. And the vapor pressure of Cu_3Cl_3 in equilibrium

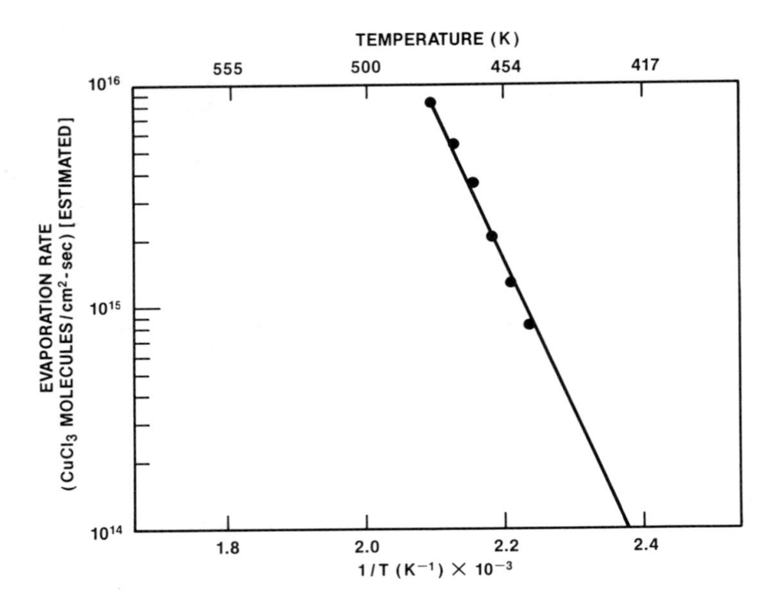

FIGURE 24. The evaporation rate of Cu_3Cl_3 is low at ordinary temperatures and shows a strong temperature dependence (high activations energy). This low product volatility is why etching of aluminum with a substantial (> 4%) copper content is inhibited at room temperature, unless plasma conditions which favor sputtering are used.

with an $Al/AlCl_3$ surface may differ from the vapor pressure of the pure chloride (because of solution and surface free energy effects). Aluminum etching is discussed in detail in Chapters 2 and 5.

IV. Process Requirements and Examples

Plasma etching processes are judged by their *rate*, *selectivity*, *uniformity*, *directionality*, *surface quality* and *reproducibility*. Production rates and cost are determined by these factors. It should be emphasized that high etching rates do not necessarily result in lower cost. Equipment size (footprint) and yield are major cost considerations because the capital and operation costs of clean rooms are a large fraction of total expense.

Uniformity really refers to two things: (1) the evenness of etching across a wafer and (2) the degree to which etch rates are maintained from wafer to wafer in the same reactor. Etch rate variations between wafers within a reactor can be caused by reactor geometry, or by plasma composition nonuniformities attributable to the gas flow or plasma operating parameters. Unevenness across a wafer can arise from localized depletion of gas phase etchant when it reacts with the film, or from the influence of a wafer on plasma boundary conditions. Sometimes the high concentrations of etchable material in specific areas of a fine pattern induce these variations. This effect is called feature sensitivity.

Selectivity defines relative material etching rates in a plasma. The degree to which differential etching of one material relative to others is possible in a plasma is one of several factors that limits the fidelity of pattern transfer (Section IV.C). Selectivities may refer to materials in the same patterned layer (level), to material in the layer immediately below, or to attack of the pattern mask relative to the substrate film. The plasma resistance of the mask is an important term in the analytical expression for the minimum attainable feature size (Eqs. (34), (40)).

Anisotropic ion-assisted straight-wall and isotropic chemical (undercut) etching profiles are the two extremes of directionality. Between these limits lie mixed combinations of these profiles, crystallographic chemical etching (Section I.E of this chapter and Chapter 2) and involuted profiles caused by yet unidentified mechanisms. Anisotropic etching is necessary to attain small feature sizes ($\lesssim 3\ \mu m$), but a slight taper is better for uniform deposition in and over the crevices and steps of a pattern.

Surface quality covers a wide range of things that can be influenced by many variables, including temperature, ion bombardment, etchant and crystallographic orientation. Sometimes etching leaves a finished surface smoothness similar to the unetched film. Under other conditions etched surfaces may be unacceptably rough, pitted or covered with cones or spikes or substrate material. Contamination from sputtering, etch residues and electrical damage caused by contamination, radiation (including ion bombardment) and charging may degrade surface quality.

At first glance, reproducibility might appear to be more a matter of process control than an inherent process characteristic. However, etching equipment and processes sometimes have insidious "memory" or hysteresis effects that are difficult to control. The initial condition of the reactor can be influenced by accumulation of etch products, exposure to the ambient atmosphere (with adsorption of air and moisture on reactor surfaces) and surface modification brought about by reaction with chemical plasmas in the equipment. Aluminum etching is sensitive to moisture, oxygen and surface condition. Thus, reproducible aluminum etching was difficult to achieve until reactor design and chemistries were developed to control these hidden variables. Another source of variability is load size, the number of wafers in a reactor. When etching reactions consume most of the etchant (e.g., etching is the major loss process for these species), the etching rate

Table 4 Typical gases, additives and radical species used for plasma etching a variety of materials.

Examples of Plasma Etching Gases and Radicals

ETCHING SPECIES	SOURCE GAS	ADDITIVE	MATERIALS	MECHANISM	SELECTIVE OVER
F	CF_4 C_2F_6 SF_6 NF_3 ClF_3 F_2	O_2 O_2 O_2 None None None	Si	Chemical	SiO_2 Resist
CF_x-film	CF_4 C_2F_6 CHF_3	H_2 H_2 None or O_2	SiO_2/Si_3N_4	Ion-energetic	Si
Cl	Cl_2 Cl_2 CF_3Cl	None C_2F_6 None	undoped Si n-type Si	Ion-energetic Ion-Inhibitor	SiO_2
Cl	Cl_2	BCl_3 CCl_4 $CHCl_3$	Al	Ion-inhibitor	Resist SiO_2

will be inversely proportional to the number of wafers in the reactor (see Chapter 2).

Plasma etching has displaced most wet etching steps because it can deliver superior results according to these criteria. For instance, plasmas routinely achieve controlled anisotropic etching of fine features; whereas wet anisotropic etching is limited to special cases where material is removed along specific crystallographic directions. Wet etching, unlike plasma etching, suffers a decrease in rate as reagent solutions are consumed, and rate variability caused by surface tension effects (lack of wetting). It also can cause resist lifting from capillarity. Finally, plasma etching is easier to automate and there are smaller amounts of hazardous material to handle.

Typical gases used for plasma etching a variety of materials are shown in Table 4. These feeds mainly contain freons, atmospheric gases and the halogens. Most are safe and easy to handle—they are nontoxic or have high threshold limit values (TLVs, see Chapter 7). There is danger at the other end, however. Although the feed gases may be benign, highly toxic substances are formed in plasma chemical reactions and should be given careful attention in process and equipment design (see Chapter 7).

A. LINEWIDTH CONTROL

Probably the single most desired result of a plasma process is faithful replication of the mask pattern in the film of semiconductor device. Submicron features can be formed with anisotropic etching chemistries and conditions. In practice the anisotropy brought about by the directed ion flux in the plasma environment allows the final etched feature to be within 10% of its dimension in the mask.

It is hard to check the accuracy of etched feature sizes in small submicron patterns. Line sizes generally must be kept within 10% of the mask dimension, and such small dimensions cannot be measured with optical microscopes. While scanning electron microscopes (SEM) are usually used

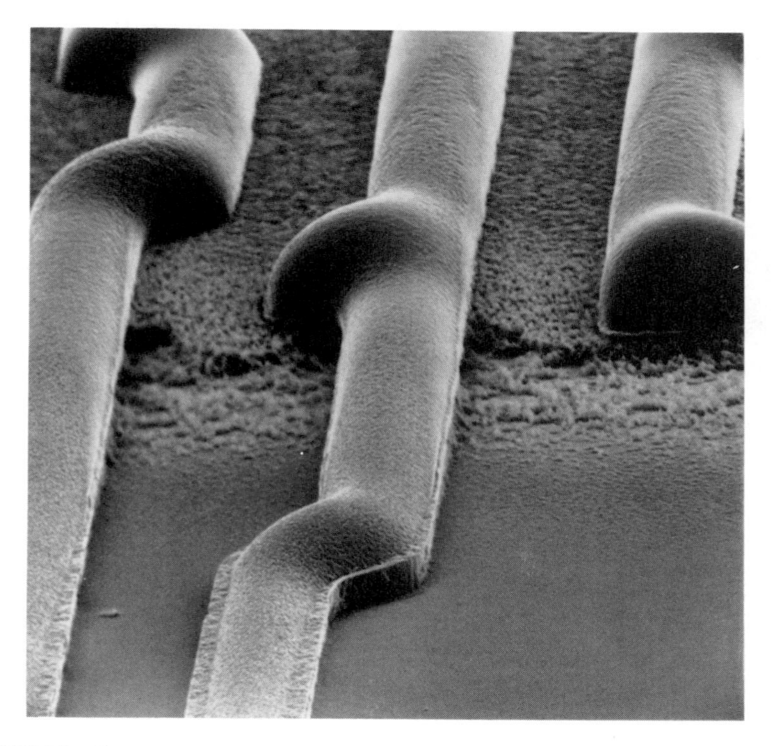

FIGURE 25. Feature size control when plasma etching aluminum. The upper portion of the wafer was protected from the etch by covering it so feature size in the resist and etched aluminum pattern can be accurately measured (from the pitch dimensions) and compared to determine the fidelity pattern transfer. After etching the aluminum, the uncovered resist was removed slightly (in an oxygen plasma) so the edge is clearly visible for measurement.

for microcircuit metrology, great care is required. Typically, dimension changes are measured by mounting etched and unetched (resist only) samples side-by-side in the SEM and comparing the two during sequential measurements. With this procedure, however, the feature pitch (line and space) should be used for comparison, since drift in the SEM may lead to error, particularly in the size scale reference marker. One way to minimize the effect of drift between sequential measurements is to cover a portion of the wafer during etching, so etched and unetched features will be visible side-by-side. Figure 25 shows such a SEM micrograph. The top of the photo shows the resist mask where it was protected from etching, while the bottom pattern was etched. The resist mask was pulled away on the bottom to unambiguously reveal the aluminum edges for measurement.

To preserve feature linewidth when highly doped polysilicon is etched by isotropic etchants such as fluorine or chlorine atoms, the sidewalls must be protected by an inhibitor film. Figure 26 illustrates the effects of adding a film-former, CCl_3F, to SF_6 [35]. The undercut in a polysilicon feature beneath the resist mask is plotted as a function of the overetch [36] time

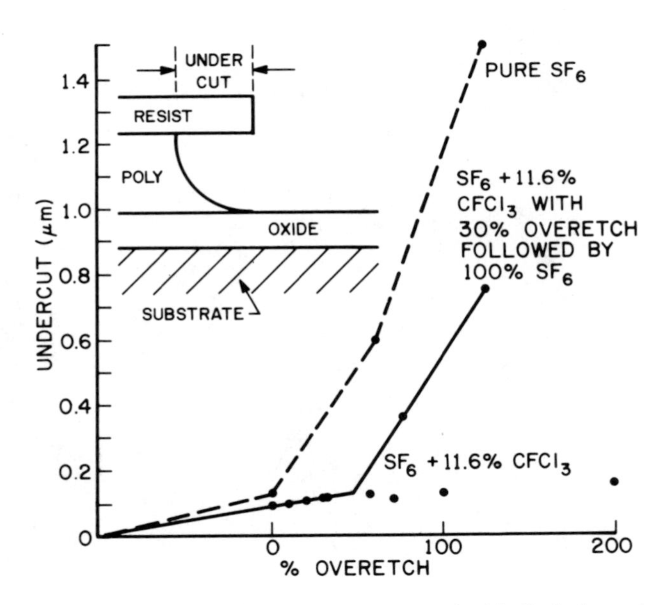

FIGURE 26. Undercut of a polysilicon film versus overetch with (dashed curve) and without (solid line) the addition of 11.6% Freon 11 ($CFCl_3$). With $CFCl_3$ there is little undercut even after long overetching (lower dotted data), indicating the formation of a protective sidewall film which inhibits lateral attack by F-atoms. When the Freon 11 is turned off after a 50% overetch, severe undercutting starts at once (solid line).

(time after the polysilicon film was cleared from the featureless planar wafer area). The upper dashed curve shows that "poly" is severely undercut by F atoms, which are formed in a pure SF_6 plasma. When 12% CCl_3F is added to the SF_6 feed, however, there is barely any undercut, even after substantial overetching. Obviously a film forms on the sidewalls of the poly feature, which inhibits attack by fluorine atoms. The protective film is apparently not present on horizontal surfaces.

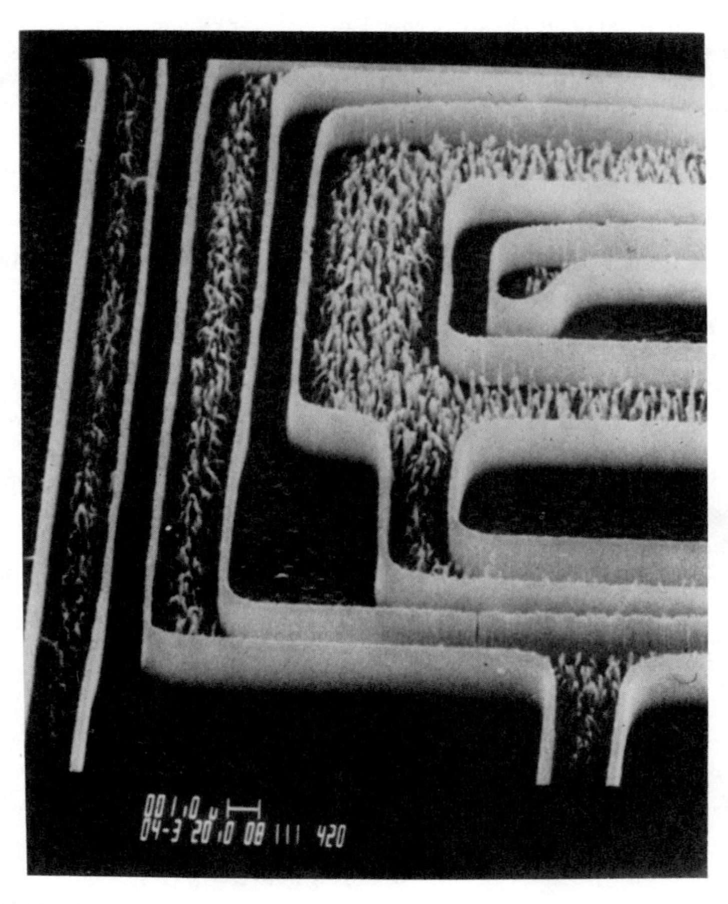

FIGURE 27. Replica of the sidewall films formed on features during anisotropic aluminum plasma etching. The micrographs made after three sequential steps: a) the masked aluminum film was plasma etched, b) the resist mask was stripped by an oxygen plasma, and c) the exposed aluminum pattern was removed by plasma etching again (see text).

If the poly is first etched in the gas mixture for a period of 50% overetch, and then the feed composition is changed to pure SF_6, there is not much undercut during the (50%) overetch period until inhibitor-forming gas is turned off. At this point the feature begins to be severely undercut. These observations have been interpreted to mean that the steady-state growth rate of the inhibitor film is fast enough to protect the poly sidewall, but the layer is so thin and easily removed that chemical attack undercuts the mask as soon as the CCl_3F flow is stopped.

Inhibitor films also prevent lateral chemical attack during the etching of aluminum in chlorine-based plasmas. Since aluminum is normally passivated by a thin ($\approx$ 50 Å) native oxide, which is impervious to chlorine, the etch chemistry must supply additional species that remove this layer. BCl_3 added to Cl_2 fulfills both rolls. As BCl_x radicals digest away the oxide layer and Cl and Cl_2 chemically etch aluminum, the BCl_x radicals also form sidewall film that delivers tight dimensional control of the feature size (refer to Fig. 25).

Figure 27 shows the result of a processing sequence designed to make the sidewall inhibitor-film visible. First, the masked aluminum was etched in the normal BCl_3/Cl_2 process. The resist mask was then stripped away in an oxygen plasma, exposing the newly patterned aluminum surface. Finally, the exposed aluminum pattern was etched in a BCl_3/Cl_2 plasma for a second time. Figure 27 shows thick standing sidewalls of a B-O-Al-contain-

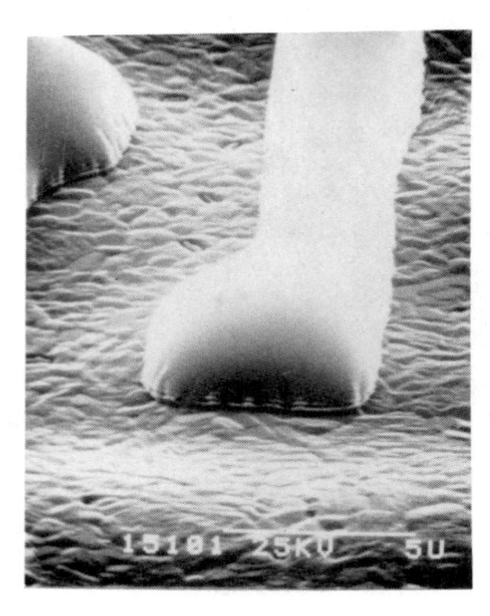

FIGURE 28. Copper agglomerates on aluminum (left) micrograph act as "micromasks," that leave "pillars" after the etch (right).

ing material remaining, which were formed by reactions between the inhibitor film, aluminum and the oxygen plasma.

This inhibitor-based anisotropic etching is capable of submicron resolution. The left-hand-side of Fig. 28 shows the top surface of a 1.0 μm thick aluminum film with a resist mask as it appears before etching. Copper agglomerates are visible on the aluminum surface. They originate from copper added to the aluminum to prevent electromigration and are formed when aluminum(copper) is deposited on the wafer at elevated temperature. When aluminum(copper) is etched, these copper areas act as "micromasks." The sidewall film protects even these small features from lateral attack forming the submicron pillars in the right side of Fig. 28. Obviously, since these pillars cannot be tolerated, copper agglomeration must be prevented.

B. APPLYING ISOTROPIC AND ANISOTROPIC ETCHING

As processes were converted from liquid chemistries to dry etching, which permitted faithful replication of a mask, anisotropy was often accompanied by a loss of selectivity over the underlying substrate film. The anisotropic character of a process sometimes has to be relaxed in order to improve selectivity over the substrate.

The need for an isotropic plasma process in pattern transfer is illustrated by Fig. 29. An anisotropically defined first level polysilicon feature (poly I) has been covered with an interlevel dielectric film, and a second level polysilicon (poly II) conductor is conformally deposited over it by chemical vapor deposition (CVD). During each patterning of the poly II film, an *anisotropic* plasma process would clear the planar portions of the circuit before the thicker poly II along the poly I feature sidewall is removed. The filament, however, must be completely removed to avoid electrical shorts between adjacent poly II runners. *Overetching* [36] for a time based on the step height of the poly I feature would eventually clear this filament from the step, but only at the expense of attack on the underlying substrate (oxide) layer. This would be intolerable, depending on the selectivity ratio for the poly II over oxide. If isotropic etching is used for the last part of the poly II removal process, the overetch time can be shortened considerably. The duration of the isotropic etch should be kept as short as possible because the poly II feature is being attacked laterally at the same time that the sidewall filament is being removed, resulting in loss of linewidth. The SEM micrograph in Fig. 29 clearly shows the poly II filament along the poly I feature, which has been "decorated" by using an oxide etch to exaggerate the relative proportions.

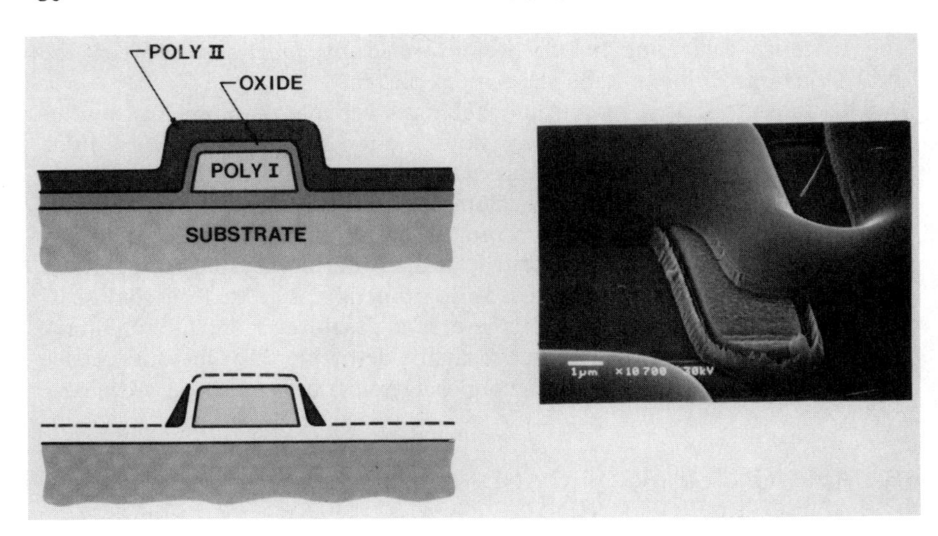

FIGURE 29. Cross section of a typical two-level conductor structure showing the poly II and interlevel oxide over a patterned poly I feature. The "poly II" thickness over planar areas is less than the vertical thickness at steps so overetching is required. The SEM micrograph (right) shows a residual poly II "filament" which is left along a poly I feature when etching is prematurely terminated. The dotted line in the bottom diagram shows where oxide was dissolved in HF to highlight the filament in the micrograph.

This problem is aggravated when the first level conductor is a multi-thin composite, such as a silicide film over a doped polysilicon (I) layer. A typical example is shown in Fig. 30, where the first level runner is a layered silicide/poly I film. Unfortunately, during the definition of the bi-layer conductor I, the etch step slightly undercuts the poly I under the silicide. Later when the interlevel dielectric and top poly II film are deposited conformally, this re-entrant structure is filled in so no void exists. During the plasma etching of the poly II film, however, the shielded filament under this ledge cannot be removed by anisotropic etching.

To deal with this problem either (1) the re-entrant ledge must be avoided by improving the first (poly I) etch process, or (2) an isotropic etch step must be provided to attack the filament laterally (and suffer the resulting poly II feature undercut). Hence, the overall sequence would consist of an anisotropic step to the planar (horizontal surface) polysilicon clearing point, followed by an isotropic etch to remove the sidewall and overhang-protected filaments. The selectivity requirement is relaxed for the first step, and since the second step should be short, even its selectivity needs may be undemanding. There are many isotropic plasma chemistries that can be used for polysilicon. Pure SF_6, which produces abundant amounts of atomic fluorine, displays good selectivity for silicon over SiO_2 (see Chapter 2) and has often been used for this sequence.

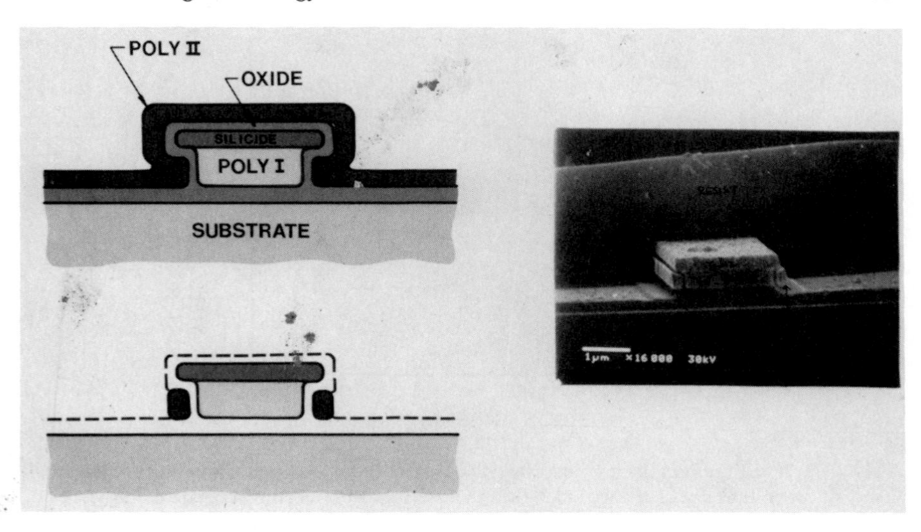

FIGURE 30. If the first level conductor has a re-entrant profile (upper diagram), interlevel oxide and second level conductor (poly II) fill the void during deposition. Thicker poly II along pattern steps can be removed by overetching in an anisotropic process, but poly II that is protected by oxide (at the re-entrate ledge) cannot be removed without isotropic etching. The micrograph shows this protected filament (highlighted by removing the oxide).

C. SELECTIVITY

1. *Selectivity in Ion Etching*

Selectivity in sputtering almost always decreases with increasing ion energy (see Fig. 31), while the yield, or sputtering rate, increases with energy. Although physical sputtering is *relatively* unselective, there are material- and ion-dependent thresholds for material removal that are related to the binding energy and, to a lesser extent, to the structure of the substrate. These differences between thresholds and the yield curves provide some selectivity. At low ion energies it is sometimes possible to etch selectively (but slowly) by maintaining the bombardment energy above one threshold and below another. At high energies, the relative rates tend to even out because differences in material binding energies are small relative to the ion kinetic energy.

There is evidence that energy-driven ion plasma etching processes have similar threshold effects and yield curves, and experiments show that high ion energies degrade plasma etching selectivity too. Moreover high energy ion bombardment is always a concern because of the mechanical and electrical damage it can do to substrates. Consequently, ion-assisted etching usually requires a careful compromise—there must be enough ion bombardment for acceptable throughput and profile control, but its energy and

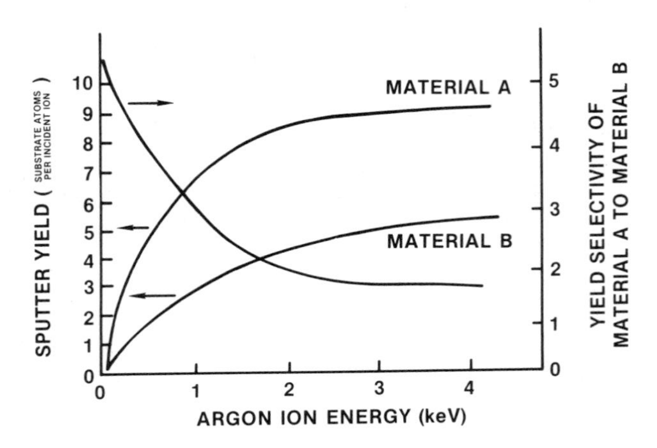

FIGURE 31. In argon sputtering, the yield (number of substrate atoms removed per incident argon ion) does not vary much (less than 10 fold with most materials) and it increases with the sputter ion energy. However the selectivity, or the ratio of the sputter yields for different materials, generally decreases with increasing ion energy. Yields for low energy sputtering (< 25 eV) are less than unity.

flux must be limited or there will be device damage and an intolerable degradation of selectivity.

In contrast to chemical or ion-enhanced plasma etching, the product leaving a sputter-etched surface may be involatile. At low pressure when the mean free path is comparable to interelectrode distances, ions sputtering "throws" material across the reactor. In fact this is the way sputter coating is done where a second substrate intercepts sputtered material from the (etched) target. When pressure is raised to the point that the mean free path is short compared with reactor dimensions, the sputter-ejected material collides with gas molecules and is reflected back to the substrate. This roughens the surface, and there is little net material removal. Thus, we see low pressure and a long mean free path are *required* for sputtering, in contrast to plasma etching where mostly volatile products are removed.

Another sputtering effect that influences selectivity is a phenomenon called "facetting." Facetting arises because the sputter yield for materials is usually a strong function of the angle at which ions are directed at the surface. The sputter yield for a few typical materials (gold, aluminum and photoresist) are given in Fig. 32. The sputter-etch rate of resist, for instance, reaches a maximum rate at an incidence angle of $\approx 60°$, which is more than twice the rate at normal (0°) incidence.

When resist is sputtered, the rapid removal rate at this critical "facet angle" causes a planar surface to develop (Fig. 33a, b) and propagate in this

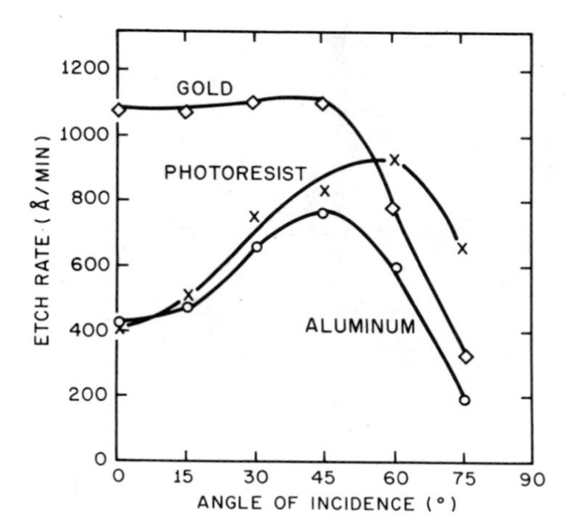

FIGURE 32. The effect of the incident angle of ions on sputter etch rates for gold, aluminum and photoresist. The sputter rate of resist at 60° incident ion flux is more than double that at normal (0°) ion incidence.

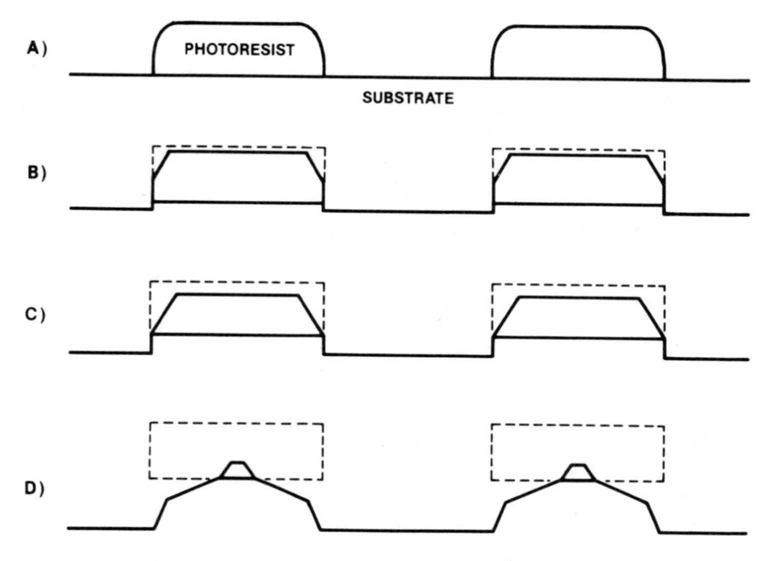

FIGURE 33. Sputtering creates angled features. If there is sputtering during the etch process, an angled surface forms on the resist mask and propagates at the facet angle ($\approx 60°$). If the mask is eroded away, this phenomena influences the feature profile even when etching is anisotropic. Sputtering the exposed feature forms a distinct facet angle.

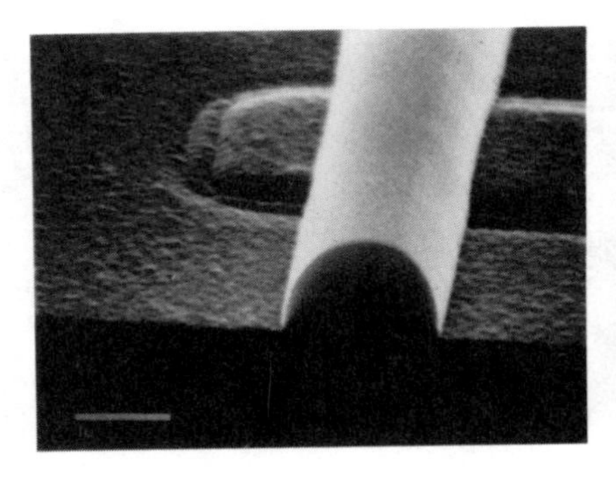
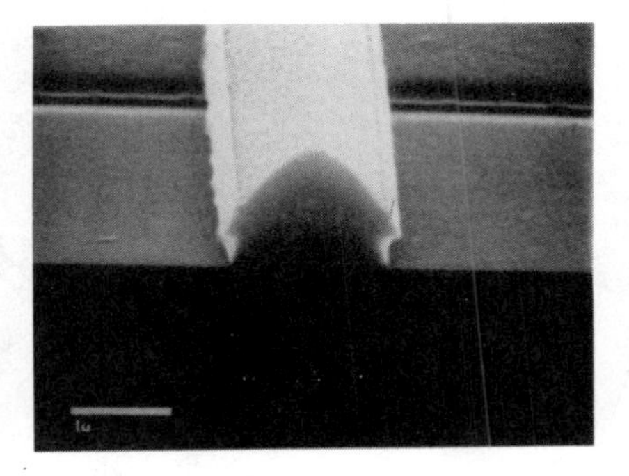

FIGURE 34. Left: a resist profile that has been rounded by post-development heat treatment ("bake") to enhance facetting. Right: micrograph shows the resist profile is facetted after etching polysilicon.

direction. As sputtering continues, this facet plane becomes larger until, if the sputtering continues long enough, it intersects the resist-substrate interface (Fig. 33c). Sputter etching beyond this point will expose the substrate under the mask edge to the plasma environment and ion bombardment may then cause a facetting of the substrate along its own preferred direction (Fig. 33d). If the mask shrinkage from this attack advances faster than the maximum sputter etch rate for the substrate (along its facet angle), sputtering of the substrate at near the normal incidence will develop a second face. The SEM micrograph in Fig. 34 (left) shows a characteristic rounded projection-exposed resist contour before etching. After an etch process the resist profile (Fig. 34 on the right) exhibits exaggerated attack at the facet angle. This effect can be minimized or eliminated by providing a resist profile with a flat top surface oriented at 0° with perpendicular sidewalls at 90°. While this type of controlled resist profile is the natural result of most direct-step-on-wafer (DSW) photolithography, postbake temperatures have to be controlled so that this profile is not degraded by flow during the resist-treatment sequence.

2. *Selectivity Analysis*

Selectivity, the etch rate ratio between materials, is a critical figure of merit. Conceptually, there are two kinds of selectivity that are needed, (1) selectivity for etching a substrate rather than the pattern mask (resist durability), and (2) etching selectivity between two material layers. The second sort of selectivity is especially critical when the film that is being patterned

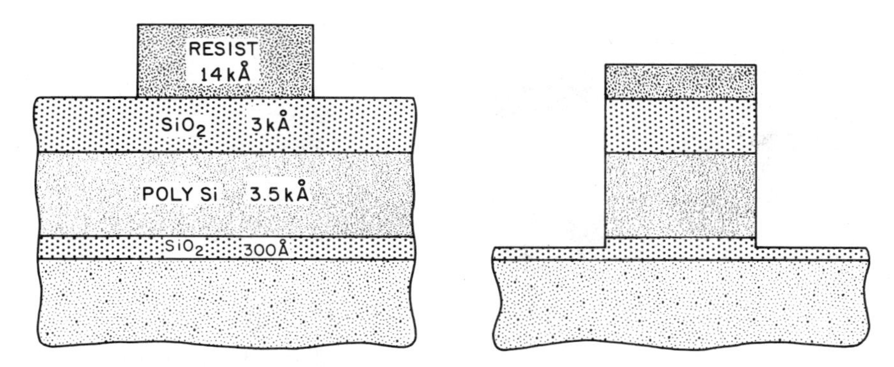

FIGURE 35. Left: cross section and typical thicknesses of structure in microelectronic device fabrication. The selectivity between each of the films that are to be etched (upper SiO_2 and poly) and the resist mask should be greater than 4 : 1, while selectivity between the poly and thin underlying gate oxide should be more than 20 : 1.

overlays a thin substrate (like a 300 Å gate oxide layer). Roughly, the minimum required selectivity is simply the thickness ratio between two material films. In Fig. 35 a representative multi-film structure is shown. The mask/film (oxide + poly) thickness ratio (2.2) and polysilicon film to lower oxide ratio (11.7) suggest minimum selectivity requirements. Selectivity is determined by the amount of material removal that can be tolerated, however, not the total film thickness. This value can range from a small fraction up to 50% of a sublayer. The topography of features introduces another variable. There are significant differences between film thicknesses at steps and over planar areas that must be taken into account. Moreover, this naive calculation does not consider film thickness and etching rate non-uniformities, which require "overetching" beyond the time to "clear" a film at the nominal film thickness and removal rate. In the following paragraphs, these factors are analyzed in detail and are used to formulate quantitative selectivity criteria.

To some extent, the resist mask thickness and selectivity for the film being etched are interchangeable degrees of freedom. Figure 36 shows a nominal 1.4 μm thick positive resist film pattern applied over a 1.0 μm film of aluminum. The features beneath the aluminum layer produce a surface topography that causes the resist to decrease from the 1.4 μm over planar areas to only about a third of this value on high steps. Figure 37 shows a wide aluminum feature that was attacked after its overlying resist was digested (because of the thin resist over the highest feature), while an adjacent narrow feature had enough mask to survive etching in the CCl_4 plasma.

FIGURE 36. The micrograph shows an undulating resist film thickness over an aluminum film, caused by step topography in the circuit patterns beneath. The resist is less than one-third as thick (< 5000 Å) where it covers the highest feature as it is over the lowest (> 1.5 microns).

We now calculate the selectivity requirements for plasma etching based on work by Mogab [37]. It is assumed that etching can have both isotropic and anisotropic components and that there is undesired plasma erosion of both the mask and the substrate underneath the layer that is patterned.

When there is an isotropic undercut, it will be necessary to make the mask opening smaller than the desired feature linewidth. The bias **B** (for mask compensation) is the difference between the mask feature size and the *top* dimension of the etched feature (see Fig. 38). For completely isotropic etching to the endpoint (film just clears with no overetching), the bias will have to be twice the film thickness

$$|\mathbf{B}| = |d_m - d_f| = 2y_f. \tag{30}$$

The absolute value sign makes this equation apply to both lines (**B** positive) and spaces (**B** negative) in Fig. 38.

The degree of anisotropy **A**, is defined as

$$\mathbf{A} = 1 - (R_h/R_v), \tag{31}$$

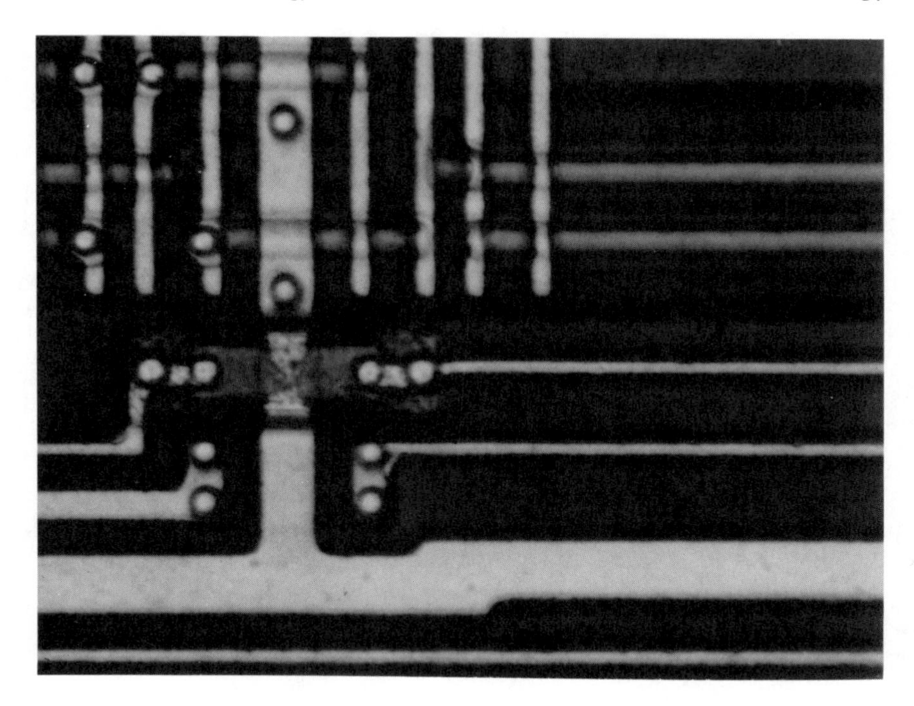

FIGURE 37. If the aluminum-to-resist selectivity is too low (as in some CCl_3-based chemistries), wide features with insufficient resist (as the high regions in Fig. 36) can be etched open, while some narrow lines with more resist are correctly patterned, as shown in this micrograph.

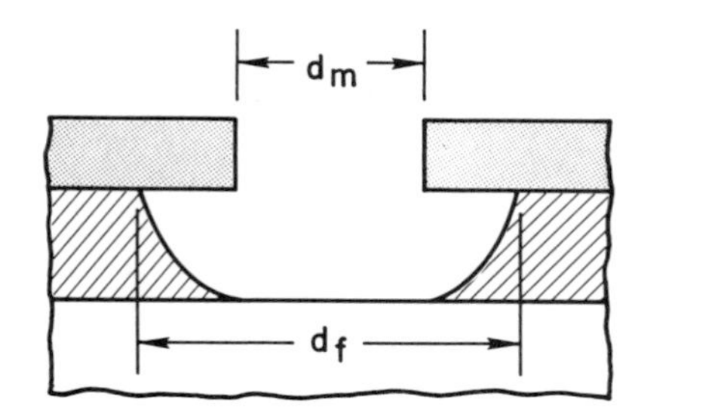

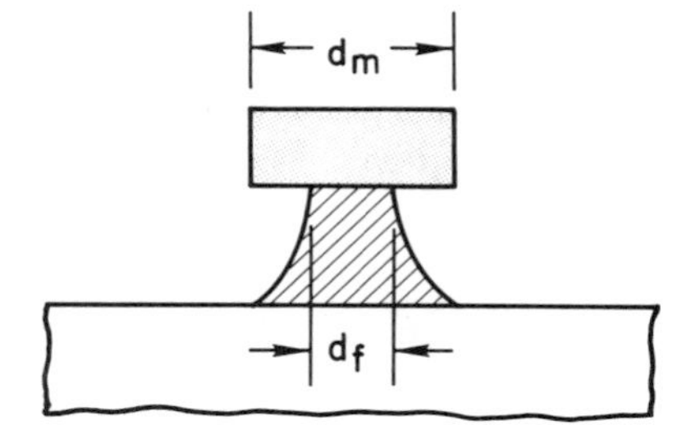

FIGURE 38. Left: the mask feature (hole) d_m is increased to d_f by a (isotropic) process with equal vertical and horizontal etch rates. Right: isotropic etching with a mask feature (line) d_m results in a reduced etched feature thickness d_f at the top of the patterned film. In both cases $|d_f - d_m|$ is twice the film thickness. Etching is stopped when the area exposed under the resist mask is removed (no overetch).

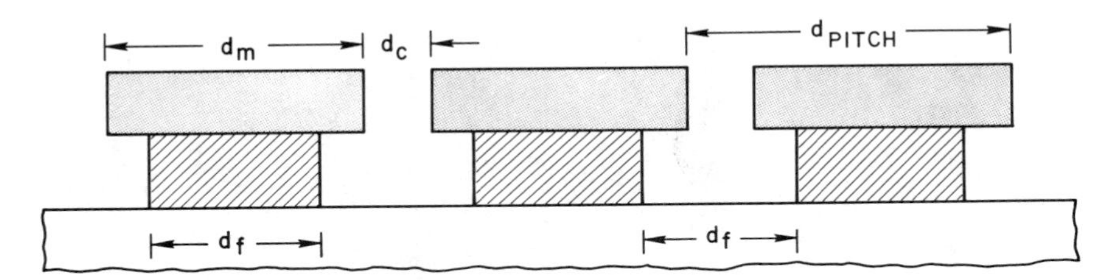

FIGURE 39. To achieve equal lines and spaces in an isotropic process, the resist mask dimension, d_m, must be "biased" larger than the desired feature size, d_f, by twice the film thickness, d_f. The minimum resolvable mask feature size, d_c, however, limits the amount of bias that can be applied. In an anisotropic process, bias is unnecessary so that the pattern can be etched with $d_f = d_m = d_c$. Sides of the etched feature are drawn vertical for clarity.

where R_h and R_v are the horizontal and vertical etch rates. For isotropic etching, R_h and R_v are equal ($\mathbf{A} = 0$) whereas for ideal anisotropic etching, $R_h = 0$ ($\mathbf{A} = 1$). **A** may be expressed in terms of the etch bias and the film thickness,

$$\mathbf{A} = 1 - \frac{|\mathbf{B}|}{2y_f}. \tag{32}$$

The lithographic constraints are illustrated in Fig. 39, where a pattern of equal lines and spaces *resulting* from an etching process is shown (recall that the dimension d_f refers to the upper width of the etched feature when there is isotropic etching). When this etching has an isotropic (chemical) component and a bias, **B**, is used to compensate for undercutting, the feature that must be printed into the resist is smaller than the final etched linewidth. In a complex integrated circuit, the smallest line or space that must be printed in the pattern is known as the "critical" feature size, d_c. From Fig. 39, an etched feature of dimension d_f, with equal lines and spaces, requires that a critical mask feature size

$$d_c = d_{\text{PITCH}} - d_m = d_{\text{PITCH}} - |\mathbf{B}| - d_f \tag{33}$$

be printed in the resist. And for an etch process with anisotropy factor, **A**

$$d_c = d_f\left[\frac{d_{\text{PITCH}} - d_f}{d_f} - 2(1 - \mathbf{A})\frac{y_f}{d_f}\right]. \tag{34}$$

Of course compensation for undercutting is limited by the smallest feature size that can be printed, and since $d_c \geq 0$ (the point at which compensation would require closing the mask opening entirely), the smallest feature that can ever be transferred when there is chemical etching (**A** = 0) is

$$d_{f,\,\mathrm{min}} = d_{\mathrm{PITCH}} - 2y_f, \tag{35}$$

which reduces to

$$d_{f,\,\mathrm{min}} = 2y_f \tag{35a}$$

for equal width lines and spaces.

However, non-uniformities in the film thickness, non-uniformities in the etching rate, and erosion of the resist mask impose more severe constraints on etching and the lithographic process. Suppose the thickness of the film that must be patterned, $y_{f,\mathrm{ave}}$, varies across the wafer surface by a fraction α ($0 \leq \alpha \leq 1$) making $y_f = y_{f,\mathrm{ave}}[1 \pm \alpha]$). Likewise, consider an etch process that removes the film with a mean etch rate $R_{f,\mathrm{ave}}$ and uniformity β, ($R_f = R_{f,\mathrm{ave}}[1 \pm \beta]$). The last remnant of the film would then be "cleared" after a total time

$$t_c = \frac{y_f[1+\alpha]}{R_f[1-\beta]}. \tag{36}$$

To allow for other sources of variation, the total etch time, t_{tot}, is usually increased to include some "overetching," δ, which is usually expressed as a percentage of the final clearing time:

$$t_{\mathrm{tot}} = t_c(1+\delta) = \frac{y_f}{R_f}\,\frac{[1+\alpha][1+\delta]}{[1-\beta]}. \tag{37}$$

At this point we must also consider the effects of resist mask erosion. There are both horizontal (R_h) and vertical (R_v) components to the mask attack. While the substrate is etched, the mask edge recedes by (refer to Fig. 40)

$$\Delta = 2t_{\mathrm{tot}}(R_h + R_v \cot\theta). \tag{38}$$

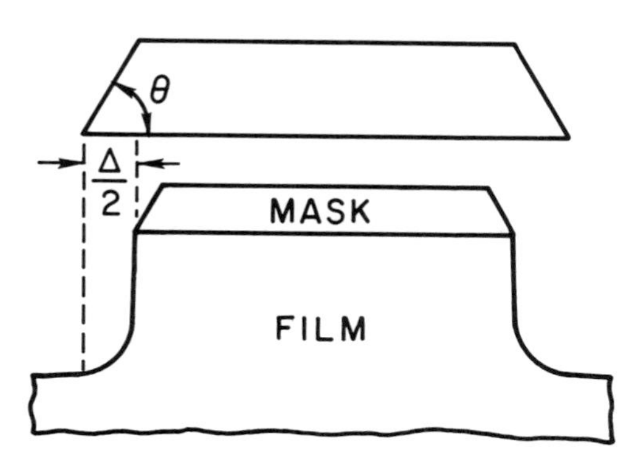

FIGURE 40. If resist mask features have an angle θ along their sides, as etching proceeds the resist mask contracts so the etched feature shrinks by $\Delta/2$ at each side.

And substituting for t_{tot} (Eq. 37),

$$\Delta = 2\frac{R_v}{R_f}y_f\frac{[1+\alpha][1+\delta]}{[1-\beta]}\left[\cot\theta + \frac{R_h}{R_v}\right]. \tag{39}$$

But non-uniformity in the plasma will also cause variability in these mask erosion rates. If $R_v = R_{v,\text{ave}}(1 \pm \epsilon)$ for the mask ($0 \le \epsilon \le 1$), a selectivity S_f^m will be required for etching the film relative to the mask (considering both mask edges):

$$S_f^m = \frac{y_f}{\Delta}U_f^m[\cot\theta + [1 - \mathbf{A}_m]], \tag{40}$$

where $\mathbf{A}_m$ is the anisotropy factor for mask attack and U_f^m is a combined uniformity factor defined as

$$U_f^m = \frac{[1+\alpha][1+\delta][1+\epsilon]}{[1-\beta]}. \tag{41}$$

Figure 41 shows the minimum selectivity S_f^m required as a function of the ratio of film thickness to the allowed linewidth loss y_f/Δ. These curves depend on the resist edge angle θ, the degree of anisotropy, $\mathbf{A}_m$, and the uniformity factor, U_f^m. As noted in Section IV.C.1, an angle of 90° (rectangular resist cross section) is common in contact and direct-step-on-wafer (DSW) printing, but it can be lowered by heat treatment before etching (postbake cycles). Angles of 60° are typical of projection exposure systems. Clearly substrate-to-resist selectivity requirements become least stringent with anisotropic etching and a vertical mask-edge profile.

Similar selectivity requirements control attack of the substrate underlying the film. An analogous calculation considers the film of thickness y_f, uniformity α, and a film etch rate $R_{f,\text{ave}}$ with variation β. The substrate is first exposed to the etching plasma when

$$t_{\min} = \frac{y_f[1-\alpha]}{R_f[1+\beta]} \tag{42}$$

and exposure stops when the plasma is turned off at t_{tot} (Eq. 37). Suppose etching into the substrate for no more than a depth, $y_{s,\max}$ is allowed. Then $y_{s,\max}$ is calculated at the maximum substrate erosion rate, R_s, and

$$y_{s,\max} = R_s(t_{tot} - t_{\min}) = R_s\frac{y_f}{R_f}\left[\frac{[1+\alpha][1+\delta]}{[1-\beta]} - \frac{[1-\alpha]}{[1+\beta]}\right]. \tag{43}$$

In terms of the film-to-substrate selectivity, $S_s^f = R_f/R_s$, Eq. 43 becomes

$$S_s^f = \frac{y_f}{y_{s,\max}}U_s^f, \tag{44}$$

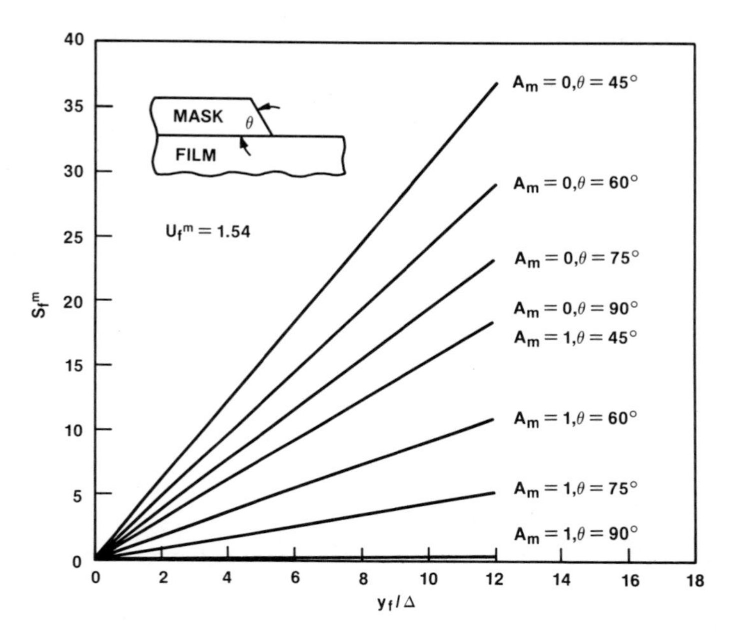

FIGURE 41. The minimum selectivity film-to-mask selectivity versus y_f/Δ, the film thickness being etched divided by the allowable dimension change, for various resist profile angles (θ), and anisotropy factors ($\mathbf{A}_m$) and a combined uniformity factor (U_f^m) of 1.54. This U_f^m was obtained from Eq. 41 assuming typical values for α (5%), β (10%), ϵ (10%) and δ (20%).

where y_s is the maximum substrate penetration allowed and U_s^f is the combined uniformity factor

$$U_s^f = \left[\frac{\beta(2 + \delta + \alpha\delta) + \alpha(2 + \delta) + \delta}{[1 - \beta^2]}\right]. \tag{45}$$

The overetch factor, δ, includes the etching necessary to clear a film from the valleys of underlying features and some margin to allow for etch rate and film thickness non-uniformities. For a perfectly uniform anisotropic etch ($\alpha = \beta = 0$) $U_s^f = \delta = h/y_f$, where y_f is the thickness of film above the highest parts of the underlying substrate and h is the top-to-bottom height of features in that layer.

Figure 42 shows film-to-substrate selectivity requirements as a function of the allowable y_f/y_s with process overetch δ as a parameter. For instance, if a 3000 Å film of polysilicon ($\alpha = 5\%$) is etched over a 350 Å oxide substrate ($\beta = 10\%$, $U_s^f = 0.886$) so that no more than 150 Å of the oxide film is removed after a 50% overetch ($\delta = 0.5$), the *minimum* selectivity is 18 : 1.

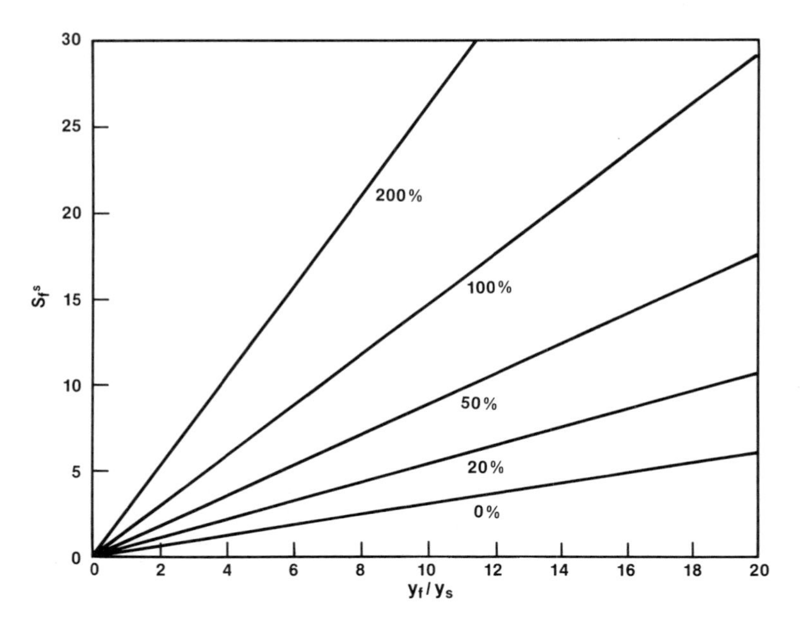

FIGURE 42. Required selectivity for the film being etched relative to the substrate as a function of the ratio of film thickness (y_f) to allowed substrate removal (y_s), with process overetch as a parameter. Film thickness and etch rate uniformities are 5% and 10% respectively.

Other parameters should also be taken into consideration to provide the best possible process for a given application. Watanabe [38] has shown that selectivity and etch uniformity often can be traded-off to advantage. In Fig. 43 polysilicon is etched over a thin (400 Å) sacrificial (gate 0) oxide layer, which had been grown on a silicon substrate. This oxide normally protects underlying silicon from attack during the plasma etch and is removed afterwards to regrow defect-free gate oxide for the device. If this gate 0 oxide layer is eroded to less than $\approx$ 300 Å, however, plasma etching degrades the underlying silicon and a poor oxide is thermally regrown from that damaged material. Experimental results in Fig. 43 show that the higher the polysilicon to oxide selectivity, the less the gate oxide removed. But if the process etch uniformity is poor ($>$ 20%), higher selectivity is necessary to guarantee the required minimum thickness of protective oxide. If the selectivity is inherently low, then great effort must be expended to develop processes with high uniformity.

As another example, Beinvogel [39] shows (Fig. 44) that high selectivity for etching polysilicon over oxide and a high degree of anisotropy in

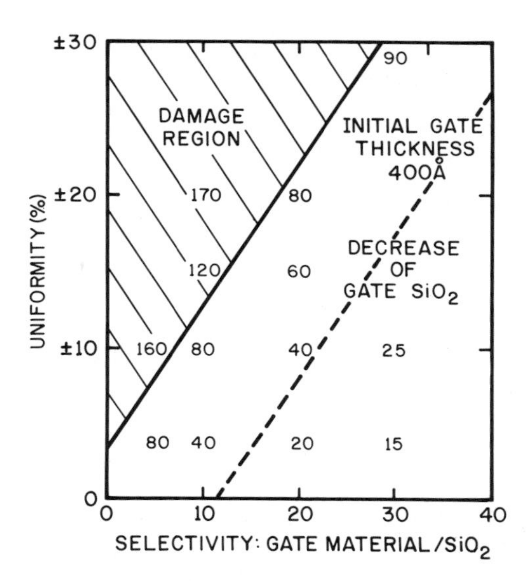

FIGURE 43. Process uniformity as a function of the selectivity for etching a polysilicon gate pattern on 400 Å of gate oxide beneath is shown. The process is unacceptable if more than 100 Å of this oxide is removed, (cross-hatched "damage region"). With improved uniformity, process selectivity requirements can be relaxed. Numbers map the oxide lost (in Å), and the dotted line is an acceptable control range based on these data (after Watanabe [38]).

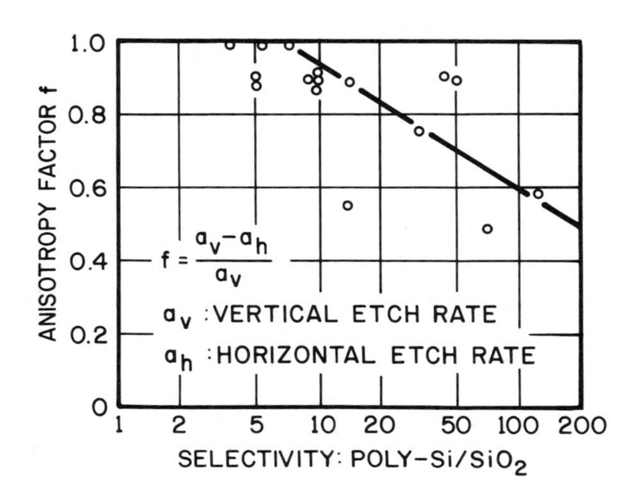

FIGURE 44. As the required selectivity for a poly/oxide etch process increases, the anisotropy deteriorates (after Beinvogl [39]).

chlorine-based plasmas are incompatible. This suggests that an empirical relationship should be developed to represent the interaction of anisotropy, selectivity and uniformity for process optimization.

V. Plasma Reactors

Although simple plasma reactors have been constructed from opposed parallel plates in a vacuum chamber (Fig. 9), a wide variety of more complex arrangements are used in commercial practice. Plasma etching reactors have been classified according to load capacity, the position of material relative to the plasma, pressure regime, electrode geometry and generator frequency.

By load capacity, we mainly refer to a distinction between *single wafer reactors* in which only one wafer is etched at a time as opposed to *batch reactors* in which many wafers are simultaneously etched in the same chamber. The choice between these alternatives is usually made on the basis of perceived requirements for economic throughput with high uniformity and reproducibility. Cost differences between batch and single wafer reactors intended for the same process step tend to be less than a factor of two, depending on the application being considered, and do not uniformly favor either type over the other. All other things equal, wafer-to-wafer reproducibility and uniformity are better when every wafer is exposed sequentially in the constant environment of a *single wafer etcher*. The drawback is that higher etch rates are required to attain the same throughput as in batch reactors, and this generally requires a more intense plasma at higher power density. These harsher conditions usually lead to higher temperatures, less selectivity and more interface damage from ion bombardment and sputtering.

A. BARREL ETCHERS

When plasma etching first found widespread use in the 1970s, *barrel etchers* or *volume loaded reactors* like that in Fig. 45 were the most common type. In barrel etching, wafers are simply stacked in a notched quartz support stand that is inserted into the reactor. External electrodes in either a "capacitively" or "inductively coupled" configuration [40] sustain the plasma, but there is no temperature control and wafers have no special orientation relative to the electrodes. As may be expected, these reactors are inexpensive, but uniformity is difficult to achieve and the rate of reaction can change erratically because of reactor heating from the RF power input

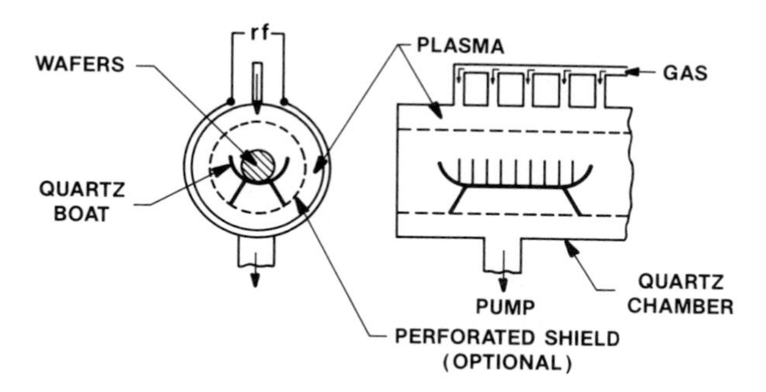

FIGURE 45. The barrel etch reactor is popular for isotropic processes such as resist stripping. The wafers are stacked on a slotted quartz holder inside the cylindrical vessel which is sometimes placed within a perforated shield, or "etch tunnel" to improve uniformity and limit exposure to charged species.

to the plasma and heat produced by the etching or stripping reactions. Nevertheless, because of their large capacity and low cost, barrel reactors are still common for noncritical processes such as stripping, cleaning, and (occasionally) large feature isotropic silicon etching.

One notable variation on this type of reactor is the incorporation of a perforated metal insert, a *tunnel*, which is intended to exclude the plasma from the wafer-containing internal volume, (e.g., an imperfect approximation of the downstream etcher). The tunnel can be a source of sputtered material and damaging contamination, however, and there are other inherent problems, such as the lack of temperature control.

B. PLANAR REACTORS AND VARIATIONS

The parallel plate or *planar* reactor and its variations is the most common type in use today. Originally introduced the early 1970s for chemical vapor deposition, the simple parallel plate "Reinberg" style reactor is shown in Fig. 46. The lower electrode supports the wafers and can be heated or cooled, while radial gas flow toward or away from the center helps uniformity. Planar reactors are available in both single wafer and larger batch sizes. Sometimes, especially in single wafer reactors, one of the electrodes contains a perforated "shower head" gas feed, which is used instead of the radial flow design.

The parallel plate reactor is often made asymmetrically with wafers placed on the smaller electrode (Figs. 20, 47). A higher potential develops across the sheath at this electrode, which increases ion energy and improves

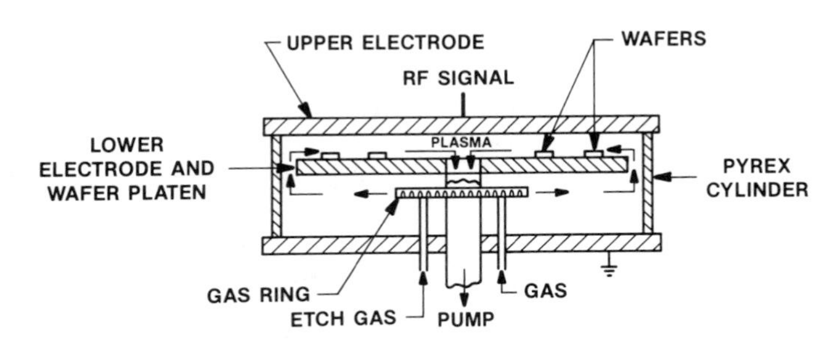

FIGURE 46. The "Reinberg reactor" is a radial flow parallel plate system with wafers placed on the "grounded" lower electrode and the electrically isolated upper electrode "driven" RF from a matching network. Gas can flow from the outside-in (as shown) or "inside-out". This configuration is sometimes referred to as plasma etching mode; by contrast, operation with the "wafer-loaded" electrode driven by RF, has been called a "reactive ion etch" (RIE) mode (after Reinberg [41]).

the directionality of the etch process. The external circuit usually has a "blocking" capacitor so that a negative DC bias voltage appears between the small electrode and ground. At low pressure, these asymmetric reactors develop still higher characteristic potentials and bias. The high-potential RF power lead is usually connected to the smaller electrode with the larger electrode grounded at common potential. This is necessary to maintain a large area ratio since there are usually other grounded surfaces, in and out of the reactor chamber, that capacitively couple to the plasma and thus add to the effective area of the grounded electrode. Trade jargon often refers to low pressure asymmetric parallel plate reactors as *reactive ion etchers*, distinguishing higher pressure operation ($\gtrsim$ 0.1 Torr) with the term "plasma etchers." Of course plasma etching is taking place in both regimes and ions are almost never the primary etchants (see Sections II.A, III.F). The term "diode reactor" has also been applied to parallel plate reactors.

A few years ago, as the diameter of parallel plate reactors was increasing to meet throughput demands, the *hexode* or Hex reactor was designed at Bell Laboratories. In this design wafer capacity is increased by folding the electrode surfaces into concentric cylinders and expanding it vertically in the third dimension. Figure 47 shows that the cathode is now surrounded by the anode (bell jar) and six wafers can be loaded onto each face on a vertical level. Typically there are between 3 and 5 vertical levels or rows. The vertical dimension and number of wafers is limited by the height of the process clean room, since the bell jar has to be raised without hitting the ceiling. The vertical orientation of wafers also helps prevent particulate matter from settling on the wafer surfaces. To date, hexode reactors have

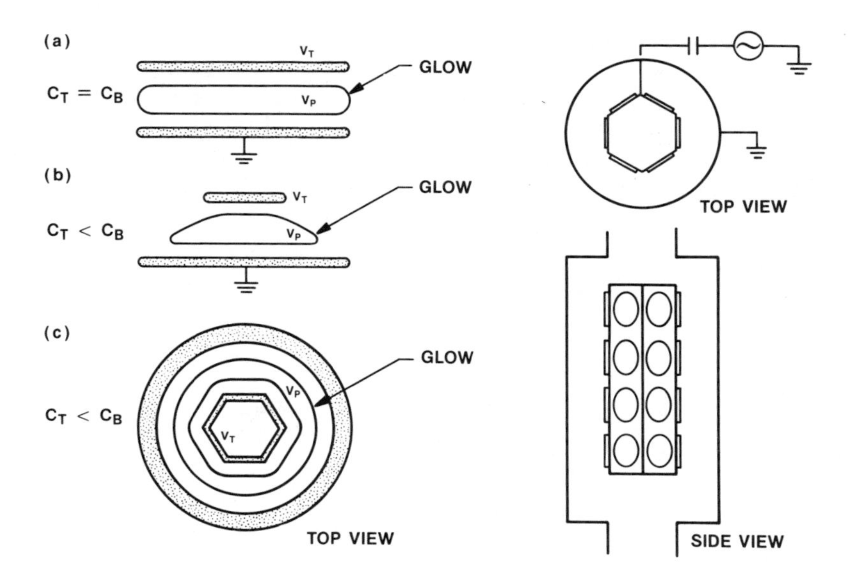

FIGURE 47. The asymmetric "diode" reactor (b) (also see Fig. 20) can be considered a modification of the symmetrical reactor (a) in which reduced sheath capacitance near the smaller "cathode" electrode (C_T) results in higher energy ion bombardment and a negative DC bias (see Section III.E.3). The "hexode" reactor [(c) and right-hand side] consists of connected electrode surfaces folded into a right hexagonal cylinder, with the larger outside vacuum vessel serving as the "anode." Wafer capacity is large because the wafers are stacked in the vertical direction.

been designed for batch operation at low pressure (around 0.005–0.100 Torr) and high frequency (13.56 MHz).

C. DOWNSTREAM REACTORS

In most plasma etching and stripping processes used today, the semiconductor substrates are immersed in the plasma where energetic charged species are accelerated toward and impinge on substrate and reactor surfaces. While ion bombardment is an integral part of some reaction chemistries, it also can sputter-deposit material from the reactor and cause device degradation by inducing chemical or structural changes. This is aggrevated when plasma and particle flux densities are high, or when ion bombardment energy exceeds the minimum level necessary for the process. An important example of structural degradation occurs when etching resist

(resist stripping) over field effect transistors. Charge is deposited on thin insulating gate oxides because of the high electron temperature, and the potentials developed can cause damaging electrostatic breakdown when the plasma is extinguished.

Because charged species recombination is more rapid than most gas phase neutral reactions, ($k_{recomb} \sim 10^{-8}$–10^{-7} cm^{-3} sec^{-1} for ions with electrons around 0.1 Torr), damage from ion bombardment can be avoided if the wafers are placed in a charge-free region outside the discharge. In a *downstream* afterglow reactor (Fig. 48a), neutral etchant radicals flow from

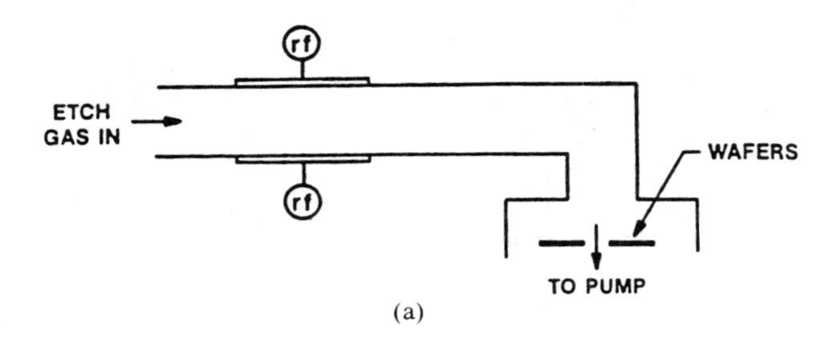

(a)

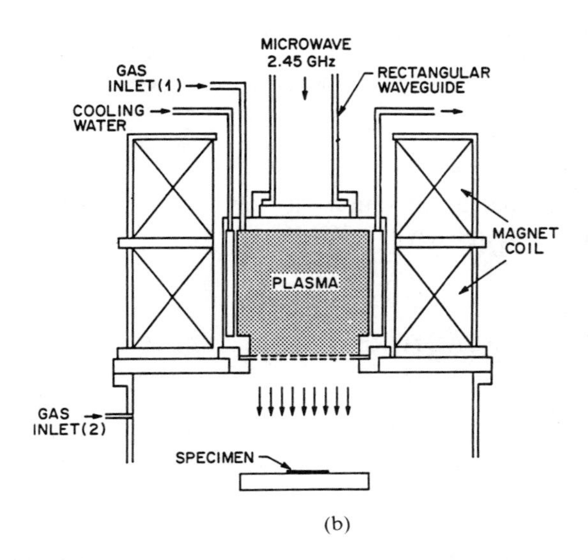

(b)

FIGURE 48. (a) In the downstream afterglow reactor, neutral etchant radicals reach the remote treatment zone and charged species are removed by recombination in transit. (b) In a low pressure ECR etcher charged species bombard the substrate and can induce anisotropic etching.

the plasma to a treatment zone that sometimes is temperature controlled. Reactors of this type have long been used for scientific studies of basic chemistry, but, except for the barrel etcher with a tunnel, commercial processing equipment based on this design is recent. Since there are no ions present to induce directional etching, applications are generally limited to the stripping of polymer resist or a layer of material where patterning is noncritical.

There is another family of etchers, closely related to these downstream etchers, which have been called "plasma transport" reactors [42, 43]. The plasma source usually is operated at microwave frequencies (2450 MHz), and sometimes with an auxiliary magnetic field to increase electron confinement (Fig. 48b). These plasma transport reactors have also been referred to as microwave or ECR (electron-cyclotron resonance) etchers, but the former nomenclature is unsatisfactory because there are in-plasma reactors and downstream etchers that also use microwave excitation and magnetically confined microwave transport devices that do not operate at resonance. These plasma transport reactors operate at low presure, as low as 10^{-5} Torr, where charged-particle recombination is slow relative to diffusion and flow. In these devices the ions may be extracted from the plasma zone by biased grids or a magnetic field and made to impinge on the substrate with a controlled, low energy. The ions sometimes are collimated and the energy can be adjusted to promote etching, yet minimize substrate damage. The flux of ions and neutral species that can be transported under these conditions is often limited, however, and consequently the etching rate may be low. This together with higher relative cost and complexity have limited their use.

D. OTHER REACTOR VARIATIONS

Some reactors have employed a magnetic field to reduce the rate of electron loss from moderate pressure plasmas, permitting higher plasma densities at lower power and characteristic potentials. These are known as *magnetron reactors*. The higher plasma (electron) density increases dissociation and the concentration of etchant species in the reactor. This, in turn, results in a faster etch rate at low bias, which can reduce (or even eliminate) ion bombardment induced device damage. In these reactors separate adjustment of the magnetic field intensity and discharge power allows some control of the plasma and etchant densities, independent of the sheath potentials and DC bias. Major drawbacks of this approach are that etching uniformity is degraded by inhomogeneities in the magnetic field, and there

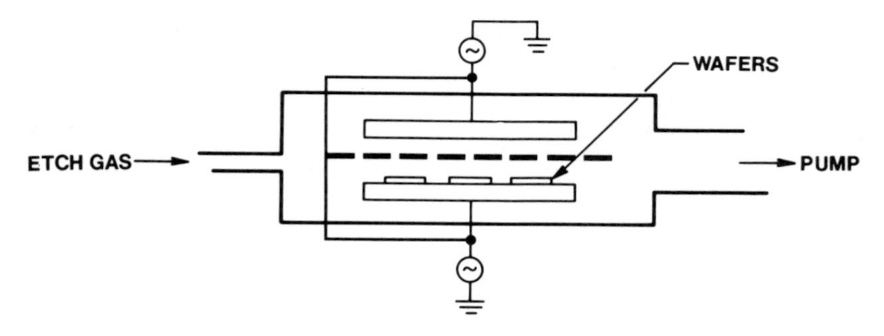

FIGURE 49. The triode system has multiple power supply connections. A third electrode serves as a grid so etch species can be generated in the upper portion of the triode (between the upper electrode and grid), while the wafers on the lower electrode can be provided with a sheath potential (bias) that is less dependent on the upper plasma sustaining power.

may be a loss of anisotropy when operating with excessive neutral-to-ion flux ratios or low ion energy.

So-called *triode reactors* employ an additional electrode and power supply. The objective here is to attain high reactant generation rates in a plasma zone that is partly shielded from the substrate (the upper zone in Fig. 49), and control ion bombardment energies in a second zone where the substrates are located (lower region in Fig. 49). In attempts to do this, a high excitation frequency has been used in the substrate-free zone with low frequency excitation enhance anisotropy in the substrate region.

Other forms of reactor geometry can be derived from these basic system configurations. An ion beam etcher (see Chapter 6) can be considered a triode, where the ions are extracted from a parallel electrode source and directed to some workplace by a third electrode, or grid arrangement. The afterglow reactor can be considered a form of the triode configuration in which wafers are placed on a remote third electrode, while the plasma is maintained between opposed driven electrodes.

E. GENERATOR FREQUENCY

Most of these reactor types can be used with either low (e.g., below 2 MHz) or high (2–30 MHz) RF excitation frequency. The effects of frequency in this range are surveyed in Section III.E.4 of this chapter and Chapters 2 and 5. Microwave excitation, on the other hand, usually requires specialized equipment.

Experimental large-area planar microwave plasma etchers have been made using surface wave technology [44], but there is controversy regarding their advantages, if any [45, 46]. To date, the chief application of microwave discharges has been as a plasma source for the downstream and plasma transport reactors, but apparently RF discharges can be used in these applications as well [27, 45, 47].

F. BATCH AND SINGLE-WAFER ETCHERS

Traditionally, wafers have been etched in a batch to maximize throughput. However, recent improvements in technology, increasing wafer size and the more stringent uniformity requirements imposed by larger wafers with smaller critical dimensions allow one-wafer-at-a-time etchers to become competitive. A single plasma etcher can cost well over half a million dollars, and considerable differences in design parameters make it impractical to convert one type of reactor onto the other. Hence, there has been concern and controversy about benefits and drawbacks of single-wafer versus batch etching equipment.

The most popular batch reactors (for example the Hex, see Section V.B) operate at high frequency and low presure (< 50 Torr) where process anisotropy is usually enhanced by energetic ions. By contrast, most single wafer equipment has been designed to provide anisotropy at higher pressure (> 100 mTorr) using inhibitor mechanism chemistry. As a result of this operating pressure difference, batch reactors often use higher ion energy than the single-wafer systems. High ion energy degrades selectivity and can cause physical sputtering at low pressure. When single-wafer systems are operated at very high etch rates, however, surface temperatures increase and cause both selectivity and linewidth loss (see below). Experience shows that both types of reactor are capable of equally good patterning when operated under favorable conditions.

Typically, single wafer etchers offer superior uniformity, can be modified to accommodate different wafer sizes with constant hourly throughput, and to maintain process control of each etched wafer (rather than using average reactor conditions or a sample wafer). Batch reactors, on the other hand, usually have higher throughput and are less sensitive to contamination from polymer forming feeds (because of their larger volume and generally lower operating pressure).

Several kinds of variability must be considered in an etching process: wafer-to-wafer, run-to-run, and within-wafer variability. With good process control run-to-run variations are small for both batch and single-wafer reactors, to the point where within-run uniformity is the main concern. This is where single wafer etchers excel. Obviously, single wafer reactors, have *no* within run wafer-to-wafer variation, and parameters contributing to within wafer nonuniformity can be compensated for by altering gas feed-distribution and the electrode shape. End point monitoring accuracy in single wafer reactors is also not subject to wafer-to-wafer variations. As a result, the uncontrolled variation for a typical batch hexode system is 7–12%, while single-wafer etchers commonly show better than 3% uniformity.

Run-to-run variation is heavily influenced by the processing sequence and chemistry. Inhibitor-forming gases, which deposit polymer films, induce changes in the RF matching conditions. These changes are greater in higher surface-to-volume ratio single-wafer chambers than in batch systems. Film buildup can be alleviated by a short in-situ cleaning step after etching each wafer, at the expense of a slight throughput loss. A cleaning step also insures that each wafer sees the same initial environment, which improves wafer-to-wafer consistency.

High uniformity makes it easy to increase etch rates in single-wafer systems by simply using more power. There has been a tendency to do this, but then the process usually suffers from abnormally high resist erosion rates and linewidth loss.

While there has been intense concern about machine throughput, in reality this figure of merit is usually confused and abused. The core of this confusion comes from neglecting wafer yield and chip complexity as part of throughput definition. There is a complex relationship between throughput and yield. The crux of the matter is that the goal of manufacturing should be to deliver the highest economic yield of *working* devices per wafer. When individual wafers have the highest possible yield, the rate of wafer processing in the slowest running (rate limiting) step limits the useful maximum EHO (estimated hourly output) of all processes. For large high quality wafers (high yield), electrically testing is commonly the slowest step and the time to test each chip increases rapidly with device complexity. For today's VLSI devices, each testing machine may be limited to as few as 5 large ($\geq$ 5 inch) wafers per hour when yields are high ($\gtrsim$ 75%). This means that an etch system that processes 25 wafers per hour (one IMS cassette) can keep five expensive testing machines busy (a typical number). On the other hand, when the yield is low, the testing machine spends much less time per chip. Hence, the number of wafers that can be tested increases inversely with yield, requiring more poor quality wafers to utilize the total production capacity.

A recent survey of commercial plasma etch machines [48] showed EHO for batch systems to range between 30–80 with single-wafer machines in the 26–60 range. Throughputs (or EHOs) of 25–40 wafers per hour should be adequate for most operations that have high yield. Hence, either system should suffice, and in this range the etch rates required of single-wafer machines can be relaxed to improve yield.

In batch reactors, patterning of multiple film layers is usually carried out sequentially in the same reactor, with appropriate changes in chemistry as layers are exposed. A major problem with this approach is that an upper layer must be *completely* removed before proceeding since residual film can act as a mask for the next, lower layer. This requires high selectivity from the etch chemistries, which sometimes means a compromise in profile control and other etch characteristics. "Memory effects" from mixing chemistries in the same reactor can make this approach difficult.

In single-wafer reactors, multiple film layers are usually dealt with continuously by processing the first layer in one chamber and the next layer in a second compartment. Volatile residues from one step can be removed by evacuation during wafer transport between chambers. This approach has the advantage that interaction between sequential etch chemistries is minimal and the system EHO is no worse than the slowest individual step.

It has been mentioned that clean room costs, which were approaching $1000 per square foot in 1987, are a major component of process economics. While the "footprints" of *simple* single-wafer systems are generally smaller than batch reactors, this is not true of continuous multi-chamber single-wafer systems. However, footprint becomes less important in some newer systems that are "bulkhead mounted," so that the body of the system is out of the high cost clean room, with the cassette loading and receiving stations protruding through the wall.

In the future most batch reactors will probably handle smaller batches of large diameter wafers, while single-wafer machines will use multiple chambers for continuous complex processing. Eventually these two schemes may evolve into a hybrid in which an automatic wafer loader places a small number of wafers on a platen that is moved through the processing sequence as a "mini-batch."

VI. Etching Endpoint Detection

It was pointed out that normal surface topography (at the edges of steps, for example, see Section IV.C.2) produces changes in film thickness that make some degree of overetch mandatory. Yet overetching must be done with great care. Planar areas, which often have the thinnest film, are particularly sensitive to excessive overetching because that is where sensitive layers such as extremely thin gate oxides ($\approx$ 150 Å) are found. These areas demand high selectivity between the film being removed and the substrate material (of course, selectivity of the film over the mask must also

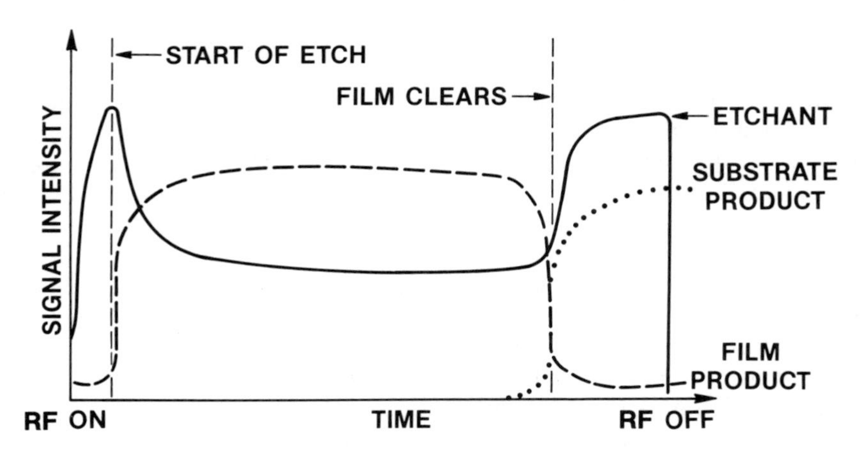

FIGURE 50. Idealized schematic of events that can be used for endpoint detection. With power on, the etchant concentration increases until etching starts. Next, etchant concentration decreases while reaction products accumulate to steady values. As the film clears, the etchant signal increases again and the product signal falls. Finally at the endpoint there is a third signal from reactions with the underlying substrate.

be high so that dimension control is not lost during the overetch). Since some anisotropy must usually be compromised for high selectivity, it is important to limit the degree of overetching, and with it, the required selectivity (see Section IV.C.2). Thus, endpoint detection is an important part of process control.

The sequence of events in a typical plasma etch process provides several ways to detect the endpoint. When RF power is first applied, the etchant species are produced. After some latency, these species react with the film to produce a volatile product. Toward the end of the etch cycle the underlying substrate begins to be exposed and reactions with the etchant produce new secondary products. The Cl etchant signal, $AlCl_3$ product signal and $SiCl_4$ signal from the underlying substrate in Fig. 50 mark the sequence of events that occur during the etching of Al by a BCl_3/Cl_2 mixture. A signal derived from etchant (Cl and Cl_2) initially rises until the less reactive surface layer is breached (e.g., contamination and Al_2O_3, which cover the Al, are removed). The signal then falls as the etchant is consumed by rapid reactions with Al to form chlorides. As aluminum finally "clears" from isolated areas of the substrate, the rate of etchant loss drops and the signal from the etchant increases rapidly. By contrast, a signal derived from the product ($AlCl_3$) reaches a steady value after etching begins, and then decreases as the film disappears at the end of the etch cycle. Finally, at the endpoint, a signal from substrate reactions emerges ($SiCl_4$).

The end of the etching period is signaled by a plateau in etchant signal intensity, when the film has cleared from the wafer. The interval between the first rise in etchant signal and its final plateau (a maximum) or between the initial fall in product signal and its final minimum, will be determined by the amount of film material that must be removed. In practice, the planar film clearing time is monitored. When these signals first approach their final plateau and the film has been cleared from the predominant planar areas, some material remains at "steps" in the circuit pattern where the film is thickest. These regions have practically no effect on the product or etchant endpoint signals because they only comprise a small fraction of the total wafer area. In practice, to allow for this, the etching time required to reach the endpoint plateau is monitored (planar clearing time) and the process is continued for a fixed fraction of this time before RF power is turned off (overetching).

There are a variety of phenomena from which signals can be derived to mark the beginning and end of an etch process. These will be discussed in the following sections.

A. PRESSURE CHANGE

Because the gas is heated and dissociated when RF power is applied, pressure tends to rise initially. When etching is fully underway, the gas-solid etching reactions usually cause a decrease in the number of gas molecules

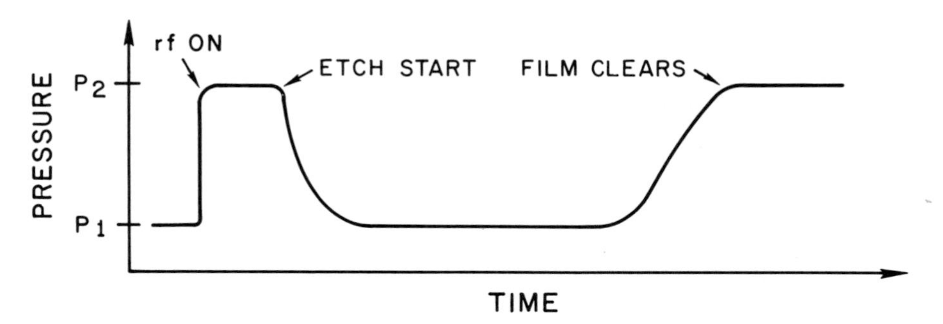

FIGURE 51. Typical pressure recording from a capacitance manometer during a plasma etch cycle (with constant pumping speed). When power is turned on, the pressure rises until the film starts to etch. Pressure falls to a lower value during etching and increases as the film clears at the endpoint.

so system pressure will fall if there is a constant resistance to flow (fixed orifice). Generally, this pressure decrease is about 5–10%. Figure 51 shows these phenomena schematically. One difficulty with pressure monitoring is that changes in the etching process are sometimes *caused* by pressure changes.

Pressure monitoring requires no special equipment, and it was used in some early batch etchers. In modern production etching reactors, however, the feed gas flow rate is fixed and an automatic "throttle" valve maintains pressure constant by adjusting flow resistance in the exhaust line (pumping speed). This arrangement improves process control, but means that pressure can no longer be used as an endpoint indicator.

B. BIAS CHANGE

As the gas composition changes during the etch cycle, the plasma impedance and sheath characteristics change. This usually causes changes in the self-bias voltage that is generated in asymmetrical reactor geometries. In some processes this effect is large—it can be more than 100% for resist stripping in a hexode batch reactor. But in other processes it can be only a few percent. Figure 52 shows the bias change during a resist strip cycle. The bias reaches a maximum as the film starts to clear, and then decreases. Normally the process is allowed to continue beyond this point while the bias signal decreases and resist stripping continues along the thicker surface topography.

Here again a major disadvantage is the need to allow the change, since power is sometimes controlled to maintain constant bias in asymmetric low pressure etchers. Nevertheless, ease of implementation and the characteris-

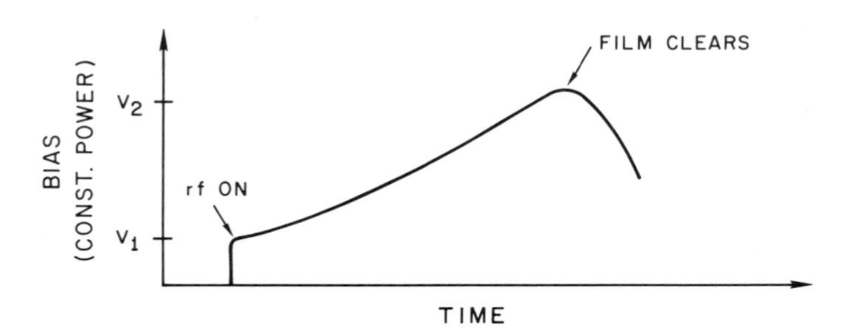

FIGURE 52. DC bias, when etching at constant power, often increases as the etch cycle progresses. The rate of increase bias of voltage falls as film clears, then bias falls during the overetch period.

tically large signals make this method attractive in automated systems. The overetch period is generally programmed to continue a constant fraction beyond the film clearing period (time of maximum bias).

C. MASS SPECTROMETRY

Since different chemical species are associated with the start and end of an etch cycle, they can be monitored by mass spectroscopy. Often, details of the chemistry suggest a particular constituent that tracks the progress of an etch. Consider, for instance, CF_4/O_2 feed chemistry that has been used to etch the plasma-deposited silicon nitride scratch-protection layers that cover aluminum interconnects. Volatile SiF_4 forms from the reaction between fluorine atoms and Si_3N_4. Figure 53 shows the mass spectrometer trace from the SiF_3^+ parent peak at mass 84 (which is formed by electron impact induced fragmentation of SiF_4 in the instrument's ionizer). A sharp drop in the intensity of the SiF_3^+ signals clearing of planar areas in the nitride film. SiO_2 is also present at the aluminum level, but it is attacked by fluorine radicals more slowly, so the mass 84 signal falls to a lower intensity at the endpoint rather than disappearing. Again, a suitable overetching interval is applied beyond the "planar clearing point" (when the film is first removed from level, horizontal surfaces) to insure that all the nitride material is removed from steps in the pattern where it is thickest. Mass spectrometry has the advantage that it can selectively detect specific species and sense air leaks and contamination.

Mass spectrometers require a high vacuum ($< 1 \times 10^{-5}$ Torr), which is provided by a separate vacuum chamber and pump. Process gas is sampled

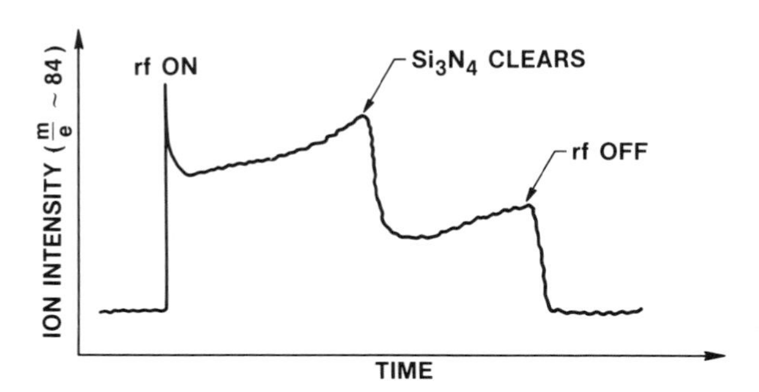

FIGURE 53. If a mass spectrometer is tuned to a product peak ($m/e = 84$ for the SiF_4 during fluorine based etching of silicon nitride) the signal level will indicate changes in chemistry, signaling the interfaces between film layers. These changes can be quite abrupt.

through a small orifice between the etching chamber and spectrometer. When an etch *product* is monitored, sensitivity can be a problem since etching is generally carried out at low depletion, where small amounts of product are highly diluted in the feed. Since product partial pressure is lowered further by the sampling process, background gas in the mass spectrometer can degrade the signal to noise ratio at low etch rates. This can be overcome to some degree by placing a sampling probe close to the wafer, but the probe may perturb the plasma and process conditions. These difficulties, along with high cost and required operator skill, usually preclude the use of general purpose analytical mass spectrometers in routine production. However, they are a powerful diagnostic tool in research and development.

Production applications are better served by small, specialized mass spectrometers that are manufactured to monitor a single mass number. These ruggedized "smart gauges" are often used to monitor water or nitrogen peaks, serving as sensitive leak detectors.

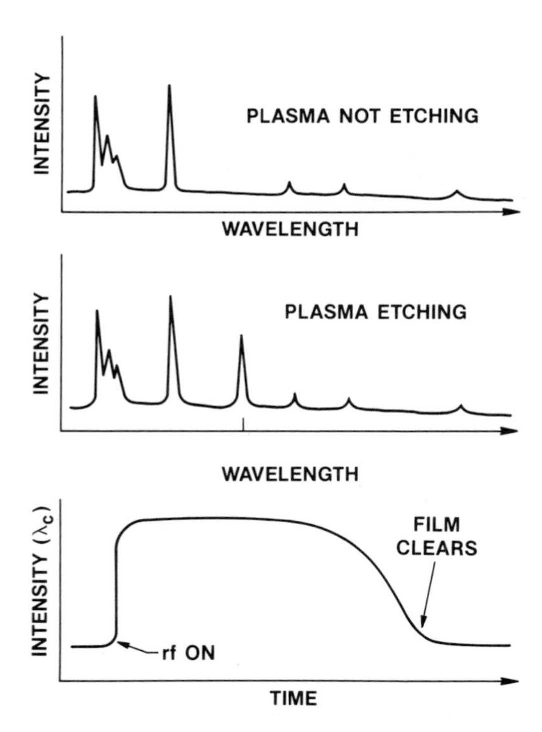

FIGURE 54. Characteristic plasma induced emissions from product species (λ_c) that can be used for etch cycle monitoring. These emissions appear at the onset of etching and fall off as the film clears.

Table 5 Characteristic wavelengths λ_c for plasma-induced-emission endpoint monitoring using selected species in various etching chemistries.

FILM	ETCHANT	λ_c (Å)	EMITTER
Al	CCl_4	2614	AlCl
		3962	Al
Cu	Ar	3248	Cu
Cr	Ar	3579	Cr
RESIST	O_2	2977	CO
		3089	OH
		6563	H
		6156	O
Si	CF_4/O_2	7037	F
		7770	SiF
	Cl_2	2882	Si
Si_3N_4	CF_4/O_2	3370	N_2
		7037	F
		6740	N
SiO_2	CHF_3	1840	CO
(P-DOPED)		2535	P
W	CF_4/O_2	7037	F

D. EMISSION SPECTROSCOPY

Optical emissions from species excited by the plasma are widely used as a sensitive and effective endpoint monitor. Emissions can emanate from etchants, etch products, or their fragments. Figure 54 shows how a spectrum from a plasma without substrate material present changes when etchable material is introduced. Wavelengths appear that are associated with the reaction product. An optical spectrometer can be set to the line of interest and follow its intensity during an etch cycle (as shown in the bottom curve of Fig. 54).

Table 5 lists characteristic emitter wavelengths that can provide useful signals in a variety of film/process chemistry combinations. An inexpensive interference filter and diode-detector can be used in place of a spectrometer, and the signal from this combination can be used to automatically turn off the plasma after the proper overetch time. These systems are small and easily transportable. Unfortunately, light from the plasma is rich in emissions from many species that cover a wide spectral range. Many times it is difficult to separate a characteristic line from the intense background.

E. LASER INTERFEROMETRY AND REFLECTANCE

When monochromatic light rays reflected from the front surface and underlying interface of a thin transparent film combine, constructive and destructive interference takes place, which is characteristic of the wavelength of the light and the film thickness. This effect can be used to monitor the thickness of materials such as polysilicon, silicon oxide and silicon nitride, which have a wavelength region where they are transparent. A laser helium-neon (632.8 nm) laser beam is directed at the material surface. At normal incidence, interference maxima and minima occur when twice the thickness of the film, d, is a multiple of the wavelength divided by the refractive index (n)

$$d = \frac{\lambda}{2} n.$$

Figure 55 shows a typical intensity trace from a laser beam being reflected from these surfaces as a film is etched. The intensity varies sinusoidally with film thickness and the etch rate can be monitored by noting the time interval between corresponding points on the signal trace (the maxima and minima are usually used).

This technique permits real time monitoring so that etch rate changes caused by variations in the composition of the film with depth, or a diffuse

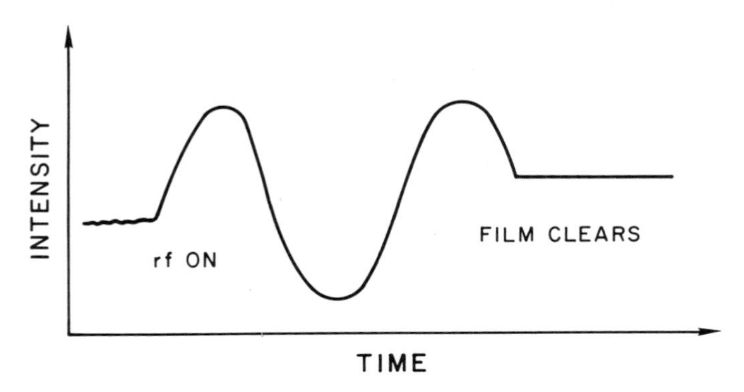

FIGURE 55. When monochromatic laser light is directed on a transparent layer, reflected wavefronts from the surface and the bottom interface interfere, causing the reflected intensity to change as etching thins the film. There is a discontinuous change in slope at the endpoint, and intensity modulation then stops or changes, depending on whether the next layer is absorbing, reflective or transparent.

interface between film layers, can be automatically detected and corrected. Laser interferometry has been used to automate etch rate control in a wide range of applications, since it is easily adapted to almost any material that is transparent to available laser wavelengths.

Metal films are opaque so their thickness cannot be monitored interferometrically. These films usually have high reflectance, however, so etching through a metal-semiconductor or metal-insulator interface is marked by a sharp change in the reflected laser light intensity.

A major limitation stems from the small spot size needed to obtain a distinct interferogram [a few mils ($\sim 10^{-5}$ cm) in diameter]. Since the endpoint determination is made from sensing a small area, non-uniformities in the film thickness or etch rate can produce an unrepresentative sample. Batch processes are especially vulnerable to these sampling errors. This is a serious problem when selectivity is low and tight control of overetch time is required. In single wafer reactors and batch processes with good uniformity, however, laser interferometry is the preferred monitoring technique. Figure 56 shows a representative laser interferogram from a complex substrate film

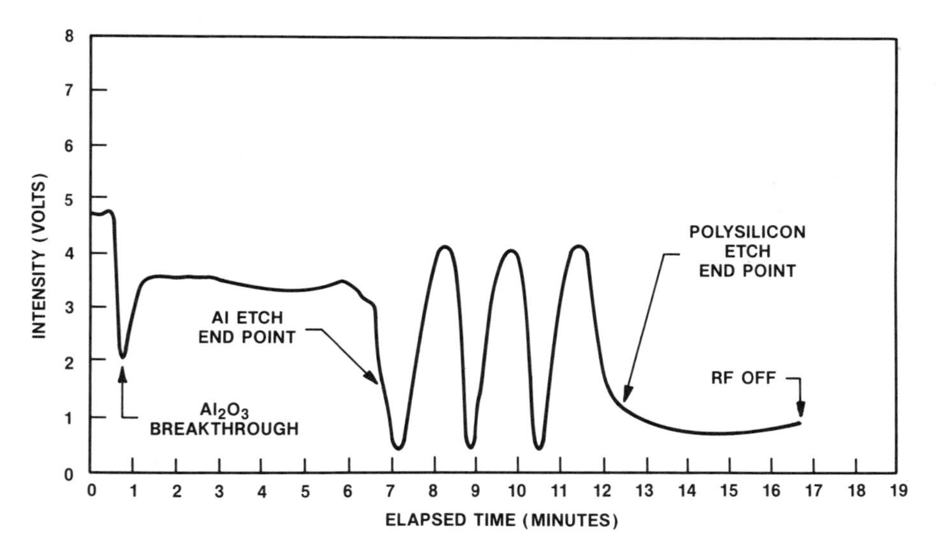

FIGURE 56. Reflected laser intensity during the patterning of aluminum (1 micron) and doped polysilicon film (0.3 micron) layers, with silicon oxide below. Intensity from unetched aluminum is high. As thin native aluminum oxide is removed the signal falls, and then increases again when the pure metal is exposed. After etching through the aluminum, into polysilicon, there is constructive and destructive interference. As the poly clears, the periodicity changes as the film is intentionally overetched 35%. The poly etch rate was about one-third that of aluminum.

combination. The wafer had a composite 1.0 μm of aluminum on a 0.3 μm layer of doped polysilicon conductor that had been deposited on a borophosposilicate glass level. When the aluminum is etched in a BCl_3/Cl_2 plasma, the laser signal intensity first decreases as the thin Al_2O_3 native surface oxide is removed, and then returns to a high value corresponding to simple non-destructive reflection from the aluminum. When the aluminum finally clears, the laser "sees" the underlying transparent polysilicon film and interference fringes are produced. Finally, the polysilicon film clears and the laser signal drops sharply, signally the endpoint. The cycle is then terminated after a suitable overetch period.

VII. Device Damage from the Plasma

Wafers in a plasma are exposed to energetic particle and photon bombardment. This *radiation* consists of ions, electrons, ultraviolet photons and soft x-rays. While most ions bombard the surface with energies below a few tens of volts, ion energy on the "tail" of the distribution can be more than 1000 V, depending on the sheath potential. Average electron energy in the plasma ranges from a few volts to 10 V or more, but electrons reaching the walls are thought to have lower energy owing to the positive plasma potential. When high energy radiation strikes the wafer surface, it can cause both shallow and deep surface disruption (damage), which alters electrical characteristics and degrades device performance. Both reversible and irreversible device damage can occur. In general, the significance of these effects increases with particle energy, and hence with peak RF potential and power. Damage appears in various forms: atomic displacements caused by ion impact, which have been observed in layers up to 150 Å deep, electron-hole pairs produced by primary ionization from the UV and X-ray photons, and secondary ionization where electrons formed by primary processes create defect centers. The damage can usually be detected as a change in the capacitance-voltage (CV) signature of MOS capacitors, or in the operating parameters of simple transistor structures, which give threshold voltage, flatband voltage, Q_{ss}, lifetime and other parameters.

A. REVERSIBLE EFFECTS

Every plasma process causes surface alteration because of the energetic radiation present in the plasma, and almost any plasma treatment causes *some* change in measured device parameters. If these changes were always irreversible, it would be hard to make useful semiconductor devices with plasma processing. Miniaturization would have been limited, since plasma processing is essential for fabricating devices with small critical dimensions. Fortunately, much of this radiation damage can be removed by thermal annealing.

Device damage from UV photons can be removed by a thermal anneal at 350°C (to the point that the original electrical parameters determined by CV measurements are restored). High energy ion bombardment damage, on the other hand, is generally removed by annealing at 650°C, while severe x-ray damage may require an annealing temperature as high as 950°C. Fortunately x-ray damage is rare, because the sheath and bias potentials in most plasma processes are too low to generate an appreciable x-ray flux. In those cases where x-rays are a problem, changing some metallic parts of the reactor to insulating materials can eliminate x-ray production. Since most device fabrication schedules include thermal cycling to at least 700–900°C after the critical gate levels, much damage from plasma processing is removed naturally, without any special processing.

However, new device designs include shallow junctions, which cannot tolerate high temperature at all. Hence, high energy ion damage is becoming more of a problem. Difficulties also arise with thin gate insulators (< 150 Å) that must have extremely good material characteristics for reliable performance. Here the x-ray flux must be reduced to an absolute minimum, so the use of metallic reactor surfaces is minimized and low potentials are used. Remote plasma systems (plasma transport reactors—see Section V) and magnetron reactors offer a less energetic ion flux and have found increasing use in these applications.

B. IRREVERSIBLE EFFECTS

Unfortunately, two "killer" effects that can occur in plasmas *are not* reversible. The first of these is contamination during critical operations such as window and/or via-step etching. If polymer formed by the plasma itself, or foreign material such as "dirt" or sputtered reactor material, enters an opened contact, it will be covered later by metallization. The result is unexpected doping or an unwanted resistance, which cannot be corrected. Hence control of reaction products from the plasma chemistry, particularly polymeric products, is essential.

A second, more subtle, source of irreversible damage is electrostatic breakdown of thin insulating films caused by charge that is deposited during the plasma process cycle. The gate insulator films in many state-of-

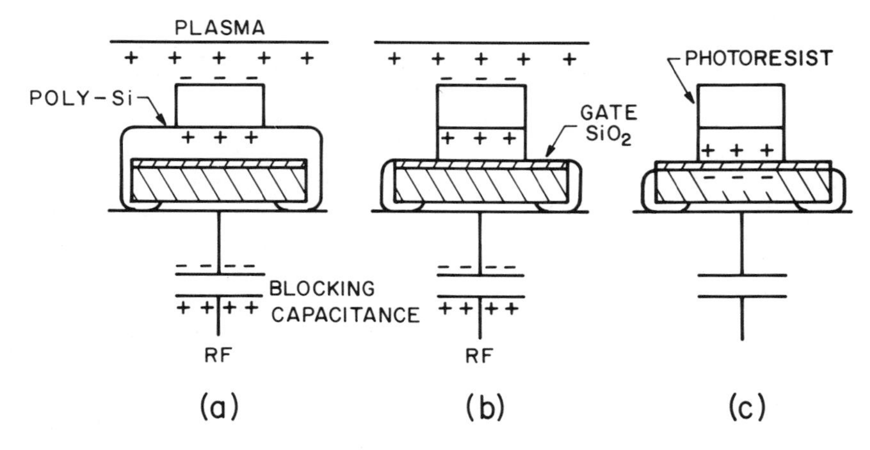

FIGURE 57. Electrostatic breakdown of gate oxide. Left: a patterned polysilicon-coated wafer is placed on the "cathode." When the plasma is turned on, the conductive poly which wraps around the wafer edge shields the gate dielectric and transfers charge to the electrode and blocking capacitor. Center: the poly is patterned and the front to back conductive path is broken. Right: charge deposited during the plasma turn-off transient is trapped on the front oxide surface with mirror charge below. If the electric field generated by this charge exceeds breakdown, the thin oxide dielectric will be mechanically disrupted.

the-art device structures are becoming extremely thin, (150–250 Å). While a high quality silicon dioxide film can withstand about 12–15 MV/cm for short periods, this amounts to only 12–15 volts across a 100 Å thick SiO_2 gate dielectric. Beyond this point there is catastrophic breakdown, and no way to repair the film.

The mechanism of plasma induced breakdown is illustrated in Fig. 57 [49]. A film of conductive polysilicon is deposited over thin gate oxide (SiO_2) and covered with a photoresist mask pattern. The wafer is then placed in a plasma to transfer the mask pattern into the polysilicon film. At the point depicted in Fig. 57a, the plasma has just been struck. As the sheath is established, the smaller electrode (cathode) is negatively charged by electrons until, at steady state, there are equal fluxes of electrons and ions (see Section III.E.1). Initially, the conductive polysilicon film extends around the wafer and forms a continuous electrical circuit from the front to the electrode underneath. Hence the negative electrode potential appears across the "blocking" capacitance, which is a part of the power supply or matching network. If the plasma were turned off at this point, there would be a transient flow of charge to the wafer and surrounding surface, but this charge would induce no potential across the gate because it is still surrounded by the conducting polysilicon layer. As the mask pattern is etched into the polysilicon, the conductive path at the edge of the wafer is finally broken (Fig. 57b), and the gate electrode is isolated. Still, at this point, there is little or no potential difference between the gate electrode and substrate below because there was electrostatic equilibrium when the path was broken.

When etching is finished and the plasma is turned off, however, positive ions from the decaying plasma will deposit small amounts of charge on the resist and gate electrode. Meanwhile, the wafer substrate is maintained at the reactor electrode potential. Since the gate oxide dielectric is thin, the charge that collects can induce high transient fields across the gate and cause breakdown. Free charge reaching the reactor electrode during this transient has little effect on its potential because of the large capacitance to ground through the blocking capacitor and stray coupling,

By contrast, if the wafer is placed on an insulating surface, substrate potential is not held at that of the large electrode and there is less danger of breakdown. In this case, voltage from the turnoff transient will be divided between the gate capacitance and the wafer/insulator/electrode capacitor. Clearly, plasma operation should be characterized with test devices to insure that transient potentials are kept well below the level that produces electrostatic breakdown.

During photoresist stripping a similar condition can occur. Wafers are put into barrel etchers where the resist collects charge as it approaches the floating potential of the plasma. For a typical oxygen plasma used for stripping photoresist, the floating potential is about -5 V. If electrical transients in the system during the resist strip cycle suddenly increase the floating potential, electrically isolated areas can charge at different rates (because of the geometrical arrangement of wafers with respect to the plasma in the reactor) so that the breakdown potential of thin insulators (12–15 V_{dc}) is exceeded.

Another experiment further illustrates these phenomena. Capacitors with 100 Å SiO_2 dielectrics fabricated on wafers were tested to determine the Fowler-Nordheim tunneling current through the capacitors as a function of impressed voltage. These wafers were divided into two groups with one group loaded into metal support "boats" (see Fig. 45), and the other into quartz boats. Both groups were placed in the same commercial resist strip system in which wafer boats are held on a "grounded" metal support system within a metal reactor chamber. After resist stripping, the capacitors were tested a second time. Wafers in the quartz boat were unaffected by the stripping cycle. Tunneling current through almost half the capacitors in the metal boat group, on the other hand, increased from 5–50 nA with about a 10 V potential difference, to 1–5 mA after the strip cycle, a level indicating a device failure (end-of-life is a tunneling current above 1 μA).

The results are simply explained. In order for breakdown to occur, a potential *difference* must exist across the film. Since both sides of the wafers reached the same floating potential in the quartz boat, there was no potential difference. By contrast, the capacitor electrodes connected to the substrate in the metal boat were maintained at "ground" potential so that transients in the floating potential during stripping were imposed across the thin SiO_2 film.

1. Testing for ESD

Plasma processing conditions should be tested to insure that devices will not suffer irreversible damage from these subtle thin-film breakdown effects. In one procedure [50], special wafers are tested for electrostatic discharge before and after plasma treatment. The chips on these wafers contain capacitors formed across 100–150 Å of "thermal" SiO_2 and arranged in a pattern similar to that used in 1 megabit dynamic random access memories (DRAMS). There are more than 1 million capacitors connected in parallel

Table 6 Failure of 150 Å capacitor test structures when stripping resist in various system configurations. Failure is signified by more than a 1 μA tunnelling current under a 10 Volt bias after stripping. In addition, there may be a time-dependent breakdown which is signalled by a slow increase of tunneling current under constant bias. Some of the failed devices (those with "yes" in the TDB column) showed a gradual increase in tunneling current until end-of-life (1 μA) was reached.

System Configuration	Percent Device Failure	Time-Dependent Breakdown
I. Barrel		
a) No etch tunnel, metal boat	~50	Yes
b) No etch tunnel, quartz boat	0	No
c) Etch tunnel	0	No
II. Parallel Plate	0.5-5.0	Yes
III. Downstream, μ-wave		
a) With grid	0	Yes
b) Without grid	0	No

on each chip, which makes the test extremely sensitive since a circuit will fail if even a single capacitor breaks down. However, these capacitors are not ordinarily stressed to breakdown (as done in other quality tests) since doing so would not provide much information about the plasma environment. Instead, a voltage in the Fowler-Nordheim regime is applied, and the tunneling current is mapped across the wafer before plasma processing. This can be compared to the tunneling current in a time-dependent breakdown (TDB) measurement after plasma processing. The applied voltage (around 10 V) on the capacitors is kept constant while the change in tunneling current is monitored. If current rises (usually it does so linearly), the time for it to reach end-of-life (1 μA) gives a measure of damage phenomena. If necessary, this test can be made more sensitive by using plasma conditions that are harsher (higher power, potential, etc.) than those expected in routine processing.

Table 6 shows results from a study where this test was applied in a variety of reactors. Wafers stripped in a barrel system with or without an etch tunnel showed no ill effects. A downstream microwave plasma stripping system also was satisfactory. In almost all other systems used in this study, however, there was some degree of damage.

References

[1] J. Bardeen and W. H. Brattain, "The Transistor, A Semiconductor Triode." Letter to the Editor, *Phys. Rev.* **74**, 230–231 (1948).

[2] W. H. Brattain and J. Bardeen, "Nature of the Forward Current in Germanium Point Contacts." *Phys. Rev.* **74**, 231–232 (1948).

[3] W. Shockley, "The Theory of p-n Junctions in Semiconductors and p-n Junction Transistors." *Bell System Tech. J.* **28**, 435–489 (1949).

[4] W. Shockley, M. Sparks, and G. K. Teal, "p-n Junction Transistors." *Phys. Rev.* **83**, 151–162 (1951).

[5] T. R. Reid, "The Chip: How Two Americans Invented the Microchip and Launched a Revolution." Simon and Schuster, New York (1984).

[6] C. J. Forsch and L. Derick, "Surface Protection and Selective Masking during Diffusion in Silicon." *J. Electrochem. Soc.* **104**, 547–552 (1957). U.S. Patent #2,802,760 issued August 13, 1957.

[7] The yield of silicon devices is related to the chip area and the number of *fatal* defects per cm^2. For a given chip area the yield increases as the D_o decreases (see Section I.B).

[8] G. E. Moore, "Proc. Caltech. Conf. on Very Large Scale Integration," California Institute of Technology, Pasadena, California (1979).

[9] W. F. Richardson, *et al.*, "A Trench Transistor Cross-Point DRAM Cell." Proc. IEDM, Washington, D.C., 714 (1985).

[10] K. Yamada, *et al.*, "A Deep-Trenched Capacitor for a 4 Mega Bit Dynamic RAM." Proc. IEDM, Washington, D.C., 702 (1985).

[11] Aubrey C. Tobey, "Semiconductor Microlithography through the Eighties." *Microelectronic Manufacturing and Testing*, 19 (1985).

[12] The "design rule" is the smallest feature dimension patterned into an integrated circuit. C. Mead and L. Conway, *Intro. to VLSI Systems*. 47–51. Addison-Wesley, Reading, Massachusetts (1980).

[13] B. T. Murphy, *Proc. IEEE* **52**, 1537 (1964).

[14] R. G. Brandes and R. H. Dudley, *J. Electrochem. Soc.* **120**, 140 (1973).

[15] K. E. Beam, *IEEE Trans. Elect. Dev.* **ED-25**, 1185 (1978).

[16] S. M. Irving, "A Plasma Oxidation Process for Removing Photoresist Films." *Solid State Technol.* **14(6)**, 47 (1971).

[17] Actually, plasmas can exist in the solid phase as well; for example, free electrons and fixed positive charge in a semiconductor or metal can be treated as a plasma.

[18] As we will see in Section III, however, chemical etching of III-V compounds by the halogens is crystal orientation dependent and anisotropic.

[19] C. J. Mogab, H. J. Levenstein, *J. Vac. Sci. Technol.* **17**, 721 (1980); U. Gerlach-Meyer, J. W. Coburn, and E. Kay, *Surface Science* **103**, 177 (1981); D. L. Smith and R. H. Bruce, *J. Electrochem. Soc.* **129**, 2045 (1982); V. M. Donnelly, D. L. Flamm, and R. H. Bruce, *J. Appl. Phys.* **58**, 2135 (1985).

[20] J. W. Coburn and H. F. Winters, *J. Vac. Sci. Technol.* **16**, 391 (1979).

[21] E. W. McDaniel, "Collision Phenomena in Ionized Gases." 23–25, J. Wiley and Sons, New York (1964).

[22] E. W. McDaniel, "Collision Phenomena in Ionized Gases." 120, J. Wiley and Sons, New York (1964).

[23] For example, in most gases the mean free path depends on ϵ_e. (Among the reasons; inelastic loss processes are a function of ϵ_e and at low ϵ_e constant time between collisions is a better approximation than constant mean free path.)

[24] E. W. McDaniel, "Collision Phenomena in Ionized Gases." 42–44, J. Wiley and Sons, New York (1964).

[25] I. Langmuir, *Gen. Elect. Rev.* **26**, 731 (1923).

[26] E. W. McDaniel, "Collision Phenomena in Ionized Gases." 35–40, J. Wiley and Sons, New York (1964).

[27] D. L. Flamm, V. M. Donnelly, and J. A. Mucha, *J. Appl. Phys.* **52**, 3633 (1981).

[28] V. M. Donnelly, D. L. Flamm, W. C. Dautremont-Smith, and D. J. Werder, *J. Appl. Phys.* **55**, 242 (1984).

[29] D. E. Ibbotson, D. L. Flamm, V. M. Donnelly, and B. S. Duncan, *J. Vac. Sci. Technol.* **20**, 489 (1981).

[30] R. Glang, in "Handbook of Thin Film Technology," L. I. Maissel and R. Glang, eds., **1-27**, 1–26, McGraw Hill, New York (1970).

[31] K. Denbigh in "The Principles of Chemical Equilibrium," 2nd ed., 197–200 Cambridge University Press, Cambridge (1968).

[32] V. M. Donnelly, D. L. Flamm, C. W. Tu, and D. E. Ibbotson, *J. Electrochem. Soc.* **129**, 2533 (1982).

[33] R. W. Perry and C. H. Chilton, *Chemical Engineer's Handbook*, 5th ed., 3–46 McGraw-Hill, New York (1973); J. T. Viola, D. W. Seegmiller, A. A. Fannin, Jr., and L. A. King, *J. Chem. Eng. Data* **22**, 367 (1977).

[34] H. Winters, *J. Vac. Sci. Technol.* **A3**, 786 (1985).

[35] M. Mieth and A. Barker, *J. Vac. Sci. Technol.* **A2**, 629 (1983).

[36] An "overetch period" is the time measured from when areas of the substrate underlying a film layer are first exposed. This overetch is usually expressed as a percentage of the time to this first exposure. Hence "50% overetch" means etching for 150% of the time needed to first expose an area of the underlying substrate.

[37] C. J. Mogab, "Dry Etching," in *VLSI Technology*, S. M. Sze, ed., 303–345, McGraw-Hill, New York (1983).

[38] M. O. Watanabe, M. Taguchi, K. Kanzaki, and Y. Zohta, *Japan. J. Appl. Phys.* Part 1, **22-2**, 281 (1983).

[39] W. Beinvogl and B. Hassler, *Electrochem. Soc. Extended Abstracts* **81-1**, 586 (1981).

[40] In the "inductively coupled" configuration, there is a coil around the vacuum chamber instead of planar electrodes. Although this arrangement is often assumed to involve magnetic coupling, in reality it is mainly the electric potential across the ends of the coil that sustains the plasma.

[41] A. R. Reinberg, U.S. Patent 3,757,733 (1975).

[42] T. Tsuchimoto, *J. Vac. Sci. Technol.* **15**, 70, 1730 (1978).

[43] T. Ono, M. Oda, C. Takahashi, and S. Matsuo, *J. Vac. Sci. Technol.* **B4**, 696 (1986).

[44] J. Musil, *Vacuum* **36**, 161 (1986); M. R. Wertheimer, J. E. Klemberg-Sapiana, and H. P. Schreiber, *Thin Solid Films* **15**, 109 (1984); J. Asmussen, *J. Vac. Sci. Technol.* **B4**, 295 (1986); S. R. Goode and K. W. Baughman, *Appl. Spectrosc.* **38**, 755 (1984).

[45] D. L. Flamm, *J. Vac. Sci. Technol.* **A4**, 729 (1986).

[46] M. R. Wertheimer and M. Moisan, *J. Vac. Sci. Technol.* **A3**, 2643 (1985).

[47] G. Lucovsky, *J. Vac. Sci. Technol.* **A4**, 480 (1986).

[48] "Dry Etching Equipment: A Special Report." *Microelectronics Manufacturing and Testing* 42–43 (1985).

[49] T. Watanabe and Y. Yoshida, *Solid State Technol.* **27-4**, 263 (1984).

[50] G. K. Herb, P. D. Cruzan, L. B. Fritzinger, G. W. Hills, L. E. Katz, and T. Kook, *Proc. 13th Annual Tegal Seminar* (Tegal Corp., May, 1987); P. H. Singer, *Semcond. International* 36 (1987).

[51] S. R. Hunter and L. G. Christophorou, "Electron-Molecule Interactions and Their Applications." **Vol. 2**, L. G. Christophorou, ed., 90–219, Academic Press, Orlando (1984).

2 Introduction to Plasma Chemistry

Daniel L. Flamm
AT & T Bell Laboratories
Murray Hill, New Jersey

ISBN 0-12-469370-9

I. Overview

This chapter discusses the mechanisms and chemistry of plasma etching. Relations between chemistry, operational parameters and mechanisms are illustrated with examples from the plasma etching of silicon, SiO_2, organic resists, metals and III-V compounds in a variety of gas mixtures. The rationales for choosing feed chemistries and parameters are discussed.

II. How Plasma Etching Takes Place

Many phenomena play a role even in the simplest examples of plasma etching. However, as pointed out in Chapter 1, we can group etching mechanisms into four functional categories.

One of these is sputtering. As shown in Fig. 1a, positive ions are accelerated across the sheath and strike the substrate with high kinetic energy. Some of this energy is transferred to surface atoms which are then ejected, leading to material removal. This process is distinguished from the other mechanisms in that the interaction is mechanical: differences in chemical bonding are important only in so far as they determine the bonding forces between surface atoms and the ballistics of dislodging them.

Chemical mechanisms, which include ion-assisted etching, form the main subject of this chapter (Figs. 1b–1d). Simple chemical etching comes about when active species from the gas phase encounter a surface and react to form a *volatile* product (Fig. 1b). As discussed in Chapter 1, high product volatility is essential—without volatility the reaction products would coat the surface and prevent gaseous species from reaching it, and cut off the etching reaction. In chemical etching, the role of the plasma is merely to supply etching species—these are usually free radicals such as fluorine atoms (F). This type of etching shares the characteristics of common chemical reactions. Etching is usually non-directional, since ion bombardment plays no role, and selectivity can be extremely high (or even "infinite") owing to large differences in an etchant's chemical affinity for various substrate materials.

Some years ago fast *directional* "dry" etching was discovered. This occurs when certain substrates are exposed to suitable neutral species in the presence of ion bombardment. The effects of neutrals and ions can be synergistic, with the resulting material removal rates exceeding the sum of separate chemical attack and sputtering. There are two general mechanisms for this ion-assisted plasma etching.

In an ideal energy-driven mechanism (Fig. 1c), neutral species cause little or no etching without ion bombardment. Ions "damage" the substrate

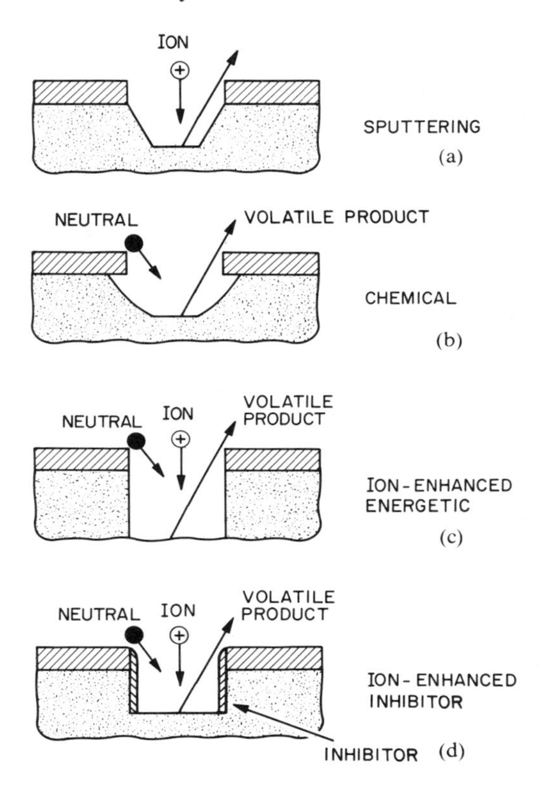

FIGURE 1. The four basic mechanisms of plasma etching are: a) sputtering in which ions *mechanically* eject substrate material at low pressure, b) purely chemical, where thermalized neutral radicals react with substrate material and form volatile products, c) ion-enhanced energetic mechanisms in which there is little or no intrinsic surface reaction with neutral radicals, until energetic ions enhance the reactivity of a substrate or product layer allowing chemical reactions to gasify the material; and d) ion-enhanced inhibitor, or protected sidewall anisotropy where inhibitor species form a sidewall film which excludes the neutral etchant. Ions disrupt this film on horizontal surfaces.

material, making it more reactive toward incident neutral radicals. We mean "damage" in a broad sense to include diverse mechanisms such as the formation of highly reactive dangling bonds, the disruption of lattice structure and formation of dislocations, forcible injection of adsorbed reactant into a lattice by the collisional cascade, or even bond-breaking in intermediate tightly-adsorbed surface compounds. We should emphasize that there probably is no universal mechanism for energy-driven ion-enhanced etching. Rather, on a case by case basis, a variety of elementary mechanisms can come into play depending on the etchant, surface being etched, and perhaps even on the rate at which radicals arrive at the surface. The common denominator to energy-driven mechanisms is that the ions supply kinetic energy, which disrupts the surface being etched. Since ions are accelerated across the plasma sheath and strike surfaces vertically, the etching induced is directional.

In the last category, inhibitor-driven ion-assisted etching (Fig. 1d), etching by neutrals is spontaneous so ion bombardment does not *cause* the etching reaction. Ions play out their role by interacting with a second ingredient—film formers, which can coat substrate surfaces and prevent etching reactions from taking place. At this point we won't be concerned about the source of inhibitor material. It has already been pointed out that the horizontal surfaces of a wafer intercept most of the normal-going ion flux. This ion flux keeps areas clear of film, while vertical feature sidewalls are coated with a thin film which inhibits chemical attack. This ion "clearing" of horizontal surfaces generally appears to be a rather mild process in which high ion energy is neither required nor desired. In some chemical systems ion bombardment may retard the growth of these films rather than literally "clearing" them.

A. SPUTTERING

Sputtering is a universal way to etch material by action of a plasma. The ballistic ejection of material occurs when positive ions are propelled into surfaces by the negative-going potentials at the edge of a plasma. When an ion transfers energy to a small region of the substrate at the point of impact, some substrate molecules receive enough of this energy to be ejected from the surface. Material is removed as it is thrown across the reactor volume under low pressure conditions (a low pressure and a long-mean-free path are required for material to leave the vicinity of the surface without being backscattered and redeposited).

Sputtering is inherently unselective because the ion energy required to eject material is large compared to differences in surface bond energies and chemical reactivity. For many applications this lack of selectivity means that sputtering cannot be used. Yet when there are no reactive etching processes available for a material, sputter etching is always a possible solution.

Sputtering suffers from other disadvantages. It is generally slow compared to other etching means, with etch rates limited to several tens of nanometers per minute compared to the hundreds of nanometers per minute and above achieved by chemical and ion-assisted etching. Sputter etching also tends to form facets and trenches near etched features as discussed in Chapter 1.

In passing, we alert the reader to the confusing use of terms such as "chemical sputtering" or "reactive sputtering" in the literature. These phrases are somewhat contradictory since "real" sputtering is basically a mechanical process. We qualify a little because details of energy exchange and material ejection are sensitive to the manner and strength by which atoms are bound to the substrate.

B. CHEMICAL ETCHING

In ideal chemical etching, the only function of the plasma is to maintain a supply of gaseous etchant species. Consequently, it is also possible to etch by this mechanism without a plasma when the feed gas is inherently reactive. For example, silicon is often etched using F atoms generated by NF_3 or CF_4/O_2 plasmas. However, silicon can also be etched spontaneously in ClF_3 gas to form the same final etch product, SiF_4. Chemical etching is inherently the most selective mechanism, since unwanted reactions will not take place at all when their thermodynamics are unfavorable. By contrast to sputtering, where involatile materials can be ejected across the reactor, a volatile reaction product is *necessary* for chemical etching. Involatile products would accumulate on the surface, and passivate it against further reaction.

Purely chemical attack, unlike other forms of plasma etching, is usually non-directional. Thus, when a silicon substrate reacts with atomic fluorine, masked areas are undercut with a characteristic isotropic circular profile (see Chapter 1). Depending on what the process is intended to do, this may be desirable or unwanted. Consider part of the sequence used to fabricate a MOS transistor gate, shown in Fig. 2. The processing sequence forms a micron-sized polysilicon plug (illustrated) over a thin oxide gate (~ 100–300 Å). This polysilicon often has a reentrant profile at its base near the oxide, which must be removed to prevent a short circuit in subsequent processing steps. While the polysilicon plug can be etched without linewidth loss by the nearly ideal ion-assisted etch available in a chlorine plasma (which is selective for silicon over silicon oxide), this etching is too vertical to remove the reentrant filament. Furthermore, the limited selectivity and ion bombardment in the anisotropic process pose the threat of electrical damage and erosion of the thin, sensitive MOS gate oxide. Isotropic etching, on the other hand, cannot maintain the small dimensions needed in this processing step.

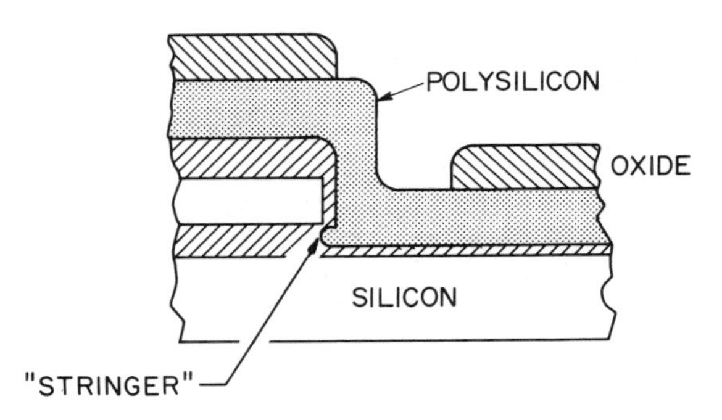

FIGURE 2. In the process of fabricating a polysilicon contact over thin gate oxide, a reentrant "stringer" is often formed and it must be removed to avoid a short circuit, later in the processing sequence. Anisotropic ion-assisted chlorine etching is used to etch the polysilicon plug almost to the gate oxide level. However, this chlorine etch will not remove the stringer and can damage the thin gate. A fluorine atom based etch gasifies the stringer and provides high selectivity over the gate oxide.

To achieve the needed process objectives, a controlled combination of anisotropic and isotropic etching can be used. First, the polysilicon plug is etched—just to the gate oxide—by the chlorine plasma. Then an isotropic etchant, such as fluorine atoms from an NF_3 discharge, can be used to remove the last bit of the polysilicon plug and the reentrant filament. Isotropic fluorine atom etching offers high selectivity over oxide and does not require ion bombardment.

Still another case where isotropic etching is required is in device fabrication by "liftoff," or additive lithography (see Chapter 1). Here a pattern is developed by depositing material over a mask with undercut features. The undercut permits the mask to be dissolved off in a later processing step. Isotropic plasma etching is also advantageous in applications where undercut-limited linewidth is not a problem. Examples include oxygen plasmas, in which O atoms are used to strip resist masks isotropically after the patterning step, and CF_4/O_2 discharges which isotropically etch silicon with fluorine atoms (for large linewidth discrete devices). Chemically-driven etching is preferred when linewidth control is not a limitation because it offers high selectivity and freedom from damaging high energy particle bombardment.

Finally, we should point out that chemical etching is not always isotropic. In fact, halogen plasmas etch III-V compounds preferentially along specific crystallographic directions. This effect can be used to obtain very smooth self-aligned vertical features or to preserve critical dimensions. We will return to this subject in Section XII.

C. ION ENHANCED ETCHING

Plasmas differ markedly from ordinary liquid and gaseous reacting media in that a directional negative-going electric field always forms along the plasma boundaries. This sheath-field, which can be quite intense, propels positive ions into boundary surfaces at normal incidence. Hence, there is an anisotropic ion flux at plasma-solid boundaries. Under the appropriate circumstances, this ion bombardment stimulates directional etching. Once this etching was thought to be caused by chemical reactions between the ions and surface material. However, for the etch rates in most practical situations, ~ 1–10 kÅ/min, this is not even a theoretical possibility because the ion flux is much lower than surface atom removal rates. Research and theory have shown that the mechanisms of anisotropic etching are substantially more complex.

It is now accepted that anisotropic plasma etching is brought about by the action of neutral gas-solid reactions, which are stimulated or directed by ion bombardment. Although ions participate in these processes, neutrals do the etching except at very low pressures where ion reactions can stimulate modest rates ($\lesssim$ 400 Å/min).

Unfortunately the term "reactive ion etching" has been applied to a variety of reactor configurations in which anisotropic etching is commonly (but not always) observed. We will avoid this confusing term.

1. Energy-Driven Anisotropy

In energy-driven ion-enhanced etching, reactions between the substrate and neutral species from the gas phase are accelerated by ion bombardment. While there is little information on the ion-energy dependence of such rates, they are thought to have a reaction-specific threshold energy, with the reaction rate increasing as ion energies surpass the threshold.

A variety of conflicting mechanistic theories have been proposed. There probably is no *single* universal mechanism, but rather there are a variety of ways in which ion bombardment can coerce different gas-substrate systems to react. Hence, the detailed mechanism of energy-driven ion enhanced silicon etching in chlorine plasmas is likely to be quite different from ion silicon etching of SiO_2 in fluorine, or in CHF_3 discharges. Various studies have attempted to simulate plasma etching in high vacuum by combining controllable fluxes of reactive neutrals with ion beams [1, 2, 3, 4, 5, 6, 7]. Figure 3 shows the dramatic result of an early experiment where silicon was exposed to a flux of Cl_2 gas in high vacuum with and without a 450 eV ion beam [1]. The ability of an ion to stimulate reaction in such experiments

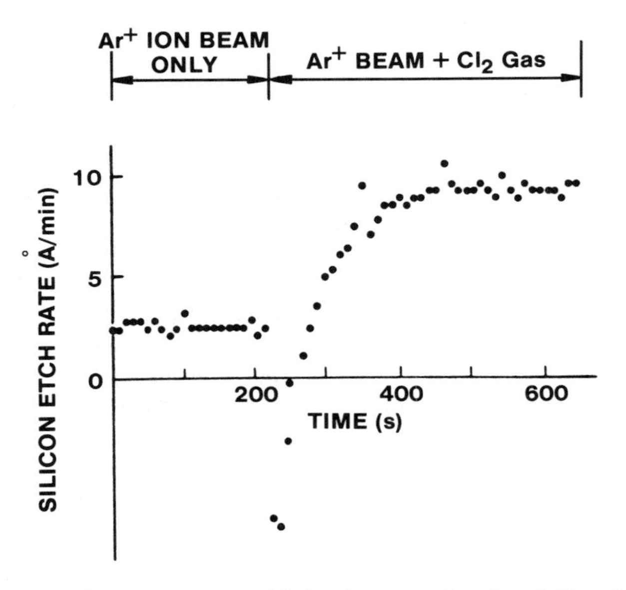

FIGURE 3. At moderate temperature, chlorine does not etch undoped silicon in the absence of ion bombardment or other energetic irradiation. A 450 V argon ion beam (left) sputters silicon slowly, but the synergistic etch rate in the presence of chlorine gas (right) is much faster. The negative transient at ~ 200 s is caused by chlorine adsorption on the surface (the etching is sensed by a qaurtz crystal microbalance)—after [1].

appears to depend more on its energy and mass than on its chemical identity [2, 3, 8]. While these simple experiments with ion beams demonstrate energy-driven etching with well-defined examples, most of the data don't appear to reproduce the etching yields [9, 10] and other aspects of in-plasma reactions [8], and should probably be regarded more as analog than a reproduction of the plasma chemistry.

2. *Inhibitor-Driven Anisotropy*

In inhibitor-driven anisotropy a surface-covering agent protects the vertical sidewalls of features from reactive neutrals, while ion bombardment suppresses inhibitor growth on the horizontal surfaces that are etched (Fig. 1d). This mechanism requires three ingredients: (1) reactive neutral etchant species, (2) an ion flux to the surface, and (3) *inhibitor* film-forming species. Inhibitor-driven anisotropy differs from energy-driven anisotropy in that the chemical etching reaction is spontaneous, even without ion bombardment. While the low flux of scattered ions to feature sidewalls has minimal effect, the main vertical ion flux either prevents the films from forming, or "clears" them as they grow. It is generally believed that low ion energies suffice to keep the horizontal surfaces free of inhibitor and allow anisotropic etching. By contrast, fairly energetic ions are usually required to drive etching which results from "damage" to the surface.

Inhibitor-forming species originate from a variety of sources. Some freon (e.g., C_2F_6, CHF_3) feed gases yield unsaturated polymer-forming species in plasmas. By unsaturated species we mean CF_2 radicals and derivatives formed by chain-building oligomerization, when fluorocarbons are involved (e.g., C_2F_4, C_3F_6), or CCl_2 radicals and derivatives (C_2Cl_4, etc.) in chlorocarbon feeds (e.g., CCl_4). These simple monomers and oligomers polymerize to form thin films on surfaces (or sometimes thick films), which induce anisotropy by the mechanism outlined above.

Aluminum etching in $CCl_4 + Cl_2$ or $CHCl_3 + Cl_2$ plasmas is a good example of an inhibitor system. Even though the Cl_2 and Cl atoms formed in these plasmas are rapid chemical etchants for clean aluminum, aluminum etching in these plasma mixtures can afford near-vertical profiles with excellent linewidth control. Similarly, heavily n-doped polysilicon (e.g., $> 10^{20}$ cm^{-3} phosphorous or arsenic) etches isotropically in pure chlorine plasmas at moderate pressure, but with CHF_3 or C_2F_6 additions, inhibitor-driven anisotropic etching is easily achieved.

While inhibitor films are usually formed from a feed gas additive, they can also arise from more subtle sources. For example, anisotropic etching of heavily n-doped silicon has been observed in *pure* chlorine plasmas at low frequency (below 1 MHz, the lower ion-transit frequency) [11]. In this case an inhibitor film was thought to be formed from sputtered reactor material (at this frequency there are very high sheath potentials and ion energies). In another situation, the resist mask catalyzed the growth of an inhibitor film [12], and material from the mask has also been suggested as a source of inhibitor [13, 14, 15].

III. Etching Characteristics and Variables

Plasma etching processes may be characterized by their *rate*, *selectivity*, *uniformity* and *surface quality*. Obviously, the etch rate must be high enough for economical process throughout (see Chapter 1). At the same time, there are selectivity constraints to insure that the etch mask is not removed or unduly eroded and that the sublayer beneath a patterned film will not be attacked during overetch. Sometimes various materials in a sublayer are exposed to the plasma during the etch step and it is required that one of these be selectively etched.

Anisotropy, as already mentioned, is an important requirement in many etch processes. However, profile requirements go beyond simple anisotropy or isotropy—a sloped profile may be necessary to guarantee adequate step coverage during a subsequent deposition.

Another key requirement is that the etched surface have good morphology and be free of physical and electrical "damage." Isotropic etching in fluorine gas (F_2) is a good example of a morphological constraint. F_2 etching can be rapid and almost totally selective for Si over SiO_2, but this etching gas is unusable because of the "moonlike" cratered surface that it leaves behind (see Section III.E.2).

A. DISCHARGE VARIABLES

Ideally we would like to know how to control etching properties by manipulating basic chemical and plasma variables. In many basic texts and literature on electrical discharges, the plasma is examined from a fundamental point of view as a function of certain quasi-dimensionless similarity variables. These variables include the ratio of electric field to number density, E/N (sometimes written as E/p since pressure, p, is proportional to number density at standard conditions), the product of number density and a characteristic length of the reactor geometry (Nd or pd), f/N, the ratio of excitation frequency (plasma RF generator frequency) to density, and reactor shape and aspect ratios. Some of these are closely related to key determinants of reaction rate constants and not all are independent. E/N, for example, is a rough measure of electron energy. n_e/N regulates the importance of charged particle-neutral relative to neutral-neutral reactions (and many other things).

Unfortunately, these variables are not generally useful for controlling most plasmas in the semiconductor industry. The reasons for this negative conclusion are threefold. The similarity variables cannot be set by a process engineer or plasma scientist—relationships between the microscopic similarity parameters are already determined by the plasma gas and apparatus. Even in somewhat ideal model systems such as the positive (glow) column of DC glow discharges or the ambipolar region (away from the sheaths) of high frequency diffusion-controlled glows, E/N is mostly fixed by gas composition and reactor geometry. Furthermore, there really is no well-defined E/N or n_e/N for many processing conditions. E and n_e often change by a factor of ten or more from one position to another in the plasma and they can also oscillate in time, in step with the applied electric field. Since E/N, n_e/N and related similarity parameters are neither constant nor at the disposal of an operator, the instrumental parameters or *discharge variables* are used instead. These discharge variables include radio-frequency (RF) input *power*, reactor *pressure*, RF excitation *frequency*, *temperature*, *flow rate*, feed *gas composition*, reactor geometry and materials of reactor construction. All of these quantities are at our disposal and, when fixed, uniquely determine the operation of a plasma process. We can't set E/N, but almost every reactor has a knob to regulate power, another to set total pressure and a controller to keep the electrodes at a chosen temperature. Many modern reactors offer control over excitation frequency as well. In the sections below we survey some effects of these variables.

B. PRESSURE EFFECTS

Pressure directly influences major phenomena that control plasma etching. Among these are: (1) the sheath potentials and energy of ions bombarding surfaces, (2) the electron energy, (3) the ion-to-neutral abundance ratio and fluxes of these species to surfaces, (4) the relative rate of higher to lower order chemical kinetics, (5) surface coverage by physisorption, and (6) the relative rates of mass transport processes. Our remarks will be restricted to the pressure range generally used for plasma etching, about 1 mTorr to 5 Torr. The influence of pressure on ion energy is treated below. Chemical kinetic pressure effects and mass transport effects are treated in Section III.D while the "loading" effect is discussed in Section V.

1. Pressure Effects on Potentials and Energy

As pressure is lowered below about 0.1 Torr, the characteristic potentials across the sheaths and the voltage applied to a discharge increase sharply, from some tens of volts up to 1000 V or more (see Fig. 4). As bias is proportional to the peak applied voltage, V_p, it rises too. Since the mean-free-paths of species are inversely proportional to pressure, the rise in potential translates into a higher energy ion flux to substrate surfaces. Sputtering does not take place until ion bombardment energy exceeds a material and ion-specific threshold [16], at which point sputtering rates increase rapidly as ion energies move above the threshold. However, the sputtering efficiency (atoms removed per incident ion) usually remains well below unity.

Similarly, it is believed that there are threshold energies and energy-dependent cross sections for damage-induced ion-assisted etching of various

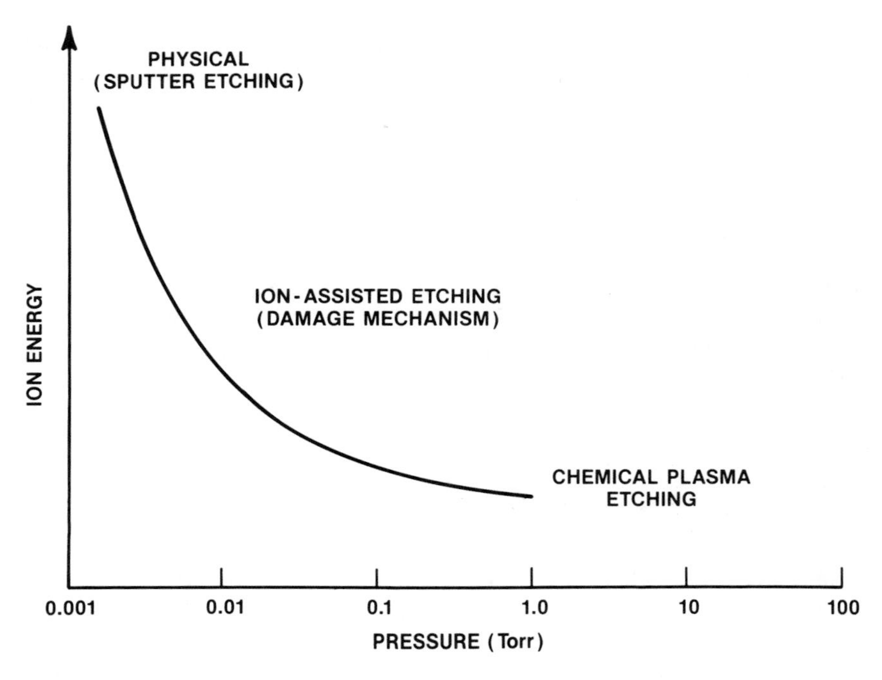

FIGURE 4. Pressure influences the sheath potential and hence the ion bombardment energy. With 13 MHz excitation frequency and the higher pressures shown, ion energy tends to be low: between ~ 10 and ~ 100 V. In this regime the neutral concentration is high and chemical etching competes strongly with ion-assisted mechanisms. At the lowest pressures shown, ion energy may be ~ 1000 V and the long mean free path permits sputtering. Most etch processes operate between these two extremes where ion-assisted etching provides anisotropy and device damage is minimal.

materials. Low pressure favors higher ion bombardment energies which promote etching by damage-induced mechanisms. However, ion energies far above threshold are undesirable because selectivity decreases with increasing ion energy, and high energy ions can cause electrical damage to devices.

Electron-molecule reaction cross sections vary with the electron energy, ϵ. For typical plasma etching conditions, the average energy, $\bar{\epsilon}$, increases with E/N (or E/p). E increases with the applied RF voltage and the applied voltage decreases with increasing pressure. Since N is proportional to pressure, this means E/N and the average electron energy decrease with rising pressure. In the range of interest, with mean electron energies $\bar{\epsilon} < \sim 5$–8 eV, many electron-molecule dissociation reaction rates tend to slowly increase with E/N because of increasing overlap between elementary reaction cross sections and electron energy distribution functions

[17, 18, 19] (the reaction rates are a convolution of these quantities—see Section III.F, Eqn. 37 and Figs. 14, 16).

As a result, electron energy and electron-molecule ionization rate constants tend to decrease with increasing pressure leaving less efficient ionization processes to regenerate the charge lost by diffusion and recombination. Because of compensating effects—the larger neutral reactant concentration (gas density) and a lower diffusive loss rate (ambipolar diffusion increases with the electron temperature, T_e, and varies inversely with pressure)—the electron number density may not have to change much with pressure to sustain a discharge.

Since energy transfer from electrons to neutrals—the thermalization of electrons—is proportional to gas pressure too, the neutral gas temperature tends to increase with pressure while the electron temperature falls. At some point, usually above 10–100 Torr, the electron and gas temperatures converge and the plasma is said to be "arc-like." At low to intermediate pressures, 1 mTorr–1 Torr, the plasma density is insensitive to total pressure, constrained to stay above 10^9–10^{10} cm^{-3} by the ambipolar limit, and below $\sim 10^{12}$ cm^{-3} by the glow-to-arc transition [20, 21]. Most measurements to date put the charge number density in etching plasmas within the $\sim 10^{10}$–10^{11} cm^{-3} range.

2. *Pressure Effects on Ion to Neutral Ratio*

Generally, we believe the average degree of ionization in most practical etching plasmas ranges about an order of magnitude $\sim 10^{10}$–10^{11}, more or less independent of the operating pressure. This heuristic may be a consequence of the apparatus, power densities and gases commonly used. To the extent that this approximation holds, it has enormous practical significance. Since the gas pressures used generally span ~ 1 Torr to 0.001 Torr, a constant plasma (ion) density implies that the surface flux of neutrals relative to ions can be varied by up to three orders of magnitude, by changing pressure alone. This should allow pressure to determine whether etching will be chiefly chemical, with little or no contribution from ion-assisted reactions, or be predominantly ion assisted.

Practically speaking these concepts are illustrated by the etching of GaAs and doped Si in chlorine plasmas. The spontaneous chemical etch rate for GaAs by both Cl atoms [22] and Cl_2 [23] is high—at 70°C, 0.3 Torr of Cl atoms etch GaAs faster than 2.5 μm/min. This makes etching predominantly chemical (and crystallographic—see Section XII.B.), even when ion bombardment energy is high (at excitation frequencies below 1 MHz—see Section III.C below). But when the pressure is reduced below approximately 0.001–0.01 Torr and the etchant is diluted, (for example with argon), it is possible to do ion-assisted vertical etching without perceptible undercutting or crystallographic dependence [24, 25, 26].

We pointed out that neutral reactants are responsible for both isotropic chemical and ion-enhanced anisotropic etching. But these two processes have different kinetics. The rate of chemical etching frequently is proportional to the concentration of neutral etchant species ("first order," see next section), while ion-enhanced etching can be independent of neutral density ("zero order," see next section) when the flux of ions limits rate. Considering chemical and ion-enhanced processes as parallel process channels appears to be a good approximation [27]. Therefore, if the rate of plasma-induced dissociation by electrons is proportional to gas pressure (see Section III.F), lowering pressure leads to a proportionate decrease in the concentration of neutral etchant while the relative rate of energetic ion-enhanced etching increases. This trend is also reinforced by the increase of sheath voltage and ion bombardment energy at lower pressure.

C. FREQUENCY EFFECTS

Much like pressure, the *RF* excitation frequency alters key discharge characteristics that have an important influence on plasma chemistry and etching [18]. We consider four separate kinds of effects, although they don't necessarily occur independently: (1) frequency can change the spatial distributions of species and electrical fields across the discharge; (2) frequency determines whether the energy and concentrations of species are constant in time, or whether they oscillate during a period of the applied field; (3) frequency affects the minimum voltage that is required to start and operate a plasma and the energy with which ions bombard surfaces; and (4) frequency may change the shape of the electron energy distribution function (EEDF) and thereby select the electron-molecule reaction channels that predominate.

Frequency changes that produce major effects on discharge characteristics can be associated with the ability or inability of a physical or chemical process to change in step with the applied voltage. Stated in another way, transitions occur when the RF period is swept through the relaxation time (τ_i) of an electrical or chemical process:

$$0.5 \lesssim \omega\tau_i \lesssim 2 \tag{1}$$

If $\omega\tau_i \ll 1$, the dynamic process state will be representative of instantaneous conditions induced by the time-varying field. On the other hand,

Table 1 Characteristic frequencies for various processes. The parameters in the table are based on a chlorine discharge at 0.3 Torr. (After [18]).

Process	Approx. Frequency $[d(2\pi \log n)/dt]^{-1}$	Rate Expression (dn/dt)	Parameters Assumed
Atom-Atom Recombination (Homogeneous)	0.3 Hz	$k_{ra}[Cl]^2N$	$k_{ra} = 3.47 \times 10^{-32}\ cm^{-6}\ sec^{-1}$ $[Cl] \sim 5 \times 10^{15}\ cm^{-3}$
Atom-Atom Recombination (Heterogeneous)	100 Hz	$\frac{\pi^2 D[Cl]}{L^2}$	$D = 250\ cm^2\ sec^{-1}$ $L = 2\ cm$ $[Cl] \sim 5 \times 10^{15}\ cm^{-3}$
Ion-Ion Recombination	250 Hz	$k_{ri}n_+n_-$	$k_{ri} = 5 \times 10^{-8}\ \frac{cm^3}{sec}$ $n_+ \cong n_- = 3 \times 10^{10}\ cm^{-3}$
Electron-Ion Recombination	500 Hz	$k_{ei}n_i ne$	$k_{ei} \lesssim 10^{-7}\ \frac{cm^3}{sec}$ $n_i = 3 \times 10^{10}\ cm^{-3}$
Ambipolar Diffusion	800 Hz	$\frac{\pi^2 D_a}{L^2} n_e$	$L = 2\ cm$ $D_a \approx 2 \times 10^3 \frac{cm^2}{sec}$
Free Diffusion	50 KHz (1 eV) 2 MHz (3 eV)	$\frac{\pi^2}{L^2} \frac{\bar{v}l}{3} n_e$	$L = 2\ cm$
Ion-Sheath Transit	LITF ≅ 200 KHz UITF ≅ 6 MHz	$\delta/v_{+,ave}$	0.3 Torr
Charge Exchange	160 KHz – 1.6 MHz	$k_{ex}n_i N$	$k_{ex} = 10^{-10} - 10^{-9}$ $N = 10^{16}\ cm^{-3}$
Attachment	2.5 MHz (0.08 eV) 500 KHz (3 eV)	$k_a n_e N$	$k_a = 1.5 \times 10^{-9}\ \frac{cm^3}{sec}$ $N = 10^{16}\ cm^{-3}$ $k_a = 3 \times 10^{-10} \frac{cm^3}{sec}$ $N = 10^{16}\ cm^{-3}$
Electron Energy Modulation	10 MHz	$\nu_u\ (\nu_u \propto N)$	experimental
Momentum Collision Frequency	160 MHz	$\nu\ (\nu \propto N)$	$\nu = 10^9$
Plasma Frequency	1.55 GHz	$\omega_p = \left[\frac{n_{eo}e^2}{m_e \epsilon_o}\right]^{1/2}$	$n_e = 3 \times 10^{10}$

when $\omega\tau_i \gg 1$, the process is too slow to respond and it reaches a static state in equilibrium with the time-average conditions.

Table 1 lists selected processes along with characteristic frequencies ($1/\tau_i$) estimated for a Cl_2 discharge at 0.3 Torr [18]. Rate expressions and assumptions are also listed. Only a few experimental studies of frequency effects have connected changes in plasma chemistry to the underlying discharge transitions. Effects of the ion-sheath transit and electron energy oscillation frequencies are discussed below. Recently, SiH_4/Ar discharges were pulsed near the attachment transition frequency to increase plasma deposition rates and manipulate film properties [28]. It should be emphasized that the mere existence of a transition (Table 1) does not mean it will have a perceptible effect on plasma etching. The table is a map of where changes, if they exist at all, may be found.

1. Ion Transit Frequency (ITF)

In studies of plasma etching at ~ 0.5 Torr [11, 29], there was a large increase in voltage and ion bombardment energy, Cl^+ ion flux to the walls, and the silicon etching rate when excitation frequency was lowered from 10 MHz to below 1 MHz (Fig. 5). Neutral reactions are too slow and electron motion is too fast to account for this transition. These effects are associated with the response of ions near the plasma-sheath boundary to the electric field.

For simplicity, consider the case where the sheath thickness, λ_s, is less than the mean-free path of an ion (collisionless sheath). When the excitation frequency is above the upper *ion transition frequency* (UITF), there is not enough time for ions to cross the sheath and reach the electrode in a single cycle. In this frequency range, ions cross the sheath over many periods and experience an electric field which is averaged over this transit time (Fig. 6a). We assume the potential across the sheath at the momentary positive electrode is small compared to the potential at the negative electrode (see Chapter 1), so ions are only accelerated toward an electrode during negative half cycles, as shown in Fig. 6a (the intervals $\pi - 2\pi$). Then the fastest ions with transit times lasting over many cycles will be accelerated by the sheath field, $E(x, t)$ to an energy

$$eV_{\max}^{\mathrm{HF}} = \frac{1}{nT}\sum_{1}^{n}\int_{0}^{T/2}\int_{0}^{\lambda_s} eE(x, t)\,dx\,dt \tag{2}$$

where the integrand in Eqn. 2 is the acceleration, T is the period of the excitation waveform and n is the number of cycles required to cross the

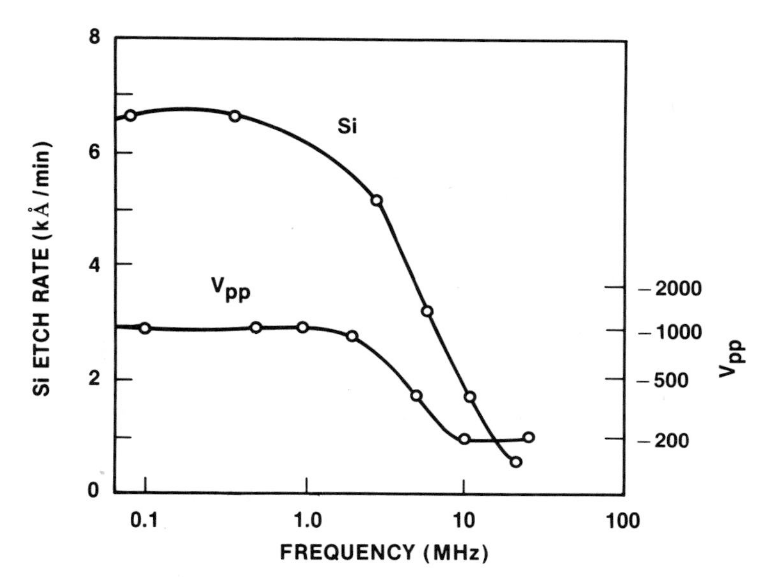

FIGURE 5. When excitation frequency is lowered from 5 MHz to below 1 MHz, ions can cross the sheath in a small fraction of a sine wave period and there is a rapid rise in peak voltage (V_p) across the sheath. As the ion energy rises, it causes a corresponding increase in the etch rate of undoped silicon in a chlorine plasma (by "damage"-induced anisotropy).

sheath. Since the entire excitation voltage, $V_0 \sin \omega t$, is applied to the sheath,

$$\int_0^{\lambda_s} E(x, t)\, dx = V_0 \sin \omega t \tag{3}$$

during the half cycles,

$$eV_{\max}^{\mathrm{HF}} = e\frac{V_0}{2\pi} \int_0^{\pi} \sin \omega t \, d\omega\, t \approx \frac{\mathrm{eV}_0}{\pi}. \tag{4}$$

At the UITF, ions entering the negative sheath when the momentary applied voltage is increasing from zero, just cross the sheath and strike the electrode at the next zero of the applied field, half period later, with an energy given by Eqn 4.

When the frequency is slightly below the UITF, some ions can cross the sheath in less than half cycle (π) and arrive at the electrode with energies greater than V_0/π. At still lower frequencies, as illustrated in Fig. 6b, there is enough time for *most* ions to cross in less than $\pi/2\omega$ (1/4 cycle). The frequency where most ions transverse the sheath in a small fraction of a

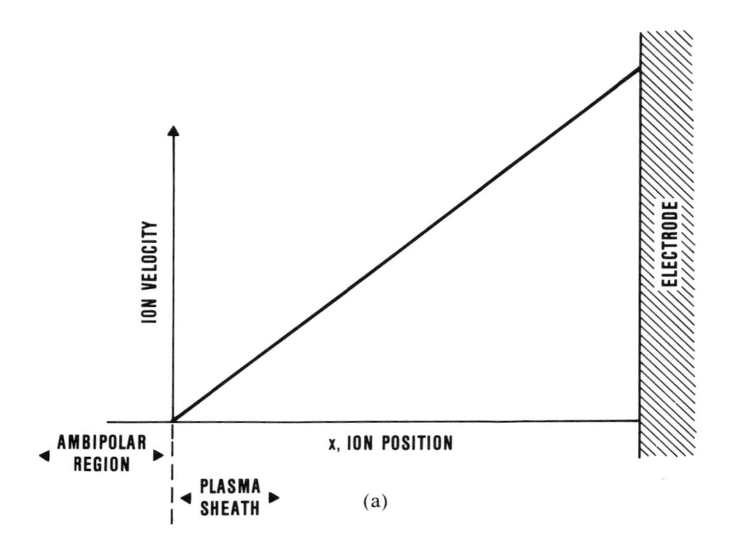

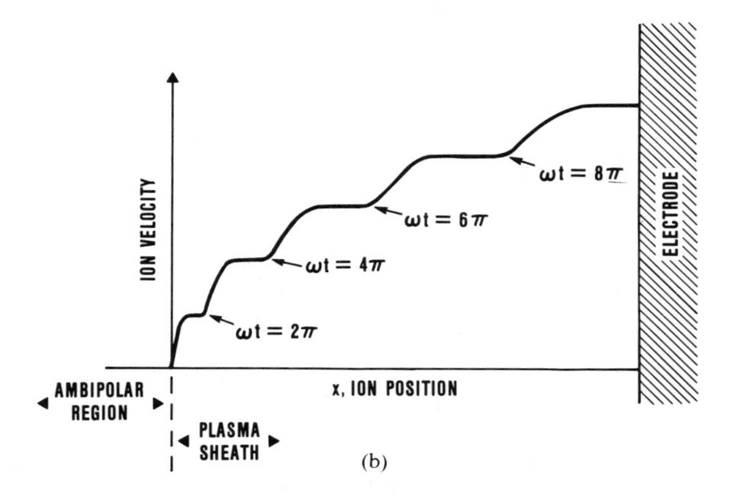

FIGURE 6. Ion velocity versus distance from the wall, in the sheath region for the most energetic ions a) well above the ion transit frequency (ITF) and b) at the UITF (see text). The electric field across the sheath is assumed to be constant.

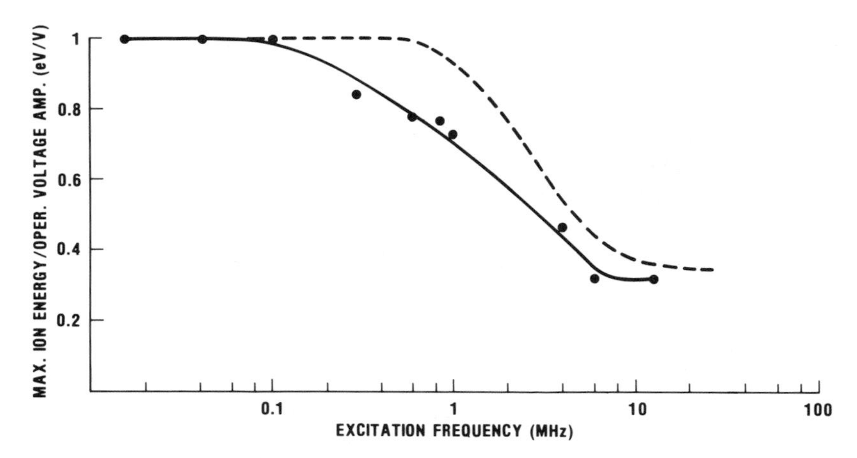

FIGURE 7. Ratio of peak ion energy to peak applied RF potential in a Cl_2 discharge at 0.3 Torr (0.6 W/cm^2). Dashed line is calculated using a collisionless model (from [11]).

period ($\approx \pi/10\omega$) has been termed the *lower ion transit frequency* (LITF). Before this point is reached, the most energetic ions move across near the instant when the electric field, $E_0 \sin \omega t$, is near its peak, E_0. These reach the wall with the maximum potential energy

$$V_{\max}^{\mathrm{LF}} \cong V_0. \tag{5}$$

Figure 7 shows the ratio of peak ion energy to peak applied voltage as frequency is swept across this transition.

These response effects account for more than a three-fold increase in the peak ion energy as frequency is lowered through the UITF at constant voltage. Experimentally, however, there is also a large increase in the discharge-sustaining voltage at this transition, and altogether the peak ion bombardment energy increases by more than a factor of ~ 10 (see Fig. 5). This voltage increase may be caused by a change in the discharge sustaining mechanism. At high frequency when ions move only a small distance during the *RF* cycle, the "glow" discharge is maintained by a balance between volume ionization, recombination and diffusive loss of charge to the walls. This does not require high potentials. Charge production at low frequency, on the other hand, is stimulated when ions bombard the momentary negative electrode and produce a secondary electron emission cascade. The secondaries must be accelerated back through a high sheath potential for there to be enough ionizing collisions to replenish charged particle losses.

2. *Electron Energy Oscillation*

While electrons in a discharge lose only a small fraction of their energy during most collisions with neutrals, gas molecules efficiently conduct translational energy to their surroundings. This makes it possible to maintain an electron temperature far above the gas translational temperature ($\epsilon_{gas} = \frac{3}{2}kT$) (see Chapters 1, 3, 4). Of course if the electric field is extinguished long enough for a great many electron-molecule collisions to take place, the electron temperature will eventually equilibrate with the gas temperature.

Under normal circumstances the sinusoidal applied field goes to zero twice each cycle, and the duration of these low field intervals is proportional to the period (1/frequency). Below some frequency, when the period is long enough, there will be enough electron-molecule collisions for the mean electron energy to respond to *instantaneous* field, changing in step with it.

At higher frequencies, oscillations of the applied *RF* field will occur faster than the electrons can lose energy. Under these circumstances the mean electron energy will assume a constant value that reflects the time-average discharge conditions. These two situations are shown in Fig. 8. The variable ν_u/ω, which is the ratio of the fraction of the average energy an electron loses per unit time to the applied frequency ω, determines whether there will be energy oscillation (ν_u is called the energy loss frequency).

ν_u/ω should have great significance for discharge chemistry because electron-molecule reaction rates are determined by electron energy. Electron-molecule reactions require a minimum collision energy (ϵ_{thresh}) to take place at all, and the reaction *cross sections* reach a maximum value at some higher characteristic energy before decreasing (the cross section for reaction resembles a bell-shaped curve as a function of electron energy). As long as the concentrations of electrons and the molecular reaction partner change slowly relative to the *RF* frequency, the effective electron-molecule reaction *rate constants*, k_i, will be determined from a time-average convolution of the instantaneous electron energy distribution function (EEDF) with the reaction cross section [30]:

$$k_i(\langle\bar{\epsilon}\rangle) = \frac{\omega}{2\pi}\int_0^{2\pi/\omega} K\int_0^{\infty} \epsilon^{1/2} Q(\epsilon) f(\epsilon,\bar{\epsilon})\, d\epsilon\, \frac{d\bar{\epsilon}}{dt}\, dt \tag{6}$$

where K is a numerical constant, $Q_i(\epsilon)$ is the reaction cross section for electrons of energy ϵ, and $f(\epsilon,\bar{\epsilon})$ is the EEDF. In general, $k_i(\langle\bar{\epsilon}\rangle)$ will depend on whether the energy remains constant over time or is scanned about an energy set by the peak applied voltage, E_o (as in Fig. 8).

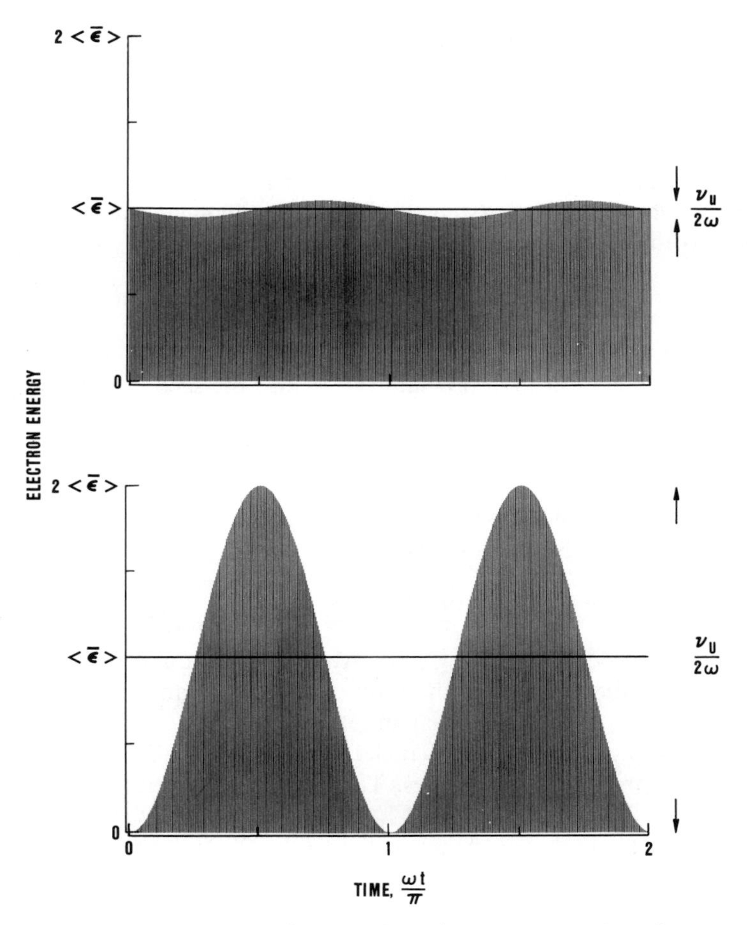

FIGURE 8. a) Electron energy as a function of ωt when $\nu_u/\omega \ll 1$ and b) when $\nu_u/\omega \gg 1$. The peak-to-peak ripple in (a) is given by $\langle\bar{\epsilon}\rangle\nu_u/2\omega$ (Eqn. 11).

Whether there is electron energy oscillation or not is determined by the balance between electron energy gain from the field and the energy loss from collisions:

$$\frac{d\bar{\epsilon}}{dt} = ev_d E_0 \sin \omega t - \nu_u \bar{\epsilon} \tag{7}$$

where v_d is the electronic drift velocity. To explicitly show that v_d depends on the electric field, E, we insert the mobility, μ,

$$v_d = \frac{dx}{dt} = \mu E_0 \sin \omega t. \tag{8}$$

Both the energy loss frequency, ν_u and $1/\mu$ are proportional to the density of the neutral gas (pressure). If ν_u and μ are independent of $\bar{\epsilon}$, the ratio of the peak electron energy (sum of the DC component plus the 2nd harmonic) to the average value (DC component) is given by

$$\frac{\epsilon_p}{\langle\bar{\epsilon}\rangle} = 1 + \frac{\nu_u/2\omega}{\left[1 + (\nu_u/2\omega)^2\right]^{1/2}}. \tag{9}$$

At low frequencies or high pressure ($\nu_u \gg 2\omega$) the second right hand term in Eqn. 9 dominates and energy follows the field:

$$\bar{\epsilon} = \epsilon_p \sin^2 \omega t \tag{10}$$

while at high frequency or low pressure ($\nu_u \ll 2\omega$) the electron energy oscillates about its mean value, $\epsilon_p/2$:

$$\epsilon = \frac{\epsilon_p}{2}\left\{1 - \frac{\nu_u}{2\omega} \sin 2\omega t\right\} \tag{11}$$

with a peak-to-peak ripple $\epsilon_p \nu_u/4\omega$. Although some approximations have been made in this derivation, the form of the results and limiting Eqns. (9) and (10) do not depend on the assumptions [30].

Data and calculations show that electron energy oscillation is common in *RF* plasma processing discharges. Time-resolved excitation and emission from Cl in a 0.3 Torr 13.2 MHz Cl_2 discharge gave a ratio $\epsilon_p/\langle\bar{\epsilon}\rangle \approx 38\%$, or $\nu_u \approx 63$ MHz [31]. This figure is divided by 2π, to obtain the transition frequency. Thus, appreciable energy oscillation is expected until the excitation frequency exceeds ~ 20 MHz ($\omega\tau = 2$). By contrast, the center of a 220 KHz Cl_2 discharge is *completely* modulated and both excitation and emission cease for an appreciable time near the voltage zero [31]. Other data and calculations show energy modulation in O_2 [32], SiH_4 [33, 34], CH_4 [34] and argon [35] discharges.

D. CHEMICAL KINETICS AND RATE EXPRESSIONS

Experimental studies show that when the stoichiometry of an *elementary* reaction between species "A" and "B" can be written as

$$A + B \rightarrow \text{products}, \tag{12}$$

the rate can usually be represented as

$$\frac{\text{product formation}}{\text{time}} = k(T)n_a n_b. \tag{13}$$

When the rate depends on the product of two concentrations in this way, the reaction is called a *bimolecular* or second order reaction. For a fixed gas feed mixture, the concentrations, n_i, are given in molar fractions by

$$n_i = x_i p, \tag{14}$$

which shows that each of the n_i is proportional to pressure. Hence the rate of a bimolecular reaction, Eqn. 12, is proportional to p^2. We should emphasize that the rate constants, $k(T)$, in Eqn. 13 and the rate equations below, depend only on temperature. They are independent of pressure.

Bimolecular reactions that produce more than one product can be quite fast—as rapid as the molecule-molecule collisions themselves. This is possible because multiple products provide enough degrees of freedom to efficiently convert some heat of reaction into kinetic energy while conserving total momentum and energy. For instance, the radical abstraction reaction

$$Cl + H_2 \rightarrow HCl + H \tag{15}$$

is almost "gas kinetic" with a rate constant of $\sim 10^{-10}$ cm^{-3} sec^{-1}. However, when two reactants combine exothermically to form a single product, the heat of reaction must somehow be taken away, or the product will decompose to the original reactants. For example, if the product of reaction 12 were "AB",

$$A + B \rightarrow AB^* \tag{16}$$

where the superscript "*" signifies an energetic activated state, "AB*" will contain enough energy to decompose via the reverse reaction

$$AB^* \rightarrow A + B \tag{17}$$

unless some of the energy from combining A + B is removed. This energy can conceivably be removed in several ways. First, the product "AB" could stablize by emitting a photon:

$$AB^* \rightarrow AB + h\nu. \tag{18}$$

However, the mean lifetime of an unstabilized AB* is only $\sim 10^{-12}$–10^{-13} s, while the average time for a (fast) radiative transition to occur is $\sim 10^{-9}$–10^{-8} s. Hence photon emission, permitting recombination, will occur in only 1 out of 1000 to 10^5 collisions, corresponding to radiative rate constants of $\sim 10^{-13}$ to 10^{-15} cm^{-3} s. This compares to $\sim 10^{-10}$ cm^{-3} s for two-body atom transfer reactions (Eqn. 15). For reaction of electrons with simple ions ($X = H^+, N^+, O^+$, etc.);

$$e + X^+ \rightarrow X + h\nu \tag{19}$$

ambient temperature radiative recombination rates are $\sim 10^{-12}\ cm^{-3}\ s^{-1}$, whereas other types of electron–ion recombination have rates up to $\sim 10^{-6}\ cm^3\ s^{-1}$ [36a] (electron–ion collision rates are larger than those between neutrals because: (1) the coulombic force gives large cross sections, and (2) the electrons have high velocities).

Collisions between AB* and a third species provide another way for removal of the excess product energy. True three-body collisions where A, B and a third species "M" come together at exactly the same time are highly improbable. However, if A reacts with B and the activated product AB* interacts with "M" almost immediately, before there is time for the reverse reaction (Eqn. 17) to occur, some of the reaction energy can be transferred to M, stabilizing the product. The probability of these three-body reactions,

$$A + B + M \rightarrow AB + M, \tag{20}$$

increases with pressure giving a rate law

$$\frac{\text{product formation}}{\text{time}} = k(T)n_A n_B n_M. \tag{21}$$

Notice that the rate for these termolecular or three-body reactions depends on the product of three concentrations which makes them proportional to the cube of total gas pressure (at fixed reactant composition).

In reality, the details of three-body kinetics are a little more complicated than Eqn. 21. The expression in Eqn. 21 is derived from the balance between formation and decomposition of AB*, as described by reactions 16, 17 and:

$$AB^* + M \rightarrow AB + M. \tag{22}$$

A steady–state analysis of these three reactions gives

$$\frac{dn_{AB}}{dt} = \frac{k_{16}k_{22}n_A n_B n_M}{k_{17} + k_{22}n_M}, \tag{23}$$

so that at pressures where $k_{22}n_M \ll k_{17}$ it is equivalent to Eqn. 21 with

$$k_{21} = \frac{k_{16}k_{22}}{k_{17}}, \tag{24}$$

while at very high pressure ($k_{22}n_M \gg k_{17}$) it behaves as a second order reaction with

$$\frac{dn_{AB}}{dt} = k_{16}n_A n_B. \tag{25}$$

Under these latter circumstances a termolecular reaction is said to be at its high pressure limit.

When reaction occurs on or near surfaces, the heat of reaction between two species, A and B, can be efficiently transferred to the surface. These *heterogeneous* reactions, where the surface acts as the third body

$$A + B + \text{Wall} \rightarrow AB + \text{Wall} \tag{26}$$

are frequently *first* order

$$\frac{\text{product formation}}{\text{time}} = k(T)n_A A_w, \tag{27}$$

where A_w is the surface or wall area. The recombination (loss) of etchant atoms in plasma discharges is governed by reactions of this type. For example,

$$F + F \xrightarrow{\text{wall}} F_2 \tag{28}$$

with a rate that is first order in F:

$$\frac{dn_F}{dt} = k_{F,\text{surf}} n_F. \tag{29}$$

This reaction is discussed in more detail in Section IV.

Gas-surface reactions, especially recombination, can be rapid, but sometimes seem to behave erratically. F atom recombination again serves as an example. As shown in Table 2, a high proportion of F atoms that strike copper, zinc or brass surfaces will recombine to form F_2, in sharp contrast to an aluminum oxide wall where only one in 10^5 impinging F atoms are

Table 2 Probabilities for F atom loss on various materials. Data from [39, 48, 49, 50].

Material	Temperature, K	Loss Probability	Process
Alumina	300	6.4×10^{-5}	Recombination
Quartz	300	1.5×10^{-4}	Total loss
Quartz	300	1.9×10^{-5}	Reaction
Pyrex	300	1.6×10^{-4}	Recombination
Steel	300–470	2.8×10^{-4}	Recombination
Molybdenum	300	4.2×10^{-4}	Recombination
Nickel	300	7.2×10^{-4}	Recombination
Aluminum (0.1% Cu)	300–560	1.8×10^{-3}	Recombination
Copper	300–570	> 0.011	Recombination
Brass	300	> 0.05	Recombination
Zinc	300	> 0.2	Recombination
Teflon	300	$< 7 \times 10^{-5}$	Recombination
BN	300–500	≈ 1	Reaction
Si	300	.0017	Reaction

Table 3 These are typical of etching, recombination and oligomerization reactions in fluorocarbon plasmas. The higher the order of a reaction (exponent of pressure) the greater is its relative importance with increasing pressure.

Overall Reaction	Effective Pressure Dependence
$F + Si_{surf} \rightarrow SiF$	p
$CF_3 + F \rightarrow CF_4$	p^2
$CF_2 + F_2 \rightarrow CF_3 + F$	p^2
$CF_2 + CF_2 \rightarrow C_2F_4$	p^2
$3\ CF_2 \rightarrow C_3F_6$	p^3

lost [37]. BN walls consume essentially all F atoms that reach them, and in doing so, form BF_3 and N_2 [38]. SiO_2 walls, on the other hand, react with only 1 in 10^5 impinging atoms (to form SiF_4 as a final product), and cause about 1 in 10^4 of the atom flux to recombine and form F_2 [39, 40]. This high sensitivity to surface composition means that reactor construction materials, reactor wall contamination and usage history can have a big effect on process chemistry.

The growth of polymer and inhibitor films during etching is partly controlled by pressure and provides a good example of the effect of reaction order. Consider the reactions shown in Table 3. This set of reactions illustrates the etching, recombination and polymerization reactions which occur in a CF_4 plasma and other fluorocarbon discharges. The higher the order of reaction (power of pressure), the more rapidly the rate and relative importance of a reaction increase with pressure. By the same token, lower order reactions become more important with decreasing pressure. Etching reactions, such as the reaction of F atoms with Si to form SiF_4, tend to have an effective *overall* rate that is first order. The reaction of F with a silicon surface in the table, can be regarded as the first order rate-determining step in a sequence of reactions between F atoms and fluorinated surface species (the other steps are very fast—see Section IV). Therefore, if the plasma etching species (F atoms in Table 3) were a *constant* fraction of the gas, and pressure was raised, the etch rate would increase directly with pressure [41] (rate $=$ const $\cdot\ p$). The reactions that form inhibitor films and polymers involve chain growth which is second or higher order in pressure. For example, the dimerization of CF_2,

$$CF_2 + CF_2(+M) \rightarrow \rightarrow C_2F_4(+M), \tag{30}$$

will have a rate law that is second or third order, $r = k_2 n_{CF_2}^2$ ($\propto P^2$) (it will be third order at pressures where M appears in the rate expression). Similarly, the effective kinetics of sequential reactions leading to growth of trimers (C_3F_6) in this example would be still higher order [42]. This is symptomatic of a general tendency to favor oligomer and polymer growth relative to etching, when the pressure increases in a fixed mixture containing F atoms and CF_2 radicals. Similarly, with decreasing pressure, the etching rate in this simplified example will rise relative to the rate of oligomer and film formation.

Increasing pressure and decreasing temperatures increase the surface concentration of physisorbed species according to their adsorption isotherms, and chemical etching rates increase in step with the surface concentration of etchant. While adsorption effects in plasma etching have not been studied, in the closely related low pressure gaseous etching (LPGE) [46, 47] they can lead to an apparent "negative activation energy." Interesting conditions were found in which decreasing temperature led to an increase in the rate of silicon etching by XeF_2, ClF_3 and other interhalogen compounds, apparently because the surface concentration of active species increased faster than the decline in rate constant with temperature [46, 47] (see Section III.1).

Finally, we note that at fixed composition and mass flow rate, the ratio of convective relative to diffusive mass transport (e.g., the Peclet number), is constant and both are independent of pressure. Therefore the ratio of mass transport rates relative to first order surface reactions will vary as $1/P$ so that lower pressure tends to overcome local reactant depletion, or mass transport limitations in chemical reaction.

E. TEMPERATURE EFFECTS

Temperature, like pressure, has a profound influence on discharge chemistry. To be clear, we really should distinguish between gas and surface temperatures. However, the gas temperature is a complex function of local power input, heat transfer and transport phenomena. Only the surface temperature is really controllable. Moreover, for the pressure and flow conditions generally encountered in low pressure plasma etching, the thermal boundary layer (e.g., the distance from the surface over which heat transfer maintains the gas close to wall temperature) is much thicker than a mean-free path, so impinging gas species are already at the surface temperature.

In our discussion of kinetics, we said that the rate constants for chemical reactions are a function of temperature. Thus temperature has a dominant effect on selectivity, etch rates and the degradation of resist masks. As we shall see, the morphology of etched surfaces is also greatly affected by temperature. Finally, physisorption and diffusion are sensitive to temperature, although these topics are beyond the scope of this chapter.

1. Effect of Temperature on Rate Constants

The rate constants for elementary chemical reactions usually vary with temperature according to the Arrhenius expression,

$$k(T) = A(T)e^{-E_A/RT}, \tag{31}$$

where A is a "pre-exponential" which is weakly dependent on temperature, and E_A is the "activation energy." The activation energy is the height of the energy barrier that reactants must overcome to approach each other and combine, (or the energy barrier for dissociation in the case of a single decomposing reactant). The exponential term is known as an "Arrhenius factor." As an example of this behavior, the etch rates for fluorine atoms etching silicon and SiO_2 are in excellent agreement with the expression

$$\text{Etch Rate} = Cn_F T^{1/2} e^{-E_A/RT}\ \text{Å/min}, \tag{32}$$

where the constants, C and E_A for Si and SiO_2 etching are shown in Table 4. Unlike the expression for the *rate constant*, Eqn. 31, the *reaction rate* Eqn. 32 depends on a concentration (n_F) as given by Eqn. 27. The weak pre-exponential $T^{1/2}$ dependence can be understood as follows: The flux of F atoms to a surface is $n_F v_F/4$ (see Chapter 1), which is proportional to $T^{-1/2}$ since n_F is inversely proportional to T according to the perfect gas

Table 4 Preexponential factors and activation energy for F atom etching of Si and SiO_2. The rate equation is ER (Å/min) $= An_F T^{1/2} e^{-E_A/RT}$.

FILM	A	E_A (kcal/mole)	RATE Å/min (298 K, $n_F = 3 \times 10^{15}$ cm^{-3})
Si	2.86×10^{-12}	2.48	2250
SiO_2	0.614×10^{-12}	3.76	55

$ER(\text{Si})/ER(\text{SiO}_2) = 4.66\ e^{1.27/RT}$
$= 44$ (at 298 K)

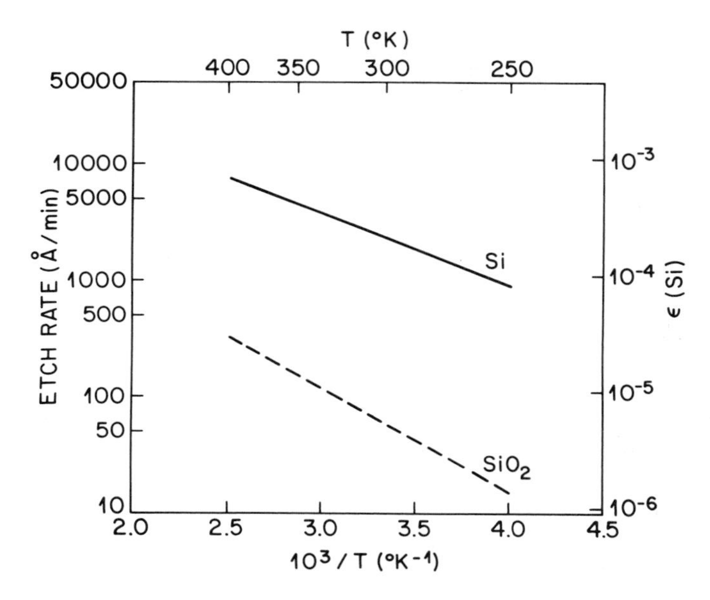

FIGURE 9. Arrhenius plot of silicon and SiO_2 etch rates, and Si reaction probability with F-atoms. The logarithms of these parameters are a linear function of $1/T$. The etch rates shown are based on a plasma F-atom concentration of 3×10^{15} cm^{-3}. Note that the reaction probability, ϵ is defined here as the probability that an impinging silicon atom leaves the surface as a silicon fluoride product (no particular product stoichiometry is assumed). Since conflicting definitions of ϵ appear in the literature, published values should be interpreted with caution.

law, and v_F is proportional to $T^{1/2}$. Hence, if the probability of an atom reacting once it is on the surface is proportional to the Arrhenius factor, the reaction rate will have the dependence shown. The logarithms of the silicon etch rate and probability of an atom reacting when it reaches a silicon surface are plotted against $1/T$ in Fig. 9. A straight line in these semilogarithmic coordinates indicates an Arrhenius dependence, and the slope of the line gives E_A. The weak $T^{1/2}$ factor has almost no effect on the slope in these coordinates.

Incidently, the final silicon-containing product of both Si and SiO_2 etching in F–containing discharges is SiF_4. The reaction rates in Table 4 belong to initial reactions between F atoms and the substrates which form intermediate species, not the final product. But they *are* also the etch rates, because the initial step is the slowest *rate determining* reaction. That is, the rate of sequential reactions is determined by the slowest step.

Since F atom etching rates of both Si and SiO_2 conform to exponential Arrhenius behavior, the selectivity, which is their ratio, is exponential in

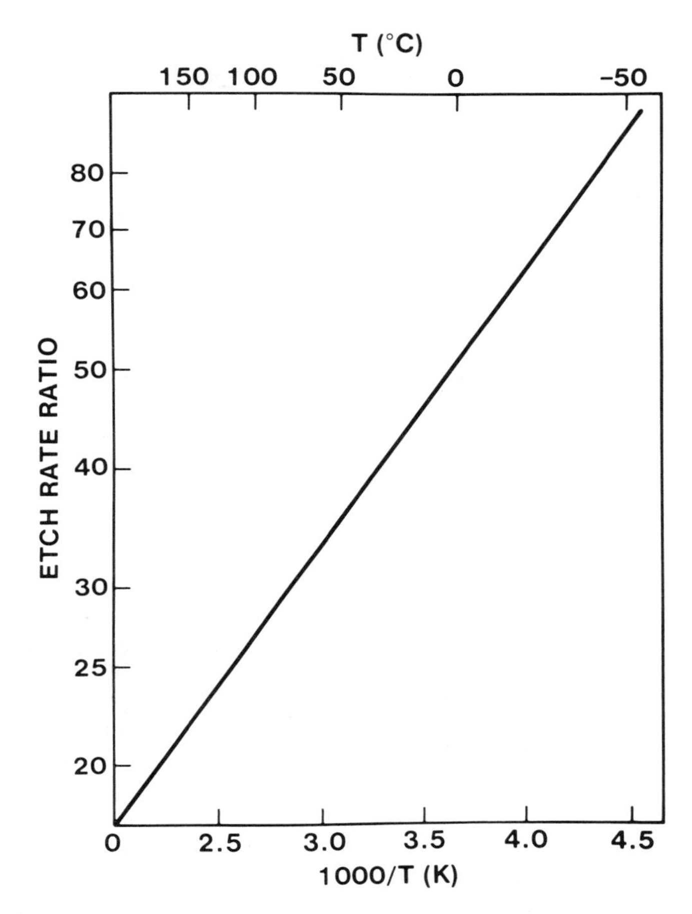

FIGURE 10. Selectivity for etching Si over SiO_2 as a function of $1/T$. The decline of selectivity with temperature is an effect of the Arrhenius dependence.

$1/T$ as well. Figure 10 shows the selectivity for etching Si relative to SiO_2 as a function of $1/T$. At room temperature, selectivity is about 44 : 1, but if the plasma heats the substrate, the selectivity will fall. Conversely, higher selectivity can be achieved by cooling the substrates to below room temperature. Notice also that these exponential factors, $\exp^{-E_A/RT}$, always approach unity when RT is much larger than E_A. Selectivities decrease with temperature and approach the ratio of the Arrhenius pre-exponential factors—in this case 4.7 : 1. Surface temperature is an essential variable.

As we will see, this example is more than an academic exercise. F atoms, made from a variety of feed gases, are widely used to etch Si, SiO_2 and

Si_3N_4. This is usually done by the purely chemical reactions discussed here, but anisotropic etching of SiO_2 with F atoms is also possible, as described in Section IX. To study the ideal etching reactions, substrates were exposed to dissociated F_2 from a discharge in a temperature-controlled cell. However, F_2 is considered too hazardous for process application so other sources of F atoms, including NF_3, mixtures of CF_4, C_2F_6 etc. with O_2, and SF_6/O_2 are used industrially. The choice between these alternatives is made on the basis of economics and side effects on the etching process. The plasma chemistry of halocarbon/oxygen mixtures is important in a rich variety of etching processes and will be discussed in Section IV.

While discussing Si etching by F atoms, we should point out that certain gaseous fluorine-bearing compounds will react even without a plasma. XeF_2 is probably the best known of these substances, and investigators have repeatedly claimed that its reaction with Si follows the same basic mechanisms as F atoms etching. This claim has, however, been repeatedly discredited. Several investigators have shown that the reactivity, the rate, the intermediates, and the temperature dependence of XeF_2 etching silicon and SiO_2 are quite different from F atoms etching. Apparently, the main reason XeF_2 has been used in so many studies is that it is commercially available (in a bottle) and stable. Measuring its reactions doesn't require the techniques that are necessary for F atom studies.

In fact, XeF_2 is but one of a family of "plasmaless" low pressure gaseous etchants (LPGE) that can be used to etch silicon selectively, making SiF_4 as a product. Other gases of this type are listed in Table 5, where atomic fluorine is included for comparison. Notice that the room temperature etch rates and apparent activation energies vary widely, consistent with the

Table 5 Reaction rates for LPGE etching of Si by various gases (at room temperature; * means not measured).

Reactant	E_a Kcal/mole	Etch Rate Å/min-Torr
XeF_2	6.1	230,000
BrF_3	*	50,000
F	2.5	9,200
IF_5	2.5	2,200
BrF_5	*	1,500
ClF_3	4.1	1,200
ClF	*	< 2
F_2	9.2	0.3

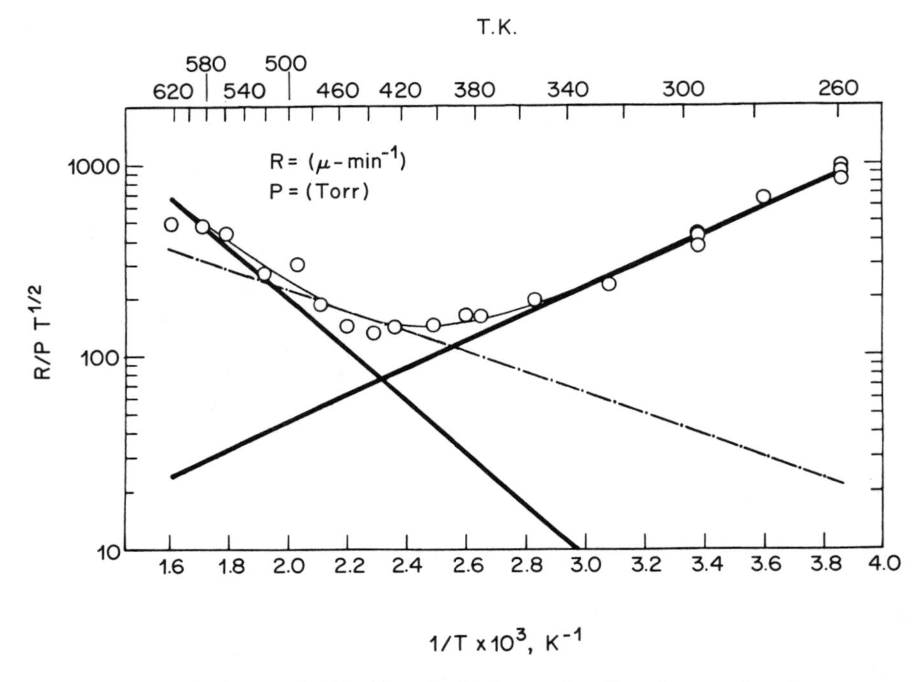

FIGURE 11. Etch rate of (100) silicon by XeF_2 as a function of temperature (upper curve and data points). The thick solid lines are asymptotes showing a normal Arrhenius behavior at high temperature (negative slope) and anomalous positive slope at low temperature which is attributed to adsorption control. For comparison, the etch rate by an equivalent pressure of F atoms is shown by the lower line (—•—•).

diverse chemical nature of these etchants. Unlike atomic fluorine, none of the plasmaless etchants attack SiO_2 at all—at least within the several hundred degree temperature range in which they have been studied. Another interesting aspect of their behavior is that many of them exhibit a "counter-Arrhenius" behavior at low temperature. Figure 11 shows this peculiarity for XeF_2, which contrasts with the temperature dependence of F atom etching. It is believed that the increasing etch rate with decreasing temperature means that adsorption of XeF_2 on the surface is rate-limiting—that is, an increase in the surface concentration of XeF_2 with falling temperature more than compensates for a decrease in the reactivity. Another way of saying this is that the concentration, n_A, in Eqn. 27 should really be a surface concentration, and that as temperature drops (below the minimum in the curve), adsorption makes the surface concentration rise faster than the rate, $k(T)$, falls.

2. *Effect of Temperature on Surface Texture*

Temperature has a systematic effect on the roughness of etched surfaces in several systems. Even when there is no plasma present (a "remote" plasma or afterglow), fluorine atom gasification reactions leave etched single-crystal silicon surfaces dull and grainy in appearance (Fig. 12a). Molecular flourine is much worse—it produces an extremely coarse texture and pitting that increases with temperature and pressure (Fig. 12b). It would be a useful, highly selective isotropic etchant for silicon over SiO_2 if it did not have this effect. By contrast, fluorine atom etching of SiO_2 preserves the initial smoothness of the oxide surface. The reason for these effects is not understood, but observations show that the surface morphology in copper-catalyzed LPGE by F_2 is much more uniform, which suggests that F_2 etching may start at defect or impurity nuclei.

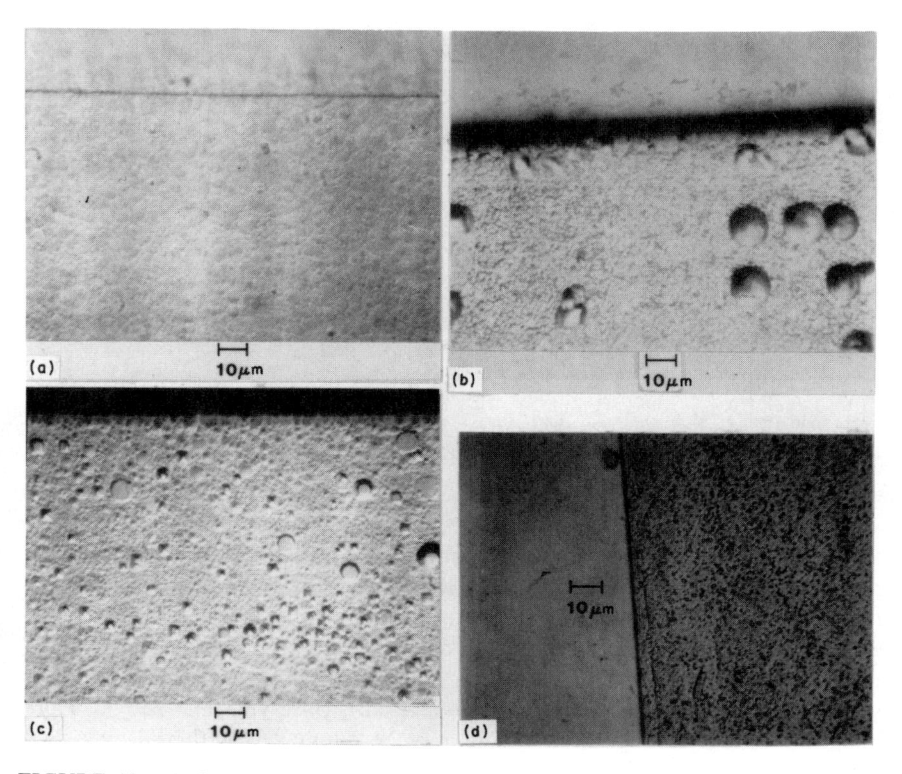

FIGURE 12. Surface texture of Si(100) etched downstream of an F_2 discharge (a) and (c) or in undissociated F_2 (b) and (d). In (a) there are 2.7×10^{15} F atoms/cm^3 at 23°C; in (b), 480 Torr of F_2 at 100°C; in (c), 2.7×10^{15} F atoms/cm^3 at 127°C; in (d) 1 torr of F_2 at 64°C.

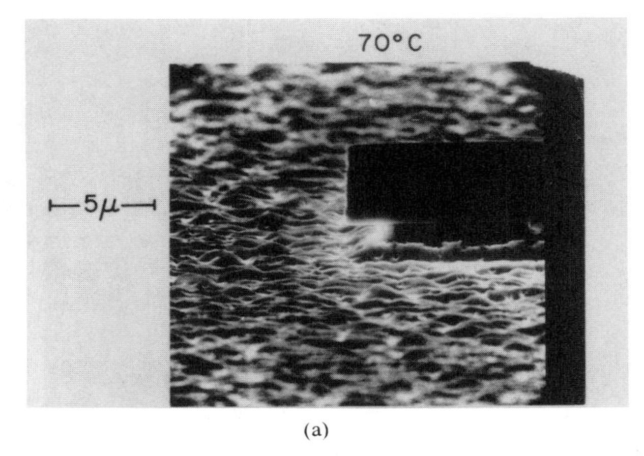

(a)

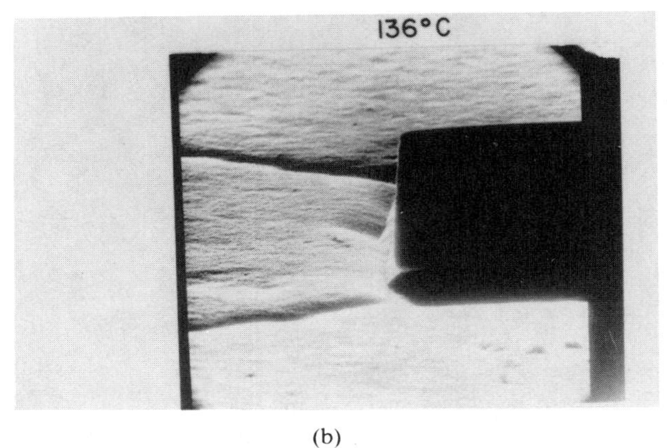

(b)

FIGURE 13. Etched surfaces of GaAs exposed to a Cl_2 discharge at (a) 70°C and (b) 130°C. The discharge operated at 40 Pa, an applied frequency of 13 MHz and 0.5 W/cm^2.

The surface morphology of InP and GaAs etched in chemically dominated Cl_2 and Br_2 plasmas is also temperature sensitive, but the effect is reversed (see Fig. 13). At high temperature the etching is smooth, but below a certain temperature range the etching becomes rough. For GaAs etching in a Cl_2 plasma this rough-smooth transition occurs at around 120°C, whereas for InP in this plasma, the transition is observed near 250°C. Here again the cause is uncertain, but it may be connected with the crystallographic nature of etching in these systems. Certain planes etch faster than others (see Section XII.B) so etching under impurity "micromasks" or slight surface roughness will tend to develop into tiny facets. At high temperatures, distinct Arrhenius factors for etching along different crystallographic planes approach unity, so there is less of an etch rate difference along different crystallographic orientations and the profiles become more isotropic. Product volatility increases rapidly with temperature in both systems, and conceivably the decreasing amounts of adsorbed product also play a role.

F. ELECTRON TEMPERATURE AND ELECTRON MOLECULE REACTIONS

Electron-molecule reactions are responsible for ionization and dissociation in the plasma. Unlike neutrals and ions which are close to translational equilibrium, electron energies are well above thermal equilibrium values, as pointed out in Chapter 1. Because of this, the reaction rate constants, k, for electron-molecule reactions, depend strongly on the average electron energy, $\bar{\epsilon}$, and on the electron energy distribution function (EEDF). For the process

$$e + A \rightarrow P(\text{products}) \tag{33}$$

P forms at a rate

$$\frac{dP}{dt} = k(\bar{\epsilon}, T_g) n_e A, \tag{34}$$

where n_e is the electron density in the plasma and we add the explicit subscript "g" to distinguish the gas temperature, T or T_g from the electron temperature T_e. $\bar{\epsilon}$ is given by

$$\bar{\epsilon} = \frac{3}{2} \mathrm{k} T_e. \tag{35}$$

Often, especially when the discharge is close to ambient temperature and electron energy is high, the dependence on T_g is weak enough to be ignored (but there are exceptions—see the discussion on dissociative attachment below). Clearly, discharge variables which change $\bar{\epsilon}$, and n_e will directly influence the rates of elementary electron-molecule reactions.

Emelius, Lunt and Meek [51, 52] showed how to calculate the rate constant, $k(T_e)$, under these conditions. Each type of reactant molecule, A, can be considered to have an effective geometric cross section for the reaction, $\sigma(\epsilon)$, which depends upon the energy, ϵ, of an incident electron. This means that for a beam of electrons, all having a single energy and velocity, $v_e(\epsilon)$, the reaction rate would be

$$\frac{dP}{dt} = n_e v_e(\epsilon) \sigma(\epsilon) n_o \tag{36}$$

where n_o is the number density of reactant molecules. However, the electrons in a discharge have a distribution of energies, the so-called electron energy distribution function or EEDF, $f(\epsilon, \bar{\epsilon})$, and Eqn. 36 must be modified to take this into account:

$$k(\bar{\epsilon}) = \int_0^{\infty} f(\epsilon, \bar{\epsilon})\sigma(\epsilon)\, d\epsilon \tag{37}$$

so that the rate of reaction is

$$\frac{dP}{dt} = k(\bar{\epsilon}) n_e n_o . \tag{38}$$

Generally, the mean energy, $\bar{\epsilon}$, or electron temperature depends mainly on E/N (which is equivalent to E/p_o, the field to reduced pressure ratio). As shown in Chapters 1 and 3, this quantity is roughly proportional to the force on an electron multiplied by the time over which it is accelerated between collisions. This means that, to a good approximation, the electron molecule rate constant can be plotted against either E/p_o or $\bar{\epsilon}$. Figure 14 shows this relationship as computed for the dissociation of oxygen,

$$e + O_2 \rightarrow O + O + e \tag{39}$$

in an oxygen discharge [53], which is used to strip resist. Both the range of

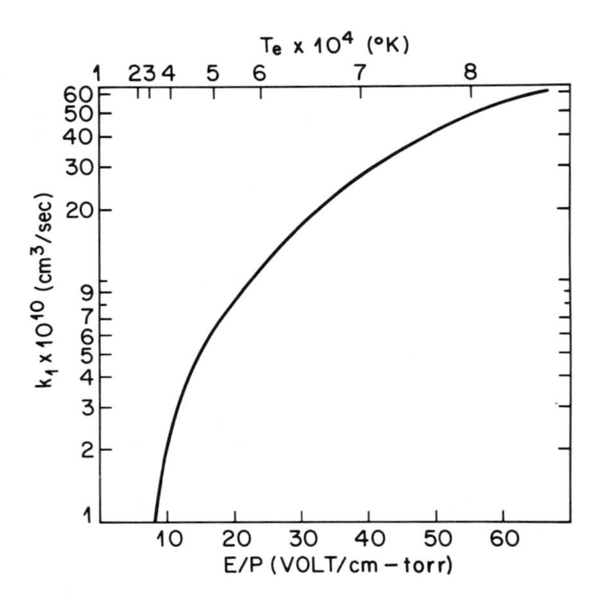

FIGURE 14. Rate constant for the electron-molecule collision-induced dissociation of O_2, (eqn. 39) versus electron temperature and E/p (rate after 53a with T_e adopted from ref. 54b).

E/p_o and k shown are typical of the orders of magnitudes expected in a low pressure glow discharge. Note that the reaction cross section, $\sigma(\epsilon)$, can peak (as does the cross section for O_2 dissociation in the 0–10 eV range shown), over an energy range where the integrated rate constant increases monotonically. The monotonic increase is caused by contributions from the tails of the EEDF (Eqn. 37) for each ϵ in the convolution. These tails broaden the dependence of $k(\bar{\epsilon})$ on $\bar{\epsilon}$ compared to that of $\sigma(\epsilon)$ on ϵ, and shift its maximum to higher energies. These rate constants eventually decline at mean energies well above the peak cross section energy.

Figure 15 shows how the distinct dependence of each electron-molecule process on E/N influences the fraction of input power that goes into

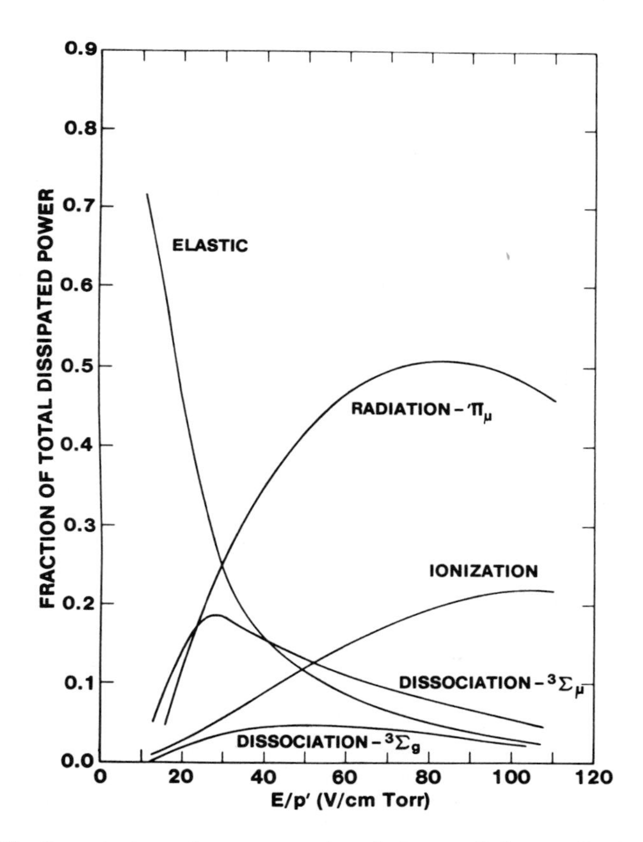

FIGURE 15. Important reaction processes in a hydrogen discharge. The input power is approportioned into different channels as E/p (E/N) is changed, owing to distinct electron energy dependencies in the underlying electron-molecule rate constants. The energy efficiency of H atom production is greatest at $E/p \approx 20$–30 (after [51, 52, 153].

different elementary processes in a low pressure *DC* discharge in hydrogen. When E/N is low, the average electron energy is too low to excite inelastic processes, so most input power is lost through elastic collisions (of course there are always *some* ionizing electrons in the tail of the distribution that sustain the discharge). As E/N and the mean electron energy is increased, a substantial amount of energy is channeled into dissociation of O_2 to create O atoms. At still higher energies (and E/N), most of the energy from electron molecule collisions goes into excited states that lose energy radiatively. Then, at the highest E/N, a growing fraction of the input power is channelled into ionization.

Another example of electron-molecule dissociation rate constants is shown in Fig. 16 [54], where the rates for dissociative attachment to various

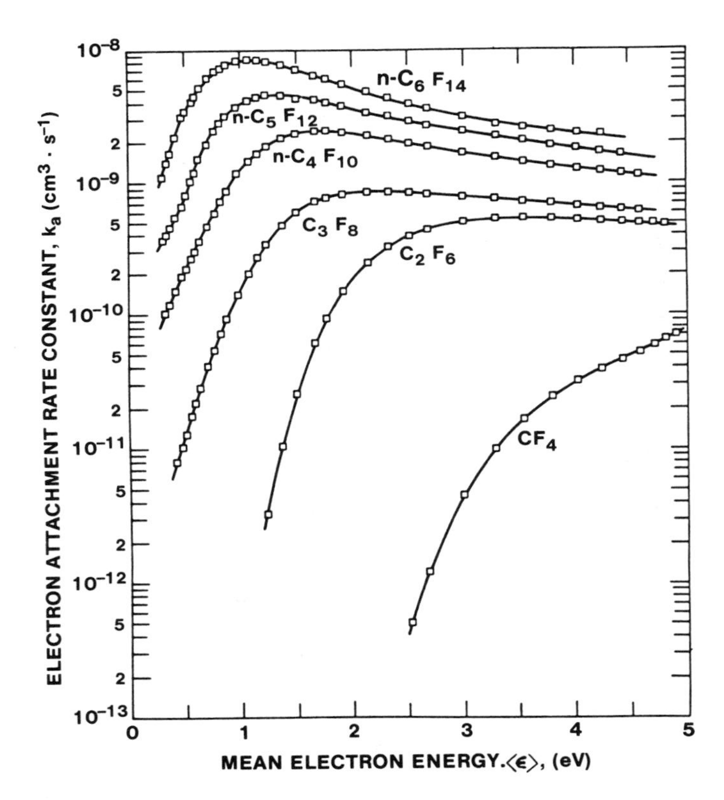

FIGURE 16. Total electron attachment rate constants as a function of mean electron energy for perfluoroalkanes (as measured in a buffer gas, from [54]).

fluorocarbons diluted in argon

$$e + C_nF_{2n+2} \rightarrow C_nF_{2n+1} + F^- \quad (40)$$

are plotted against $\bar{\epsilon}$. These dissociative attachment processes are an important source of etching species in fluorocarbon plasmas. The threshold energies for dissociative attachment decrease with chain length in the series, and maximum $\sigma(\epsilon)$ for the longer chain members ($C_3 - C_6$) are at low energy so that their $k(\bar{\epsilon})$'s reach broad maxima and decline at the higher energies in this range. Dissociative attachment cross sections and rates are often sensitive to gas temperature. For example, as gas temperature is raised from 300°C to 700°C, the energy integrated dissociative attachment cross section for C_2F_6 (according to Eqn. 40) increases by 25%, while dissociative attachment to CF_3Cl increases more than four-fold over this temperature interval [55].

In CF_4 discharges, the parent gas can be dissociated into atoms by these electron impact processes:

dissociative ionization

$$e + CF_4 \rightarrow CF_3^+ + F + 2e, \quad (41)$$

dissociation

$$e + CF_4 \rightarrow CF_3 + F + e \quad (42)$$

$$e + CF_4 \rightarrow CF_3^* + F + e \rightarrow CF_2 + 2\,F + e, \quad (43)$$

and *dissociative attachment*

$$e + CF_4 \rightarrow CF_3 + F^- \quad (44)$$

$$e + CF_4 \rightarrow CF_3^* + F^- \rightarrow CF_2 + F + F^-, \quad (45)$$

followed by *detachment*

$$e + F^- \rightarrow F + 2e. \quad (46)$$

Ionization in CF_4 is a high energy process with a threshold of 15.5 eV [56]. Electrons produced by this reaction make up for recombination losses and sustain the (relatively low) concentrations of ions and electrons in the discharge (10^{10}–10^{11} cm^{-3}). But for the range of E/N found in CF_4 etching plasmas, the mean electron energy is only around 6–8 eV [57], so ionization is driven by the small number of electrons in the tail of the distribution. Dissociation (Eqn. 42) also requires high electron energies—the threshold is 12.5 eV. Although dissociative attachment (Eqn. 44) has a smaller cross section, it reaches a maximum at low energy [54] (7.3 eV), and can be excited by a much larger segment of the electron population. Furthermore, room temperature cross section data may underestimate dissociative attach-

Table 6 Mechanism for a CF_4 discharge [58].

Reaction Number	Reaction	Rate Coefficient at 0.5 Torr
1.	$CF_4 \xrightarrow{e} CF_3 + F$	6
2.	$CF_4 \xrightarrow{e} CF_2 + 2\,F$	14
3.	$CF_3 \xrightarrow{e} CF_2 + F$	20
4.	$CF_2 \xrightarrow{e} CF + F$	20
5.	$CF_3 + CF_3 \xrightarrow{M} C_2F_6$	8×10^{-12}
6.	$CF_3 + F \xrightarrow{M} CF_4$	1.3×10^{-11}
7.	$CF_2 + CF_2 \xrightarrow{M} C_2F_4$	5×10^{-14}
8.	$CF_2 + F \rightarrow CF_3$	4.2×10^{-13}
9.	$CF + F \xrightarrow{M} CF_2$	5×10^{-15}
10.	$C_2F_6 \xrightarrow{e} CF_3 + CF_3$	20
11.	$C_2F_4 \xrightarrow{e} CF_2 + CF_2$	20
12.	$F + C_2F_4 \rightarrow CF_3 + CF_2$	4×10^{-11}
13.	$CF_2 + CF_3 \xrightarrow{M} C_2F_5$	8.8×10^{-13}
14.	$C_2F_5 + F \rightarrow CF_3 + CF_3$	1×10^{-11}
15.	$CF + CF_2 \rightarrow C_2F_3$	1×10^{-12}
16.	$C_2F_3 + F \rightarrow C_2F_4$	1×10^{-12}
17.	$F + Si_{surf} \rightarrow SiF$	10
18.	$SiF + F \rightarrow SiF_2$	1×10^{-10}
19.	$SiF_2 + F \rightarrow SiF_3$	1×10^{-10}
20.	$SiF_3 + F \rightarrow SiF_4$	1×10^{-10}

ment in plasmas since the rate increases with temperature and most discharges are "warm" (~ 100–200°C or so).

G. MODELLING

The plasma chemistry of most discharges cannot be modelled because the principal reactions and rate constants are not known. However, CF_4 and CF_4/O_2 plasmas are exceptions. Recent models of the chemistry [58, 59] reproduce the effects of pressure, flow and comparison, in good agreement with experiment. Table 6 gives a reaction subset that includes all of the important reactions in a CF_4 plasma. Several of these deserve further comment. The first dissociation reaction in Table 6 represents a combination of reactions 42, 44 and 46 above, while the second reaction in the table represents the combined effects of Reactions 43, 45 and 46. These electron molecule reactions are probably the most uncertain part of the model because the space- and time-dependent electron energy distribution in plasmas still cannot be calculated from first principles, and data on the dissociation cross sections for CF_4 are sketchy. The etching reactions, presented as 17–20 in the table, are also a representation of more detailed steps. The rate constant for Reaction 17 governs the overall etching rate, with fast reactions 18–20 fulfilling the stoichiometric requirement that SiF_4 is the final product. This constant was selected to represent Eqn. 32 (and Table 4) under the conditions modelled. The detailed etching steps represented by the reactions in this table, and Eqn. 32 are discussed in the next section.

IV. Etching Silicon in Fluorine Atom Based Plasmas

The isotropic etching of silicon (Si) and polycrystalline silicon (poly-Si) by fluorine atoms produced in plasmas is one of the oldest, best understood and most widely used plasma processes. A variety of studies have led to a detailed understanding of how the etching takes place.

Data show that when clean silicon is exposed to atomic fluorine, it quickly acquires a fluorinated "skin" that extends about one to five monolayers into the bulk (depending on the gas and exposure). Hence when etching, fluorine atoms encounter a fluorinated silicon surface which may be as different from silicon as "Teflon" is from carbon. Although some details are currently under investigation, evidence supports the overall model shown in Fig. 17. F atoms penetrate into the fluorinated layer and attack subsurface Si—Si bonds. Many studies have confirmed that the attack liberates two gaseous desorption products—the free radical SiF_2 and the stable end product SiF_4. Although one might expect these two reaction channels to have a different temperature dependence, in fact they have precisely the same activation energy, possibly because there is a common activated state that undergoes dissociation (to form SiF_2, channel *a* in the figure), or stabilization (channel *b* in Fig. 17). The stabilized species formed through channel *b* are further fluorinated (steps II-IV) to form SiF_4. The total etching rate represents the sum of both channels. SiF_2 can be separately followed since it reacts in the gas phase with F or F_2 to form an excited state of SiF_3, which emits a broad visible chemiluminescence, peaking around 500 nm:

$$F + F{-}Si_{\overline{surf}} \rightarrow SiF_2 \tag{47a}$$

$$SiF_2 + \begin{cases} F \\ F_2 \end{cases} \rightarrow SiF_3^{*}\,(+F) \tag{47b}$$

$$SiF_3^{*} \rightarrow SiF_3 + h\nu_{continuum}. \tag{47c}$$

The degree to which the rate-limiting step branches into channel *a* or *b* is

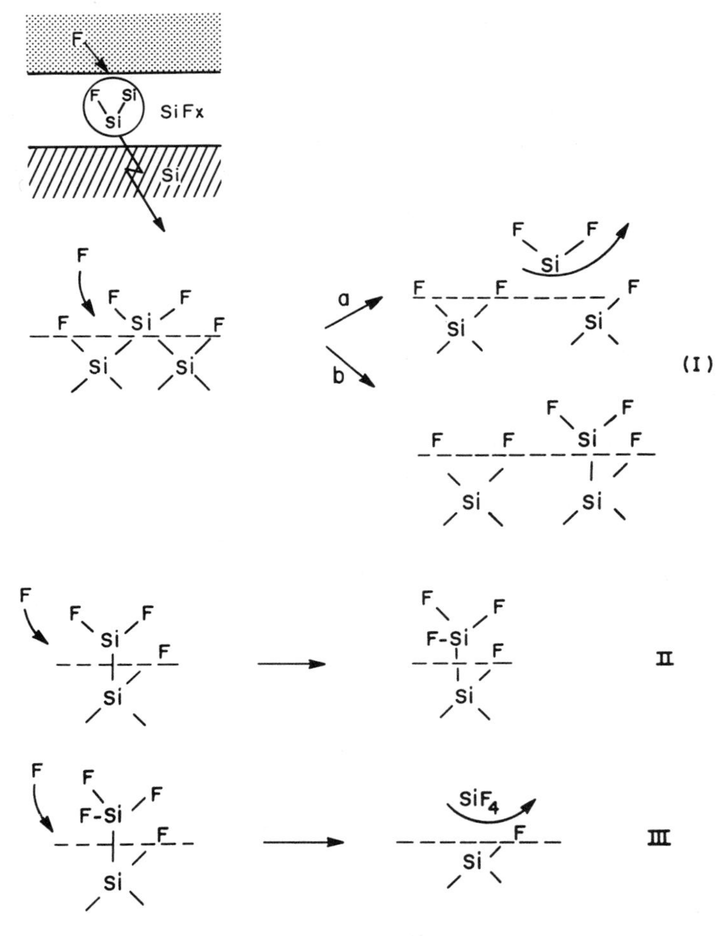

FIGURE 17. When silicon is exposed to F atoms or F_2, it acquires a fluorinated skin which extends about 1–5 monolayers below the surface (depending on the gas and exposure). Of the F atoms reacting with the surface, a small fraction attack underlying bonds to liberate SiF_2 (Ia), while the major reaction pathway proceeds via fluorinated moieties (Ib) which are saturated through a series of additions (II, III) before desorbing as SiF_2. Experiment shows that the fractions of product desorbing as SiF_2 (Ia) and SiF_4 (III) are temperature independent, which suggests that (Ia) and (Ib) are two channels of a single reaction.

statistical, depending more on the heat of reaction than the substrate temperature. Studies have shown that most of the material leaves the surface as SiF_4, with SiF_2 probably amounting to between 5 and 30% of the etch product.

Although the fundamental concept of etching with a branching precursor, slowed by a layer that must be penetrated for gasification to proceed, is well founded, there is controversy about chemical details of the fluorinated layer. This layer appears to contain $—SiF—$, $—SiF_2—$, $—SiF_3—$ and even some SiF_4. It is not totally clear that a bound $—SiF_2—$ group, shown in Fig. 17, rather than a $—SiF—$ group, is the precursor to the branching reaction. Furthermore, it has been suggested many times that silicon etching by F_2, XeF_2, and by inference other LPGE or "plasmaless etchants," goes by the same basic process as F atom etching. However, this supposition disagrees with experimental evidence. Independent studies show the kinetics and product distribution of XeF_2 etching are remarkably different, and F_2 etching proceeds with yet another set of kinetics and a much less fluorinated surface (approximately a monolayer).

Many different plasma feed gas mixtures produce F atoms as the dominant etching species. These include F_2 [40], CF_4 [60, 61], CF_4/O_2 [60, 61], SiF_4/O_2 [62], SF_6 [63], SF_6/O_2 [63], NF_3 [64, 65] and ClF_3 [64, 65]. These are all highly selective etchants for Si over SiO_2 and Si_3N_4. The CF_4 and SF_6 feeds are preferred over pure F_2 because of their low toxicity; however, they form unsaturated species (oligomers derived from CF_2) and fluorosulfur radicals, S_xF_y, in the discharge which can react with free F atoms and sometimes form polymeric residues. Oxygen is frequently added to these plasmas with at least two different effects. First, as suggested by the etchant-unsaturate model in Section VIII, O atoms react with unsaturates making F atoms, while depleting these polymer-forming species. This enhances the silicon etch rate and improves selectivity compared to pure CF_4, because unsaturated species selectively etch SiO_2 when there is ion bombardment. Second, when enough O_2 is added to the feed, O chemisorbs on the silicon surface making it more "oxide-like" which slows the etching.

These effects can be seen in the etch rate curves plotted in Figs. 18 and 19. In Fig. 18, the etch rate of both Si and SiO_2 increase dramatically as oxygen is added to the feed mixture. This is the result of increasing F atom concentrations. Superficially, oxygen can be considered to "burn" fluorocarbon radicals

$$\left.\begin{matrix} O \\ O_2 \end{matrix}\right\} + CF_x \rightarrow \left\{\begin{matrix} COF_2 \\ CO \\ CO_2 \end{matrix}\right. + F, F_2, \qquad (48)$$

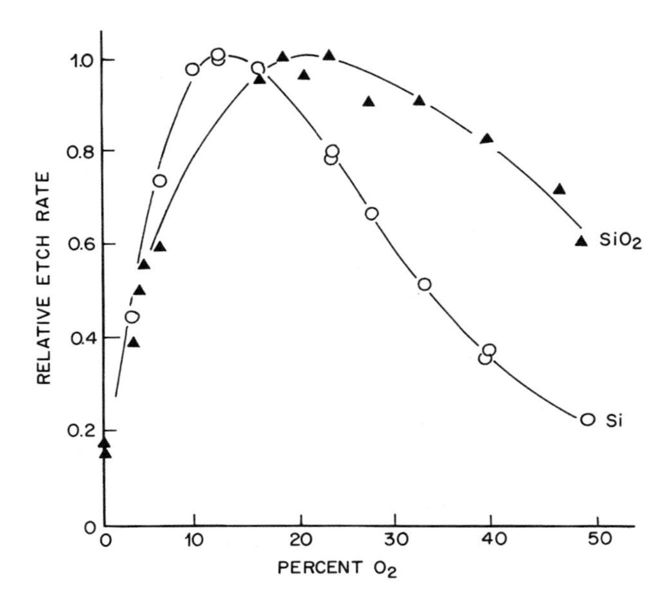

FIGURE 18. *Normalized* etch rates for Si and SiO_2 wafers in a parallel plate reactor using CF_4/O_2 mixtures at 0.35 Torr, 0.16 W/cm^2, 200 sccm feed and a substrate temperature of 100°C. These rates are shown in the same plot for comparison, but the Si maximum etch rate was ~ 4600 Å and the SiO_2 maximum was ~ 325 Å. The SiO_2 etch rate peaks at the same concentration as 704 nm emission from F-atoms and F atom concentration (after [60]).

and this scheme is in agreement with mass spectral analyses of the effluent from CF_4/O_2 discharges, as shown in Fig. 20a. In Fig. 20b, a piece of silicon has been placed in the discharge zone which lowers the total amount of F observed (F and F_2 in the figure) and reacts to form SiF_4, the final product (CO, CO_2 and COF_2 curves are virtually the same as in Fig. 20a, but have been suppressed for clarity). With large oxygen additions (in Figs. 18 and 20) the etch rates and product concentrations fall because of dilution by the oxygen.

The peak etching rates of Si and SiO_2 in Fig. 18 are different because of the effects of O on silicon, depicted in Fig. 21. When the silicon etch rate is plotted against F atom concentration (Fig. 19) it forms a "hysteresis curve" in which there are two etching rates for most F atom levels. The higher etching rate at small O_2 additions reflects the increase in etching with F, brought about by the virtually complete consumption of oxygen by Reactions 48. However, the etching rate is lower at the same concentration of F atoms with larger oxygen additions. This is caused by the competition between F and O for surface sites, as depicted in Fig. 21.

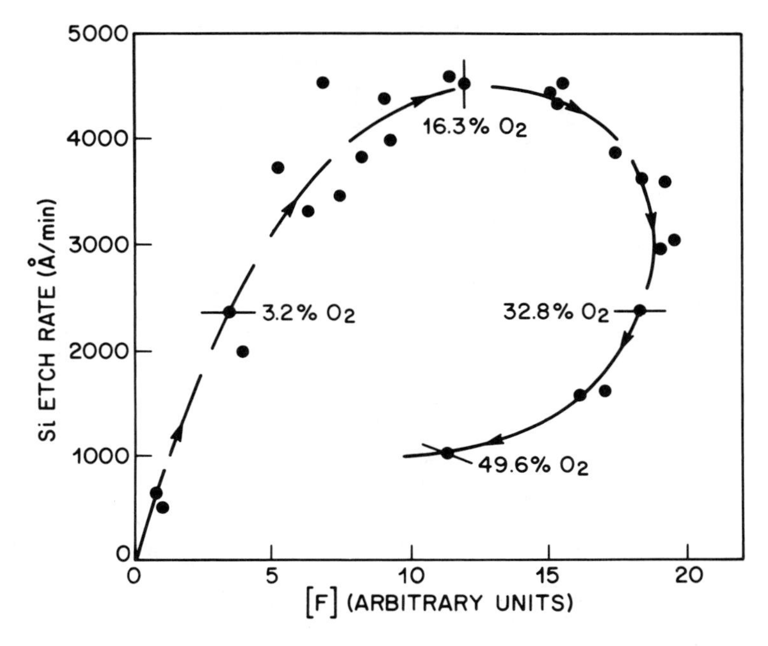

FIGURE 19. Silicon etch rate in a CF_4/O_2 plasma is plotted as a function of F-atom concentration, with oxygen content of the feed as a parameter. As oxygen raises the F-atom concentration by reacting with fluorocarbon radicals, the etch rate increases. However, adsorbed oxygen atoms depress the etch rate and at large O_2 additions these counteracting effects cause "hysteresis." Similar results are obtaind in SF_6/O_2 (from ref. [60]).

It is of interest to note that the detailed chemistry of CF_4/O_2 discharges has been modelled and the results are in close agreement with experimental data. The core of the mechanism is schematically shown in Fig. 22, where "e" represents electron impact dissociation reactions (see Section III.F) and F represents combination with F atoms. Model predictions are compared with data in Fig. 23. Here, as in modelling of the CF_4 discharge, the rate of the electron impact reactions cannot be calculated directly because of uncertainty about the electron energy distribution function in the discharge.

Anisotropic etching of silicon in fluorine-containing plasmas is practically impossible under most plasma etching conditions because of the rapid spontaneous chemical reaction. Low pressure plasmas, with high substrate "bias" in CF_4 and SF_6 are an apparent exception, but in this regime the gas phase concentration of F atoms is lower relative to the adsorbed halocarbon species and the ion bombardment flux, and the selectivity for Si over SiO_2 is well below 10 : 1 for these anisotropic conditions (compare with 40 : 1 for

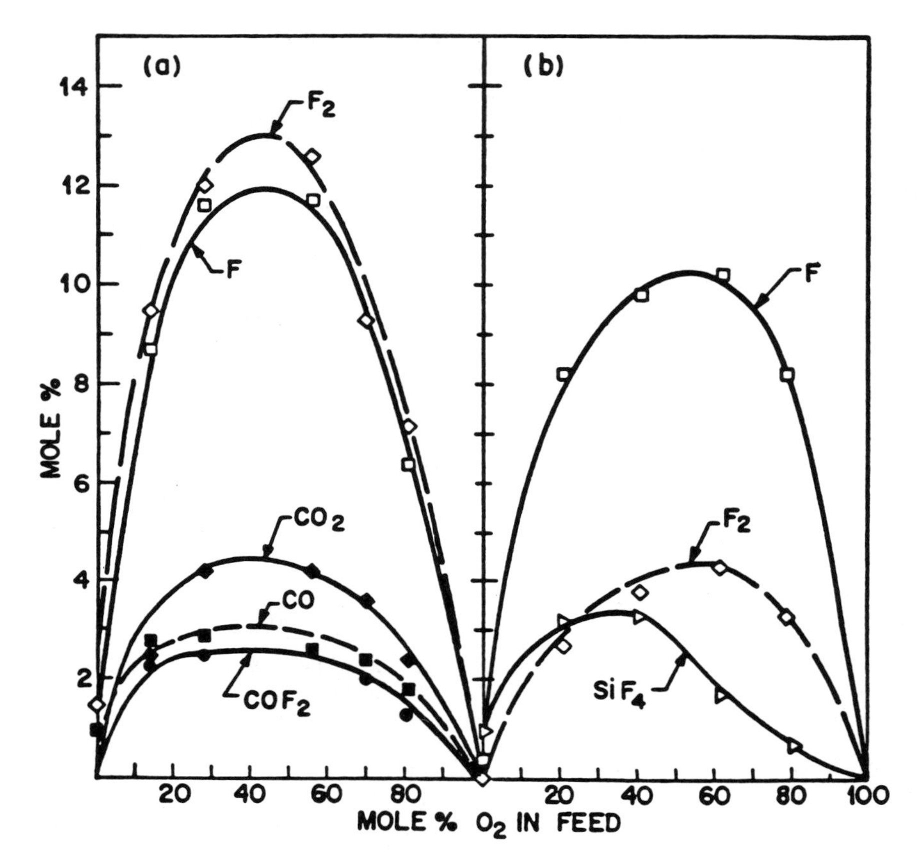

FIGURE 20. Mole percent of products found in effluent of 0.5 Torr, short residence time (~ 7 msec) CF_4/O_2 discharges with (b-right) and without (a-left) silicon present. Curves for CO, CO_2 and COF_2 have been omitted in (b) but were similar to those in (a). The F_2 curves include a major contribution from F atoms that recombined during the sampling procedure (hence $[F]_{actual} \approx [F] + 2[F_2]$) after [61].

the isotropic process at room temperature, Table 4). Hence, it is possible that adsorbed halocarbon or sulfuryl species are the etchant in these low pressure systems where the mechanisms and etchant species have not been identified. Adsorbed halocarbons and halocarbon radicals are an effective etchant for SiO_2 (see Sections VIII–IX).

While F atom producing plasmas are an isotropic etchant for silicon, they can be a good anisotropic etchant for other substrates. Table 7 lists some of the other substrates that have been etched with F atom producing plasmas.

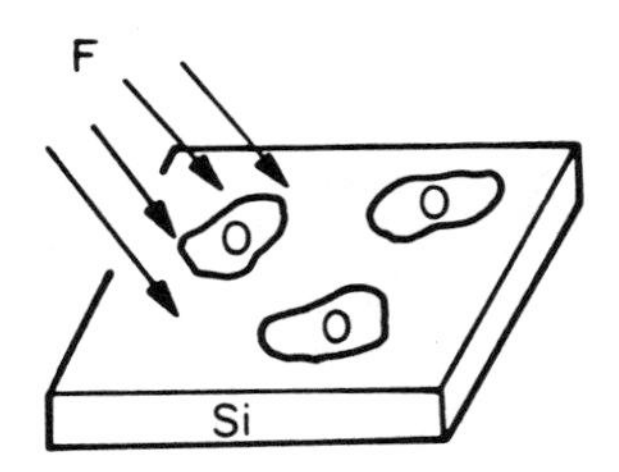

FIGURE 21. Schematic depicting the mechanism of F-atom attack on silicon when oxygen is present. F-atoms chemisorb on the exposed Si and etch. When oxygen chemisorbs on the surface, F-atoms must attack an oxidized layer and desorb the oxygen.

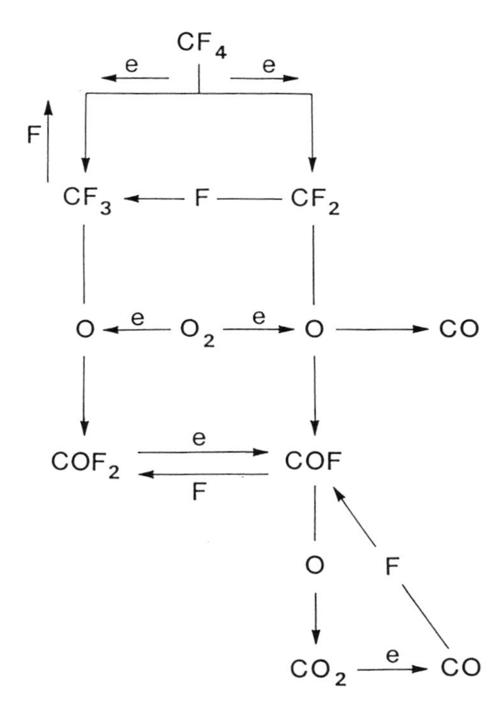

FIGURE 22. A reduced reaction set contains the essential reactions for modelling CF_4/O_2 discharges (from [58]).

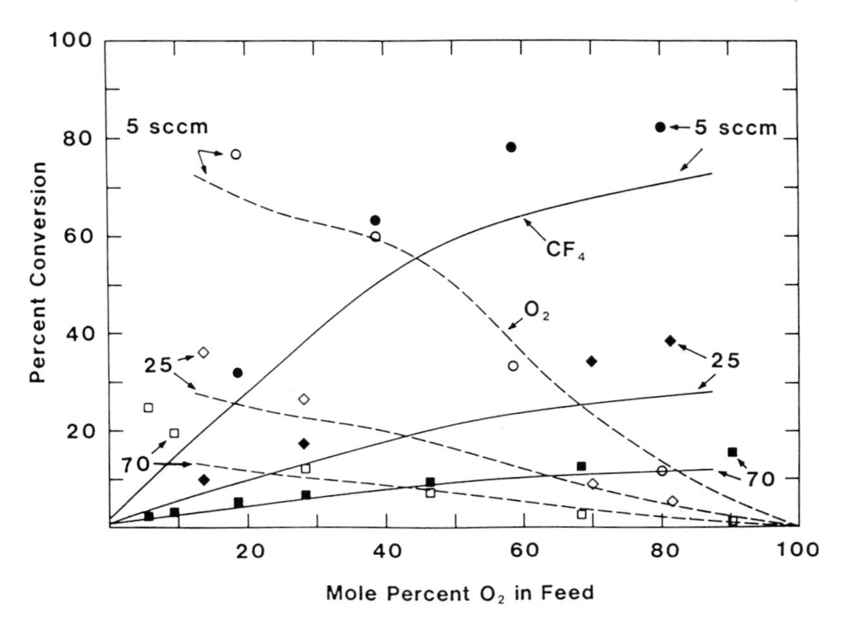

FIGURE 23. Comparison between conversion of feed to products obtained by computer simulations of CF_4/O_2 plasmas (solid lines for CF_4, dotted for O_2) and experimental data at 0.5 Torr (open points for O_2, solid points for CF_4). Flow rates are shown by each data set.

III-V semiconductor materials don't form volatile fluorides so NF_3 and other F atom plasmas are effective for the anisotropic patterning of SiO_2 on these materials; this is discussed more in Section IX. Similarly, fluorine-containing plasmas have been used to anisotropically etch silicides, some metals and nitrides.

V. The Loading Effect

Etching rates often decrease as more etchable substrate material is placed in a reactor. This *loading effect* is the result of gas phase etchant being depleted by reaction with the substrate material. The etch rate is usually proportional to etchant concentration, and if a significant portion of these reactive species are consumed in etching reactions, their concentration decreases with the area of etchable surface in the plasma. Most isotropic etchants exhibit this effect.

Table 7 Feed gases and mechanisms for plasma etching various materials with fluorine atoms.

Source Gas	Additive	Materials Etched	Mechanism	Selective Over
CF_4	O_2			
C_2F_6	O_2			SiO_2
SF_6	O_2	Si	Chemical	Resist
NF_3	None			III-V's
ClF_3	None			
F_2	None			
		SiO_2/Si_3N_4	Chemical	III-V's
				Resist
		$TiSi_2$, $TaSi_2$,	Ion-energetic	SiO_2
		$MoSi_2$, WSi_2		Si_xN_y
				Resist
		Ti, Ta, Mo	Ion-energetic	SiO_2
		W, Nb		Resist
		Ta_2N	Chemical	

If a species F is the etchant, and there are m wafers in the reactor, a mass balance on F can be written as

$$r_F V = A k_{sF}^* n_F + k_{vF} n_F V + m A_w k_{wF} n_F, \tag{49}$$

where r_F is the rate at which F is generated per unit volume, A is the area of the empty (unloaded) reactor, A_w is the area of a single wafer or substrate, n_F is the concentration of F in the gas, k_{sF}^* is the net rate constant for loss of F on reactor walls, k_{vF} is the loss rate of F per unit volume of the gas, and k_{wF} is the loss of F from etching substrate material. The etch rate R_m is

$$R_m = \alpha k_{wF} n_F \tag{50}$$

where α is the constant of proportionality between units of linear etch rate (Å/min) and material consumption (molecules/cm^3). Equation 49 is solved for n_F and substituted into Eqn. 50 to derive the loading effect equation [64, 66]:

$$\frac{R_o}{R_m} = 1 + m\Phi_F = 1 + m\frac{A_w k_{wF}}{A k_{sF}}, \tag{51}$$

where R_o is the etch rate in an empty reactor ($m = 0$), R_m is the etch rate with m wafers present, and Φ_F is the ratio of F consumed by etching a wafer (rate k_{wF}) to that lost by recombination ($Ak_{sF} = Ak^*_{sF} + Vk_{vF}$). Φ_F is the slope of the loading effect curves (e.g. $1/R_m$ vs m) shown in Fig. 24. Use of a large reactor volume (V), with high surface area (A) will minimize the sensitivity of etch rate to the area of etchable material ($\Phi_F \lesssim 1$). Equation 51 provides a good description of silicon etching in CF_4/O_2 and certain other discharges (see Fig. 24). Etchant loss in reaction systems with little or no loading effect is dominated by volume and heterogeneous radical recombination reactions, rather than by etching.

Loading can be avoided by using plasmas in which the principal etchant loss process is insensitive to the etching reaction ($k_{\text{loss}} \gg k_{\text{etch}}$) or where ion bombardment flux, rather than the etchant supply, controls the reaction

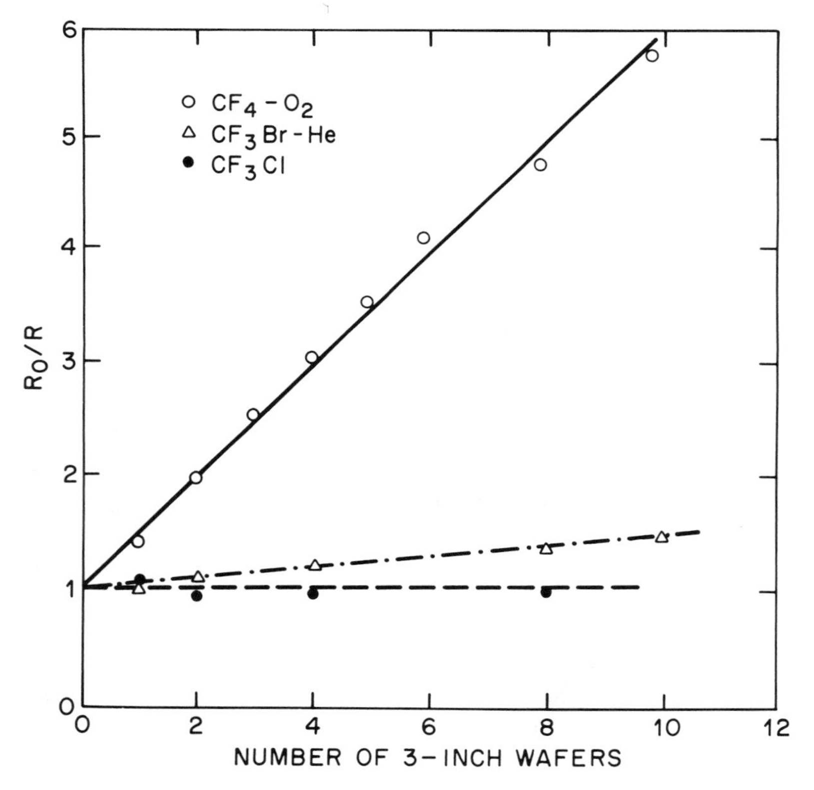

FIGURE 24. The loading effect in CF_4/O_2 (top), CF_3Br (center) and CF_3Cl (bottom) plasmas (after [66]). The etch rate in an empty reactor (R_0), divided by the etch rate with m units of area, forms a straight line with an intercept at 1.

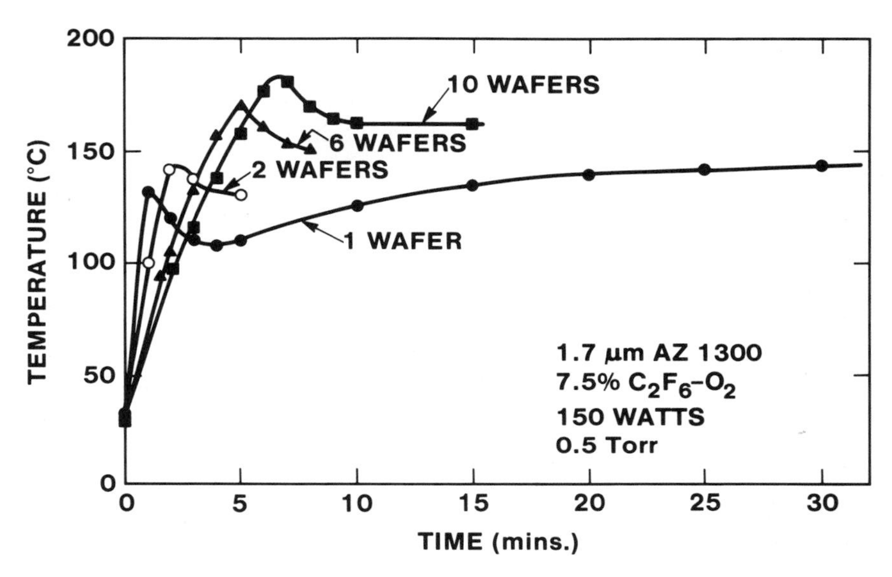

FIGURE 25. Variation of wafer temperature with number of wafers and time when plasma stripping 1.7 μm of AZ 1300 photoresist in a "barrel" reactor. The temperature maxima occur close to when the resist is "cleared" (based on [154]).

rate (this situation is not covered by Eqn. 51). Silicon etching in fluorine atom plasmas usually shows a loading effect because etching reactions dominate the loss of fluorine. However the concentration of atoms in most chlorine and bromine containing plasmas tends to be limited by atom–atom recombination [67], so the etch rates are less sensitive to load size and circuit topology (Fig. 24).

Figure 25 shows an interesting case of loading during resist stripping with oxygen in a barrel reactor. The resist is etched mainly by O atoms, but a small amount of fluorocarbon (C_2F_6) in the feed increases the O atom concentration and generates low levels of F atoms. This fluorine hastens resist degradation (see Section XI).

Wafers in a barrel reactor are thermally isolated, and wafer temperature increases during the etch cycle. The temperature rise shown is caused by two effects. First, the *RF* excitation heats the plasma gas, and in turn the wafers and the reactor. Second, the reaction between oxygen atoms and resist is exothermic. As the oxidation process forms CO, CO_2 and water, the enthalpy of reaction heats the wafer surfaces, and in turn this heat is transferred to the gas and reactor. The curve marked "1-wafer" shows the wafer temperature rise from these combined effects. At first the resist oxidation causes a steep increase in wafer temperature. After some time, the resist begins to "clear" from the wafer and the heat release from "combustion" tapers off. This "endpoint" coincides with the maximum on the temperature-time curve; beyond the "endpoint," there is no heat from oxidation and the wafer cools as it equilibrates with the surrounding plasma gas temperature.

The barrel reactor itself is slowly heated by the reactor gas while the wafers are stripped. When stripping is complete, the gas temperature continues its slow rise toward steady state and carries the wafer temperature with it as it does so. This cooling and heating produces a local minimum in the temperature-time history.

When many wafers are stripped together, the concentration of oxygen atoms is reduced by the loading effect. Hence the rates of stripping and heat release on each wafer are lower, wafer temperature increases more slowly and it takes longer to reach the endpoint. However, the total enthalpy released by oxidation is proportional to the number of wafers. Hence the wafer temperature at endpoint and the reactor temperature just afterwards go up with the number of wafers stripped.

For isothermal etching, the time to reach endpoint should be inversely proportional to the stripping rate. The loading effect relation predicts stripping times proportional to the number of wafers under these conditions. In fact the data for 1, 2 and 6 wafers in Fig. 25 are in fair agreement with this relation, despite the temperature variations.

When more than one etchant species is present, for instance in ClF_3 plasmas where both Cl and F atoms participate in surface gasification reactions (see Section VII.A.3 below), a two-component loading effect applies [64]:

$$\frac{R_o}{R_m} = \frac{1 + \Phi_{F,Cl}}{1/(1 + m\Phi_{Cl}) + \Phi_{F,Cl}/(1 + m\Phi_F)}. \tag{52}$$

This loading curve depends on three dimensionless parameters: Φ_F, the ratio of the specific etch rate for F atoms to their recombination loss; Φ_{Cl}, the equivalent ratio for Cl atoms; and $\Phi_{F,Cl}$, the ratio of F concentration in an empty reactor times specific F etch rate, to Cl concentration in an empty reactor times its etch rate, which is a dimensionless measure of the F to Cl atom reactivity ratio. Again, m is the number of wafers. Figure 26 shows the etch rate as a function of the number of wafers in a ClF_3 plasma. For the plasma conditions where the data in Fig. 26 were taken, $\Phi_F = 2.19$, $\Phi_{Cl} = 0$ and $\Phi_{F,Cl} = 6.14$. A single component loading effect curve with the same initial slope is also drawn for comparison. The lack of loading for Cl ($\Phi_{Cl} = 0$) is characteristic of a variety of anisotropic etchants and leads to curvature in the two component loading effect plot. As the "loaded" etchant (F in Figs. 24, 26) is depleted by increasing amounts of reactive

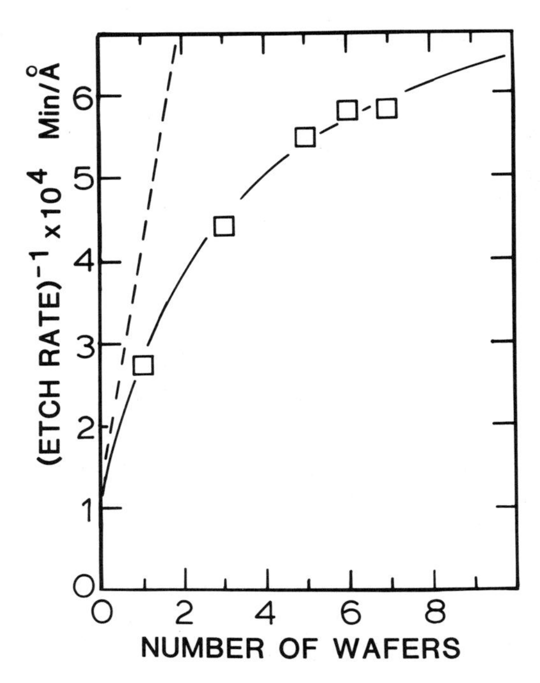

FIGURE 26. Multiple etchant loading effect in a ClF_3 discharge. Both Cl and F atoms participate in surface gasification reactions. Solid line: two component loading effect model, eqn. 52; dashed line: initial slope of the loading curve (points are data) [from [64]].

surface material, the unloaded component (Cl) does most of the additional etching and R_0/R_m approaches an asymptotic limit $(1 + \Phi_{F,Cl})$. The theory has also been developed for the case of n simultaneous etchants [64].

In principle, etchant species can also be depleted by fast convective flow (Q) through the reactor. To account for this loss process, an additional term, Qn_F, is added to Eqn. 49:

$$r_F V - Qn_F = Ak^*_{sF}n_F + k_{vF}n_F V + mA_w k_{wF}n_F. \tag{53}$$

If this convective term dominates so that the reactive losses on the right hand side can be neglected, then substitution of Eqn. 53 into Eqn. 50 gives

$$\frac{1}{R} = \frac{Q}{k_F V r_F}. \tag{54}$$

In this limit the etch rate is inversely proportional to flow. This regime is probably undesirable for process applications, since etching rates should ideally be insensitive to process parameters like flow. Fortunately, this limit usually does not apply. However, with low wall recombination rates and

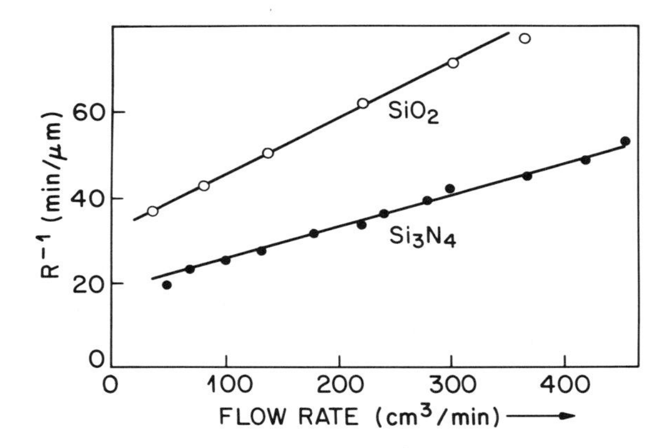

FIGURE 27. Reciprocal etch rate ($1/R$) vs. flow for SiO_2 and Si_3N_4 in a CF_4/O_2 plasma at 1 Torr. In this example, convection controls reactant loss and the etch rate, R is proportional to residence time; $1/R$ is proportional to flow rate—see eqn. 54 (from [155]).

slow etching substrates in a small reactor, fast flows can make convective loss controlling. Figure 27 shows an example where the rates of F atoms etching SiO_2 and Si_3N_4 conform to Eqn. 54.

VI. The Role of Gas Additives

A variety of feed gas mixtures are listed in the tables and discussed throughout this chapter. It is helpful to list and group feed additives according to their function in these mixtures. Most additive effects can be placed in one of the classes shown in Table 8 and outlined below. Many of these mixtures are described more fully in connection with specific etching processes.

Oxidants are used to increase etchant concentrations or suppress polymer. The addition of O_2 or Cl_2 to CF_4, CCl_4 and the other freons in the table are examples. The interaction between etchant radicals and unsaturated polymer-formers is described more fully in Section VIII.

Radical-Scavengers such as hydrogen increase the concentration of film formers or reduce etchant concentration. Hydrogen is added to fluorocarbon feeds to promote CF_x-film growth for selective SiO_2 etching as discussed in Section IX. Unsaturated film-formers (such as CF_2, C_2F_4) also scavenge oxidants and etching radicals.

Most of the freons listed supply *Inhibitor-Formers*, which are usually unsaturated radicals like CF_2 and CCl_2 that oligomerize to form the

Table 8 Gas additives for plasma etching.

Additive	Purpose	Example (Additive-Etchant Gas : Material)
Oxide Etchant	Etch through material oxide to initiate etching.	C_2F_6—Cl_2 : SiO_2; BCl_3—Cl_2 : Al_2O_3; CCl_4—Cl_2 : Al_2O_3
Oxidant	Increase etchant concentration or suppress polymer.	O_2—CF_4 : Si; N_2O—CHF_3 : SiO_2; O_2—CCl_4 : GaAs, InP
"Inert" Gas, N_2	Stabilize plasma, dilute etchant, improve heat transfer.	Ar—O_2 : organic material; He—CF_3Br : Ti
Inhibitor-Former	Induce anisotropy, improve selectivity.	C_2F_6—Cl_2 : Si; BCl_3—Cl_2 : GaAs, Al; H_2—CF_4 : SiO_2
Radical-Scavenger	Increase Film-Former, improve selectivity.	H_2—CF_4; CHF_3—C_2F_6 : SiO_2; H_2—CF_4 : SiO_2
Water/Oxygen-Scavenger	Prevent Inhibition improve selectivity.	BCl_3—Cl_2 : Al; H_2—CF_4 : SiO_2
Volatilizer	Form a more volatile product, increase etch rate.	O_2—Cl_2 : Cr, $MoSi_2$

sidewall films which induce anisotropy. Ions promote desorption of these species, so that the surface concentration is larger in regions not exposed to ion irradiation (i.e., sidewalls of features). Some inorganic gases (like BCl_3, O_2 and H_2O) have the same effect. BCl_x radicals from BCl_3, and $SiCl_x$ from $SiCl_4$ can induce inhibitor-type anisotropy when etching Al or III-V compounds. Although GaAs and GaAlAs have nearly equal etch rates in a pure chlorine plasma, when small amounts of water or fluorinated gases are present, surface layers form on GaAlAs (Al_2O_3, $Al(OH)_3$, AlF_3) that strongly inhibit etching. This inhibition can be used to intentionally etch various $Ga_xIn_yAs_uP_v$ alloys selectively over the aluminum-containing III–V's.

Some etchants, such as Cl atoms, do not readily etch the thin oxide films that form on Si, GaAs, InP and Al. These layers prevent the onset of etching unless small amounts of *native oxide etchants*, like C_2F_6 (for

SiO_2—Section VII.A.2), BCl_3 (to remove Al_2O_3 or the hydroxide [68, 69, 70],—Section VII.B, or Ga_2O_3 on GaAs [71]—Section XII) are added. Furthermore, the plasma decomposition products of BCl_3 and $SiCl_4$ *scavenge* water and oxygen which helps prevent oxide regrowth.

"*Inert*" *Gases* (usually Ar or He) can help stabilize a plasma, enhance anisotropy, or reduce the etching rate by dilution. Since helium has high thermal conductivity, it also improves heat transfer between wafers and the supporting electrodes. Stabilization may be due to an effect on the thermal properties of the discharge gas [36b] (especially with helium additions) or to a shift in the electron energy and density resulting from altered electron balance processes. Sometimes the "inert" addition can enhance anisotropic etching by its effect on the sheath potentials and by providing non-reactive ion-bombardment that stimulates etchant-surface gasification [1, 3].

In a few systems, etch rates can be increased by additives that combine with the primary etchant and substrate to form more volatile ternary products. O_2 is a good *volatilizer* for etching either chrome or $MoSi_2$ in Cl_2 plasmas. Cr etch rates are very low in Cl_2 plasmas, owing to the low volatility of the chloride. However with $< 20\%$ oxygen added to the feed, the formation of volatile $Cr_2O_2Cl_2$ enables etching [72]. The same feed chemistry has been applied to $MoSi_2$ etching in Cl_2 where molybdenum chlorides have a lower vapor pressure than the oxychlorides [73].

VII. Chlorine Plasma Etching

As shown in Table 9, chlorine- and bromine-containing plasmas can be used to etch silicon, III-V compounds, aluminum and a variety of other substrates. Depending on the substrate material, the reactions can be "energetic" ion enhanced, chemical and isotropic, or can be made anisotropic with ion enhanced inhibitor chemistry. Chlorine and bromine atoms have similar etching characteristics with respect to these substrates, but chlorine is more widely used because it has a higher vapor pressure, is less corrosive and tends to form somewhat less toxic byproducts. In the sections below, we will discuss silicon, aluminum and III-V compound semiconductor etching. The other processes in the table are discussed in reference [74].

A. ETCHING SILICON WITH CHLORINE

Practically all chlorine based plasma etching processes for silicon can be understood in terms of a few basic principles. First, Cl and Cl_2 etch undoped silicon very slowly (~ 100 Å/min below 100°C at 0.1 Torr), or

Table 9 Feed gases and mechanisms for plasma etching various materials with chlorine atoms.

Source Gas	Additive	Materials Etched	Mechanism	Selective Over
Cl_2	None	heavily *n*-doped Si	Chemical	SiO_2
	C_2F_6		Ion-inhibitor	
	$SiCl_4$		Ion-inhibitor	
Cl_2	None	Si	Ion-energetic	SiO_2
CCl_4	O_2			
$SiCl_4$	O_2			
	$SiCl_4$			
	CCl_4			SiO_2, Some
Cl_2	$CHCl_3$	Al	Ion-inhibitor	resists,
	BCl_3			Si_3N_4
Cl_2	O_2	$MoSi_2$	Ion-energetic	SiO_2
Cl_2	None		Chemical-	SiO_2,
		III-V	Crystallographic	Resists
	BCl_3	Semiconductors	Ion-inhibitor	
	CCl_4			
CCl_4	O_2			
$SiCl_4$	O_2			
Cl_2	O_2, H_2O	III-V Alloys	Chemical-	$AlGa_xAs_y$,
		without Al	Crystallographic	$AlIn_xP_y$,
				SiO_2, Si_3N_4,
				Resists
CF_2Cl_2	None			Ion-inhibitor

not at all (depending on the crystallographic orientation, or whether it is polycrystalline). However, Cl will etch undoped silicon in the presence of energetic ion bombardment ("damage" ion-enhanced etching). Second, heavily n-type doped silicon and polysilicon are rapidly and spontaneously etched by Cl atoms. Thus, features are often severely undercut when heavily As- or P-doped layers are etched in Cl atom plasmas, but this can be prevented by sidewall inhibitor chemistry. Cl_2 will also attack the doped material, but at a much slower rate [75]. There is good evidence that these same circumstances are true of bromine plasma etching of silicon.

The effect of frequency on undoped silicon etching in a Cl_2 plasma, shown in Fig. 5, illustrates the effect of ion bombardment. The applied voltage is a good indicator of sheath potential, and the large voltage increase below ~ 3–5 MHz is a characteristic effect associated with the ion transit frequency (see Section III.C). The close correlation between the increase in etch rate and voltage with decreasing frequency is symptomatic of an energy-driven ion enhanced etch process.

1. The Doping Effect

The chemical etching of silicon in halogen-based discharges is affected by the type and concentration of electrically active dopants. In F atom systems, *p*-type doping (boron) suppresses silicon etch rates slightly (by as much as a factor of two) [76, 77, 78, 79, 80], while high concentrations of *n*-type dopants (As or $P \gtrsim 10^{19}\,cm^{-3}$) enhance etching [76, 77, 78, 79, 81] by a factor of 1.5–2. By contrast heavily *n*-doped (100) and (111) silicon [82, 83, 84] and polysilicon [67, 85, 86, 87, 88, 89] ($\sim 10^{20}\,cm^{-3}$) in Cl atom plasmas (Cl_2, Cl_2/Ar, CCl_4/Ar, CF_3Cl, $SiCl_4/O_2$, CF_3Br/Cl_2, C_2F_6/Cl_2) etch as much as 15–25 times faster than undoped substrates. This enhancement is related to the concentration of *active n*-type carriers (e.g., the Fermi level), rather than the chemical identity of the dopant [67, 79, 90, 91, 92]. Unannealed or electrically inactive dopant implants have a minimal influence on etching [67, 90].

The detailed mechanisms through which chlorine-silicon etch rates depend on doping levels are still being studied. However, it is generally agreed that *n*-type doping raises the Fermi level and thereby reduces the energy barrier for charge transfer to chemisorbed chlorine [67, 93, 156]. As depicted in Fig. 28, chlorine and/or bromine atoms are covalently bound to specific sites on an undoped silicon surface. Steric hindrance impedes impinging etchant from penetrating the surface to reach subsurface Si-Si bonds. The formation of a more ionic silicon–halogen surface bond, due to the *n*-type doping and enhanced electron transfer, opens additional chemisorption sites and facilitates etchant penetration into the substrate lattice. This makes it possible for impinging Cl atoms to more readily chemisorb, penetrate the lattice, and react. This notion of steric hindrance is reinforced by a variety of data. The less densely packed [100] planes of undoped single crystal are slowly etched by Cl, even under conditions where etching along the densely packed [111] plane is undetectable [94]. Evidence from x-ray photoelectron spectroscopy shows more etchant fragments on n-type than on undoped silicon when etching in a $CFCl_3$ plasma is interrupted [90].

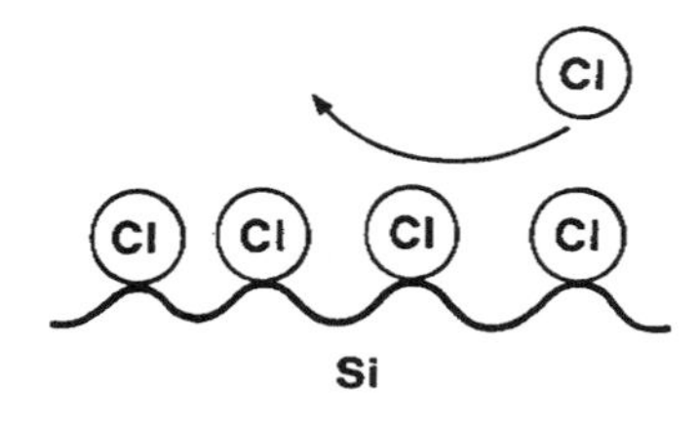

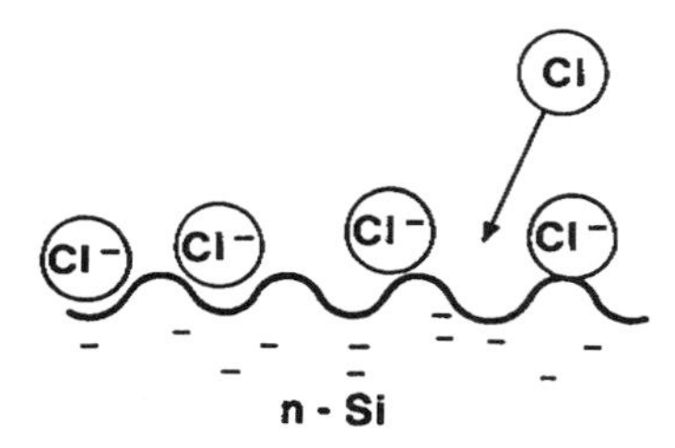

FIGURE 28. Chlorine atoms spontaneously etch heavily *n*-doped silicon, but on the undoped material they chemisorb and reaction stops. This difference is attributed to steric hindrance which impedes etchant from reaching subsurface Si-Si bonds. The higher Fermi level in the doped material promotes charge transfer, and a more ionic Cl-Si bond allows the Cl access to additional sites (from [156]).

Finally, recent results on molecular chlorine/polysilicon etch rates (Cl_2) as a function of n-dopant level, show the increasing etch rate is attributable to an increase in the preexponential (frequency) factor rather than the Arrhenius activation energy (see Section III.E), consistent with an increasing number of adsorption sites [75].

2. *Sidewall Inhibitor Chlorine Etching*

An example of inhibitor-based chlorine plasma etching is shown in Fig. 29. As chlorine is added to a pure C_2F_6 feed, both undoped and phosphorous-doped polysilicon etch rates initially increase. Heavily-doped polysilicon etching then shows a rapid, almost linear rise with increasing amounts of chlorine in the feed, while the undoped polysilicon rate remains low. This difference reflects the fast reaction of chlorine atoms with doped polysilicon, which should increase in direct proportion to the atomic chlorine concentration. The reaction of undoped polysilicon is probably connected with other species in the plasma, and its low etch rate in chlorine reflects the small proportion of energetic ions at high frequency and moderate pressure.

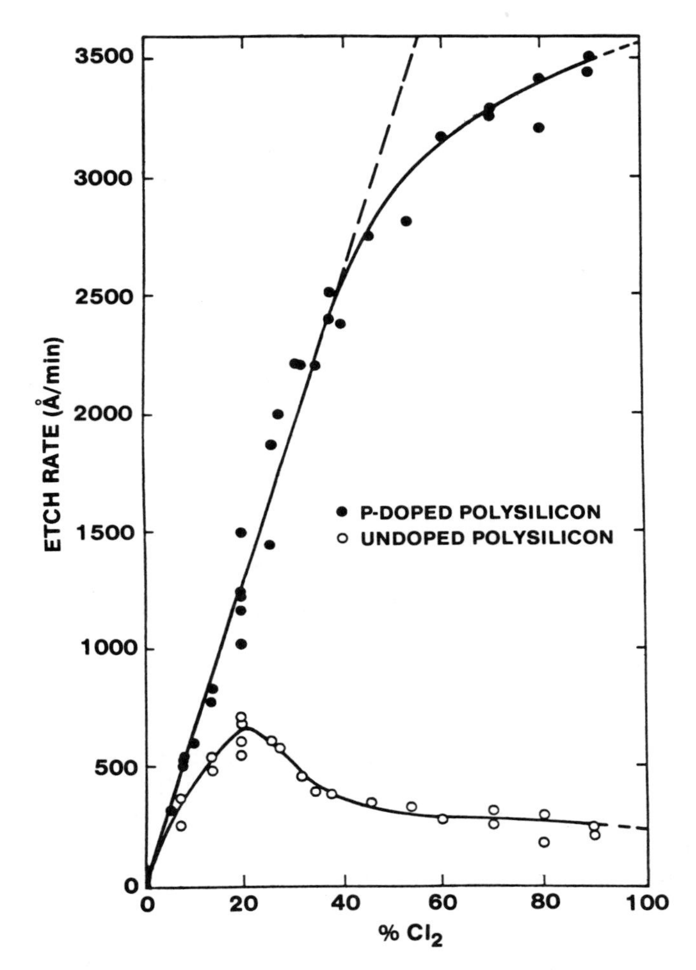

FIGURE 29. Etching of doped and undoped polysilicon using a Cl_2/C_2F_6 feed in a parallel plate reactor at 0.35 Torr, 200 sccm feed, 0.32 W/cm^2 and a 25°C electrode temperature. Cl atoms in chlorine-rich plasmas chemically etch *n*-type *P*-doped polysilicon, while the etch rate of undoped silicon is low. Species from C_2F_6 form a sidewall inhibitor layer which provides anisotropy. Beyond about 20% Cl_2 etching becomes isotropic because there is not enough inhibitor to prevent Cl-atoms from attacking doped polysilicon sidewalls. Species from C_2F_6 also remove native etch native oxide from the Si surface so etching can start (after [157]).

Significantly, the doped polysilicon etch profile is anisotropic until about 15–20% chlorine is added to the feed. Beyond this point there is not enough inhibitor-forming material to adequately protect the sidewalls of etched features. It can be difficult to initiate the etching process in nearly pure chlorine feed mixtures because Cl atoms will not etch native oxide on the silicon without energetic ion bombardment. This problem can be overcome with a brief dip in hydrofluoric acid solution before etching, or by briefly running a pure C_2F_6 plasma before switching to the selected feed composition (C_2F_6 plasmas are a source of unsaturated monomer that etches SiO_2 —see Sections VIII, IX).

The C_2F_6/Cl_2 system is one member of a family of etching mixtures that can be used to etch doped polysilicon anisotropically via the inhibitor mechanism. The results of etching by related gas mixtures are shown in Fig. 30. Various profiles are possible, including isotropic etching, anisotropic

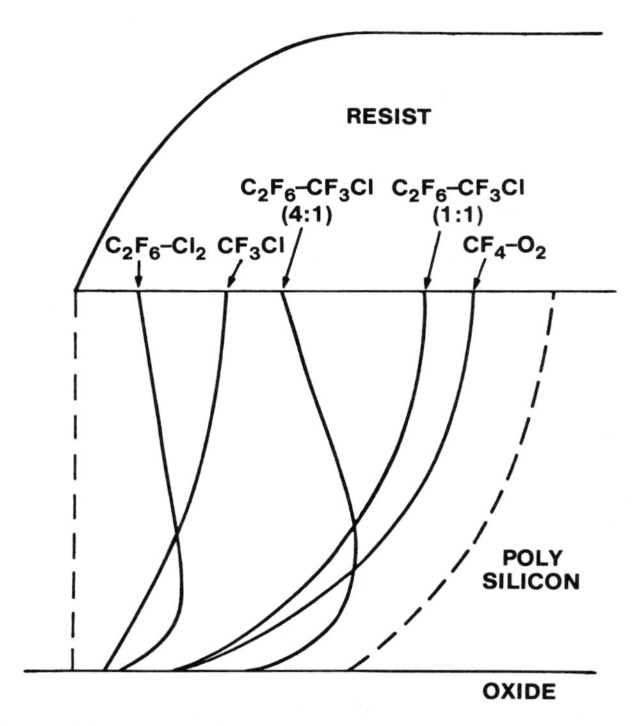

FIGURE 30. Profile control by the ion-enhanced inhibiting mechanism is sensitive to chemistry and plasma conditions. Details of the processes involved are poorly understood and a variety of unusual contours have been observed, as illustrated by these profiles found on 1.2 μm polysilicon lines after 25% overetching in various mixtures (~ 0.4 Torr, after [158]).

etching, and even involuted profiles with various degrees of undercut. The mechanisms responsible for producing these complex profiles have not been explained. Mainly they show how manipulation of the feed mixture composition can be used to adjust the profile contour and degree of anisotropy.

3. *Mixing F and Cl for Profile Control*

We have seen that fluorine etching of silicon is isotropic, while chlorine etching of undoped silicon is usually anisotropic. These facts suggest that a plasma containing a mixture of chlorine and fluorine atoms could be used to form intermediate tapered profiles, between these two limits. Figure 31

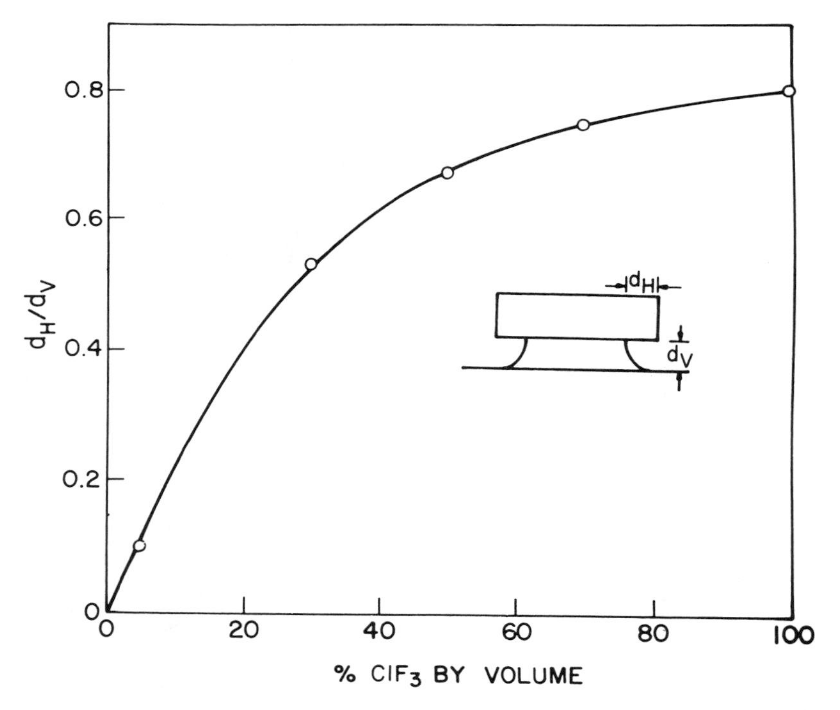

FIGURE 31. Effect of ClF_3 concentration on anisotropy in a ClF_3/Cl_2 plasma. Conditions were 100 W, 5 sccm total flow, and 0.02 Torr with six wafers in the reactor (from [64]). Etching in pure ClF_3 plasmas is mostly by F-atoms (in the absence of heavy loading). Since plasma etching by F atoms is isotropic and ion-assisted etching of undoped silicon by Cl-atoms from Cl_2 is anisotropic, mixtures of these etchants in controlled proportions achieve various degrees of anisotropy.

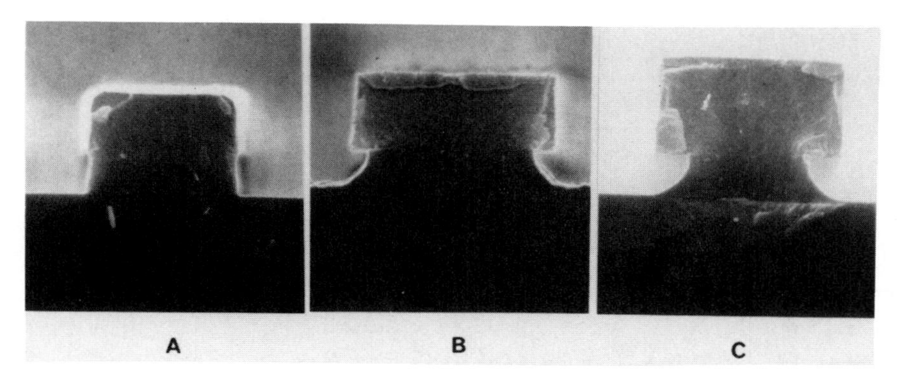

FIGURE 32. Electron micrographs showing profiles at (a) 5% ClF_3, (b) 30% ClF_3 and (c) 50% ClF_3 for conditions of figure 31 (from [64]).

shows how this can be done. A plasma in pure ClF_3 dissociates the gas mostly into fluorine atoms and a much smaller fraction of Cl. More chlorine can be added to the feed mixture as Cl_2. By varying feed composition between pure chlorine and pure ClF_3, the ratio of chlorine to fluorine atoms in the discharge can be continuously changed from zero to almost all fluorine, which is characteristic of ClF_3. With this, the horizontal to vertical etching ratios vary between anisotropic etching, characteristic of chlorine, to isotropic profiles from fluorine atoms. Resulting profiles are plotted, and profiles at selected points on this curve are shown in Fig. 32.

B. ALUMINUM ETCHING

Since aluminum is an essential interconnect material in virtually all integrated circuits, micron- and even submicron-sized aluminum features must be patterned anisotropically. Aluminum will not etch in fluorine atom plasmas because aluminum fluoride is completely involatile. Fortunately, aluminum chlorides and aluminum bromides are both volatile. Al_2Cl_6 appears to be the etch product from chlorine attack near room temperature [95], while $AlCl_3$ is the main species at higher temperatures ($> 200°C$).

A small number of factors dictate the etching conditions for aluminum. First, both atomic and molecular bromine and chlorine all attack bare aluminum vigorously. Figure 33 compares total silicon and aluminum chloride product evolution when these substrates are etched by chlorine atoms and molecules at low pressure in the presence of positive ion bombardment. Ion bombardment does not affect the aluminum etching rate because spontaneous chemical attack is very fast. By contrast, the reaction

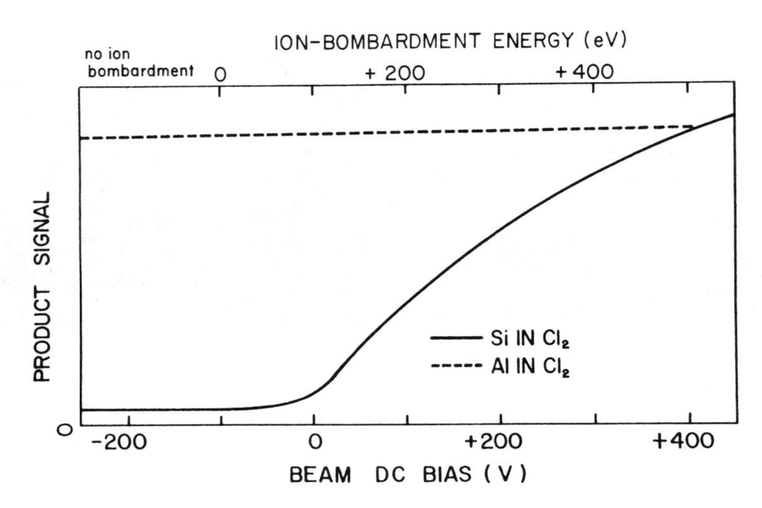

FIGURE 33. Etching by species extracted from a Cl_2 discharge as a function of bias (Si etch rate, solid line; Al etch rate, broken line). The same aluminum etch rate is obtained with the discharge extinguished (from [93]). While ion bombardment is required for etching of undoped silicon by Cl_2 or Cl-atoms at room temperature, both Cl and Cl_2 chemically etch aluminum, rapidly and spontaneously, without ion bombardment.

between undoped silicon and chlorine *is* enhanced by energetic ions. This fast chemical etching means that inhibitor chemistry is *necessary* for linewidth control in aluminum etching. CCl_4, $CHCl_3$, BCl_3 and $SiCl_4$ are among the inhibitor-forming additives that promote anisotropy in aluminum etching.

A second factor is the unreactive oxide, which is always formed on aluminum surfaces exposed to air; the oxide forms especially fast when there are traces of moisture present. This means that a thin (~ 30 Å) native aluminum oxide layer must be etched before chlorine can reach the aluminum, and that aluminum etching is sensitive to traces of oxygen or water vapor in the reactor. This oxide can be sputtered off or etched with species that consume the oxide. Unsaturated radicals from CCl_4 or $CHCl_3$, CCl_x, or radicals derived from BCl_3 and $SiCl_4$, BCl_x, or $SiCl_x$ are capable of etching this layer. It etches about 100 times slower than aluminum in a BCl_3 discharge and about 25 times slower in a CCl_4 discharge [68].

Moisture and oxygen are readily adsorbed on many materials of construction. Anodized aluminum is often used for electrodes because it is lightweight and corrosion-resistant, but it also is porous and has a high effective surface area to adsorb oxygen and water vapor. Oxygen and water desorbed during aluminum etching will seriously interfere with the etching process if appropriate measures are not taken [96]. Desorbed water and oxygen can be minimized by excluding them—for example, by using a "load lock" so the reactor is not exposed to air and water vapor between runs. Luckily, some of the additives which form inhibitor films and etch Al_2O_3, BCl_3 and $SiCl_4$, are also effective scavengers for water and oxygen [69, 97].

Another problem is that up to 5% copper is commonly alloyed with aluminum to improve electromigration resistance. This makes etching more difficult because copper chlorides have low volatility. Three complementary approaches are used to overcome this problem—the substrate temperature can be raised (but after a point, photoresist erosion and hardening are problems); aluminum with a larger grain size and less copper can be used; and etching conditions with more ion bombardment can be used, which promote copper chloride desorption (but again this is limited by resist erosion).

Photoresist degradation is a serious problem in chlorine-based plasma etching of aluminum. The etch product, $AlCl_3$, is a Lewis acid, and well-known catalyst for halogenation and "Friedel-Crafts" reactions of organic compounds [98]. As such, it accelerates degradation of resist. The deterioration increases rapidly with etching temperature and in practice sets the upper limit for aluminum etching in a chlorine plasma to around 100–150°C.

After aluminum etching, residual $AlCl_3$ and chlorine-containing residues from the resist remain on the surface. These are hygroscopic, and hydrolyze in the presence of atmospheric humidity to form HCl, which corrodes the aluminum. Rinsing with water or exposure to an oxygen plasma after etching reduces the amount of chlorinated residue, but sometimes these procedures are not enough to prevent post-etch corrosion. Another effective treatment is to expose the etched aluminum film to a fluorocarbon plasma [99],—a brief exposure to a discharge in C_2F_6 converts the chlorides into fluorides, which are neither hygroscopic nor corrosive. A dilute nitric acid rinse can then be used to remove the fluorides and regrow a protective oxide, if desired. Low temperature thermal oxidation in dry oxygen is also effective in some situations [100].

VIII. Etchant-Unsaturate Concepts

Halocarbons and their mixtures with oxidants such as O_2, Cl_2, NF_3, etc. are used for practically all plasma processing. Unsaturated halocarbon radicals and oligomers derived from the feed gas usually are the precursors to inhibitor sidewall protection films. When inhibitor ion-assisted etching is taking place, these films can be extremely thin, often no more than ~ 30–150 Å at steady state [74]. However, if excessive concentrations of unsaturated species are formed in the gas phase, they may leave contamination or produce gross amounts of polymer that coat all surfaces and stop etching. In halocarbon plasmas, there is ordinarily a balance between unsaturated species and the etchant/oxidant atoms such as F or O. The *etchant-unsaturate* concept provides a framework for understanding this balance in general terms and is a guide to predicting the effects of composition changes on halocarbon/oxidant plasmas.

In this reaction scheme, unsaturated fluoro- and chlorocarbon polymer precursors derived from the CF_2 or CCl_2 radicals are saturated during reactions with atoms and reactive molecules. The most reactive species are preferentially removed by the saturation reactions. An ordering of this reactivity can be used to predict the dominant atomic etchants as a function of halocarbon and additive gas compositions. The general formulation is given by Eqns. 55–58:

$$\text{e} + \text{Halocarbon} \rightarrow \begin{matrix}\text{Saturated}\\ \text{Species}\end{matrix} + \begin{matrix}\text{Unsaturated}\\ \text{Species}\end{matrix} + \text{Atoms}, \tag{55}$$

$$\left.\begin{matrix}\text{Reactive Atoms}\\ \text{Reactive Molecules}\end{matrix}\right\} + \text{Unsaturates} \rightarrow \text{Saturates}, \tag{56}$$

$$\text{Atoms} + \text{Surfaces} \rightarrow \left\{\begin{matrix}\text{Chemisorbed Layer}\\ \text{Volatile Products},\end{matrix}\right. \tag{57}$$

$$\text{Unsaturates} + \text{Surfaces}\left(+ \begin{matrix}\text{initiating}\\ \text{radicals}\end{matrix}\right) \rightarrow \text{Films}. \tag{58}$$

In CF_3Br, for instance, Br and F atoms are formed by electron-impact dissociation of the halocarbon, and there is usually a negligible steady state concentration of F atoms because of reaction with CF_2 radicals and unsaturated species such as C_2F_4. In heavily chlorinated plasmas (e.g., CCl_4 as the halocarbon in Eq. 56), CCl_2 and C_2Cl_4 will play analogous roles. The relative reactivities of atoms and molecules in these saturation reactions are assigned a hierarchical order: $F \sim O > Cl > Br$; $F_2 > Cl_2 > Br_2$. This ordering is used to predict the predominant etchant species and reaction products as a function of feed composition. The most reactive atoms (F, O) do not coexist with unsaturates at appreciable concentrations; thus, either these atoms or the unsaturates are substantially depleted. When atoms that can gasify a substrate predominate, etching takes place. Film formation is observed when unsaturates are present in excess and are adsorbed on surfaces where polymerization proceeds by reaction with saturated fluorocarbon radicals [101]. Alternatively, the unsaturated radicals may form sidewall films that result in anisotropic etching in the presence of ion bombardment. Polymerization of unsaturates is inhibited by surfaces that react with fluorocarbon radicals to form entirely volatile products [101, 102] (e.g., SiO_2 surfaces).

Oxidant additions to a plasma alter the balance between halogen atoms and unsaturates. More reactive oxidants will be preferentially consumed by unsaturates, tending to increase the relative concentration of less reactive halogen atoms, and while doing so they will suppress polymer formation. The effect may be illustrated by the addition of oxygen to a CF_3Cl plasma. Oxygen atoms and molecular oxygen in the discharge will result in the following overall reactions:

$$\left.\begin{matrix} O \\ O_2 \end{matrix}\right\} + C_xF_{2x} \rightarrow \left\{\begin{matrix} COF_2 \\ CO \\ CO_2 \end{matrix}\right. + F, F_2 \tag{59}$$

and

$$\left.\begin{matrix} O \\ O_2 \end{matrix}\right\} + C_xF_{2x-y}Cl_y \rightarrow \left\{\begin{matrix} COF_2 \\ COFCl \\ CO \\ CO_2 \end{matrix}\right. + F, F_2, Cl, Cl_2. \tag{60}$$

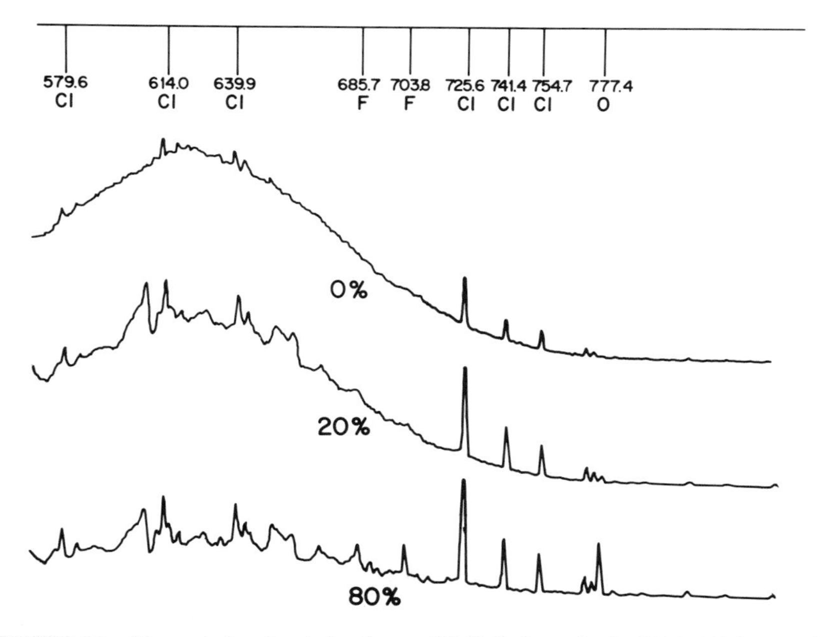

FIGURE 34. Plasma induced emission from a CF_3Cl discharge in the 566.5–833.5 nm region as a function of oxygen additions (selected atomic lines are identified). When oxygen is added the broad feature (on the left), which is attributed to fluorocarbon and unsaturated compounds, disappears and emissions from Cl increase. Little or no free O or F are apparent in the spectrum until the unsaturated feature is suppressed (curve for 80% O_2), from [159].

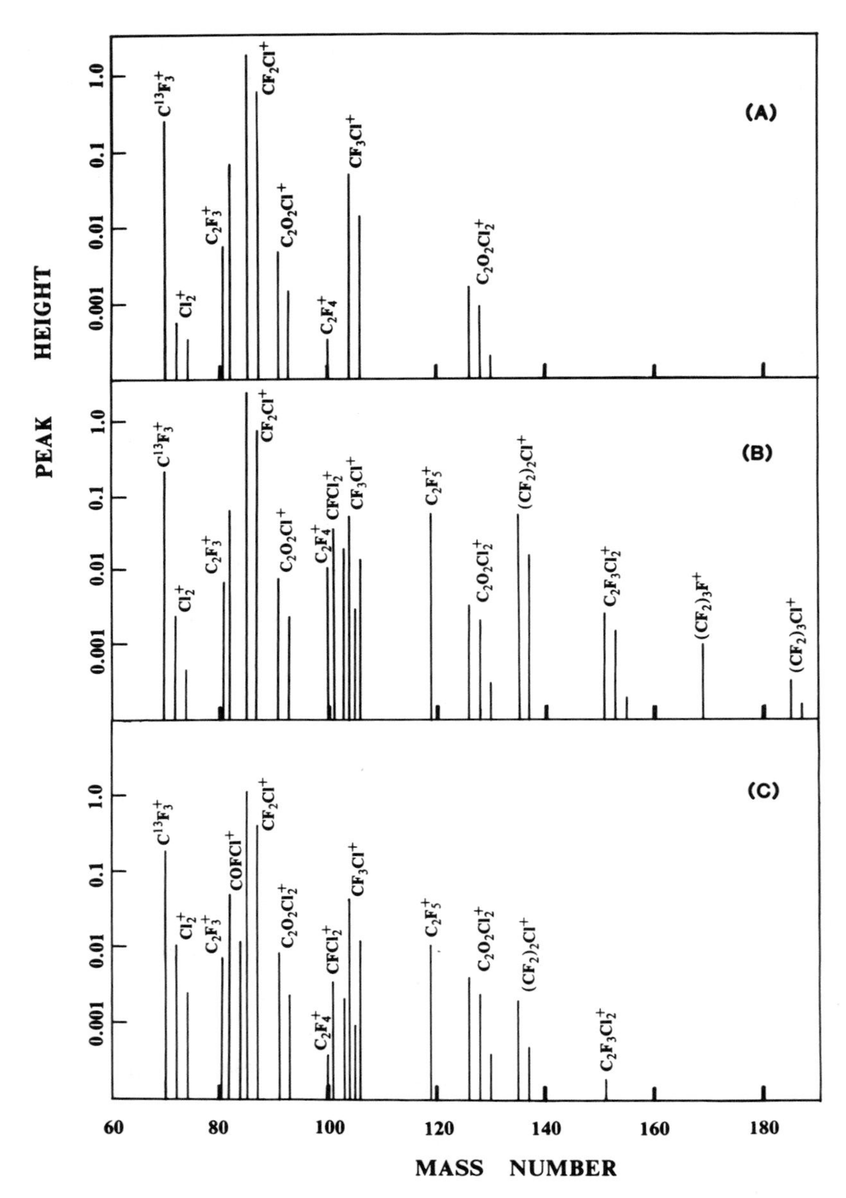

FIGURE 35. Principal peaks in the high range ($\geq$ 70 mass units) of the mass spectrum of products from a) CF_3Cl, b) a CF_3Cl discharge, and c) a CF_3Cl 20% O_2 discharge (100 W, 0.5 Torr, 100 sccm total flow rate in a parallel plate reactor, ionizer energy was 20 eV). Oxygen in (c) suppresses the high mass peaks in (b) which are derived from unsaturate-like $(CF_2)_n$ compounds formed in the discharge (after [105]).

These reactions cause an increase in F, F_2,Cl, and Cl_2 concentrations in the plasma, as shown in Fig. 34, and depletion of unsaturated compounds in the plasma (see Fig. 35) [103, 104, 105]. Fluorine and oxygen combine with unsaturated species at similar rates; hence if unsaturated materials are present in excess, the fluorine is consumed by saturation reactions such as:

$$\left.\begin{matrix} F \\ F_2 \end{matrix}\right\} + C_xF_{2x} \rightarrow \left\{\begin{matrix} C_xF_{2x+1} \\ C_xF_{2x+2} \end{matrix}\right. . \tag{61}$$

Chlorine-unsaturate reactions are slower, and chlorine removal is proportional to the unsaturated species concentration. Consequently, the consumption of unsaturated compounds by oxygen and fluorine lowers the rate of chlorine-unsaturate reactions, leading to an increase in free chlorine within the discharge. Yet there is little or no free O or F. Finally, if an excess of oxygen is added, unsaturated compounds are substantially depleted so fluorine and oxygen atoms will appear in the discharge along with the increased concentration of Cl and Cl_2. This is usually undesirable since F atom reactions with the substrate will lead to mask undercutting, while O atoms attack photoresist as well as form surface oxides that retard etching [60]. Note that oxygen itself is only one example of an oxidant that can be added. For instance, NF_3 added to CF_3Br plasma (or a CF_3Cl plasma) will also increase the concentration of Cl atoms by liberating F and F_2 which participate in reactions analogous to Eqn. 60, and will suppress unsaturated species (Eqns. 59, 61).

IX. Etching Silicon Oxide in Unsaturated and Fluorine-Rich Plasmas

SiO_2 can be etched in plasmas that produce fluorine atoms or, alternatively, using feed mixtures that generate unsaturate-rich fluorocarbon species. In silicon microelectronic processing, it is usually necessary to etch SiO_2 preferentially over silicon. Since F atoms etch Si faster than SiO_2 (see Section III.E), the unsaturated gas feeds are usually employed to pattern oxide on silicon devices. However, F atoms *are* a selective etchant for SiO_2 layers on III-V semiconductors, thanks to the involatility of group III fluorides. F atom-producing plasma feeds such as NF_3/Ar, have the advantage over unsaturated plasmas in that they leave no carbonaceous residue on surfaces, which lessens the possibility of contaminative device degradation. At higher pressures and excitation frequency ($\gtrsim$ 0.1 Torr, $\gtrsim$ 5 MHz), where ion bombardment energy is low, F atom etching of SiO_2 is chemical, and therefore isotropic. However, at low frequency ($\lesssim$ 1 MHz) and/or low pressure, damage-driven ion-enhanced attack dominates [27]. Oxide etching in unsaturated plasmas can *only* go by the anisotropic

Table 10 Plasmas etching SiO_2 under various conditions.

Additive	Etchant	Conditions	Mechanism	Selectivity	Rate
O_2	F	High Pressure High Frequency	Isotropic (Chemical)	Over III-V's	Low
		High Pressure Low Frequency	Energetic Ion-Enhanced	For Silicon	Moderate
H_2	C_xF_y	Low Frequency Low Pressure	Energetic Ion-Enhanced	Over Silicon	Moderate

damage-driven route. Unfortunately, this means there is no way now known to selectively plasma etch SiO_2 over Si with an isotropic profile. These alternatives are summarized in Table 10.

CF_3 radicals were once thought to be the selective SiO_2 etchant in unsaturate-forming, "fluorine-deficient" plasmas [106, 107]. However, it turns out that CF_3 does not spontaneously etch SiO_2 [108, 109]. Instead, it appears that halogen atoms are transferred to the substrate from a thin ($\lesssim$ 10 Å) fluorocarbon layer that is formed when unsaturated fluorocarbon species impinge on the oxide surface (see Fig. 36) [110, 111, 112]. As fresh CF_x radicals are added from the gas phase, material in this carbonaceous layer is continuously gasified by ion-induced reaction with the SiO_2 substrate and sputtering. Ion damage produces dangling bonds and radical groups at the silicon dioxide interface, where Si bonds are eventually

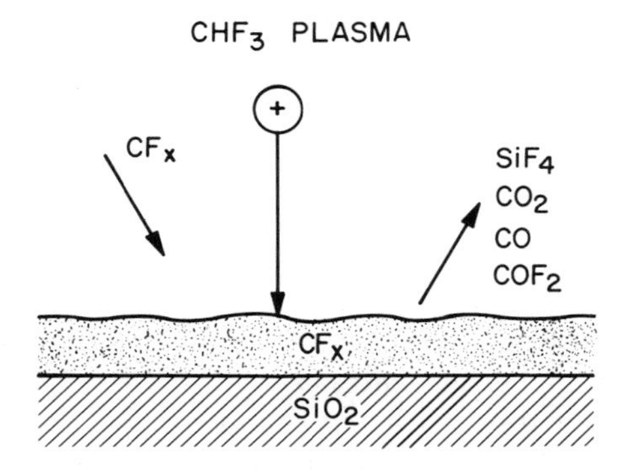

FIGURE 36. During anisotropic etching of SiO_2 in unsaturated plasmas, halogen atoms are transferred to the substrate from a thin fluorocarbon layer.

converted into SiF_x groups, and exposed defects are covered with new polymer. For example, the following types of reactions have been proposed [93]:

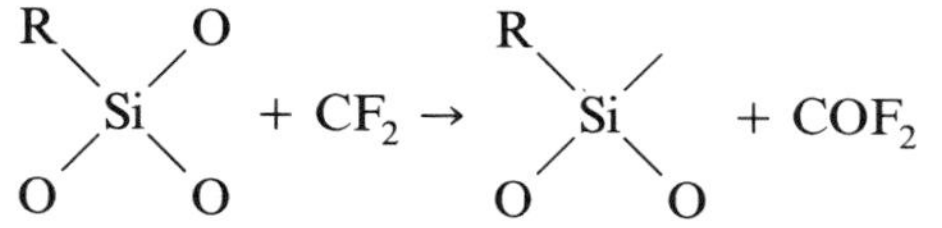

$$\begin{array}{ccc} R & & O \\ & \diagdown\ \diagup & \\ & Si & \\ & \diagup\ \diagdown & \\ O & & O \end{array} + CF_2 \rightarrow \begin{array}{ccc} R & & \\ & \diagdown\ \diagup & \\ & Si & \\ & \diagup\ \diagdown & \\ O & & O \end{array} + COF_2 \qquad (62)$$

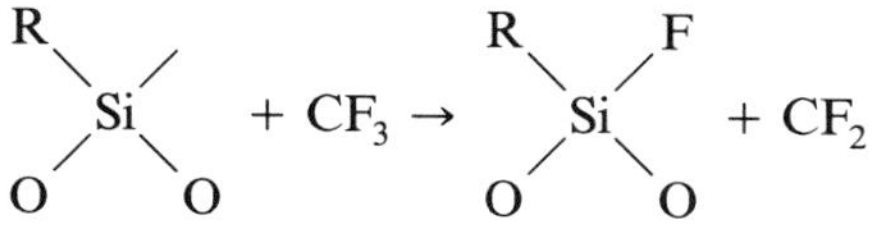

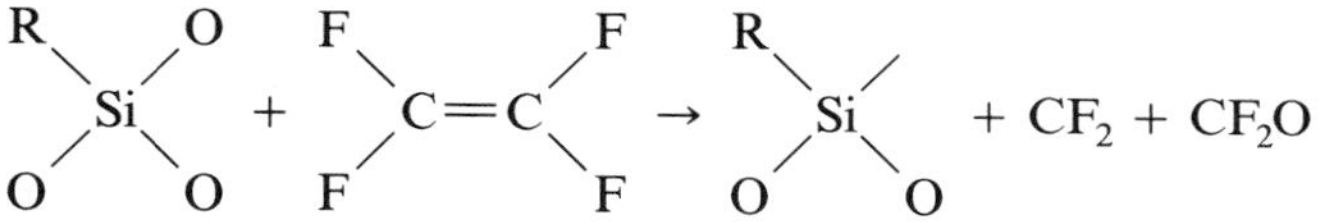

$$\begin{array}{ccc} R & & \\ & \diagdown\ \diagup & \\ & Si & \\ & \diagup\ \diagdown & \\ O & & O \end{array} + CF_3 \rightarrow \begin{array}{ccc} R & & F \\ & \diagdown\ \diagup & \\ & Si & \\ & \diagup\ \diagdown & \\ O & & O \end{array} + CF_2 \qquad (63)$$

$$\begin{array}{ccc} R & & O \\ & \diagdown\ \diagup & \\ & Si & \\ & \diagup\ \diagdown & \\ O & & O \end{array} + \begin{array}{ccc} F & & F \\ & \diagdown\ \ \diagup & \\ & C{=}C & \\ & \diagup\ \ \diagdown & \\ F & & F \end{array} \rightarrow \begin{array}{ccc} R & & \\ & \diagdown\ \diagup & \\ & Si & \\ & \diagup\ \diagdown & \\ O & & O \end{array} + CF_2 + CF_2O \qquad (64)$$

$$\begin{array}{ccc} & & SiF_3 \\ & & \diagup \\ R & & O \\ & \diagdown\ \diagup & \\ & Si & \\ & \diagup\ \diagdown & \\ O & & O \end{array} + CF_3 \rightarrow \left\{ \begin{array}{l} SiF_4 + COF_2 + \begin{array}{ccc} R & & \\ & \diagdown\ \diagup & \\ & Si & \\ & \diagup\ \diagdown & \\ O & & O \end{array} \\ SiF_3 + COF_2 + \begin{array}{ccc} R & & F \\ & \diagdown\ \diagup & \\ & Si & \\ & \diagup\ \diagdown & \\ O & & O \end{array} \\ SiF_4 + COF + \begin{array}{ccc} R & & F \\ & \diagdown\ \diagup & \\ & Si & \\ & \diagup\ \diagdown & \\ O & & O \end{array} \end{array} \right. \qquad (65)$$

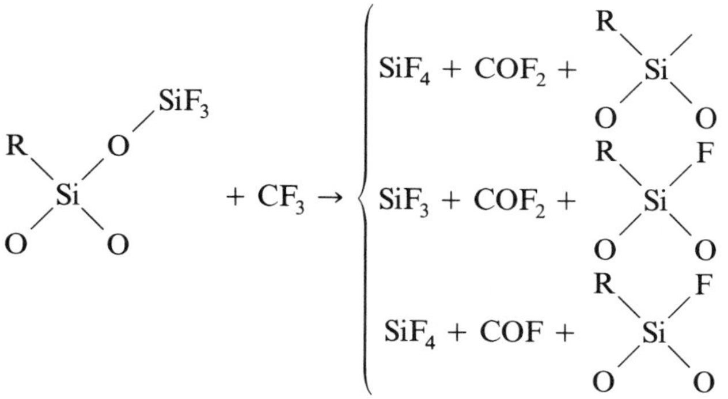

where R denotes F, O, or bound fluorocarbon chains ($C_xF_y \cdots$) and the reactants shown (CF_2, CF_3, C_2F_4) represent classes of functional groups in the fluorocarbon layer. The thickness of the layer on SiO_2 depends on the feed gas and etching conditions—in the extreme of high concentrations of unsaturated material in the gas phase, visible polymer forms; in ~ 10 mTorr C_2F_6/C_2H_4 plasmas Auger analysis and ellipsometry showed a thickness of 10–20 Å [113], while in hydrogen-lean CF_4/H_2 mixtures at 25 mTorr, the film was too thin to detect using the same technique [114].

Naturally, fluorocarbon layers also form on exposed silicon during oxide etching [46, 101]. Since there are no reactions with silicon that can gasify the carbonaceous film, a fairly thick layer (~ 150 Å or more) accumulates on the silicon and helps block attack by sputter and chemical etching. Although the protective film on silicon provides selectivity, silicon surfaces are contaminated by carbon in this process, and ion mixing causes amorphoti-

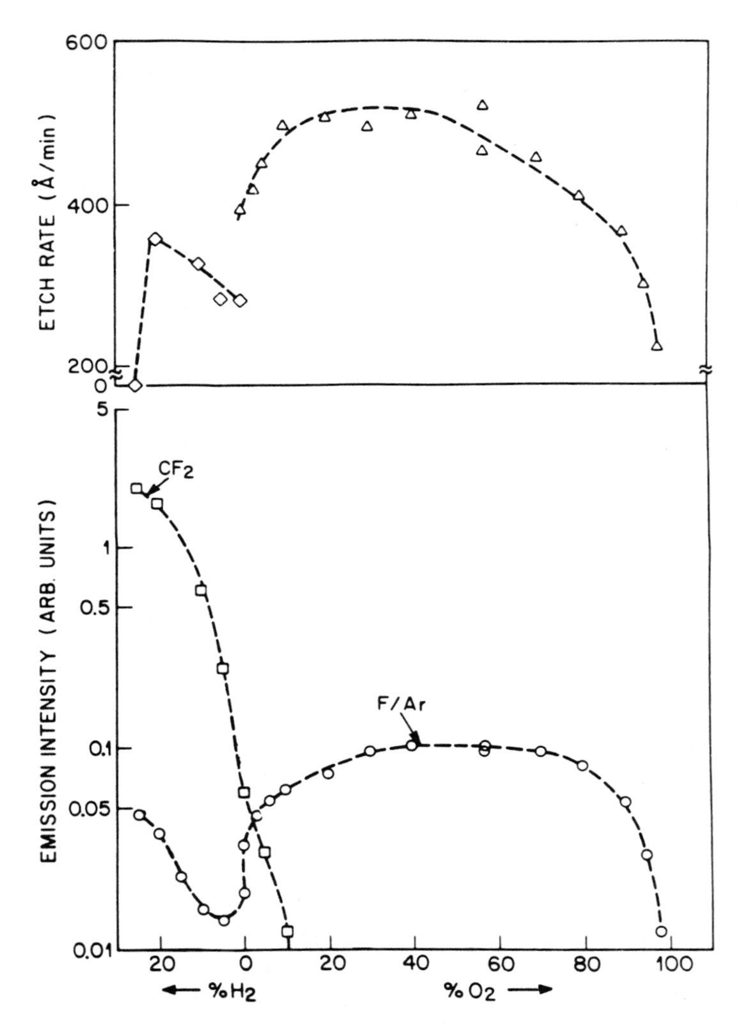

FIGURE 37. SiO_2 etch rate (top) and plasma-induced emission intensities (bottom) as a function of H_2 and O_2 additions to a 100 kHz CF_4 plasma (parallel plate reactor, Si-covered lower electrode, 0.35 Torr, 0.35 W/cm^2). 704 nm F-atom emission intensity was divided by the emission from small amounts of added Ar (< 5%) to measure the relative F-atom concentration actinometrically (see Chapter 4). With more than ~20% H_2 a thick polymer film forms on the oxide and etching stops.

zation, silicon carbide formation and damage, sometimes extending more than 100 Å below the silicon surface [115, 116, 117].

Starting with CF_4, adding either O_2 or H_2 increases the etch rate of SiO_2 under anisotropic ion energy enhanced etching conditions—low frequency and high pressure, or low pressure. However, the chemistry under these two conditions is drastically different according to the scheme depicted in Table 10. Figure 37 shows the etch rate and emissions from F atoms and CF_2 in the discharge versus additives. When oxygen is added, F atoms are the etchant, while added H_2 results in the formation of CF_2 and derivatives that cause thin CF_x films to form on surfaces and etch SiO_2 in the presence of ion bombardment. This is reflected in the emission spectrum (Fig. 37). Note that adding too much hydrogen (at the leftmost side of the composition curve) causes a thick plasma polymer to coat all surfaces until etching stops. The onset of this unselective thick film growth has been called the "polymer point."

The remaining condition in Table 10, adding oxygen at high frequency, results in slow, isotropic, purely chemical SiO_2 etching by fluorine atoms which is not selective over Si (discussed in Sections III.E and IV).

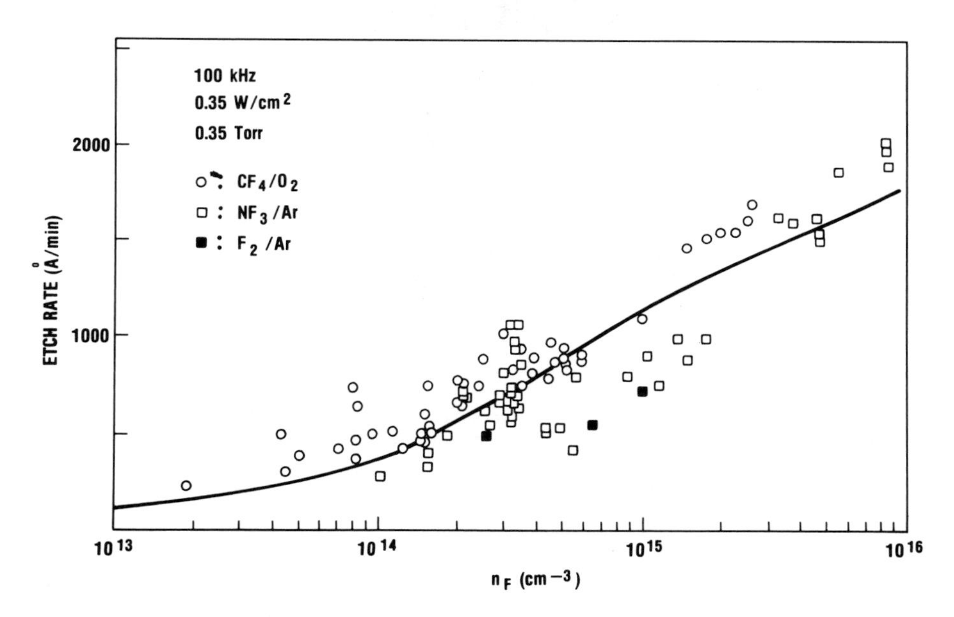

FIGURE 38. SiO_2 etch rates vs. fluorine atom concentration, n_F, in CF_4/O_2 plasmas (circles), NF_3/Ar plasmas (squares), and F_2/Ar plasmas (triangles) at 100 kHz, 0.35 W/cm^2, 100 sccm and 0.35 Torr total pressure. The solid points were achieved by loading the plasma with a silicon-covered lower electrode. The solid curve is a fit to Eqn. 63 (after [27]).

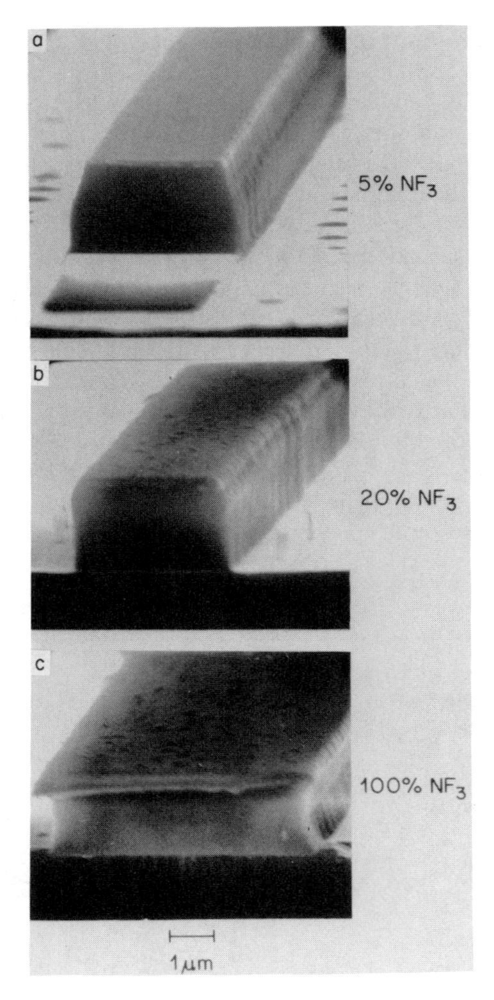

FIGURE 39. Profile in masked SiO_2 films etched at 100 kHz, 0.35 W/cm_3, and 0.35 torr using (a) 5% NF_3 in Ar, (b) 20% NF_3 in Ar, and (c) 100% NF_3. The large F-atom concentration in 100% NF_3 leads to fast chemical etching and a noticeable undercut. At 5% NF_3 extended etch times are required and features are tapered because of edge erosion of the mask (from [27]).

It is reasonably well-established that rapid anisotropic SiO_2 etching in F atom-rich plasmas is induced by ion damage, which makes the surface richer in silicon and more reactive [27, 29, 118, 119, 120]. Hence, the most apparent difference between the various feed gas mixtures that can be used lies in the concentration of atoms they make. Figure 38 shows SiO_2 etch rates in a low frequency plasma plotted against a three orders of magnitude change in F atom concentration. Although there is scatter, the points group about a "universal" curve which is largely independent of the feed gases and conditions (e.g., Si, F atom consuming electrodes) used to cover this range. In the presence of ion bombardment, the etch rate can be represented by [27]:

$$r_{SiO_2} = j_+\phi_s + \frac{n_F\bar{v}_F}{4}\epsilon_{F(SiO_2^*)}\Theta + \epsilon_{F(SiO_2)}(1 - \Theta) \tag{66}$$

where j_+ is the positive ion flux, ϕ_s is the sputtering efficiency, Θ is the fraction of the surface that has been sensitized by ion bombardment, $1 - \Theta$ is the unsensitized fraction that is chemically etched, and $n_F\bar{v}_F/4$ is the fluorine atom impingement rate. Increasing ion flux increases the sensitized fraction, Θ, while at large n_F, Θ is reduced in proportion to n_F as the faster-etching sensitized material is depleted relative to the unconditioned oxide surface. Thus, at the largest n_F in Fig. 38, Θ is small and the (second) chemical etching term makes a substantial contribution to the rate. This means that profiles should have an appreciable isotropic component under these conditions. On the other hand, if n_F is small (toward the left hand side of the figure), the first term representing purely physical sputtering will be a factor. The electron micrographs in Fig. 39 confirm these effects. In a pure NF_3 plasma, where high fluorine atom concentrations lead to fast chemical etching, there is significant undercutting (Fig. 39, top). The undercutting is eliminated at 20% NF_3/Ar because the rate of chemical etching (third term) relative to sensitized etching slows in proportion to the fluorine atom concentration. However, with too little NF_3, etching time must be prolonged to compensate for the reduced fluorine atom concentration and extended sputter-erosion of the mask leads to a tapered feature.

X. Silicon Nitride Etching

Silicon nitride is etched by both fluorine atom containing plasmas and CF_x/unsaturate-rich plasmas, and shows etching characteristics intermediate between those of silicon and SiO_2. Since "nitride" is widely used for oxidation masks, there is a demand for etching nitride over oxide and silicon with high selectivity. There are really two distinct kinds of "silicon nitride" in use: (1) thermal nitride is the stoichiometric compound Si_3N_4

that is usually grown by high temperature chemical vapor deposition (CVD); (2) plasma silicon nitride, sometimes called "plasma nitride" or p-Sinh [121], refers to a range of Si : N : H alloys which are plasma-deposited at low temperature ($\lesssim 400°C$). The etch rate of p-Sinh is usually much higher than that of thermal CVD nitride.

A. ISOTROPIC PROCESSES

Many chemistries that etch SiO_2 will also gasify Si_3N_4 and p-Sinh. CF_4/O_2 and fluorine atom-producing feeds such as NF_3/Ar are widely used to etch silicon nitride under high frequency, high pressure conditions, although these processes are neither selective (over silicon) nor anisotropic. The underlying F atom—Si_3N_4 reaction rates have not been measured, but some data hint that the intrinsic $Si_3N_4 : SiO_2$ F atom etch rate ratio may be as high as 5–10, depending on temperature [122, 123]. In CF_4/O_2 plasmas at $\sim 100°C$, the measured selectivity is only ~ 2–3 [124, 125]. Perhaps some selectivity loss is caused by atomic oxygen in the plasma (see Section IV), which may create slow-etching Si-O sites on the Si_3N_4 surface. Adding N_2 to CF_4/O_2 improves the selectivity for Si_3N_4 over oxide to about 10 : 1, but the mechanism is not understood. One speculation is that N atoms extract nitrogen from the Si_3N_4 surface, making its etch behavior more Si-like [123]. Another way to improve selectivity in CF_4/O_2 plasmas is to add CF_3Br to the feed [125], which can increase the selectivity to 10 or more as it depresses the etch rate. Since CF_3Br by itself tends to form an unsaturate-rich plasma, its main role is probably to lower the concentration of O atoms, but it could also convert some of the F atoms into a more selective etchant such as BrF or BrF_3 (see below). The apparent activation energy for etching nitride by F atoms is reportedly ~ 0.17 eV [122].

ClF_3 [126] (and by analogy, other interhalogen fluorides such as BrF_3 [47]) offer a better approach to selectively etching silicon nitride over silicon oxide. SiO_2 isn't attacked by these gases whereas Si_3N_4 and p-Sinh are gasified in *undischarged* ClF_3 even at room temperature [126]. This information has been exploited to achieve a completely selective etching process for Si_3N_4 on thin (100 Å) gate oxide. A feed mixture of NF_3 and Cl_2 ($\approx 1:2$) is discharged with etching done in a downstream reaction zone (this is a "CDE," or "remote" plasma chemical dry etching reactor). Although ClF was proposed as the active etchant—and this cannot be completely ruled out—it more likely is ClF_3 or a (synergistic) mixture of ClF and F atoms. The discharge effluent can supply ClF, ClF_3, Cl, Cl_2 (and possibly some F atoms), but ClF appears to be the least reactive of these, since it does not etch *n*-type (100) Si perceptibly, even at 100°C [47] (compare Cl reactions with n-type Si, Section VII.A).

B. ANISOTROPIC SILICON NITRIDE ETCHING

Unsaturated fluorine-deficient plasmas (in feed gases like CHF_3, C_2F_6, CHF_3/N_2O or CF_4/H_2) anisotropically etch both Si_3N_4 and SiO_2, with high selectivity over Si. Etching is done under conditions that favor energetic ion bombardment (e.g., low frequency or low pressure), and it is difficult to get much selectivity for Si_3N_4 (or *p*-Sinh) over SiO_2. These plasmas typically etch Si_3N_4 faster than SiO_2, but there are feed mixtures and etching conditions where SiO_2 etches at a higher rate. These shifts in selectivity are achieved by adjusting gas stoichiometry so that the CF_x surface film selectively thickens on the nitride (which has no oxygen to help gasify the CF_x layer), and possibly by operating with low ion bombardment energies near the thresholds where relative rates change rapidly (see Section III.B).

Conditions that parallel anisotropic ion-enhanced silicon oxide etching by atomic fluorine have not been explored for nitride etching.

XI. Oxygen Plasma Etching of Resists

Oxygen plasmas are used to etch resists both isotropically and anisotropically. Isotropic resist stripping with O atoms is one of the oldest plasma applications, while anisotropic pattern transfer was developed to permit micron- and submicron pattern transfer using the "trilevel" process [127] (see Chapter 5). Oxygen chemical treatment is also used to passivate metal- and silicon-bearing resists for bilevel lithography and to remove epoxy smears from holes in multilayer printed circuit boards [128].

At moderate pressure and high frequency, a discharge in pure oxygen produces oxygen atoms. These atoms attack organic materials to form CO, CO_2 and H_2O as the final end products [12]. The main degradation mechanism appears to be random chain scission [129]. Stripping rates in pure oxygen plasmas are proportional to oxygen atom concentration [130, 131]. Inert gas additions can help stabilize O_2 plasmas. Chemical attack of resist in oxygen plasmas is heavily influenced by the structure and substitutional groups on the polymer, and by physical variables such as temperature. The temperature dependence of attack in an oxygen plasma may also depend on the resist's glass transition temperature (T_g). For example, the apparent activation energy of poly(methyl methacrylate) (PMMA) etching in O_2 plasmas increased sharply from 4.4 Kcal/mole below the glass transition temperature, T_g (60–90°C), to ~ 9.7 Kcal/mole above T_g [132]. Activation energies between ~ 5–15 Kcal have been measured for resist oxidation under chemical attack by O atoms (with O_2 present), depending on the particulars of the resist and other physical conditions.

Stripping rates are enhanced when fluorinated gases are added to an oxygen plasma. Even a few percent of C_2F_6 added to O_2 can bring about a large increase in the removal rate for many resists. The enhancement seems to be caused by two effects [130, 133]: (1) fluorine atom reactions produce reactive sites on the polymer backbone, and (2) small amounts of fluorine increase the concentration of atomic oxygen in an oxygen plasma [134, 135, 136, 137]. Atomic fluorine *abstracts* hydrogen from organic polymers, leaving unsaturated or radical sites. These sites are much more susceptible to oxygen than a saturated polymer chain. When oxygen attacks these sites there is chain scission, carbonyl groups are formed, and the polymer is "burned" into volatile oxidation products. The initial steps in fluorine-assisted degradation are probably similar to

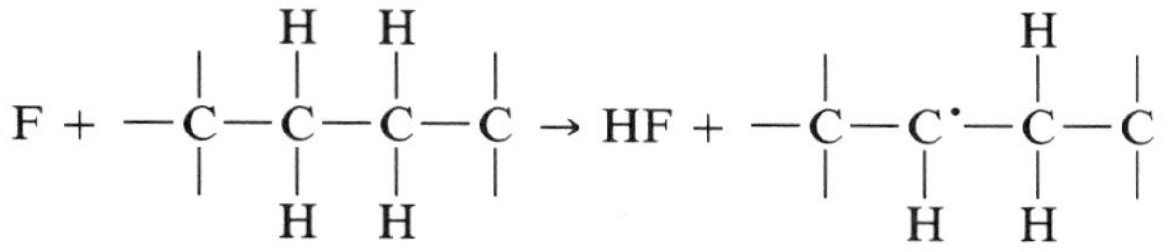

$$\mathrm{F} + \begin{array}{cccc} & \mathrm{H} & \mathrm{H} & \\ | & | & | & | \\ -\mathrm{C}- & \mathrm{C}- & \mathrm{C}- & \mathrm{C} \\ | & | & | & | \\ & \mathrm{H} & \mathrm{H} & \end{array} \rightarrow \mathrm{HF} + \begin{array}{cccc} & & \mathrm{H} & \\ | & & | & | \\ -\mathrm{C}- & \mathrm{C}^{\bullet}- & \mathrm{C}- & \mathrm{C} \\ | & | & | & | \\ & \mathrm{H} & \mathrm{H} & \end{array} \tag{67}$$

followed by

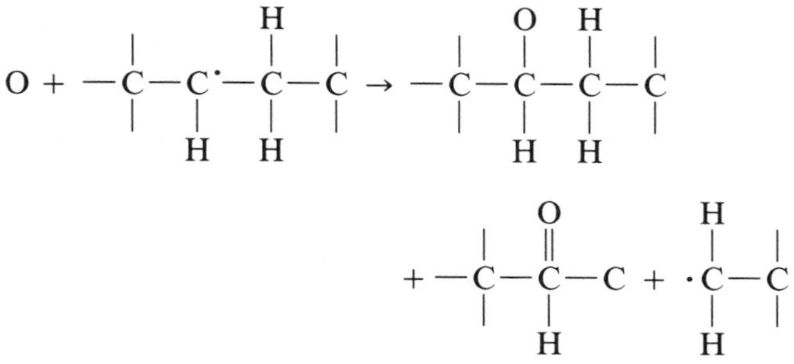

$$\mathrm{O} + \begin{array}{cccc} & & \mathrm{H} & \\ | & & | & | \\ -\mathrm{C}- & \mathrm{C}^{\bullet}- & \mathrm{C}- & \mathrm{C} \\ | & | & | & | \\ & \mathrm{H} & \mathrm{H} & \end{array} \rightarrow \begin{array}{cccc} & \mathrm{O} & \mathrm{H} & \\ | & | & | & | \\ -\mathrm{C}- & \mathrm{C}- & \mathrm{C}- & \mathrm{C} \\ | & | & | & | \\ & \mathrm{H} & \mathrm{H} & \end{array}$$

$$+ \begin{array}{ccc} & \mathrm{O} & \\ | & \| & \\ -\mathrm{C}- & \mathrm{C}- & \mathrm{C} \\ | & | & \\ & \mathrm{H} & \end{array} + \begin{array}{cc} \mathrm{H} & \\ | & | \\ \cdot\mathrm{C}- & \mathrm{C} \\ | & | \\ \mathrm{H} & \end{array} \tag{68}$$

The energy liberated when fluorine abstracts hydrogen can cleave carbon-carbon bonds [138] and is probably a primary degradation mechanism in plasmas containing a high concentration of fluorine (e.g., CF_4/O_2 or SF_6/O_2) [139].

With polymers that contain unsaturated backbone groups, fluorine addition reactions may contribute to the degradation:

$$\mathrm{F} + -\overset{|}{\underset{|}{\mathrm{C}}}-\overset{\mathrm{H}}{\overset{|}{\mathrm{C}}}=\underset{\underset{\mathrm{H}}{|}}{\mathrm{C}}-\overset{|}{\underset{|}{\mathrm{C}}} \rightarrow -\overset{|}{\underset{|}{\mathrm{C}}}-\underset{\underset{\mathrm{H}}{|}}{\mathrm{C}^{\bullet}}-\overset{\mathrm{F}}{\overset{|}{\underset{\underset{\mathrm{H}}{|}}{\mathrm{C}}}}-\overset{|}{\underset{|}{\mathrm{C}}} \tag{69}$$

The radical sites are then oxidized as before. Curiously, there is a shift from an oxygen-dominant etch mechanism to fluorine-dominant degradation in the etching of polyimide by SF_6/O_2. This leads to two maxima in the etch rate, one at low oxygen additions (at ~ 20% O_2) and one for large additions (at ~ 60% O_2) [139].

The fluorine also tends to lower the activation energy. For example, the apparent activation energy for stripping AZ340B resist decreases from about 11 Kcal/mole in a pure O_2 plasma to 5 Kcal/mole using 1% CF_4 in O_2 [140]. For plasma etching epoxy printed circuit boards in 50% CF_4 in O_2, the Arrhenius coefficient is 4.3 ± 0.2 Kcal/mole.

Two mechanisms have been proposed to explain the increase in atomic oxygen and both may play a role. Normally, recombination of O atoms to reform O_2 on reactor walls is a dominant loss process for O atoms. It is known that F_2 lowers the rate of Cl recombination on walls [141], and it seems likely that fluorine may similarly reduce the rate of O atom wall loss. A second proposal is that the added halocarbon changes the electron energy distribution function [103] in a way that favors O production. In any case, the increased oxygen atom concentration is apparent from the intensity of optical emission from excited O atoms [103, 105] and has been confirmed by laser fluorescence measurements [136].

In purely chemical etching or stripping of resist, plasma exposure may be undesirable since impinging ions can cause electrical damage to devices and charging from the plasma-induced potential can cause electrostatic breakdown of thin gate oxides (see Chapter 1). Downstream plasma strippers are increasingly used to circumvent these problems.

Directional, ion-assisted, resist etching is also in wide use, especially for multilevel resist processing. The "trilevel" process is the oldest and probably best known. Conventional resist is deposited and patterned over a thin SiO_2 layer, and the resist pattern is transferred to the SiO_2 using a CHF_3 based plasma. In turn, the patterned SiO_2 is used as a mask to etch a high aspect ratio pattern anisotropically in a *thick* polyimide or conventional resist layer using an oxygen plasma. This layer becomes a highly resistant fine-featured mask for further plasma etching steps. Ion bombardment flux stimulates the vertical etch rate of horizontal polymer surfaces by oxygen.

To achieve high aspect ratios, the process is done under plasma conditions that provide ion bombardment with low gas phase oxygen atom concentrations—at low pressure (around a few mTorr) or at higher pressure (~ 0.1 Torr) using low frequency (e.g., ~ 200 KHz–1 MHz) and high dilution with an inert gas (usually argon). It has been proposed that ion bombardment assists etching by driving off H and partially oxidized carbon groups in the polymer (C—O, C=O, C—O—C) as CO [142] and that sputtering of the remaining C-rich surface may be rate limiting. While details of the ion-assisted mechanism are uncertain, clearly the gas phase concentration of O must be low relative to the flux of ions to avoid undercutting by purely chemical reaction.

Recently, bilevel lithography schemes using silicon or metal containing organic resists as the first layer have been studied as a means to simplify making the thick mask. The thin metal-containing resist is deposited directly on the thick polymer layer rather than on an intervening layer of SiO_2. The basic idea is for the oxygen plasma to form a protective oxide skin on the metal (silicon) containing mask layer that will slow or stop further erosion of the mask. Problems with developing this scheme are that passivation cases shrinkage and linewidth loss in the metal-containing resist, and the ion bombardment needed for anisotropic etching of the underlying thick polymer level erodes the mask passivation. Increasing the concentration of oxygen atoms favors growth of the passivation layer, but oxygen atoms also undercut the thick polymer layer by chemical etching.

XII. III–V Etching Chemistries and Mechanisms

A. DEVICES AND CHEMISTRIES

Technologies based on III-V compound semiconductor devices, made of GaAs, AlGaAs, GaP, InP, InGaAsP and related alloys, have become increasingly important. Techniques for etching these materials are sometimes complicated by the need to penetrate multiple layers of binary, ternary and even quaternary alloys in a single structure. High selectivity between related compounds is required at times, yet in other cases it is necessary to etch different III-V materials at the same rate. For example, laser structures in $GaAs/Al_xGa_{1-x}As$ require smooth, anisotropic and non-selective etching, while fabrication of heterostructure transistors demands highly selective etching of GaAs over AlGaAs. Group III fluorides are involatile at practical processing temperatures, but the chlorides and bromides have adequate volatility. For the most part, chlorine-containing plasmas have been used, but bromine and iodine act similarly as etchants.

H_2 plasmas etch some of the III-V compounds slowly and are effective for removing native oxides and carbonaceous contaminants from III-V surfaces. A new chemistry based on H_2/hydrocarbon mixtures appears to etch GaAs and InP at high rates with good surface morphology.

B. CHLORINE AND BROMINE ETCHING OF III-V COMPOUND SEMICONDUCTORS

The group V halides (e.g., $InCl_3$,$GaCl_3$), have high vapor pressures while the group III halides are much less volatile and have relatively high latent heats of vaporization, ~ 37 Kcal/mole for $InCl_3$ and 12 Kcal/mole for $GaCl_3$. Removing these group III halide products (e.g., $InCl_3$, $GaCl_3$) from the surface is often rate limiting, so these substrates are commonly heated to increase product vapor pressure (chemical InP etching usually requires substrate temperatures above ~ 150°C). Rate limiting product volatility means that the activation energy for etching, E_A, equals the heat of sublimation. Elevated temperature can also improve the morphology of etched III-V substrates—(see Section III.E.2).

Etching in moderate pressure Cl_2 plasmas ($\gtrsim$ 0.1 Torr) is generally chemical, and both GaAs and InP etching rates show an Arrhenius type dependence with activations energies close to the heats of sublimation [22]. Because of the high heat of sublimation, InP etching in a 250 kHz Cl_2 plasma at 0.3 Torr increases by more than two orders of magnitude, from ~ 1 μ/min, to ~ 100 μ/min as the substrate temperature is raised from 200°C to 300°C [143] (Fig. 40). Etch rates and activation energies for etching GaAs at 13 MHz also are shown in Fig. 40. As with Si etching (Chapter 1), we can enumerate the etching process as a series of sequential steps:

(I) ion and electron formation (which balance loss of charged species and are part of a more complex reaction set)

$$e + Cl_2 \rightarrow Cl_2^+ + 2e, \tag{70}$$

(II) etchant formation

$$e + Cl_2 \rightarrow 2Cl + e, \tag{71}$$

(III) adsorption of etchant on the substrate

$$\left\{\begin{matrix} Cl \\ Cl_2 \end{matrix}\right. \rightarrow InP_{surf} - nCl, \tag{72}$$

(IVa) chemical reaction to form product

$$InP - nCl \rightarrow \left\{\begin{matrix} InCl_{x(ads)} \\ PCl_{y(ads)} \end{matrix}\right. \tag{73a}$$

(IVb) ion enhanced etching to form product

$$InP - nCl \xrightarrow{(ions)} \begin{cases} InCl_{x(ads)} \\ PCl_{y(ads)} \end{cases} \tag{73b}$$

and, product desorption either by evaporation

$$\begin{aligned} InCl_{x(ads)} &\rightarrow InCl_{x(gas)}, \\ PCl_{y(ads)} &\rightarrow PCl_{y(gas)}, \end{aligned} \tag{74a}$$

or by ion-assisted desorption

$$\begin{matrix} InCl_{x(ads)} \\ PCl_{y(ads)} \end{matrix} \xrightarrow{(ions)} \begin{matrix} InCl_{x(gas)} \\ PCl_{y(gas)} \end{matrix}, \tag{74b}$$

where the slowest step, usually removal of the group III halide product, will limit the etching rate.

The relative importance of the spontaneous (Eqns. 73a, 74a) and ion-assisted (Eqns. 73b, 74b) reactions depends on the etchant flux and substrate. At moderate pressure (~ 0.3 Torr), etching in chlorine and bromine plasmas is rapid, spontaneous and crystallographic. However, at low pressure with intense ion bombardment (~ 0.001–0.01 Torr), ion-enhanced reactions are more important.

Surprisingly, GaAs etching appears to be controlled by the same factors. As with InP, the latent heat of vaporization for $GaCl_3$ is close to the apparent activation energy for etching [22] (see Fig. 41), but the vapor pressure reported for $GaCl_3$ is so high (10.4 Torr at 77.8°C [144]) that the product evaporation rate should be orders of magnitude larger than Cl_2 plasma GaAs etch rates [22] and therefore not limiting. Initially the agreement seemed fortuitous. However, recent experiments [145] show that $GaCl_3$ *does* accumulate on GaAs surfaces and limit its etch rate, just as $InCl_3$ limits InP etching. So either (1) the published vapor pressure data are in error, or (2) the volatility of *thin* layers of $GaCl_3$ is anomalously low.

Halogen and halogen atom etching of III-V's differ from most other examples of chemical plasma etching in that the etching is crystallographically selective. The relative rate at which GaAs crystal planes etch in Cl_2 and Br_2 plasmas is $[111]_B > [100] > [110] > [111]_A$, with the exact ratios depending on substrate temperature and plasma composition. The subscript "B" denotes As-rich [111] planes while subscript "A" are the Ga-rich planes. Figure 41 shows the result of etching of [100] plane of GaAs masked by an SiO_2 rectangle oriented along the [110] cleavage planes. The $[111]_A$ facets form because they are the slowest etching planes, but disappear as etching proceeds (not shown) because of the converging {100} planes. Rates are high in a pure Br_2 plasma. Etch rates in the $\langle 100 \rangle$ direction are 20–80 $\mu m/min$, which correspond to ~ 0.1 Ga or As atoms removed per

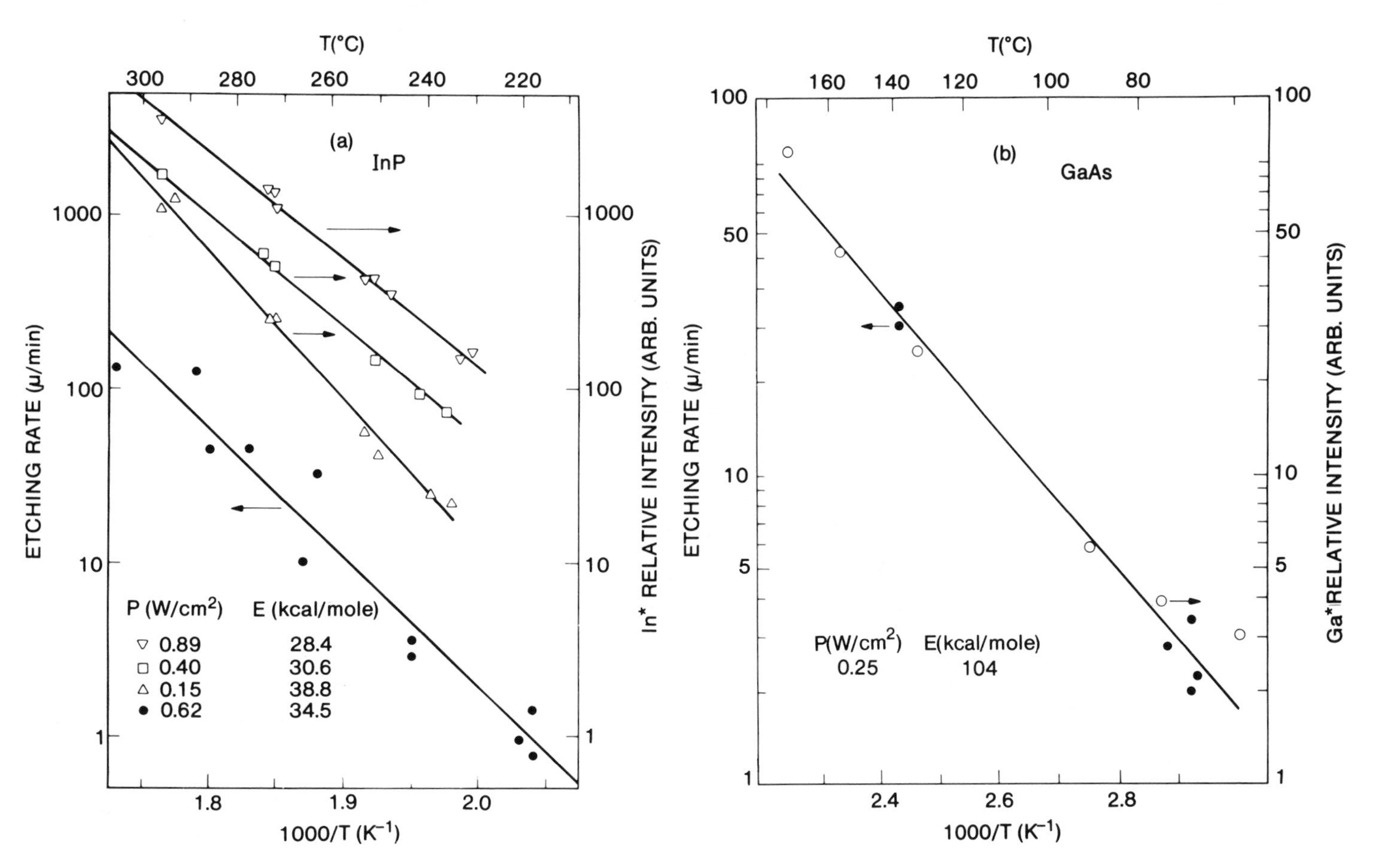

FIGURE 40. Arrhenius plot of the etch rate and plasma-induced group III emission for (a) InP in a 0.3 Torr Cl_2 plasma at 250 kHz (left), and (b) GaAs etching under the same conditions with 13 MHz excitation (right). Since the product concentration in the gas phase is low, emission intensity is nearly proportional to the etch rates (a is from [22]).

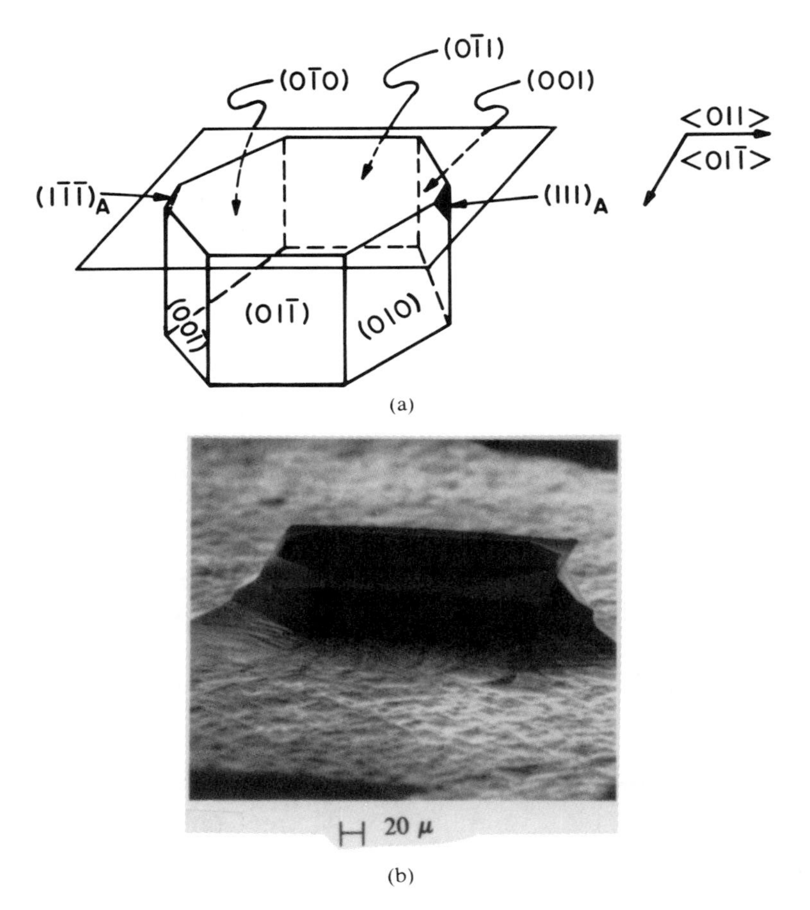

(a)

(b)

FIGURE 41. a) Schematic of the crystallographic planes that are exposed when the (100) surface of GaAs is etched in a Cl_2 or Br_2 plasma under a mask aligned along the ⟨011⟩ directions. The relative etch rates are $(111)_B > (100) > (110) > (111)_A$ where the "B" planes are As-rich and the "A" planes are Ga-rich (adapted from [160]). b) A (100) GaAs surface etched crystallographically in a low power, 0.3 Torr, Br_2 plasma under a square SiO_2 mask oriented along the ⟨110⟩ and ⟨111⟩ directions.

incident Br atom. This etch rate is so high that it dominates ion-bombardment induced etchingwell into the low frequency regime.

However, ion bombardment does seem to play a role in InP etching at low frequency and moderate pressure (~ 0.3 Torr). This process shows the characteristic "signature" of an energetic ion enhanced process—the etch rate increases rapidly as frequency is lowered from 14 MHz to below the ion transit frequency (250 kHz). Yet the etch rate shows a high activation

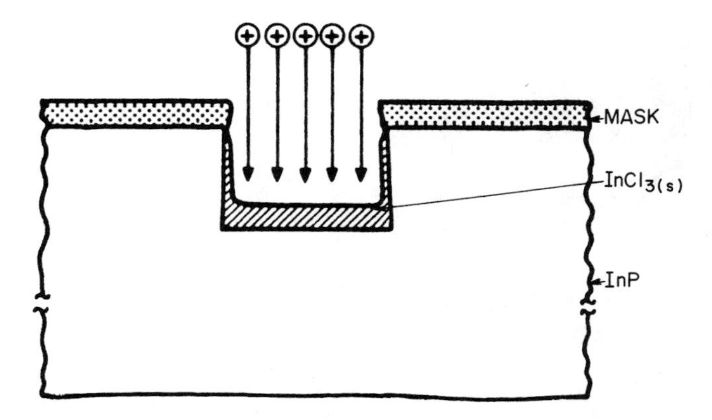

FIGURE 42. In ion-assisted anisotropic etching of InP by Cl_2 plasmas, the etch rate is limited by volatilization of an $InCl_3$ product layer (see text).

energy, which is uncommon for energetic ion enhanced etching, and sputter Auger analysis shows that a multilayer film of $InCl_3$ is present during etching. One proposed explanation is that "damage" by ions penetrating the $InCl_3$ film induce reaction. The $InCl_{3(s)}$ layer will thicken, attenuating ion bombardment until the ion-assisted $InCl_{3(s)}$ formation rate balances evaporation. Clearly, the details of energetic ion-enhanced etching mechanisms may vary widely from system to system.

Since the ratio of ion to neutral flux increases with decreasing pressure, a point finally is reached where ion induced processes overwhelm purely chemical etching. Energetic ion-enhanced III-V etching ordinarily becomes dominant below $\lesssim$ 10 mTorr [146, 147], where anisotropic etching *without* undercutting can be achieved independent of crystallographic orientation. A disadvantage of operating in this regime is that ion induced damage can degrade device structures.

BCl_3 or chlorocarbons (CCl_4, $CHCl_3$ CCl_2F_2) are often added to chlorine discharges because they form species which remove surface oxides and promote anisotropy by the inhibitor mechanism. Oxygen additions are sometimes used to control polymerization and regulate the balance between unsaturated chlorocarbons and Cl. Since oxygen also increases the concentration of Cl in chlorocarbon plasmas (see Section VIII), it is sometimes added to chlorocarbons in place of Cl_2 to increase the etching rate.

Removing the Ga and In chlorides or bromides from the surface is the bottleneck for chemical and low pressure ion-enhanced etching alike. This is still true when ternary or quaternary aluminum-containing alloys are etched because aluminum chloride is more volatile than these heavier compounds. It follows that chlorine or bromine plasmas etch GaAs and

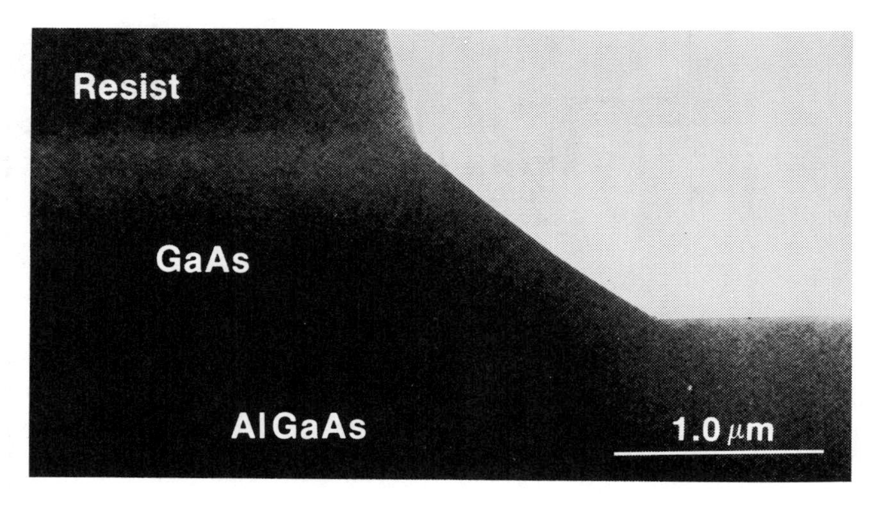

FIGURE 43. This 1 μm width space was faithfully transferred to form a GaAs trench on GaAlAs by crystallographic etching in bromine with controlled water vapor additions. The photoresist mask on the GaAs is marked by the steep vertical slope, while the sidewall of the GaAs layer is bounded by a slow-etching $(111)_A$ plane that is sloped at $\approx 54°$. Etching stops at the horizontal GaAlAs interface. Similar results are obtained in Cl_2 and Br_2 plasmas.

AlGaAs at nearly equal rates, providing the involatile aluminum oxides and fluorides are not allowed to form. To prevent this, the reactor and feed systems must be kept rigorously free of oxygen, moisture and fluorinated species. Conversely, if selective etching is desired, small, controlled amounts of water (or equivalently H_2 and O_2 which react to form water) or fluorinated gases such as CCl_2F_2 may be added to intentionally grow a protective aluminum-containing skin on the surface of AlGaAs (for instance) to selectively etch GaAs. This type of process is illustrated by Fig. 43, where a submicron line was chemically etched through GaAs over a AlGaAs sublayer, with a Br_2 feed to which a controlled partial pressure of H_2O had been added. The mask was oriented along the $\langle 0\bar{1}1 \rangle$ GaAs direction so that the slow etching Ga-rich planes "formed" on the sides of the GaAs until the GaAlAs level was reached where water passivation made the etching stop.

C. HYDROGEN / HYDROCARBON ETCHING OF III-V'S

Hydrogen plasma III-V etching was reported some years ago [48], but there were difficulties in obtaining adequate rates, surface morphology and reproducibility [149], and the underlying chemical processes were poorly under-

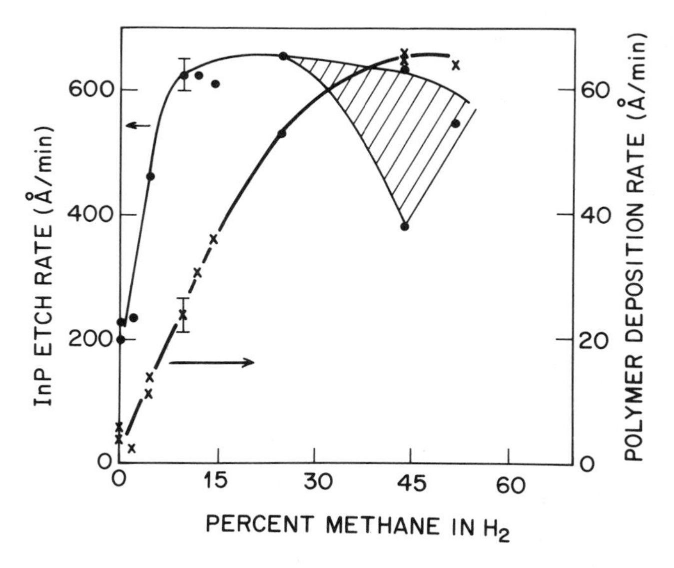

FIGURE 44. InP etch rate and polymer deposition rate on a silicon surface as a function of feed composition during anisotropic etching of InP in CH_4/H_2 mixtures at 90 mTorr (parallel plate reactor, 13.6 MHz, 0.6 W/cm^2, 230 sccm total flow rate, from [152]). Beyond the plateau at ~ 25% CH_4 the etch rate becomes highly variable (shown by the shaded area) and surface hydrocarbon levels on the InP increase.

stood. However, a hydrocarbon chemistry in which 5–25% CH_4 is added to a H_2 plasma appears to offer high etch rates, anisotropic etching and less surface damage [150, 151, 152].

An understanding of the chemistry is still evolving, but there seems to be an analogy with SiO_2 etching in unsaturated plasmas (Section IX). Both CF_4/H_2 and CF_4/H_2 plasmas form polymerizing radicals. Fluorocarbon films are deposited when SiO_2 etches in CF_4/H_2, while hydrocarbon films accompany InP etching in CH_4/H_2. Figure 44 shows the InP etch rate in this system with the rate of film deposition on an unetchable surface (Si) as a function of feed composition. With up to 30% CH_4 no polymer is detected on InP. Beyond this point, as the feed is made richer in methane, some polymer forms on the InP as its etch rate decreases slightly [152]. Apparently ion bombardment stimulates reaction between adsorbed CH_x and the GaAs or InP substrate. The substrate elements are likely to be volatilized as group III organometallic compounds (e.g., $Ga(CH_3)_x$, $In(CH_3)_x$ and group V hydrides (i.e., PH_x, AsH_x), since Ga and In hydrides are unstable. This mechanism suggests that other hydrocarbon feeds such as C_2H_6 will also work and might be used in combination with oxidant additions to regulate CH_x film growth and etching (see Section VIII).

Acknowledgment

I gratefully acknowledge help and encouragement from my close colleagues, Dale E. Ibbotson, John A. Mucha, and Vincent M. Donnelly who worked with me and made this chapter possible. Equally, I thank my wife Dr. Lois E. Flamm for her editorial hand and support.

References

[1] J. W. Coburn and H. F. Winters, *J. Appl. Phys.* **50**, 3189 (1979).
[2] Y. Y. Tu, T. J. Chuang, and H. F. Winters, *Phys. Rev. B.* **23**, 823 (1981).
[3] U. Gerlach-Meyer and J. W. Coburn, *Surf. Sci.* **103**, 177 (1981).
[4] T. Mizutani, C. J. Dale, W. K. Chu, and T. M. Mayer, *Nucl. Instr. Methods B* **7/8**, 825 (1985).
[5] A. W. Kolfschoten, R. A. Haring, A. Haring, and A. E. de Vries, *J. Appl. Phys.* **55**, 3813 (1984).
[6] E. L. Barish, D. J. Vitkavage, and T. M. Mayer, *J. Appl. Phys.* **57**, 1336 (1985).
[7] H. P. Gillis and W. J. Gignac, *J. Vac. Sci. Technol.*, **A4**, 696 (1986).
[8] D. J. Oostra, A. Haring, R. P. van Ingen, and A. E. de Vries, *J. Appl. Phys.*, submitted for publication, 1987.
[9] T. M. Mayer and R. A. Barker, *J. Vac. Sci. Technol.* **21**, 757 (1982).
[10] N. A. Takasaki, E. Ikawa, and Y. Kurogi, *J. Vac. Sci. Technol. B* **4**, 806 (1986).
[11] V. M. Donnelly, D. L. Flamm, and R. H. Bruce, *J. Appl. Phys.* **58**, 2135 (1985).
[12] S. E. Bernacki and B. B. Kosicki, *J. Electrochem. Soc.* **131**, 1926 (1984).
[13] M. Oda and K. Hirata, *Jpn. J. Appl. Phys.* **19**, L405 (1980).
[14] R. H. Bruce and G. P. Malafsky, *J. Electrochem. Soc.* **130**, 1369 (1983).
[15] C. J. Tracy and R. Mattox, *Solid State Technol.* **25**(5), 83 (June, 1982).
[16] C. M. Melliar-Smith and C. J. Mogab, "Thin Film Processes." J. L. Vossen and W. Kern, eds. Academic, Orlando, (1978), pps. 521–525.
[17] D. L. Flamm and G. K. Herb, in "Plasma Etching: An Introduction." D. M. Manos and D. L. Flamm, eds. Academic, New York (1989) Chap. 1.
[18] D. L. Flamm, *J. Vac. Sci. Technol.* **A4**, 729 (1986).
[19] A. T. Bell, in "Techniques and Applications of Plasma Chemistry." J. R. Hollahan and A. T. Bell, eds. John Wiley & Sons, New York (1974), Chap. 1.
[20] A. M. Howatson, in "An Introduction to Gas Discharges." Pergamon Press, New York (1965), Chap. 4.
[21] W. P. Allis and D. J. Rose, *Phys. Rev.* **93**, 84 (1964).
[22] (a) V. M. Donnelly, D. L. Flamm, C. W. Tu, and D. E. Ibbotson, *J. Electrochem. Soc.* **129**, 2533 (1982); (b) D. E. Ibbotson, D. L. Flamm, V. M. Donnelly, and B. S. Duncan, *J. Vac. Sci. Technol.* **20**, 489 (1982); (c) D. E. Ibbotson, V. M. Donnelly, and D. L. Flamm, *Extended Abstracts*, **81-2**, 650 (1981).
[23] J. H. Ha, E. A. Ogryzlo, and S. Polypronopolos, *Unpublished results*, (1988).
[24] H. Yamada, H. Ito, and H. Inaba, *J. Vac. Sci. Technol.* **B3**, 884 (1985).
[25] J. Z. Li, I Adesida, and E. D. Wolf, *J. Vac. Sci. Technol.* **B3**, 406 (1985).
[26] E. L. Hu and R. E. Howard, *Appl. Phys. Lett.* **37**, 1022 (1980).
[27] V. M. Donnelly, D. L. Flamm, and W. C. Dautremont-Smith, D. J. Werder, *J. Appl. Phys.* **55**, 242 (1984).
[28] L. J. Overzet, "Investigation of RF Discharges in Silane/He Mixtures and α-Si:H Deposition," presented at AT & T Bell Laboratories, Murray Hill, New Jersey (April 29, 1988).

[29] V. M. Donnelly, D. E. Ibbotson, and D. L. Flamm, in "Ion Bombardment Modification of Surfaces: Fundamentals and Applications." O. Auciello and R. Kelley, eds. Elsevier, New York (1984), Chap. 8.
[30] D. L. Flamm, R. F. Baddour, and E. R. Gilliland, *I & EC Fundamentals* **12**, 276 (1973).
[31] D. L. Flamm and V. M. Donnelly, *J. Appl. Phys.* **59**, 1052 (1986).
[32] R. M. Barnes and R. J. Winslow, *J. Phys. Chem.* **82**, 1869 (1978).
[33] G. de Rosny, E. R. Mosberg, Jr., J. R. Adelson, G. Devaud, and R. C. Kern, *J. Appl. Phys.* **54**, 2272 (1983).
[34] H. Tagashira, K. Kitamori, M. Shimozuma, and Y. Sakai, in "Proc. 7th Intl. Symp. on Plasma Chem." C. J. Timmermans, ed. Eindhoven Univ. Technol., Eindhoven, The Netherlands (1985), p. 1337.
[35] T. Makabe, in "Proc. 7th Intl. Symp. on Plasma Chem." C. J. Timmermans, ed. Eindhoven Univ. Technol., Eindhoven, The Netherlands (1985), p. 1331.
[36] (a) B. E. Cherrington, in "Gaseous Electronics and Gas Lasers." Pergamon, New York (1979) pps. 130–133; (b) *ibid.* pps. 156–158.
[37] P. C. Nordine and J. D. LeGrange, *AIAA Journal* **14**, 644 (1976).
[38] D. A. Winborne and P. C. Nordine, *AIAA Journal* **14**, 1488 (1976).
[39] D. L. Flamm, C. J. Mogab, and E. R. Sklaver, *J. Appl. Phys.* **50**, 6211 (1979).
[40] D. L. Flamm, V. M. Donnelly, and J. A. Mucha, *J. Appl. Phys.* **52**, 3633 (1981).
[41] While this statement is literally true, usually the percentage of etchant decreases with pressure owing to rapid gas phase recombination reactions (typified by the second entry, Table 3). The reaction of CF_3 with F is in its high pressure limit[59] above 0.1–0.3 Torr where it conforms to second-order kinetics.
[42] The kinetics of polymerization are considerably more complex than this discussion implies. Although the sequential reactions $CF_2 + CF_2 \rightarrow C_2F_4 + CF_2 \rightarrow C_3F_8$ are third order in CF_2 concentration when CF_2 conversion is low, a complete set of reactions to simulate plasma polymerization kinetics would have to include branching, termination, and reactivation reactions. Also, equations for a steady state open system (CSTR) would be more appropriate for modelling a plasma reactor. Using the simplistic reactions given, chain growth in a CSTR would be high order only if first-order loss processes dominated CF_2 consumption. Experimental evidence and recent models seem to suggest that plasma polymer deposition rates and the gas species chain lengths increase with pressure[44,45,46] when reactant supply and initiation are not limiting. However, there are few data and these effects have not been studied experimentally in a realistic etchant gas.
[43] K. Nakajima, A. T. Bell, and M. Shen, *J. Appl. Poly. Sci.* **23**, 2627 (1979).
[44] H. Yasuda, "Plasma Polymerization." Academic Press, Orlando (1985), pp. 286–287.
[45] R. d'Agostino, F. Cramarossa, and F. Illuzzi, *J. Appl. Phys.* **61**, 2754 (1987); R. d'Agostino, F. Cramarossa, V. Colaprico, and R. d'Ettole, *J. Appl. Phys.* **54**, 1284 (1983); R. d'Agostino, S. De Benedictis, and F. Cramarossa, *Plasma Chem. Plasma Proc.* **4**, 1 (1984).
[46] D.L. Flamm, D. E. Ibbotson, J. A. Mucha, and V. M. Donnelly, *Solid State Technol.* **24**(4), 117 (1983).
[47] D. E. Ibbotson, J. A. Mucha, D. L. Flamm, and J. M. Cook, *J. Appl. Phys.* **56**, 2939 (1984).
[48] D. A. Winborne and P. C. Nordine, *AIAA Journal* **14**, 1488 (1976).
[49] P. C. Nordine and J. D. LeGrange, *AIAA Journal.* **14**, 644 (1976).
[50] P. S. Ganguli and M. Kaufman, *Chem. Phys. Lett.* **25**, 221 (1974).
[51] K. G. Emeleus, R. W. Lunt, and C. A. Meek, *Proc. Roy. Soc. A*, **156**, 394 (1936).
[52] R. W. Lunt and C. A. Meek, *Proc. Roy. Soc. A*, **157**, 146 (1936).
[53] (a) B. Eliasson and U. Kogelschatz, *J. Phys. B: At. Mol. Phys.*, **19**, 1241 (1986); (b) A. T. Bell and K. Kwong, *Ind. Eng. Chem. Fund.* **12**, 90 (1973).

[54] S. R. Hunter and L. G. Christophorou, *J. Chem. Phys.* **80**, 6150 (1984); L. G. Christophorou, D. R. James, R. A. Mathis, I. Sauers, S. R. Hunter, M. O. Pace, D. W. Bouldin, S. M. Spyrou, A. Fatheddin, V. K. Lakdawala, J. L. Adcock, C. E. Easterly, and G. D. Griffin, "Gaseous Dielectrics Research and Applications, Semiannual Report," ORNL/TM-8368, Oak Ridge National Laboratory, Oak Ridge, July, 1982.

[55] L. G. Christophorou, D. L. McCorkle, and A. A. Christodoulides, in "Electron Molecule Interactions and Their Applications." Vol. 1, "Electron Attachment Processes," p. 477 L. G. Christophorou, ed. Academic Press, Orlando (1984); L. G. Christophorou, R. A. Mathis, S. R. Hunter, and J. G. Carter, "Effect of Temperature on the Uniform Field Breakdown Strength of Electronegative Gases," in "Gaseous Dielectrics V." ed. L. G. Christophorou and D. W. Bouldin, Pergamon Press, New York (1987) p. 88; S. M. Spyrou and L. G. Christophorou, *J. Chem. Phys.* **6**, 2620 (1985).

[56] H. M. Rosenstock, K. Draxl, B. W. Steiner, and J. T. Herron, "Energetics of Gaseous Ions," *J. Phys. Chem. Ref. Data* **6**, Suppl. No. 1, Amer. Chem. Soc., Amer. Inst. of Phys., and U.S. Nat'l Bur. Stds, (1977).

[57] M. S. Naidu and A. N. Prasad, *J. Phys. D.* **5**, 983 (1972).

[58] I. C. Plumb and K. R. Ryan, *Plasma Chem. Proc.* **6**, 205 (1986).

[59] D. Edelson and D. L. Flamm, *J. Appl. Phys.* **56**, 1522 (1984).

[60] C. J. Mogab, A. C. Adams, and D. L. Flamm, *J. Appl. Phys.* **49**, 3796 (1979).

[61] G. Smolinsky and D. L. Flamm, *J. Appl. Phys.* **50**, 4982 (1979).

[62] H. Boyd and M. S. Tang, *Solid State Technol.* **25**(4), 133 (1979).

[63] R. d'Agostino and D. L. Flamm, *J. Appl. Phys.* **52**, 162 (1981).

[64] D. L. Flamm, D. N. K. Wang, and D. Maydan, *J. Electrochem. Soc.* **129**, 2755 (1982).

[65] D. L. Flamm, D. Maydan, and D. N. K. Wang, U.S. Patent No. 4,310,380 (January 12, 1982).

[66] C. J. Mogab, *J. Electrochem. Soc.* **124**, 1262 (1977).

[67] C. J. Mogab and H. J. Levenstein, *J. Vac. Sci. Technol.* **17**, 721 (1980).

[68] D. W. Hess, *Solid State Technol.* **24**(4), 189 (1981); K. Tokunaga and D. W. Hess, *J. Electrochem. Soc.* **127**, 928 (1980).

[69] K. Tokunaga, F. C. Redeker, D. A. Danner, and D. W. Hess, *J. Electrochem. Soc.* **128**, 851 (1981).

[70] K. Tokunaga and D. W. Hess, *J. Electrochem. Soc.* **127**, 928 (1980).

[71] L. A. D'Asaro, A. D. Butherus, J. V. DiLorenzo, D. E. Iglesias, and S. H. Wample, in "Gallium Arsenide and Related Compounds 1980." Conf. Ser. No. 56, H. W. Thim, ed. Inst. of Physics, Bristol, England (1980), pp. 267–273.

[72] H. Nakata, K. Nishioka, and H. Abe, *J. Vac. Sci. Technol.* **17**, 1351 (1980).

[73] T. Watanabe, Japanese Patent No. 4,478, 673, (October 23, 1984); K. Nishioka, H. I. Takura, M. Yoneda, M. Nagatome, H. Abe, and H. Nakata, in "1983 Symposium on VLSI Technology. Digest of Technical Papers." Tokyo (1983), p. 24.

[74] D. L. Flamm, V. M. Donnelly, and D. E. Ibbotson, in "VLSI Electronics: Microstructure Science." N. G. Einspruch and D. M. Brown, eds. Academic Press, New York (1984) Vol. 8, Chap. 8.

[75] E. A. Ogryzlo, D. L. Flamm, D. E. Ibbotson, and J. A. Mucha, *Appl. Phys. Lett.* in press (1988).

[76] Y. H. Lee, M. M. Chen, and A. A. Bright, *Appl. Phys. Lett.* **46**, 260 (1985).

[77] Y. H. Lee and M. M. Chen, *J. Vac. Sci. Technol.* **B 4**, 468 (1986).

[78] T. Makino, H. Nakamura, and M. Asano, *J. Electrochem. Soc.* **128**, 103 (1981).

[79] L. Baldi and D. Beardo, *J. Appl. Phys.* **57**, 2221 (1985).

[80] R. C. Hanson, *J. Electrochem. Soc.*, in press (1988).

[81] A. F. Borghesani and F. Mori, *Jpn. J. Appl. Phys.* **22**, 712 (1983).

[82] G. C. Schwartz and P. M. Schaible, *J. Vac. Sci. Technol.* **16**, 410 (1979).
[83] G. C. Schwartz and P. M. Schaible, *Solid State Technol.* **23**, 85 (1980).
[84] G. C. Schwartz and P. M. Schaible, *J. Electrochem. Soc.* **130**, 1898 (1983).
[85] M. F. Leahy, *Proc. 3rd Symp. on Plasma Processing*, Electrochemical Society, Pennington, New Jersey, **82-6**, 176 (1982).
[86] M. F. Leahy and D. J. Tanguay, *Extended Abstracts*, Electrochemical Society, Pennington, New Jersey, **83-1**, 244 (1983).
[87] S. M. Cabral, D. D. Rathman, and N. P. Economou, *Extended Abstracts*, Electrochemical Society, Pennington, New Jersey, **83-1**, 246 (1983).
[88] M. Shibagaki, T. Watanabe, H. Takeuchi, and Y. Horiike, *Proc. 3rd Symp. on Dry Processes*, IEE, Tokyo, pp. 39–45 (1981).
[89] N. Awaya and Y. Arita, *Proc. 6th Symp. on Dry Processes*, IEE, Tokyo, pp. 98–103 (1984).
[90] S. Berg, C. Nender, R. Buchta, and H. Norstrom, *J. Vac. Sci. Technol.* **A5**, 1600 (1987).
[91] H. Okano, Y. Horiike, and M. Sekine, *Jpn. J. Appl. Phys.* **24**, 68 (1985).
[92] M. Sekine, H. Okano, and Y. Horiike, *Proc. 5th Symp. on Dry Processes*, IEE, Tokyo, p. 97 (1983).
[93] D. L. Flamm and V. M. Donnelly, *Plasma Chem. Plasma Proc.* **1**, 317 (1981).
[94] D. E. Ibbotson, D. L. Flamm, J. A. Mucha, and E. A. Ogryzlo, *J. Appl. Phys.* in press (1988).
[95] H. F. Winters, *J. Vac. Sci. Technol.* **B3**, 9 (1985).
[96] K. Donohoe, *7th Annual Tegal Plasma Seminar* (May 1981); *Proc. 5th Int. IUPAC Symp. Plasma Chem.*, Heriott-Watt Univ., Edinburgh (1981), pp. 13–19.
[97] G. K. Herb, R. A. Porter, P. D. Cruzan, J. Agraz-Guerena, and B. R. Soller, *Extended Abstracts*, Electrochemical Society, Pennington, New Jersey, **81-2**, 710 (1981).
[98] D. W. Hess, *Plasma Chem. Plasma Processing* 2, 141 (1982).
[99] Y. T. Fok, *Extended Abstracts*, Electrochemical Society, Pennington, New Jersey, **80-1**, 301 (1980).
[100] W. Y. Lee, J. M. Eldridge, and G. D. Schwartz, *J. Appl. Phys.* **52**, 2994 (1981).
[101] D. L. Flamm, P. L. Cowen, and J. A. Golovchenko, *J. Vac. Sci. Technol.* **17**, 1341 (1980).
[102] J. W. Coburn and E. Kay, *IBM J. Res. Develop.* **23**, 33 (1979).
[103] D. L. Flamm, *Solid State Technol.* **22**(4), 109 (1979).
[104] J. Heicklen, *Adv. Photochem.* **7**, 57 (1969).
[105] D. L. Flamm, *Plasma Chem. Plasma Proc.* **1**, 37 (1981).
[106] R. A. Heineke, *Solid State Electronics* **18**, 1146 (1975).
[107] R. A. Heineke, *Solid State Electronics* **19**, 1039 (1976).
[108] R. M. Robertson, M. J. Rossi, and D. M. Golden, *J. Vac. Sci. Technol.* **A5**, 3351 (1981).
[109] N. Selamoglu, M. J. Rossi, and D. M. Golden, *J. Chem. Phys.* **84**, 2400 (1986).
[110] J. W. Coburn, *J. Appl. Phys.* **50**, 5210 (1979).
[111] M. Shima, *Surf. Sci.* **86**, 858 (1979).
[112] S. Joyce, J. G. Langan, and J. I. Steinfeld, *J. Chem. Phys.* **88**, 2027 (1988).
[113] J. A. Wenger, AT & T Bell Laboratories Internal Technical Memorandum (January, 1981).
[114] M. A. Jaso and G. S. Oehrlein, *J. Vac. Sci. Technol.* **A6**, 1397 (1988).
[115] N. Yabumoto, M. Oshima, O. Michikami, and S. Yoshii, *Jpn. J. Appl. Phys.* **20**, 893 (1981).
[116] D. J. Vitkavage and T. M. Mayer, *J. Vac. Sci. Technol.* **B4**, 1283 (1986).
[117] G. S. Oehrlein and Y. H. Lee, *J. Vac. Sci. Technol.* **A5**, 1585 (1987).
[118] H. J. Tiller and G. Rudakoff, *Kristall und Technik* **15**, 865 (1980).

[119] S. Thomas, *J. Appl. Phys.* **45**, 161 (1974).
[120] H. J. Tiller and J. Krausse, *Crystal Research and Technology* **17**, K87 (1982).
[121] C. P. Chang, D. E. Ibbotson, D. L. Flamm, and J. A. Mucha, *J. Appl. Phys.* **62**, 1406 (1987).
[122] E. P. G. T. van der Ven P. A. Zilstra, *Proc. Electrochem. Soc.*, **81-1**, 112 (1981).
[123] Y. Horiike, in "Application of Plasma Processes to VLSI Technology." T. Sugano, ed. John Wiley & Sons, New York (1985), p. 138.
[124] Y.Horiike and M. Shibagaki, "Semiconductor Silicon 1977," *Proc. Third Int'l Symp. Silicon Matl. Sci. and Tech.*, H. R. Huff and E. Sirtl, eds. Electrochem. Soc., Princeton (1977), Vol. **77-2**, pp. 1071–1081.
[125] F. H. M. Sanders, J. Dieleman, H. J. B. Peters, and J. A. M. Sanders, *J. Electrochem. Soc.* **129**, 2559 (1983).
[126] D. E. Ibbotson, J. A. Mucha, D. L. Flamm, and J. M. Cook, *Appl. Phys. Lett.* **46**, 794 (1985).
[127] J. B. Kruger, M. M. O'Toole, and P. Rissman, in "VLSI Electronics, Microstructure Science." N. G. Einspruch and D. M. Brown, eds. Academic Press, Orlando (1984), Chap. 5, pp. 91–136.
[128] (a) D. A. Niebauer, *Electronic Packaging and Production*, p. 153 (September, 1980); (b) B. Kegel, *Circuits Manufact.* **21**, 27 (1981); (c) R. D. Rust, R. J. Pachter, and R. J. Rhodes, Extended Abstracts, **83-1**, Electrochemical Society, Pennington, New Jersey, 182–184 (1983).
[129] B. J. Wu, D. W. Hess, D. S. Soong, and A. T. Bell, *J. Appl. Phys.* **54**, 1725 (1983).
[130] J. M. Cook and B. W. Benson, *J. Electrochem. Soc.* **130**, 2459 (1983).
[131] J. F. Battey, *IEEE Trans. Electron. Dev.* **ED-24**, 140 (1977).
[132] K. Harada, *J. Appl. Polym. Sci.* **26**, 1961 (1981).
[133] G. N. Taylor and T. M. Wolf, *Polym. Eng. and Sci.* **20**, 1086 (1980).
[134] J. J. Hannon and J. M. Cook, *J. Electrochem. Soc.* **131**, 1164 (1984).
[135] C. J. Mogab, unpublished results (1980).
[136] R. E. Walkup, K. L. Saenger, and G. S. Selwyn, *J. Chem. Phys.* **84**, 2668 (1986).
[137] Comparing the effects of fluorocarbon, hydrocarbon, and chlorocarbon, additions to an oxygen plasma, it was found that fluorine increased the O concentration while hydrocarbon and chlorocarbon oxidation products tended to decrease the O concentration. (D. L. Flamm and D. E. Ibbotson, unpublished data, 1986).
[138] L. A. Pederson, *J. Electrochem. Soc.* **129**, 205 (1982).
[139] G. Turban and M. Rapeaux, *J. Electrochem. Soc.* **130**, 2231 (1983).
[140] J. M. Cook, J. J. Hannon, and B. W. Benson, in "Proc 6th Int. Symp. on Plasma Chem., ISPC-6." M. I. Boulos and R. J. Muntz, eds. Univ. Scherbrooke, Montreal, Canada (1984), p. 616.
[141] P. Nordine, private communication (1981).
[142] H. Gokan, Y. Ohnishi, and K. Saigo, *Microelectron. Eng.* **1**, 251 (1983).
[143] V. M. Donnelly, D. L. Flamm, C. W. Tu, and D. E. Ibbotson, *J. Electrochem. Soc.* **129**, 253 (1982).
[144] N. N. Creenwood and K. Wade, *J. Inorg. Nucl. Chem.* **3**, 349 (1957).
[145] M. Balooch, D. R. Olander, and W. J. Siekhaus, *J. Vac. Sci. Technol.* **B4**, 794 (1986).
[146] (a) D. E. Ibbotson and D. L. Flamm, *Solid State Technology* **31**(10), 77 (October, 1988); (b) *ibid*, 105 (November, 1988).
[147] D. E. Ibbotson, *Pure and Appl. Chem.* **60**, 703 (1988).
[148] R. P. H. Chang and S. Darack, *Appl. Phys. Lett.* **38**, 898 (1981).
[149] C. W. Tu, R. P. H. Chang, and A. R. Schlier, *Appl. Phys. Lett.* **41**, 80 (1982); D. T. Clark and T. Fok, *Thin Solid Films* **78**, 271 (1981).

[150] R. Cheug, S. Thoms, S. P. Beamont, G. Doughty, V. Law, and C. D. W. Wilkinson, *Electronics Lett.* **23**, 16 (July, 1987).

[151] S. Thoms, S. P. Beaumont, C. D. W. Wilkinson, J. Frost, and C. R. Stanley, *Microelectron. Eng.* **5**, 249 (1986).

[152] T. W. Hayes, M. A. Dreisbach, W. Dautremont-Smith, P. Thomas, and L. A. Heimbrook, unpublished results, (1988).

[153] A. T. Bell, "Chemical Reaction in a Radiofrequency Discharge. The Oxidation of Hydrogen Chloride." Sc.D. Thesis, Massachusetts Institute of Technology, Cambridge, Massachusetts (1967).

[154] C. J. Mogab, in "VLSI Technology." S. M. Sze ed., First ed., McGraw-Hill, New York (1983), pps. 303–345.

[155] H. Kalter and E. P. G. T. Van de Ven, *Philips Tech. Rev.* **38**, 200 (1978/79).

[156] D. L. Flamm and V. M. Donnelly, *Solid State Technol.* **24**(4), 161 (April, 1981).

[157] C. J. Mogab and H. J. Levenstein, *J. Vac. Sci. Technol.* **17**, 1721 (1980).

[158] A. C. Adams and C. D. Capio, *J. Electrochem. Soc.* **128**, 366 (1981).

[159] D. L. Flamm, *J. Appl. Phys.* **51**, 5688 (1980).

[160] D. E. Ibbotson, D. L. Flamm, and V. M. Donnelly, *J. Appl. Phys.* **54**, 5974 (1983).

3 An Introduction to Plasma Physics for Materials Processing

Samuel A. Cohen

Plasma Physics Laboratory
Princeton University
Princeton, New Jersey

Plasma Etching:
An Introduction

ISBN 0-12-469370-9

I. Introduction

The processing of materials by plasmas requires detailed knowledge in several scientific and technological areas. This is particularly true in the field of semiconductor fabrication where the continuing development of denser arrays with finer features has demanded the combination of various techniques into a highly specialized art. Perhaps the basic foundation for this art is plasma physics, though chemistry, electrical engineering, and vacuum technology have defensible claims. Each must be understood and well-practiced for the material processing to succeed.

In this chapter we will present the fundamental concepts in plasma physics which underlie the operation of plasma processing equipment. This will include discussions of the types of particles present in processing plasmas, their energies and fluxes, and an elucidation of the characteristic lengths, time scales, excitable modes (both stable and unstable), and atomic and surface processes important in the initiation and maintenance of plasma discharges. To discuss these topics in a practical way, we include information from a wide range of plasma configurations used in plasma processing, presenting material on dc- and rf-driven discharges with and without externally applied magnetic fields.

The understanding of plasmas that are unmagnetized, isothermal, isobaric, and isotropic is already rather difficult. The configurations used in all processing devices do not have even this simplicity, in large part due to the boundary between the plasma and the solid surfaces. It is at the boundary that our ultimate interests lie. However, the reader should find that the simplified situations described here will form a good understanding of the often counter-intuitive behavior of plasmas and will encourage improvements in existing equipment or processes.

Plasmas are usually created in metal vacuum vessels, commonly used to attain the low pressures essential for particular plasma properties. Plasmas have a propensity to fill every crevice in these vacuum vessels. (The word "plasma" originates from a Greek root meaning deformable.) And though efforts are made to constrain the plasma to particular sections of the vessel, these are not completely successful. Device operation is considerably affected. To emphasize this, we shall use the label "containment" vessels to fully appreciate that some plasma reaches everywhere in them.

We assume familiarity with college physics (especially Maxwell's equations) and introductory calculus. Most equations will be presented both in cgs and practical units to aid their easy application. Section II presents most of the basic ideas and definitions concerning plasmas. These are developed in the later sections. Section III concerns single-particle motion;

Section IV gives details of plasma parameters; Section V is devoted to plasma formation; and Section VI applies the previous four sections to the magnetron device.

II. The Plasma State

Plasmas are a state of matter that consists of a large group of electrons and ions with nearly equal numbers of opposite charges, each particle moving at a high rate of speed relative to the others. It is the precise electric field of the individual charged particles that gives the plasma its unique properties. The electric field of each particle influences the motion of distant particles, whether they have like or opposite charge. This action-at-a-distance causes a wide variety of waves and instabilities to be possible in a plasma. And because each particle is influenced by the electric and magnetic fields of many particles, the term used to describe the kinematics is *collective motion*.

The electric field of a single isolated electron is proportional to r^{-2}. The volume of a spherical shell a distance r from that electron increases proportional to r^2. Thus, the product of the electric field times the volume, a measure of the effectiveness of the field at a distance, is constant (Fig. 1a). It is the same near the electron as it is far away, showing how the action-at-a-distance arises.

What differentiates a plasma from a group of neutral atoms that also has equal and large numbers of electrons and ions? It is the distance at which the electric field is felt strongly. Neutral atoms (and molecules) have an electric field no stronger than a dipole. This falls-off proportional to the distance cubed or faster. Hence at large distances, it is weak compared to the Coulomb electric field of the bare electrons found in a plasma. Because of the very short range of their electric and magnetic fields, molecules interact with each other only by "hard" collisions, meaning close encounters, typically at separations of about 1 Å. Free electrons and ions in a plasma interact over much greater distances, typically 1000 Å or more, as well as less, of course!

Numerous distant interactions will change a charged particle's trajectory more than the infrequent hard collisions (Fig. 1b). For this reason close encounters may be unimportant to the charged particles in a plasma. (This is related to another reason why a group of neutral atoms does not behave like a plasma. The quantal nature of the electronic energy levels in an atom precludes the small changes in energy required by distant encounters so important to plasma behavior.) Hence plasmas are frequently termed *colli-*

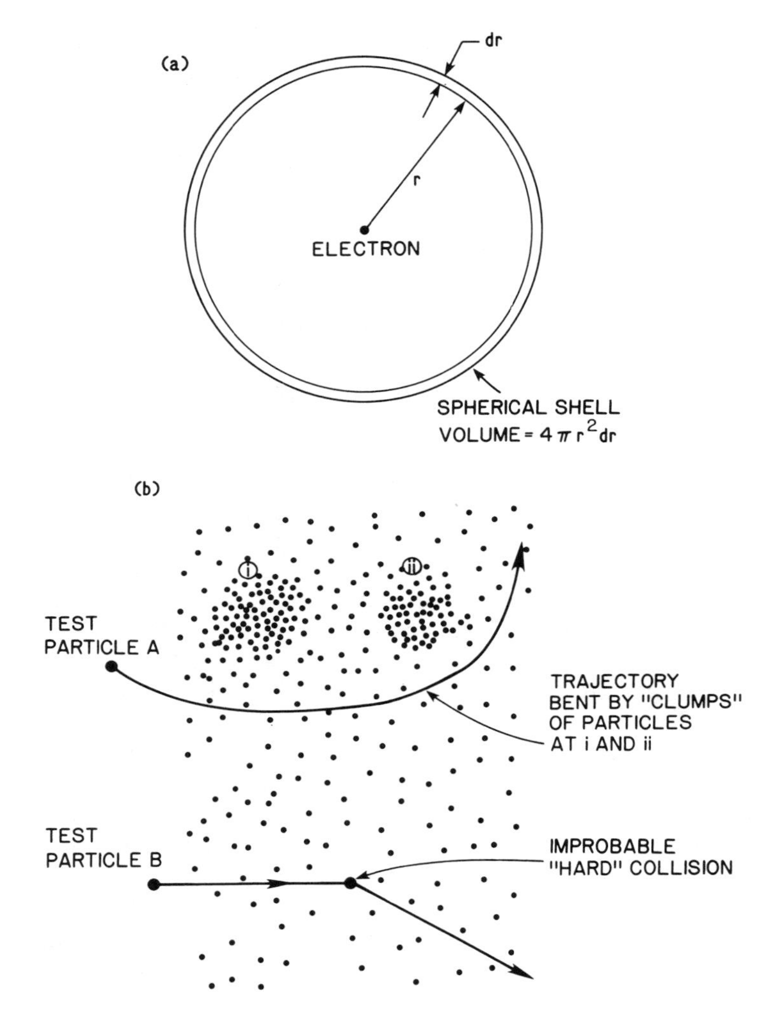

FIGURE 1. a) The electric field of an isolated electron falls off proportional to r^{-2}. The volume of a spherical shell around that electron increases as r^2. This shows the importance of distant particles to the motion of that single electron. b) When a charged test particle moves through a cloud of charged particles, the electric field of the distant particles alters the trajectory of the test particle more than the infrequent hard collisions with nearby particles.

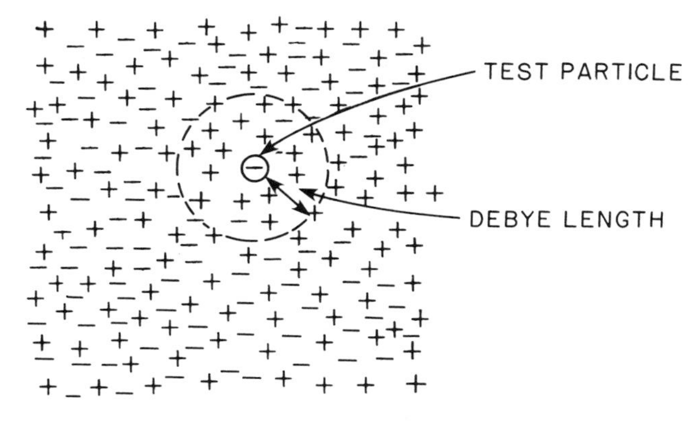

FIGURE 2. When a negative test charge is placed in a plasma, a volume around it, of radius equal to the Debye length, is partially depleted of negative charges. This results in the shielding of that test particle's electric field at distances larger than the Debye length. Outside that volume the negative and positive charges are more nearly equal in abundance.

sionless, meaning that individual hard collisions are unimportant compared to the numerous distant soft ones. Exact criteria, which hinge on the characteristic lengths in the problem, must be examined before a particular plasma can be correctly labeled collisionless. For example, is the size of the containment vessel larger than the mean-free-path between hard collisions?

When many charged particles are present, they alter their positions, like-charged particles being repelled and oppositely charged ones being attracted, to reduce the distance over which an applied electric field is effective. The source of this field could be external metal plates attached to a battery, or a single electron placed in the plasma as a *test particle*. The *shielding* (Fig. 2) occurs in a distance called the *Debye length*, whose size determines many properties of the plasma relevant for material processing. The plasmas typically used in materials processing have a Debye length in the range of .01 to 1.0 mm. Within a sphere of this radius there are still many (typically more than a million) charged particles to influence and to be influenced by the test charge.

Shielding does not prevent the penetration of all fields into a plasma. Certain electrostatic and electromagnetic waves, for example, can penetrate into and propagate through plasmas. This is essential to many schemes for plasma heating.

By extrapolation it is clear that, at too high a density, plasma particles may be too close together and thus "appear" like dipoles to the distant particles. Also, at room temperature electrons and ions will rapidly recombine to form neutral atoms and molecules. Hence, to sustain a plasma, its temperature must be kept above some minimum, about 10,000 K (about 1 eV), which depends on the density. The higher the temperature, the higher is the allowed density. From this, one can estimate that most laboratory plasmas have densities in the range of 10^8 to 10^{12} cm^{-3}.

In astrophysical situations, plasmas exist at much lower densities (10^{-3} cm^{-3}) as in the interstellar medium, and at much higher densities (above 10^{20} cm^{-3}) as in certain stars (Fig. 3). Other systems, such as electrons in metals or ions in liquids, also have certain properties like those of our gaseous plasmas.

The approximate equality between oppositely charged particles is termed *quasineutrality*. It is one of the most basic tenets of plasma physics. A 1% excess of either charged species in a plasma with parameters like a magnetron planar etcher would cause an electric field in excess of about 1000 volts/cm. This large field would cause the electrons to rearrange their positions to restore a more balanced distribution of charges. Small electrical imbalances do occur. The resulting restoring force causes the plasma electrons to oscillate internally at a frequency (naturally enough) called the *plasma frequency*. These occur, again for the magnetron etcher, at about 10^{10} Hz.

Three phases of matter—gas, liquid, and solid—are commonly experienced first hand, i.e., they can be readily touched if not too hot or cold. Plasmas cannot. They are generally too hot, too tenuous, and too fragile, almost like a soap bubble. A further similarity to soap bubbles is that the plasma possesses at its boundaries a skin, called the plasma or Debye *sheath*. The sheath is about 5 Debye lengths in thickness. Trying to touch a plasma by penetrating the sheath can destroy the plasma or, if the plasma has enough stored energy, the finger! Hence one of our best ways to test material properties is unavailable. But visual inspection with the naked eye can reveal the presence and dimensions of the sheath, properties of which it has taken scientists decades to confirm and quantify accurately with other diagnostic equipment.

As might be inferred from their densities alone, plasmas behave more like gases than solids. But their modes of internal motion are complex because of collective motion. One similarity is that both support longitudinal compressional (sound) waves. Another is that both expand to fill their containers. On encountering the walls of the containment vessel, the charged particles of a plasma are neutralized by attachment to charged particles from the solid, often metallic, surfaces. Thus, the rate at which the plasma expands will determine how long, and if, it can be sustained. The charged particles lost by contact with the wall must be replaced if the plasma is to continue in its existence (Fig. 4). The steady state thus achieved will depend

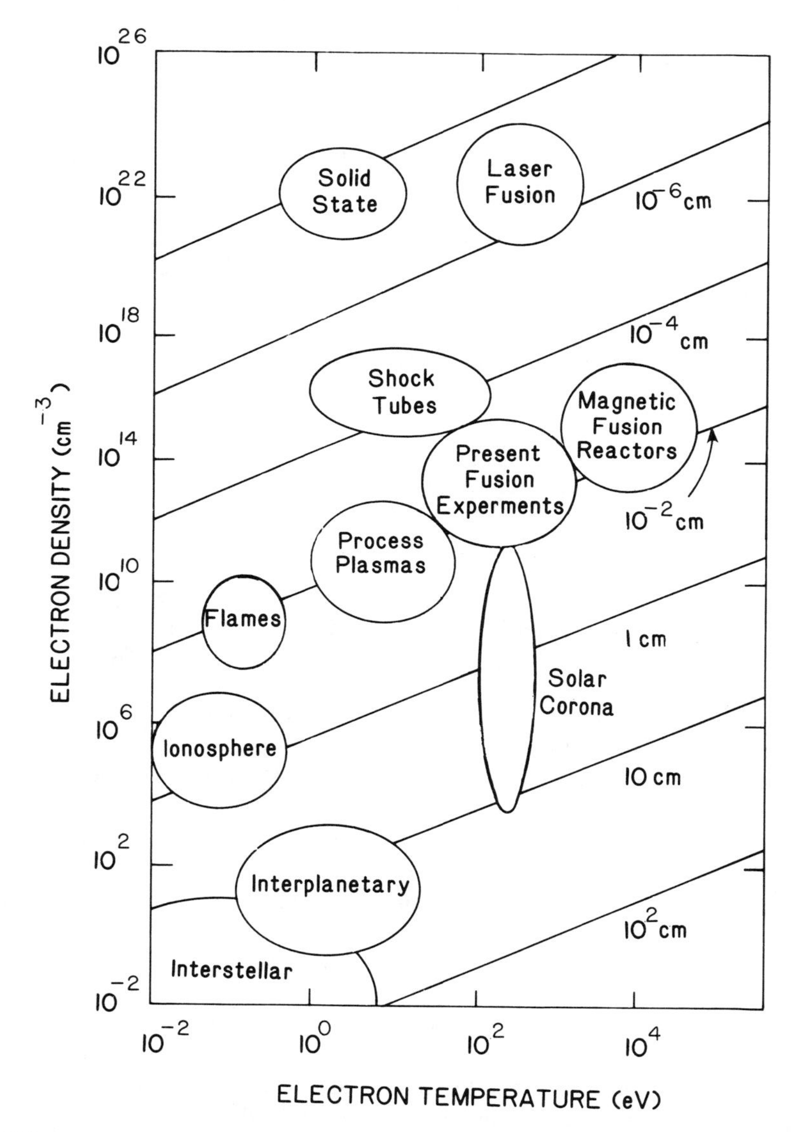

FIGURE 3. Types of plasmas, categorized by their temperatures and densities. The corresponding Debye lengths are the diagonal lines.

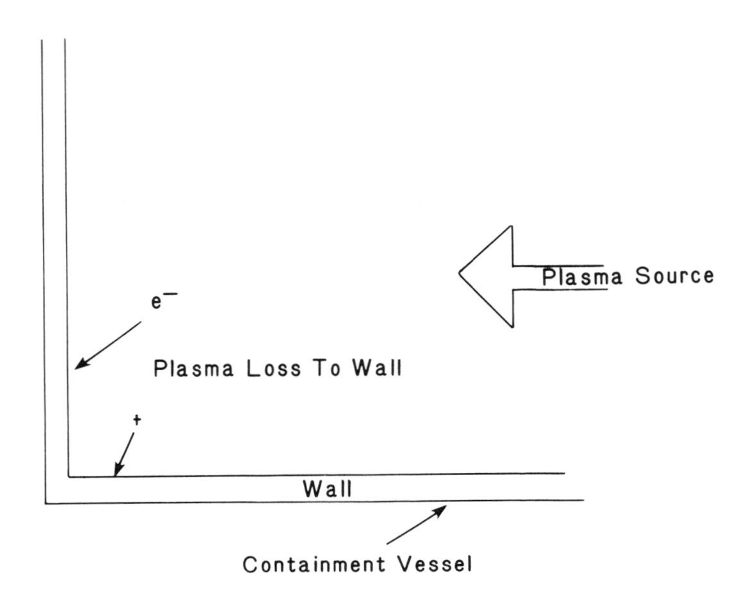

FIGURE 4. Plasma particles are transported to the walls of the containment vessel. There they are absorbed. The lost plasma must be replenished by an internal or external source.

strongly on the properties of the containment vessel, e.g., will it resupply electrons and the right ions when and where needed.

There are numerous ways to replenish the lost particles. Some devices rely on generation of plasma in a separate volume and its injection into the containment vessel. In these apparatus, a plasma "gun" may be used to produce "fresh" plasma by applying a high voltage across a low density gas. Free electrons in the gas are accelerated by the electric field, picking up sufficient energy to ionize any gas atoms they impact. The more usual way to replenish lost plasma is to resupply gas atoms directly into the containment vessel to replace the lost ones and to rely on the impact of the plasma's own electrons to ionize them (Fig. 5). In either case, the ions thus formed may be of several species, depending on the gas feedstock and the plasma parameters. For example, helium gas will form only two types of ions, He^{+} and He^{++}, whose abundances will depend on the plasma temperature; in contrast, CF_4 will be fragmented into nearly a dozen different species. Thus a plasma contains neutral atoms as well as electrons and ions.

Neutral collisions are very important, not just for the chemical processes described in other chapters of this book, but for their effects on plasma behavior. Many properties of the plasma will depend on the relative

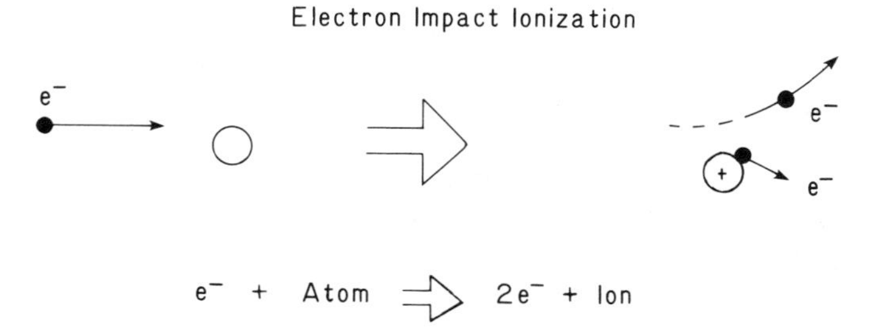

FIGURE 5. The impact of a plasma electron on a neutral atom may result in ionization of that atom.

temperatures and densities of these different species. Though the interaction between the charged particle pairs is much stronger than between a charged particle and a neutral, if the neutral density is very high, then electrons will collide more often with them, and, for example, the electrical resistivity will be affected. The neutral particles in materials processing plasmas typically outnumber the charged ones by more than 10^4 to 1. At the extreme, if the neutral to charged particle ratio exceeds about 10^{11}, the neutral collision frequency will exceed the plasma frequency, collective aspects of the motion will be lost, and the plasma state destroyed.

III. Single-Particle Motion

Bulk plasma motion is dominated by collective effects. Yet the motion of each charged particle under the influence of the local electric and magnetic fields is the correct description of what is happening at a microscopic level. And so the Lorentz force law and Maxwell's equations provide the proper way to predict a single test particle's motion. What this approach lacks is the back-effect of the test particle's own motion on the local fields as caused by its fields acting on the distant particles. To improve the single-particle picture, one could start with a description of the plasma as a fluid or as an ensemble of particles. Then fluid equations or the kinetic equations would be used to describe the plasma's evolution. These approaches are usually reserved for the more advanced students. Several references are listed at the end of this chapter that should satisfy the more ambitious. Many results achieved by a kinetic or fluid analysis can be reproduced in a single-particle description by choosing the proper initial conditions based on the known answer. This is often done for pedagogical reasons because the single particle picture is so easy to visualize and hence to remember. We use this approach here.

In the single-particle approach to plasma physics, the motion of an individual charged particle under the influence of externally applied electric and magnetic fields is examined. The fields are allowed to vary in space and time but do not change to reflect the subsequent motion of the charged particle. The motion of the single particle is readily obtained from the Lorentz force law, which (in cgs units) is

$$\mathbf{F} = \frac{m\,d\mathbf{v}}{dt} = q\left(\mathbf{E} + \frac{\mathbf{v} \times \mathbf{B}}{c}\right), \tag{1}$$

where c is the speed of light (3×10^{10} cm/s), m is the mass of the charged particle (in gm), q its charge (-4.8×10^{-10} statcoul for a single electron), v its velocity (cm/s), and E (in statv/cm) and B (in gauss) are the applied electric and magnetic fields, respectively. We now discuss several simple cases of this equation.

A. $\mathbf{E}$ = CONSTANT, $\mathbf{B} = 0$

The application of a constant and homogeneous electric field, but no magnetic field, results in the constant acceleration of a charged particle. The particle gains energy from the field at an ever increasing rate. If allowed to continue, the particle would eventually reach relativistic speeds. Then the simple Newtonian description fails and the particle gains mass rather than speed, and also energy and momentum. This relativistic limit is beyond the scope of processing plasma physics. Integrating Eqn. (1) for E parallel to the x-direction gives

$$x(t) = x_0 + v_{x0}t + \frac{qEt^2}{2m}, \tag{2}$$

where x_0 and v_{x0} are the initial x-position and x-velocity of the particle. If the initial velocity v_0 is zero or is in the direction parallel to E, this motion is in a straight line. Otherwise it is parabolic (Fig. 6). Though in most

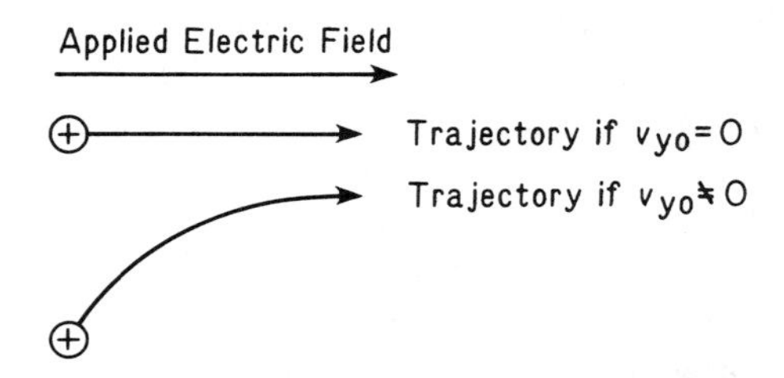

FIGURE 6. Trajectories of charged particles in a constant electric field.

situations it is adequate to ignore a particle's past history, i.e., x_0 and v_0, this is not true in the initiation of a discharge or in most microwave-driven discharges.

B. $\mathbf{E} = 0$, $\mathbf{B}$ = CONSTANT

As seen from the Lorentz law, if there is no electric field and no initial velocity, the particle remains at rest. More precisely, the magnetic force acts only if the particle's motion is perpendicular to B. The result is that a particle moving initially along B is unaffected by B while a particle moving perpendicular to B has its trajectory turned into a circle. For a particle moving at an angle to B the net result is a corkscrew or helical motion. There is no gain or loss of total particle kinetic energy from a static or slowly varying magnetic field (Fig. 7). Integrating Eqn. (1) for this choice of parameters gives

$$
\begin{aligned}
z(t) &= z_0 + v_{z0}t \\
x(t) &= x_0 + r_L \sin(qBt/mc) \\
y(t) &= y_0 + r_L \cos(qBt/mc)
\end{aligned}
\tag{3}
$$

where

$$r_L = \frac{mv_\perp c}{qB}, \tag{4}$$

$v_\perp$ is the component of the velocity perpendicular to B, and we assumed B parallel to z, $B = B_z$.

Notice that the circular motion has a characteristic radius, r_L, called the *Larmor radius*, and a characteristic frequency, $\omega_c = 2\pi f = qB/mc$, called

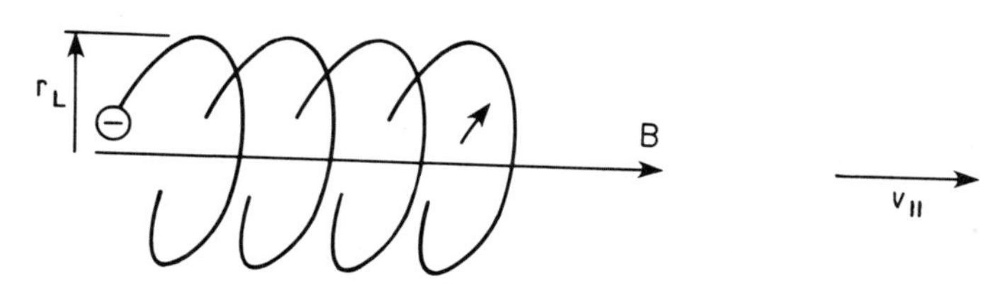

FIGURE 7. Trajectory of a (negatively) charged particle in a constant magnetic field.

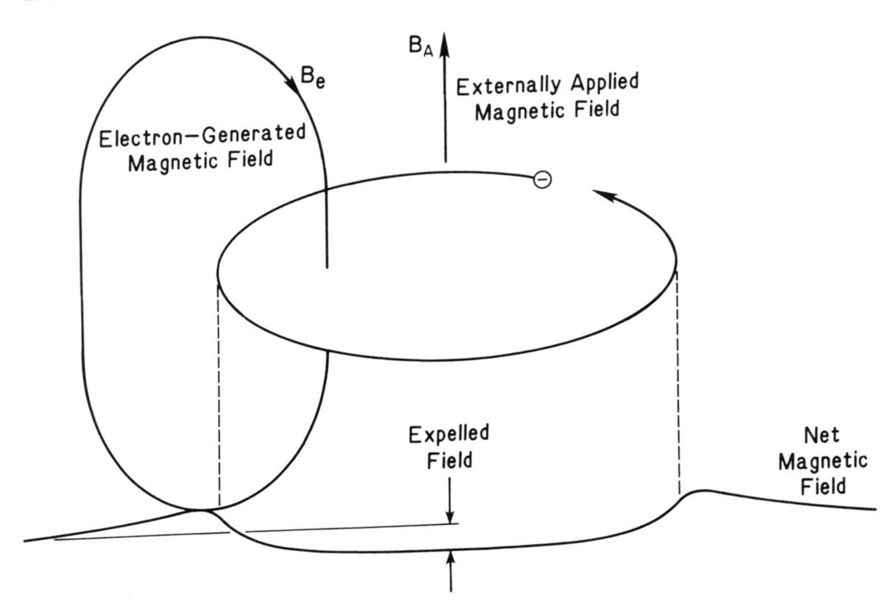

FIGURE 8. The magnetic field generated by the circling electron is oppositely directed to the applied B inside the orbit and in the same direction outside. The net field is thus reduced inside the orbit.

the *Larmor-*, *cyclotron-*, or *gyro-frequency*, which are related by

$$r_L = \frac{v_\perp}{\omega_c}. \tag{5}$$

The circular motion of the charged particle produces a dipole magnetic field in such a way as to reduce the strength of the field inside the circular orbit and to increase it outside (Fig. 8). Then one can view the circular motion as being caused by a higher magnetic pressure on the outside of the orbit. This property of a charged particle or a plasma to partially expel magnetic fields from their vicinity (interior) is called *diamagnetism*. The dipole field has a strength called the *magnetic moment* equal to

$$\mu = \frac{qv_\perp r_L}{2c} = \frac{mv_\perp^2}{2B}. \tag{6}$$

Another convenient picture of the motion is to separately consider the circular (cyclotron) and linear motions. By ignoring the cyclotron motion one has the *guiding center motion* remaining. This nearly linear motion is a worthwhile concept if the cyclotron radius is small compared to the other characteristic dimensions. And so it is of most use for electrons whose Larmor radii are small because of their small mass.

C. E-PERPENDICULAR TO B

If **E** and **B** are parallel to each other, there is little change in a particle's motion from the above description. That is, a particle will accelerate along **E** and **B** and have an unchanging circular motion around them.

If **E** and **B** are perpendicular, however, a new effect takes place that is contrary to most everyday experiences, except that of the common toy top. What happens is that the charged particle starts to accelerate parallel to E until its velocity is large enough for the magnetic force to bend it. The bend becomes so large that the particle ends up moving in a direction that is perpendicular to both **E** and **B** (Fig. 9). Similarly, a spinning toy top, if pushed by a finger or by gravity, responds by precessing perpendicular to both the applied force and its axis of rotation, and not by falling over.

So the motion of the charged particle is the sum of three distinct motions: cyclotron motion around **B**; linear motion along **B**; and a drift

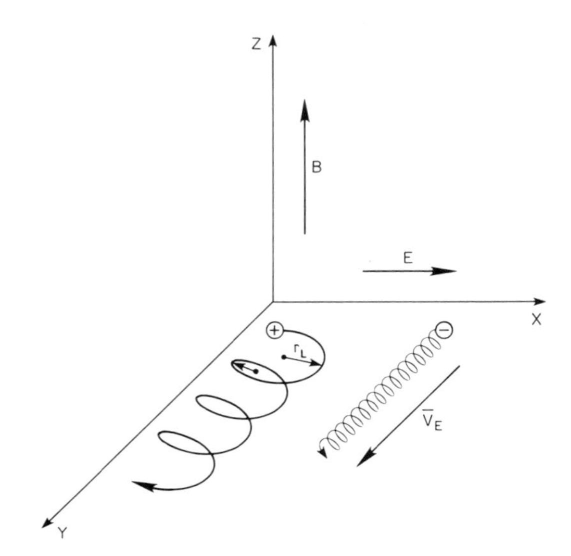

FIGURE 9. Drift of a charged particle in the presence of crossed **E** and **B** fields. Both electrons and ions drift in the same direction and at the same speed. Note that the Larmor radius changes as the particle gains and loses energy from the electric field. It is this difference in Larmor radius that causes the drift.

perpendicular to both **B** and **E**. The **B**-parallel and **B**-perpendicular (drift) motions now comprise the guiding-center motion.

Integrating the equation of motion for this case is more complicated than in the previous paragraphs. Instead, one can take the vector cross-product of **B** with the Lorentz law, Eqn. (1), ignoring the $d\mathbf{v}/dt$ term, which describes the cyclotron motion. Then one obtains for the transverse component of the velocity (the drift velocity)

$$v_E = c\left(\frac{\mathbf{E} \times \mathbf{B}}{B^2}\right). \tag{7}$$

This drift velocity is independent of the particle's energy, charge, and mass because the electric and magnetic forces are both proportional to a particle's charge and independent of a particle's mass. Though a heavy particle has a larger gyroradius, its cyclotron frequency is slower by the same amount so the gain and loss of energy from the electric field during each cycle are balanced.

D. NON-UNIFORM FIELDS AND OTHER FORCES

The same derivation could have been carried out for a different force **F** acting on a charged particle in a **B** field. The result can simply be obtained by replacing **E** by $\mathbf{F}/q$, the equivalent electric field. The classic example most quoted is that of a charged particle in crossed magnetic and gravitational fields (Fig. 10a). The result is a drift velocity of magnitude cmg/qB, where g is the gravitational acceleration. This drift does depend on charge and mass because the gravitational force does not depend on charge but does depend on mass. Charges separate and the heavier particles drift faster. Hence, a net current flows, in contrast to the $\mathbf{E} \times \mathbf{B}$ case.

Another drift would occur if the particle experiences a force due to a pressure gradient, ∇p. This would arise in a plasma with temperature or density gradients. The resulting drift is called the diamagnetic drift and has a velocity equal to

$$v_D = \frac{c\nabla p \times \mathbf{B}}{qnB^2}. \tag{8}$$

Similarly, the drift resulting from gradients or curvature in the **B** field itself can be written down immediately after identifying the forces they cause. A charged particle moving along a curved field line (Fig. 10b) feels a centrifugal force of magnitude $mv_{\|}^2/R_c$, where R_c is the radius of curvature of the field. And magnetic field gradients (Fig. 10c) cause a force equal to

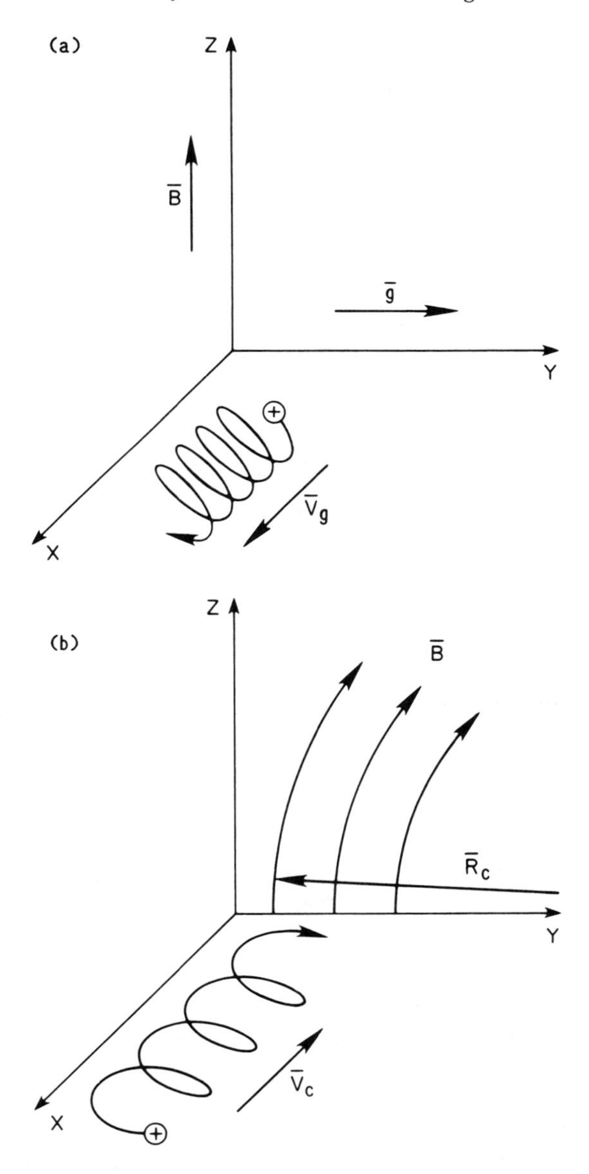

FIGURE 10. Drift motion of positively charged particles under the combined influences of a magnetic field and a) gravity, b) curved magnetic field, and c) magnetic field with a gradient (continued on next page).

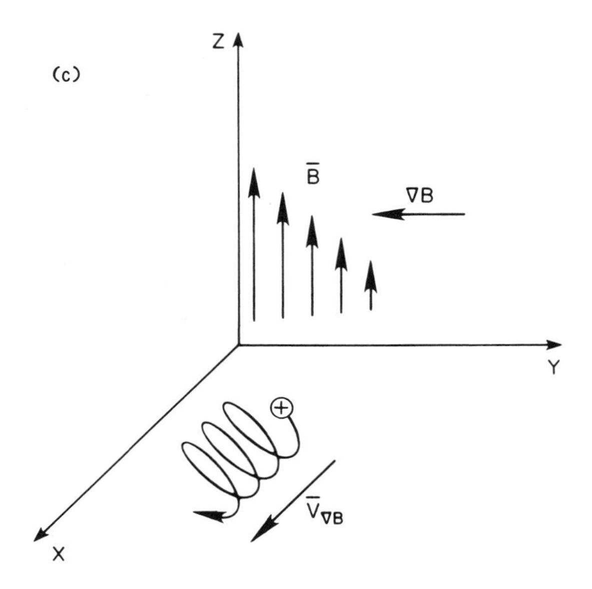

FIGURE 10c. Continued from previous page.

$\mu\nabla$ **B**. Hence the drifts are:

$$v_G = \pm \frac{v_\perp r_L \mathbf{B} \times \nabla \mathbf{B}}{B^2} \tag{9}$$

and

$$v_c = \frac{m v_\parallel^2 \mathbf{R}_c \times \mathbf{B}}{q B^2 R_c^2}. \tag{10}$$

In contrast to the **E** × **B** drift, these depend on a particle's mass, charge, and energy. The more massive a particle, the faster it drifts. This is another counter-intuitive property of plasmas. Generally one expects the electrons, the lighter particle, to carry all the electrical currents. Not so. The ions can play a dominant role in current flows.

A non-uniform electric field will alter the drift velocity by giving weight to the time the particle spends in the regions of different field strength and direction. The result for a sinusoidally varying electric field with a periodicity distance d is that the drift is proportional to the usual **E** × **B** drift times $(1 - r_L^2/4d^2)$. This is called a finite Larmor radius correction because of the presence of r_L in the equation. This is different for each species and may cause charge separation and thus plasma waves.

E. TIME-VARYING FIELDS

When the electric field varies in time the drift is modified. One can again understand the motion from a microscopic picture. Each time the field "turns-on" the particle slowly accelerates parallel to **E** and then the **E** × **B** drift develops. When the electric field reverses the processes again occur, but with the drifts in the opposite directions. This can be seen quantitatively from the following derivation (Fig. 11a and b). The magnetic field is along the z-axis; the electric field is along the y-axis. The equations of motion for the x and y directions are:

$$\ddot{x} = \omega_c \dot{y} \tag{11a}$$

and

$$\ddot{y} = \frac{qE}{m} - \omega_c \dot{x} . \tag{11b}$$

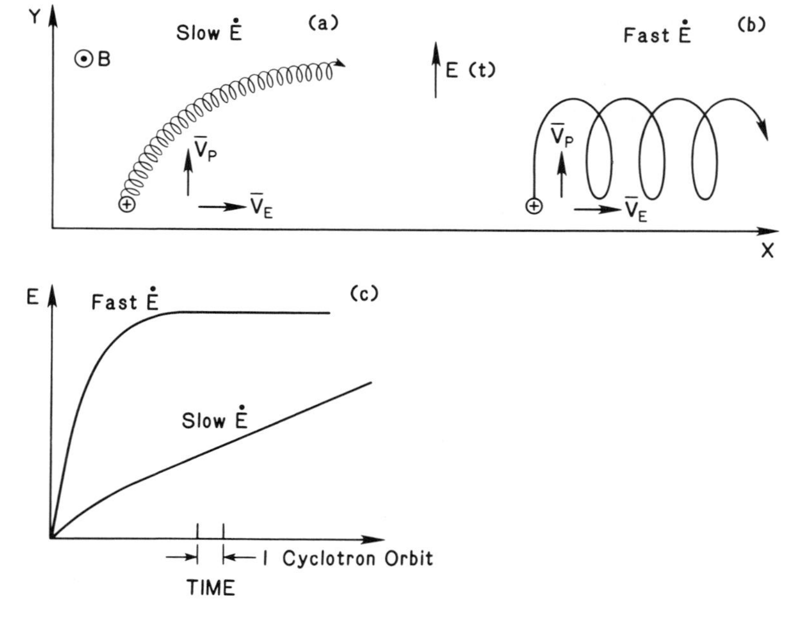

FIGURE 11. Polarization drifts for positively charged particles when **E** increases monotonically with time. If **E** increases slowly, the Larmor radius is unchanged. Two drifts develop: the **E** × **B** drift and the polarization drift (parallel to **E**). When **E** increases at a fast rate, relative to the period of the cyclotron motion, the Larmor radius grows as the particle gains energy from the field. Again two drifts develop. Note that when $d\mathbf{E}/dt = 0$ the polarization drift also is 0.

By differentiating (11a) and substituting in (11b) and vice-versa, the equations can be made separable yielding,

$$\dddot{x} + \omega_c^2 x = \frac{qE\omega_c}{m} \tag{12a}$$

and

$$\ddot{y} + \omega_c^2 y = \frac{q\dot{E}}{m}, \tag{12b}$$

which have solutions

$$\dot{y} = -v_\perp \sin(\omega_c t) + \frac{mc^2\dot{E}}{qB^2} \tag{13a}$$

and

$$\dot{x} = +v_\perp \cos(\omega_c t) + \frac{cE}{B}. \tag{13b}$$

It is now easy to identify the $\mathbf{E} \times \mathbf{B}$ drift motion in the x-direction and the polarization drift in the y-direction, $mc^2\dot{E}/qB^2$. The orbit shapes will vary depending on the rate of change of the electric field. Figure 11 shows two cases, both for monotonically increasing fields, not oscillating ones. In Fig. 11b the rapidly changing intense field increases the Larmor radius because it adds an appreciable amount of energy to the particle in a time faster than the cyclotron period.

The polarization drift is in opposite directions for oppositely charged particles. Hence charge separation occurs. Again the more massive particle carries the current.

F. ADIABATIC INVARIANTS

A similar derivation carried out for slowly varying magnetic fields shows that no additional drift arises. Instead the Larmor radius grows or shrinks, depending on whether B decreases or increases. This is associated with a decrease or increase of the particle's transverse energy (see Eqn. (4)). The change is such that the magnetic moment is unchanged. So $v_\perp^2$ increases proportional to B.

The magnetic moment is called the *first adiabatic invariant*. It is constant as long as changes in **E** or **B** occur slowly compared to a cyclotron orbit. Its constancy reflects the symmetry and periodicity of the cyclotron orbit.

Sitting in the particle's frame-of-reference, a change of **B** can come from a change in its position as well as a change in the local field strength. Recall that static magnetic fields add no energy to a particle. It is then clear that the change in perpendicular energy must be accompanied by an equal but opposite change in parallel energy. So if a particle moves into a region of

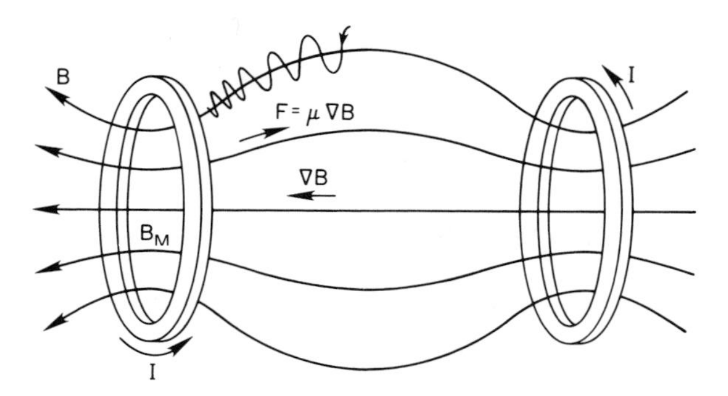

FIGURE 12. Magnetic mirror formed by two coaxial coils with co-directed currents, I. As a charged particle approaches a region of higher B, its perpendicular energy grows at the expense of its parallel energy. Reflection will occur if B reaches a high enough value. Note that the Larmor radius shrinks as the particle moves into a region of higher B.

increasing **B**, it will continue to gain perpendicular energy at the expense of its parallel energy. At some point it may have lost all its parallel energy. With $\nabla \mathbf{B}$ still causing a force on the particle it is reflected. This property of spatially varying magnetic fields is called the *mirror effect* and forms the basis for many plasma confinement configurations (Fig. 12). For the mirror to "work," the particle must start with sufficient perpendicular energy because the mirror force only acts on the magnetic moment ($\mathbf{F} = \mu \nabla \mathbf{B}$). From conservation of energy and μ we can show that a particle starting from a region of magnetic field strength B_0, will only be reflected in the region of higher field if its pitch angle, $\theta = v_{\|}/v_{\perp}$, is less than

$$\frac{B_0}{B_M} = \sin^2\theta = (R_M)^{-1} \tag{14}$$

where

R_M = the mirror ratio, and

B_M = the maximum field strength.

The underlying principles for these statements are the laws of conservation of energy and angular momentum. These are very powerful tools when a physical situation has symmetry. For example, calculations of cosmic ray trajectories in the earth's vicinity are easy because of the size of the earth coupled with the symmetry of its magnetic field. The symmetry is such that other invariants of the motion, the so-called second and third adiabatic invariants, help to solve the problem. They are based on the symmetry and periodicity of the large-scale motion around the earth (bouncing back and forth between the magnetic mirrors at the poles and circulating around the equator), not just of the cyclotron orbit. In plasma processing equipment, however, collisions are too frequent and the size of the apparatus too small to allow undisturbed orbits around the material structures. Thus, the only useful invariant is the first.

G. SUMMARY OF PARTICLE DRIFTS

It is easiest to use the information just presented on the drift motions by rewriting their equations expressed in practical units. The parameters, symbols and practical units for this are listed next.

These drift velocities should be compared with the thermal velocities of the ions and electrons, which are

$$v_t = (2W/m)^{1/2} = 5.9 \times 10^7 (W)^{1/2}\ cm/s \quad \text{for electrons} \tag{16a}$$

$$= 1.4 \times 10^6\ (W/M)^{1/2}\ cm/s \quad \text{for ions.} \tag{16b}$$

Parameter	Symbol	Practical Units
magnetic field	B	gauss
electric field	E	volts/cm
gravitational constant	g	cm/s^2
ion mass	M	amu (proton mass = 1)
charge	q	-1 for an electron
radius of curvature	R_c	cm
gradient scale length	R_G	cm
temperature	T	electron volts
parallel energy	$W_\parallel$	electron volts
perpendicular energy	$W_\perp$	electron volts
total kinetic energy	W	electron volts
cyclotron frequency	ω_c	/sec

Drift	Symbol	Value (cm/s)
$E \times B$	v_E	$10^8\ E/B$
diamagnetic	v_D	$10^8\ T/R_G B$
gravitational	v_g	$10^{-4}\ gM/B$
curved B	v_c	$2 \times 10^8\ W_\parallel/R_c B$ (15)
grad B	v_G or $v_{\nabla B}$	$10^8\ W_\perp/R_G B$
polarization	v_p	$v\ (\omega/\omega_c)$ for $E = E\cos(\omega t)$

IV. Plasma Parameters

The preceding section described the motions of individual particles under the joint influences of electric and magnetic fields. Nowhere was there a need to consider the presence of a group of particles. In this section we describe the salient properties of a group of charged particles. Some of these properties are so general, e.g., temperature and pressure, that they apply to all states of matter; others concern the plasma state only, e.g., plasma oscillations.

A. TEMPERATURE, DENSITY, AND PRESSURE

The concept of temperature is a very precise one. Only if a collection of particles exists *and* only if that collection has a certain distribution of particle energies, may a temperature be defined. That distribution is called a Maxwellian and its shape is shown in Fig. 13. It is not necessary—in fact it is a rarity—for plasmas to be exactly Maxwellian. Even the average energy of particles may dramatically vary between different locations in the containment vessel. In spite of all this, the concept of temperature, or at least of an average energy, is useful in a local description of plasmas.

Note in Fig. 13 that there are no particles with zero energy. There is also a clear maximum in the distribution. Most particles have energies within a few times the most probable energy. It is usually the particles in the tail, however, with about four times the most probable energy, that dominate the physical processes. There are two reasons for this. First, these particles are twice as fast-moving as the others, hence they interact more frequently; and more importantly, most physical processes, ionization for example, are near threshold at the low energies (5 to 10 eV) characteristic of plasma processing discharges. Hence small changes in energy make exponential changes in reaction rates. This is one reason why great efforts are made to increase the temperature of plasmas. A two-fold increase in the electron temperature can increase the reactivity ten-fold. In Section IV.D, we evaluate this quantitatively to see if there is an upper limit to the desired temperature.

A second important parameter describing a plasma is its density. This too is of critical importance in determining reaction rates. As stated earlier, the densities of positive and negative charges are locally balanced, usually to better than 1%. Across a plasma device, though, the densities can vary a hundred-fold or more. One usually aims at keeping the density high near the work surface if it is desired to have plasma bombardment of that surface. But there are applications where plasma bombardment is undesirable, such as when lattice damage or arcing might occur. Figure 14 shows

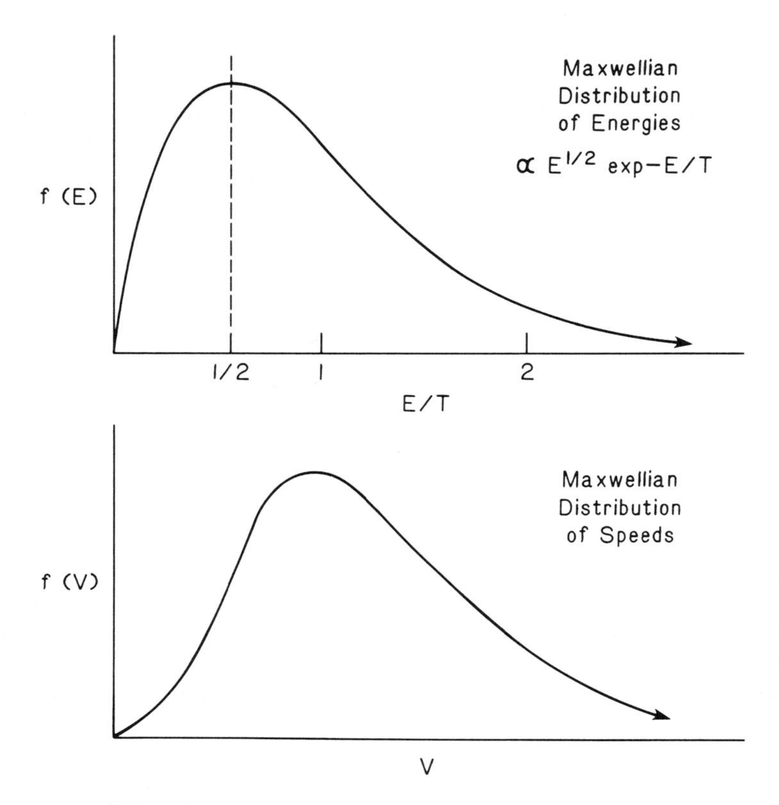

FIGURE 13. Maxwellian distributions of energy and speed.

typical ranges for the temperatures and densities of the various species in processing plasmas. Note that the ion energies in the sheath may be very high.

The product of plasma density times temperature gives the plasma pressure. This quantity is a measure of how well used is the energy provided to form the plasma. The *energy confinement time* is the ratio of the stored energy (plasma pressure times the plasma volume) to the input power. As will be seen later by an application of this concept, different methods of plasma formation and confinement place quite different requirements on the power supplies used to generate and to maintain the plasmas and also affect where the input power is ultimately deposited.

In the ensuing sections the reader should carefully note the use of k_B denoting Boltzmann's constant and k denoting the wavenumber of a periodic disturbance in the plasma.

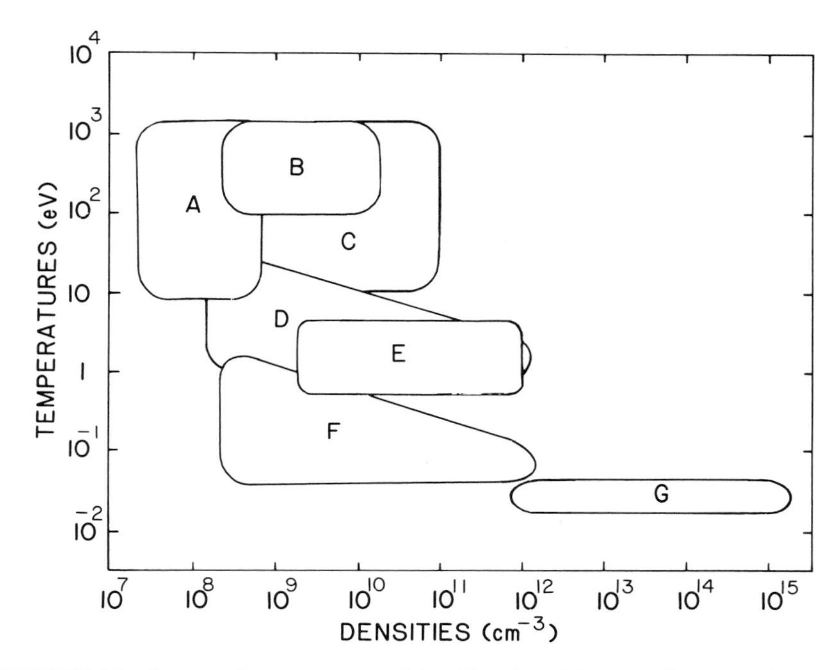

FIGURE 14. Ranges of temperatures and densities of particles typically present in process plasmas: A = secondary electrons accelerated through sheath (cathode), B = ions backscattered from cathode (possibly neutralized), C = ions accelerated towards cathode, D = electrons in main plasma, E = Franck-Condon ions and neutrals, F = ions in main plasma, G = neutral atoms and molecules.

B. DEBYE LENGTH AND PLASMA FREQUENCY

Consider a collection of charged particles arranged in a slab, with equal numbers of each charge. Try to displace the positive charges to the right and the negative ones to the left (Fig. 15). Energy is required to effect the separation. An electric field develops, which makes the separation of charges progressively more difficult. Now consider the particles as they try to separate by simply using their own thermal energy, not some externally supplied source. The distance they can separate depends on their thermal energy. When their thermal energy is all converted into potential energy they can separate no further. The potential energy, $U = q\Delta\phi$, may be derived from Poisson's equation for the potential

$$\nabla^2\phi = 4\pi(n_i - n_e)q, \tag{17a}$$

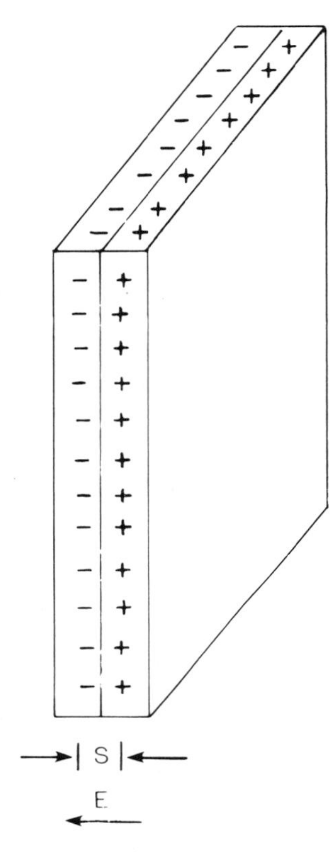

FIGURE 15. Collection of charged particles in a slab. The positive particles were displaced from the negative. A strong electric field results, which will force the two sets of particles back toward each other.

which, for a separation s, gives

$$\Delta\phi = -2\pi nqs^2. \tag{17b}$$

This is the potential energy change, $q\Delta\phi$, as a function of the separation of the charges. By equating the available thermal energy ($k_BT_e/2$ because of the 1-dimensional motion) with the potential energy, we can solve for the maximum distance, λ_D, by which the charges may separate from each other before the restoring force of the electrostatic field pulls them back together again. This distance is the Debye length, λ_D.

$$\lambda_D = \left(\frac{k_BT_e}{4\pi n_e q^2}\right)^{1/2}. \tag{18}$$

Another way to calculate the Debye length is by asking how charges rearrange themselves around a test particle in response to its electric field. After finding the distribution of charges, we again would use Poisson's equation to see how much the Coulomb field is reduced. The answer we would obtain is that the electric field is reduced to $1/e$ of its Coulomb value in a distance equal to the Debye length. Thus, the electric field does penetrate into a plasma further than one Debye length. We shall discuss this more in regards to the presheath and to magnetized plasmas.

We are now able to find out how many particles are still affected by the shielded field of the test particle. This can be estimated by calculating the number of particles within a sphere of radius λ_D.

$$N_D = \frac{n_e 4\pi\lambda_D^3}{3} = \frac{\left(k_B T_e/q^2\right)^{3/2}}{3(4\pi n_e)^{1/2}} \tag{19}$$

This number, commonly called the *plasma parameter*, must be large (more than 100) for collective effects to be important.

There are two reasons for showing the first derivation. First, the mathematics is simpler. And second, it naturally leads into the next topic, the plasma frequency, ω_{pe}. As just stated, when the two slabs of charge are separated (Fig. 15), a restoring force is developed. This pulls the charges back toward each other. They accelerate and then pass through their equilibrium positions and separate in the opposite sense. Now the positive charges are to the left and the negative to the right. It is easy to estimate the frequency of this motion. It is simply the inverse of the time it takes particles to move the Debye length at their thermal velocity.

$$\omega_{pe} = \frac{v_{e\perp}}{\lambda_D} \tag{20}$$

Consider the influence of a material object on the plasma with a sheath developed at the interface. For convenience, choose a metal sphere floating freely in the plasma. At the instant of its introduction it has no net charge. But the electrons flow rapidly into it because they are much faster than the more massive ions. The electron flow progressively charges the sphere to a negative potential. When the value of this potential exceeds a few times the electron temperature, most electrons are repelled by the field. Soon a steady state is reached where the greatly reduced flow of electrons is balanced by the flow of ions. This is called the *floating potential*,

$$\phi_f = 0.5(k_B T_e/q) \ln\left[\left(2\pi\frac{m_e}{m_i}\right)\left(1 + \frac{T_i}{T_e}\right)\right]. \tag{21}$$

The ions that provide the charge to balance the electron flow originate deep in the plasma from a region called the *presheath*. Its length may be 100s of Debye lengths. The condition of charge neutrality in the plasma requires that the ions are accelerated from the presheath into the sheath up to a velocity, c_s, called the *sound speed*,

$$c_s = \left(\frac{k_B T_e + k_B T_i}{m_i} \right)^{1/2}. \tag{22}$$

To arrive at the sheath with this velocity, the potential drop, $\Delta\phi_{ps}$, integrated along the entire presheath, must be $\sim 0.5\ T_e$. (More detailed theories of c_s and ϕ_f change the coefficients of Eqns. (21) and (22) by up to 40%, depending on geometry, the plasma equation-of-state, and other factors.)

It is clear from this discussion that the plasma is severely perturbed in the immediate vicinity of the object. For example, there is no return flow of electrons and ions from the sphere. The electron and ion densities in the sheath do not balance because of the electric field that has developed (Fig. 16). The distance of the perturbation is approximately five times the Debye length. And the size of the potential drop is 4.7 times the electron temperature for an argon plasma (Eqn. (21)). Thus, the ions that reach the surface will have an energy equal to what they have in the plasma plus what they have gained in passing through the presheath and sheath, $(4.7 + 0.5)k_B T_e$.

With these three concepts now in hand, it is instructive to review the single-particle approach to see what other effects (beyond increasing the ion energy) the plasma presheath and sheath may have on ions hitting surfaces. For example, consider a magnetized isothermal plasma with **B** parallel to a metal surface. Ions that enter the sheath—later we will describe how they might be transported across the magnetic field—will find themselves in a region of crossed electric and magnetic fields. Will they be rapidly accelerated and hit the surface at normal incidence before the $\mathbf{E} \times \mathbf{B}$ drift develops, or will they be so deflected by the drift that they skim along the surface only to hit it at a grazing angle? This has consequences to both the sputtering rate and erosion profile. Though a detailed calculation of particle trajectories through the sheath is necessary to answer these questions accurately, we can get an approximate answer by simply comparing the $\mathbf{E} \times \mathbf{B}$ drift velocity and the thermal velocity. Setting Eqns. (7) and (16) equal gives

$$v = \left(\frac{k_B T}{2m} \right)^{1/2} = v_E = \frac{cE}{B}.$$

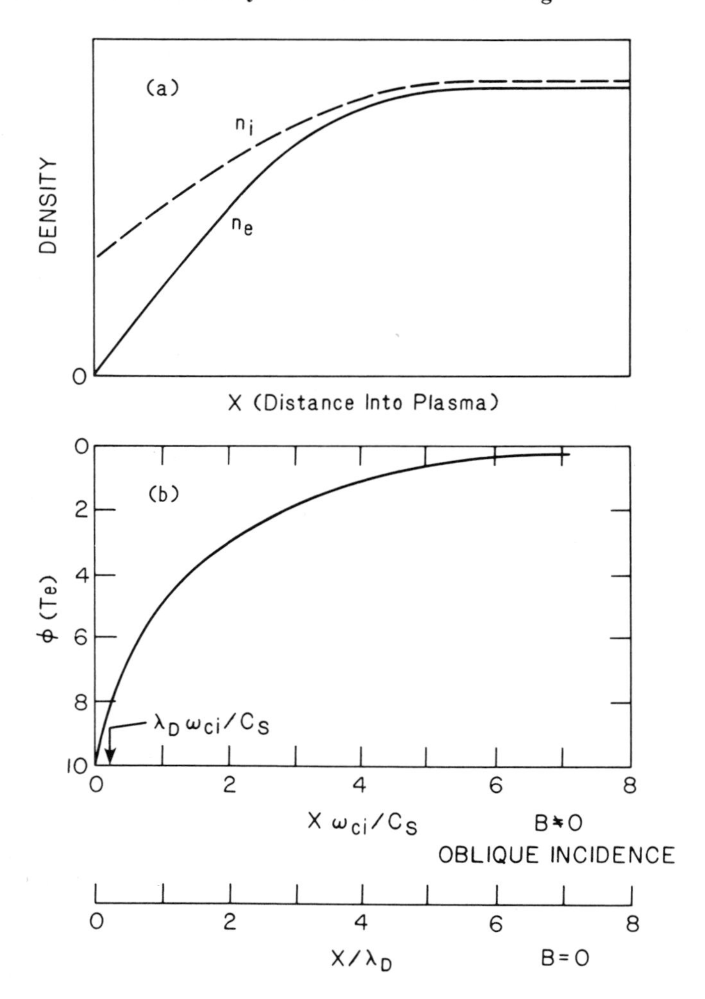

FIGURE 16. a) The electron and ion densities in the sheath region of plasma. b) The potential distribution is shown with two different distance scales: one is to be used for unmagnetized plasmas (or when *B* is normal to the surface); the other is for when *B* intersects the surface at an oblique angle. The value of the potential at the surface depends on the ion mass and temperature (see Eqn. 21).

In the sheath the electric field is approximately the potential energy change, $U \cong 5kT_e$, divided by the sheath thickness, $\sim 5\lambda_B$. So using Eqn. (19) for λ_D, this can be rearranged to yield

$$\frac{B^2}{8\pi} = nm_i c. \tag{23}$$

From Eqn. (23), we can determine the conditions for which the $E \times B$ drift velocity will be more important than the thermal motion or the sheath-induced velocity (since both have the same dependance on T). This occurs when the magnetic energy, $B^2/8\pi$, is less than the Einstein self-energy, $nm_i c^2$. Using standard parameters for a magnetron discharge, i.e., $B = 1000$ g and argon ions, we find that the critical density above which v_E may cause glancing ion impacts is about 10^6 cm^{-3}. The single-particle picture is open to question if the orbit size, r_L, exceeds the sheath size. The comparison is made using Eqns. (4) and (19),

$$\lambda_D = r_L,$$

which upon substitution and rearrangement gives:

$$\frac{B^2}{8\pi} = nm_i c^2.$$

That is, the same conditions satisfy both requirements. If the density is below the critical value, then the ion cyclotron radius is smaller than the sheath thickness and the single particle picture is correct. Then the **E** $\times$ **B** drift is less important than the thermal motion. But magnetrons operate at densities of about 10^{11} cm^{-3}, well above the critical value. So the application of the drift equations is not as simple. The ions do not stay in the sheath long enough for the **E** $\times$ **B** drift to fully develop. All this was based on the assumption of an isothermal plasma. Usually the ions are colder than electrons in these devices. Hence their gyroradii may be smaller. Again, to answer these questions accurately, one should calculate the forces on an individual ion as it enters and moves through the sheath. This could be done with the Lorentz force law, Eqn. (1). One calculation, see Chodura in Ref. 10, shows the angle of incidence (with respect to the surface-normal) of ions is about 1/2 the angle of incidence of the magnetic field. (The presheath has much larger dimensions; hence, the **E** $\times$ **B** drift does develop and ions are accelerated parallel to the surface.)

In summary, evaluation of ion trajectories based on the single particle equations of motion in a magnetized plasma shows that ions are accelerated in both the sheath and presheath to a net velocity, which is oblique to metal surfaces when the plasma density exceeds about 10^6 cm^{-3}. Detailed calculations are required for a particular geometry and set of plasma parameters to quantify the distribution of angles of impact.

Before leaving the subject of magnetized plasmas, it is important to note the effects of the magnetic field on the sheath. If the magnetic field is normal to the surface there is no change in the sheath. But an obliquely incident magnetic field does change things considerably. Mainly, this is due to the fact that the electric field, set up by the initial electron flow to the object, now has to overcome the Lorentz force on the ions in addition to the ion inertia. The net result is that the floating potential changes little, but the sheath extends further into the plasma by a distance equal to several ion gyroradii. However, the appropriate ion gyroradius to use is that of the ion with an energy obtained from the $\mathbf{E} \times \mathbf{B}$ drift motion. This second part of the sheath is termed the magnetic part; the first is the electrostatic (see Fig. 16b). This reduces the rigidity of the requirement we arrived at in Eqn. (23).

C. SKIN DEPTH AND DIELECTRIC CONSTANT

We have just seen that a plasma will shield its interior from an externally applied static electric field. It also reduces the distance over which the electric field from individual charges in the plasma has an effect. But what about the penetration of time-varying fields, i.e., waves? There are two general classes of waves: electrostatic (also called longitudinal) and electromagnetic (also called transverse). The first have their electric field parallel to their direction of propagation. These types of waves cannot propagate in empty space. They need a medium, like a plasma, to support their motion. The second have their electric field perpendicular to their direction of propagation. Light waves are a member of this class, and they can obviously propagate through vacuum.

It is instructive to compare the derivations of the penetration depth, the *skin depth*, for these two different types of waves. For the longitudinal waves, we will start with the force law. To Eqn. (1) we must add a term that denotes the pressure caused by the bunching of charges. After all, the longitudinal wave is a pressure wave generated when the applied electric field causes electrons to bunch. Magnetic forces are unimportant. For simplicity we will consider only motion in one direction, the x-direction. Then the force equation is

$$\frac{mn\,\partial v_1}{\partial t} = -qn_0E - 3T\frac{\partial n}{\partial x}. \tag{24}$$

We now look for wave solutions with the standard form, $\exp(i(kx - \omega t))$.

(Perturbed quantities have subscript 1). Poisson's equation gives

$$\nabla^2\phi = 4\pi(n_i - n_e)q$$

or

$$-\nabla E = ikE = 4\pi q n_1 \tag{25}$$

and the equation of continuity gives

$$\frac{\partial n}{\partial t} = -n_0 \nabla v_1 \tag{26a}$$

or

$$i\omega n = n_0 ikv. \tag{26b}$$

Substituting (25) and (26) into (23) gives

$$\omega^2 = \omega_{pe}^2 + \frac{3}{2}k^2 v_t^2 \tag{27}$$

or

$$k = 2\omega_{pe}\frac{\left(\dfrac{\omega^2}{\omega_{pe}^2} - 1\right)}{3v_t}. \tag{28}$$

For $\omega \ll \omega_{pe}$,

$$k \simeq \frac{2i}{3\lambda_D} \tag{29}$$

or $\delta_{\|}$, the skin depth for low frequency longitudinal waves, is

$$\delta_{\|} = \frac{3\lambda_D}{2}. \tag{30}$$

So longitudinal waves with frequencies less than the electron plasma frequency only penetrate about a Debye length into a plasma. This is because the electrons in the plasma are able to move fast enough to shield out the relatively slowly varying longitudinal electric field.

We now consider transverse waves, again with frequencies less than the plasma frequency. The difference between this case and the previous can be seen in Fig. 17. The electrons are pushed up and down in sheets in the plasma. No bunching develops to shield the electric field. Instead the flowing electrons generate a magnetic field to counter that of the electromagnetic wave. To analyze this situation, start with Maxwell's equations in a dielectric medium,

$$\nabla \times D = -\frac{1}{c}\frac{\partial H}{\partial t} \tag{31}$$

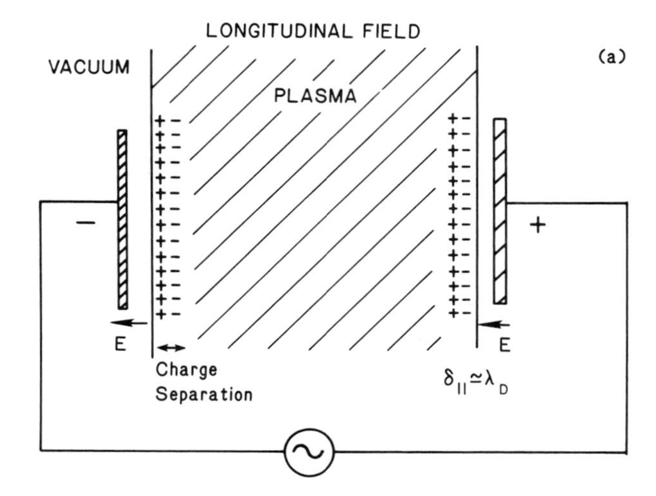

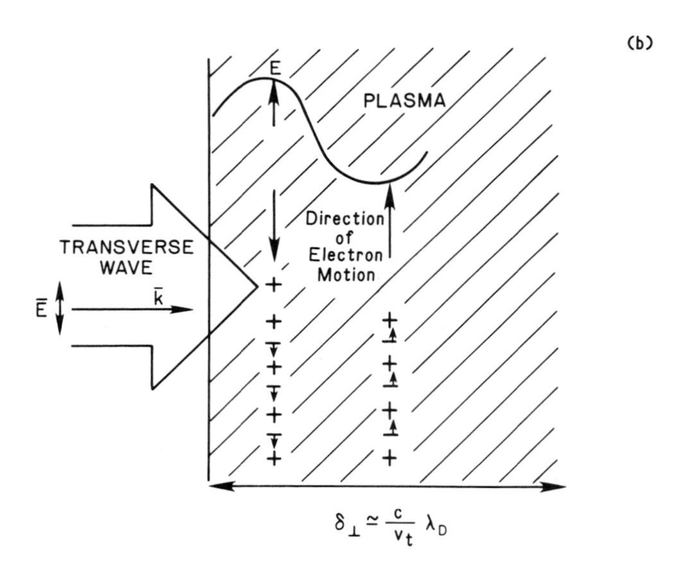

FIGURE 17. Penetration of low frequency longitudinal a) and transverse b) waves into plasmas. A longitudinal field causes charge separation at the plasma surface. The transverse wave does not and hence may penetrate further.

and

$$\nabla \times H = -\frac{1}{c}\frac{\partial D}{\partial t} + \frac{4\pi J}{c}, \tag{32}$$

where $D = \epsilon E$ and $B = \mu_D H$. ϵ is the dielectric constant and μ_D is the magnetic permeability $\simeq 1$. Take the curl of both sides of Eqn. (31). Use the facts that

$$J = -n_0 e v_1 \tag{33}$$

and

$$\frac{m\,\partial v_1}{\partial t} = \frac{qE_1}{im\omega}. \tag{34}$$

This gives

$$k = \omega_{pe}\frac{\left(\dfrac{\omega^2}{\omega_{pe}^2} - 1\right)^{1/2}}{c}, \tag{35}$$

which for $\omega \ll \omega_{pe}$ yields

$$k \simeq \frac{i\omega_{pe}}{c} \tag{36}$$

or

$$\delta_{\perp} = \frac{c}{\omega_{pe}} = \frac{c\lambda_D}{v_t}. \tag{37}$$

So the penetration of transverse waves into plasmas is much further than longitudinal waves. Typically the ratio c/v_t is 1000, so transverse waves may penetrate tens of centimeters into the plasma before they are absorbed (or reflected). This offers an important choice when waves are being considered for heating or sustaining plasmas. For example, does the process require a high energy content in the edge (plasma sheath) for sputtering, or is it better to deposit the energy further in the plasma where it may preferentially be used for ionization and molecule fragmentation?

The inclusion of a magnetic field into the preceding calculations increases the complexity of the analysis. Some results will be presented in a later section on waves. Here, however, a more basic concept is presented, the dielectric constant. This has close ties to the results just obtained on the strength of the electric field inside plasmas and on the drift motions described in the previous section.

Let us guess at the result. Because a magnetic field hinders the movement of charged particles, we expect that for large B both the charge bunching and the sheets of current (formed in response to the longitudinal and transverse waves respectively) will be reduced; hence the shielding ability of the plasma for both longitudinal and transverse waves will decrease. To see if this is true start again with Maxwell's equation,

$$\nabla \times B = \frac{4\pi J}{c} + \frac{\dot{E}}{c}. \tag{38}$$

The current can be divided into two terms: one that is the polarization current due to the changing E field, J_p; the other containing all other currents, J_0. From Eqn. (14) one can write an expression for the polarization current (which is due mostly to ions)

$$J_p = n_0 q v_p = \frac{n_0 m_i c^2 E}{B}. \tag{39}$$

So Eqn. (38) becomes

$$\nabla \times B = \frac{4\pi J_0}{c} + \frac{\epsilon \dot{E}}{c} \tag{40}$$

where ϵ is defined as

$$\epsilon = 1 + \frac{4\pi n m_i c^2}{B^2}. \tag{41}$$

This is nearly the same as Eqn. (23). Equation (41) was derived for low frequencies and **E** perpendicular to **B**. The point to be remembered is that ϵ represents how much the strength of a time-varying transverse electric field of a wave is reduced inside the plasma.

D. COLLISIONS

Collisions take place when charged and/or neutral particles interact in the plasma and also when they impact the walls and structures of the containment vessel. Though most textbooks ignore the second type, these are so critical to plasma processing that a short review of the most important surface effects is included at the end of this section. This is a prerequisite for understanding the sustenance of plasma discharges presented in Section V.

In the following discussion we consider only positive ions. In many plasmas these overwhelmingly dominate. There are cases, however, particularly at high density or when electronegative atoms or molecules are present, that copious numbers of negative ions will exist. They modify the species that impact the anode and also substantially change plasma properties such as mobilities, diffusivities, sheaths, floating potentials, and so forth.

Collisions in the plasma have contrary roles. They can result in the birth or loss of ions, the transport or confinement of plasma, and the gain or loss of heat. In the next paragraphs qualitative arguments will be presented to support this statement followed by quantitative formulae.

The impact of an energetic electron on a neutral atom may result in the creation of an ion-electron pair. This process is energy dependent. As shown in Fig. 18, the ionization rate coefficient increases rapidly with electron energy to a maximum at about 100 eV and then falls slowly. Since most plasma processing devices have a low electron temperature, it is clear that a slight increase in T_e will result in a great increase in plasma production rate. The ionization rate coefficient, R_I, may be approximated by

$$R_I = \frac{2.5 \times 10^{-6} \eta_e (T_e/\chi) \exp(-\chi/T_e)}{(1 + T_e/\chi)} \text{ cm}^3/\text{s} \tag{42}$$

where

T_e = electron temperature (eV)

η_e = number of equivalent outer electrons (1 for H, 2 for H_e)

χ = ionization potential (eV).

The exponential term in Eqn. (42) arises from the exponential in the Maxwellian velocity distribution function of the impacting electrons. This formula, a phenomenological fit, is only valid for $\chi \gtrsim T_e/30$.

As hinted earlier, there may be unwanted effects if the plasma is too hot and too dense. Consider the case of Al sputtering by Ar. If the plasma density exceeds 10^{13} cm^{-3} and the temperature exceeds about 50 eV, the sputtered Al will be ionized close (within 1 cm) to where it has been sputtered away. Then, as positive ions, it would be attracted toward and redeposited on the negatively biased Al target plate, reversing the desired effect.

Electrons may reattach themselves to ions. The simple radiative recombination process is slow at plasma temperatures above 1 eV. Even at this low temperature, the recombination time for a plasma of density 10^{11} cm^{-3} is about 10 sec, which is much longer than the lifetime of particles in the plasma due to flow to the electrodes, for example. If the density is much higher or the temperature much lower, as might occur in high-pressure plasma processes, then recombination can be sped up to play an important role in plasma loss. An approximate formula to estimate the recombination rate coefficient, R_R, is

$$R_R = 1 \times 10^{-13} Z \frac{\chi}{T_e} \left(0.4 + 0.5 \ln \frac{\chi}{T_e} + 0.4 \frac{\chi}{T_e} \right) \text{ cm}^3/\text{s}. \tag{43}$$

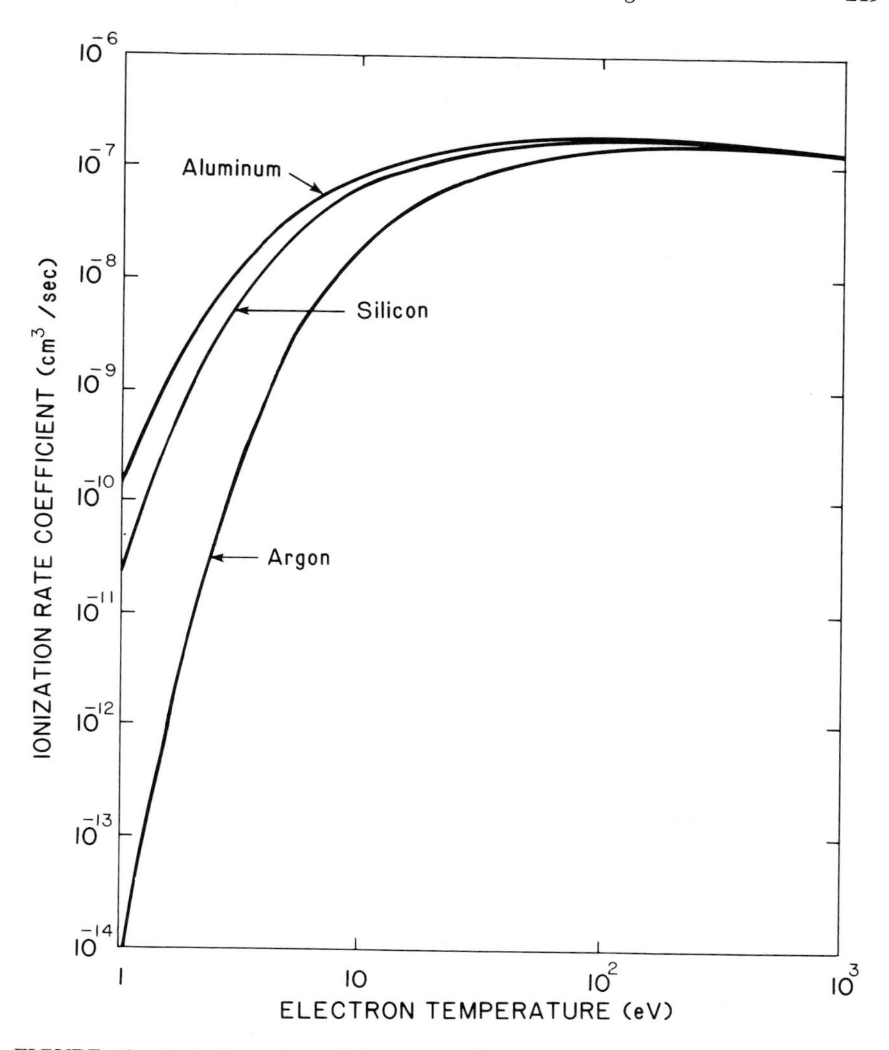

FIGURE 18. Ionization rate coefficient versus electron temperature for electron impact on Al, Ar, and Si.

The collision of an ion with a neutral will rarely result in the creation of a free electron. However, the exchange of an electron is highly probable. The rate coefficient of this process is nearly independent of energy and is about the same as for electron impact ionization, 10^{-9} cm^3/s. This phenomena can lead to the presence of energetic neutrals whose trajectories are unaffected by electric or magnetic fields. The importance of this to the etch profile is clear.

The second set of contrary effects deals with plasma confinement. For simplicity consider the case of an unmagnetized plasma (i.e., there is no external magnetic field applied) in a metal containment vessel. If the plasma density is low then the charged particles can freely stream to the walls where neutralization takes place (Fig. 19a). One way to reduce the rate of plasma loss is to increase the plasma density. Then collisions will convert the free-streaming situation into a diffusion-dominated one (Fig. 19b). One need not rely on charged-particle collisions to effect this improvement in confinement; neutral-charged particle collisions will do as well. The diffusion coefficient is approximately given by

$$D \cong \frac{\lambda_m^2}{\tau_m}, \tag{44}$$

where λ_m is the mean-free-path between collisions and τ_m is the time between collisions. The mean-free-path will depend on whether it is charged-particle pair scattering or charged-neutral scattering that is more important.

The inclusion of a magnetic field into the problem completely alters the outcome. Consider an infinitely long cylinder with a uniform magnetic field parallel to its axis. The ions and electrons may flow freely along B but are constrained to remain within one Larmor radius of the field line on which they originate—unless, of course, they collide with other particles. Then they may jump one Larmor radius inward or outward (see Fig. 19c). So diffusive motion occurs again. But this time the "perfect" confinement of the magnetic field is degraded by collisions. If the collision is with a neutral, the neutral may move in an arbitrary direction relative to the field and the ion (or electron) then bounces in such a way to conserve total momentum. In short, the ion will probably bounce one Larmor radius and the ion density profile will expand diffusively. When two similar ions collide, the result may again be a single Larmor radius shift for each particle. But when one particle goes inward the other goes outward. There is a detailed balance and no net change in particle density. The classical collisions between charged particles of equal mass and charge do not contribute to particle diffusion. Ion-electron collisions do lead to diffusive transport, but at a slow

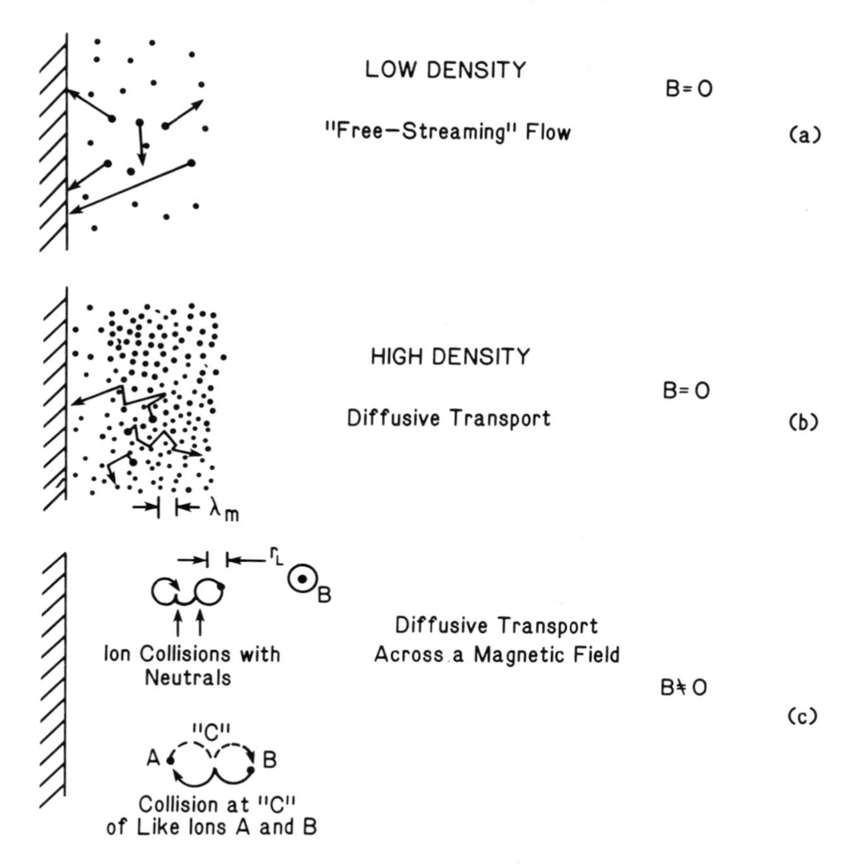

FIGURE 19. Plasma transport dominated by a) "free-streaming" flow, b) diffusive flow, and c) diffusion across B due to collisions. Note that in case c), collisions between like ions do not result in the net transport of particles. The inward displacement of one ion is exactly balanced by the outward displacement of the other.

rate because of the poor momentum transfer between particles of such disparate mass.

There is yet another type of "collision" that leads to transport. It is more accurately described by the term particle-wave interaction. Waves, or fluctuations in particle densities, result in fluctuations in electric field. The magnitude of the electric field is directly proportional to the electron energy, hence its temperature. This electric field may lead to an $\mathbf{E} \times \mathbf{B}$ drift and particle loss. Surprisingly, a simple expression for the diffusion coefficient, derived by Bohm using very elementary arguments based on this picture, proved to be more accurate than the more complicated and detailed theories derived over the next 40 years.

The final pair of contrasting effects caused by collisions concerns heat transport. These can be understood by arguments similar to those just presented for particle transport and thus will not be repeated.

In addition, there is another very important effect of collisions on energy content, the extraction of energy from electric fields. If a DC electric field is applied to a plasma, the electrons will be accelerated. Without collisions, the electrons would simply be lost to the anode. Collisions allow the randomization of the directed energy into a thermal-type distribution. This is called *ohmic heating*. In a related fashion, when an oscillatory field is applied, though the electrons first gain energy, they lose the same amount of energy when the field reverses, unless a collision occurs first. Then again, their momentum is redirected and heating occurs. One new aspect of the collision process is that as temperatures increase, the ease with which charged particles share energy diminishes because the interaction time, hence energy transfer, is smaller at higher velocities. The rate of energy equilibration is given in practical units in the next section.

Let us now turn to a quantitative summary of these phenomena. First consider a weakly ionized unmagnetized plasma with sufficiently high neutral density that they dominate the ion and electron collisions.

Solving the force equation with the addition of a collision term resulting in momentum loss, $-nmv\nu_m$, gives an expression for the particle flux, Γ_j,

$$\Gamma_j = \mu_j n_j E - D\nabla n_j, \tag{45}$$

where

$\mu_j = |q|/m_j\nu_m$, the mobility of the jth species,

$D = k_0 T/m_j\nu_m$, the diffusivity and

$\nu_m = \tau_m^{-1}$, the collision frequency.

Because the electron mobility (Fig. 20) is so much higher than the ion mobility due to the mass dependence, the electrons would tend to run away from the ions. However, then the electric field generated ensures that the ions are pulled along. The net effect is that the transport is ambipolar, and that the effective diffusion coefficient is the mass-weighted mean of the two,

$$D_{eff} = \frac{\mu_e D_i + \mu_i D_e}{\mu_i + \mu_e} \cong 2D_i, \tag{46}$$

for equal ion and electron temperatures.

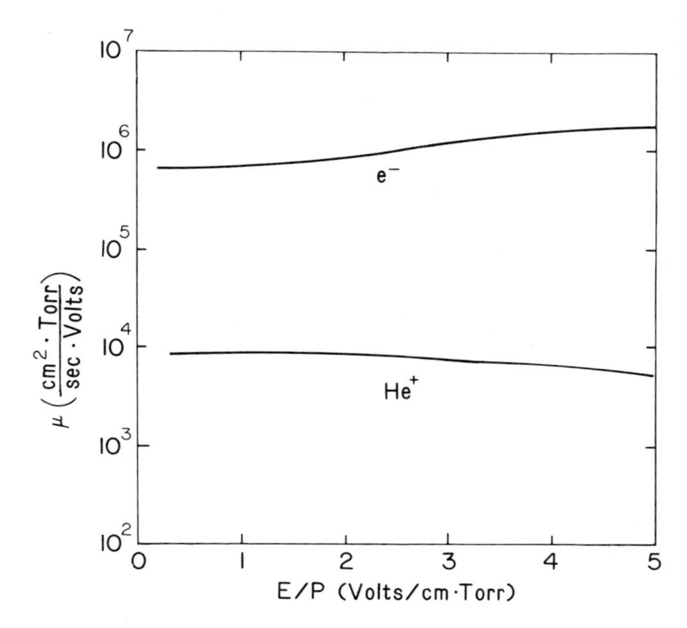

FIGURE 20. Mobilities of electrons and helium ions in helium gas (adapted from Ref. 12).

If the external electric field is shut off, the density distribution will decay in time. The loss processes, being mostly at the wall, ensure that the density there is lowest (Fig. 21). The detailed shape of the density profile depends on where the new plasma is created. The more central the plasma "fueling," the more peaked the density profile.

The inclusion of a magnetic field into this situation requires the inclusion of the $\mathbf{v} \times \mathbf{B}$ term in the force equation. The result is that the mobilities and

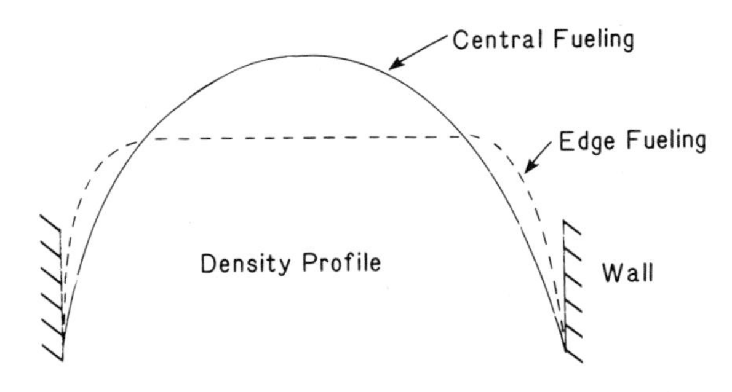

FIGURE 21. Density profile in a plasma device dominated by diffusive losses to the walls. The detailed shape of the profile depends on where the new plasma is created to replenish that which is lost.

diffusivities are modified to

$$\mu_{\perp} = \frac{\mu_i}{1 + \omega_c^2 \tau_m^2} \tag{47}$$

and

$$D_{\perp} = \frac{D}{1 + \omega_c^2 \tau_m^2}. \tag{48}$$

Note the limiting cases for $D_{\perp}$ show the expected dependence on ν_m.

$$D_{\perp} = \frac{k_B T \nu_m}{m \omega_c^2} = r_L^2 \nu_m \qquad \text{for } \omega_c^2 \tau_m^2 \gg 1 \tag{49}$$

and

$$D_{\perp} = \frac{k_B T}{m \nu_m} = \lambda_m^2 \nu_m \qquad \text{for } \omega_c^2 \tau_m^2 \ll 1. \tag{50}$$

For this situation, i.e., $B > 0$ and neutral-dominated collisions, particles will diffuse parallel to the gradients and also flow perpendicular to them (and to B) at the v_D and v_E drift speeds, which are reduced somewhat by neutral-particle drag.

To evaluate the effects of collisions in fully ionized plasmas, we must know the energy-dependent cross section or collision frequency for Coulomb scattering between electrons and ions. An alternate, but equivalent, parameter is the resistivity, $\eta_{\parallel}$. This is related to the collision frequency by

$$\eta_{\parallel} = \frac{m_e \nu_{ei}}{n q^2}, \tag{51}$$

and is equal to

$$\eta_{\parallel} = \pi q^2 m^{1/2} \frac{\ln \Lambda}{(k_B T_e)^{3/2}} \text{ ohm-cm}. \tag{52}$$

The resistivity is the impedance presented by an unmagnetized plasma to a flow of current driven by a static electric field. Note that $\eta_{\parallel}$ is independent of density. This is because, though the number of charge carriers increases with density, so does the number of scattering centers. The effects exactly cancel.

The term $\ln \Lambda$ is due to Spitzer, who showed that the correct way to perform the integration over impact distances was to stop at a separation equal to the Debye length at large distances and at the classical radius of the electron at short distances. The value of $\ln \Lambda$ is about 10 for most laboratory plasmas, ranging from the cold and tenuous glow discharge to hot and dense pinches.

The temperature dependence of $\eta_{\parallel}$ reflects the fact that as a charged particle's velocity increases, it spends less time near enough to another charged particle for its trajectory to be greatly altered. So the hotter a plasma, the less effective is ohmic heating.

When a magnetic field is applied, the resistivity parallel to B is the same, but that perpendicular to B increases. $\eta_\perp$, as defined by the collision rate (c.f. Eqn. (51)), is about 3.3 $\eta_\parallel$. But the perpendicular resistivity, as defined by a flow of current in response to an applied electric field, increases much more, about $3.3(1 + \omega_c^2\tau_{ei}^2)$. Remember, in the absence of collisions (e-i, i-i, or quasi-), the application of an electric field to a magnetized plasma results in a drift perpendicular to **E** and **B**, not current parallel to **E**.

Solving for the diffusion of particles, not a flow of current, in this case gives

$$D_\perp = \frac{3.3\eta_\parallel nkT}{B^2}. \tag{53}$$

In contrast to diffusion caused by neutral impacts, $D_\perp$ depends on the plasma density. So the higher the density, the more rapid the loss. Also note the dependence on k_BT_e/B^2. This is just the step size, r_L^2, of the random walk induced by the Coulomb collisions.

The diffusion coefficient in Eqn. (53), however, only represents a lower limit on how quickly plasma will be lost. Experiments usually show a loss rate 10^4 times faster. One explanation of this was that presented by Bohm. He estimated that the transport could be driven by self-generated electric field fluctuations in the plasma of strength k_BT_e/R, where R is a characteristic distance in the plasma. These would then cause a drift velocity that can be converted into an effective diffusion coefficient

$$D_B = \frac{k_BT_e}{16qB}. \tag{54}$$

The coefficient, 1/16, does not come from Bohm's theory, but is a factor that fits many experiments.

We have seen that there are five different diffusion coefficients (Fig. 22), which may be used to describe the loss of plasma to the walls and other structures of plasma devices. These are: $D_\perp$ and $D_\parallel$ for ion/neutral collisions; $D_\perp$ and $D_\parallel$ for ion/electron collisions; and D_B. Which is the correct one for a particular situation will depend on the detailed plasma parameters. These will be presented in practical units in the next section. But as rules of thumb, the following guidelines can be used:

1. If the magnetic field is zero and the pressure exceeds 10^{-2} Torr, then the neutral particle collisions dominate plasma transport and resistivity.
2. If the neutral pressure is less than 10^{-3} Torr and a magnetic field > 100 g is present, use D_B for perpendicular transport of ions and $2D_{\parallel ii}$ for parallel transport.

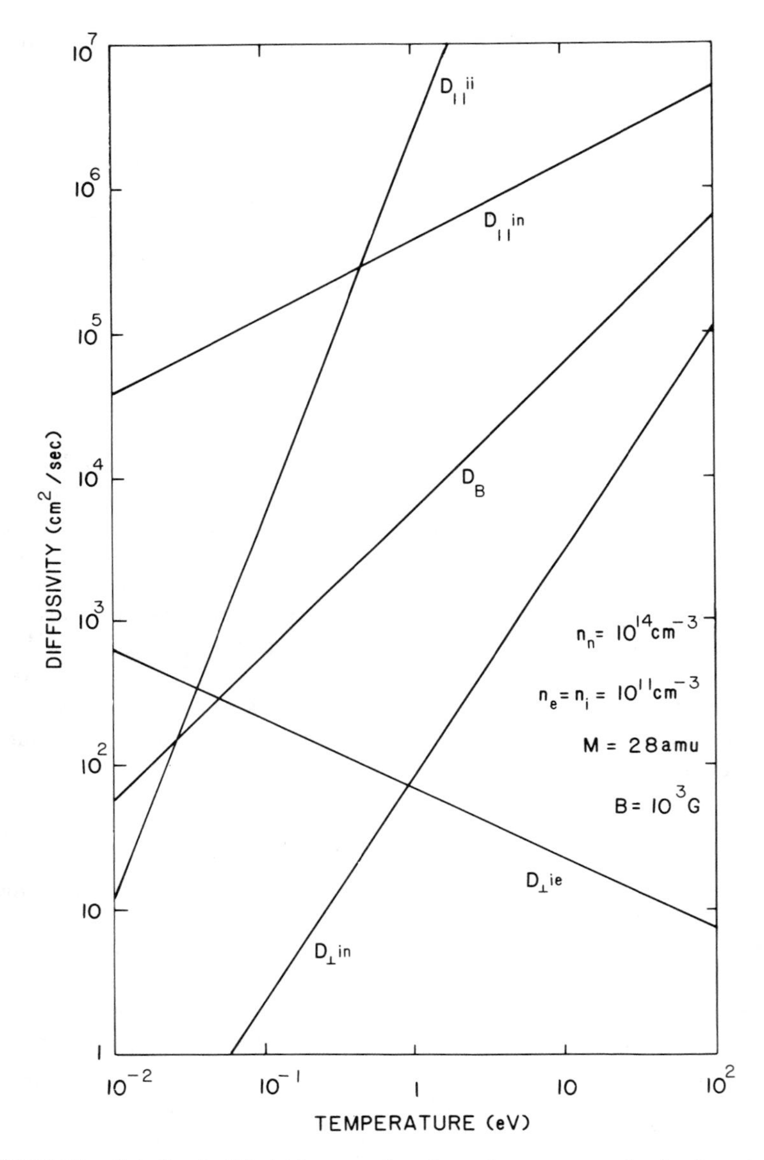

FIGURE 22. Calculated diffusivities as a function of temperature for fixed neutral and electron densities, ion species, and magnetic field.

Collisions of plasma particles with surfaces can be divided into three types depending on the impacting species: neutrals, ions, and electrons. Here we consider only the physical, not chemical, processes. The latter are more appropriately handled in the other chapters of this book.

Neutral atoms and molecules that hit surfaces generally accommodate thermally with the surface and then desorb. Their residence time is typically 10^{-13} sec. This can be altered by cooling the surface or increasing its chemical affinity for the gas atoms. In plasma processing equipment, most neutral atoms have energies of only 1/40 eV. Hence they do not sputter or cause the emission of secondary electrons or ions. A counter-example to this guideline is charge-exchange neutrals formed by interactions of cold neutrals with energetic ions. This would predominantly occur in the sheath where the ions are most energetic. Other counter-examples are molecular fragments (Franck-Condon neutrals) and sputtered atoms. Yet another will be noted shortly.

Ions that impact metal surfaces are rapidly neutralized by the free electrons in the metal unless special conditions are present, one example of which is Cs atoms impinging on solid hot W. The Cs desorbs as ions because the work function of W is lower than the ionization potential of Cs.

The impact of ions can result in their immediate reflection, implantation, and/or delayed release. The immediate reflection is a process that has not

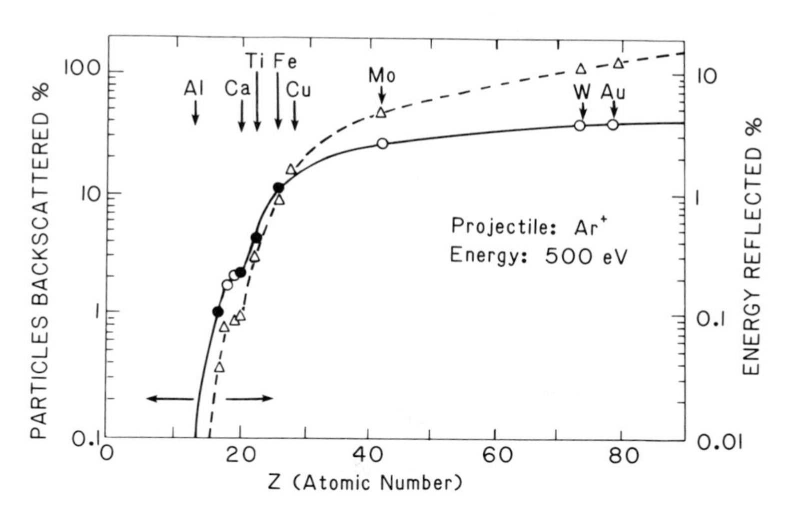

FIGURE 23. Calculated particle and energy reflection coefficients for 500 eV Ar bombardment (normal incidence) of various solids.

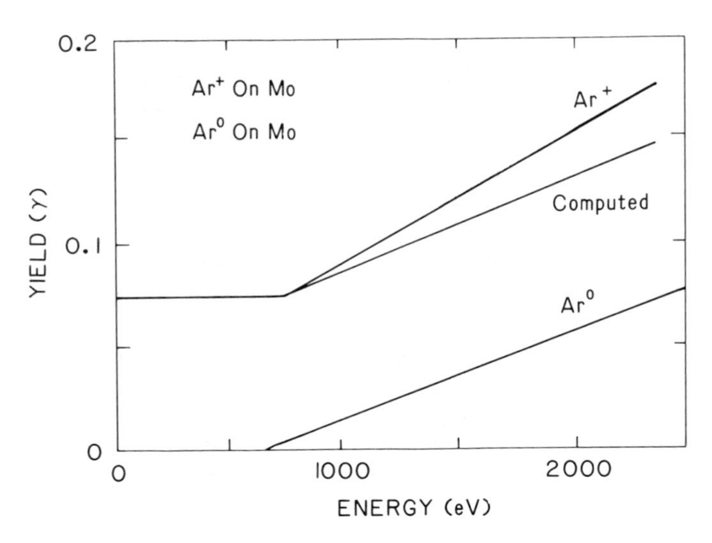

FIGURE 24. Secondary electron yield from molybdenum due to argon atom and ion bombardment (adapted from Ref. 13).

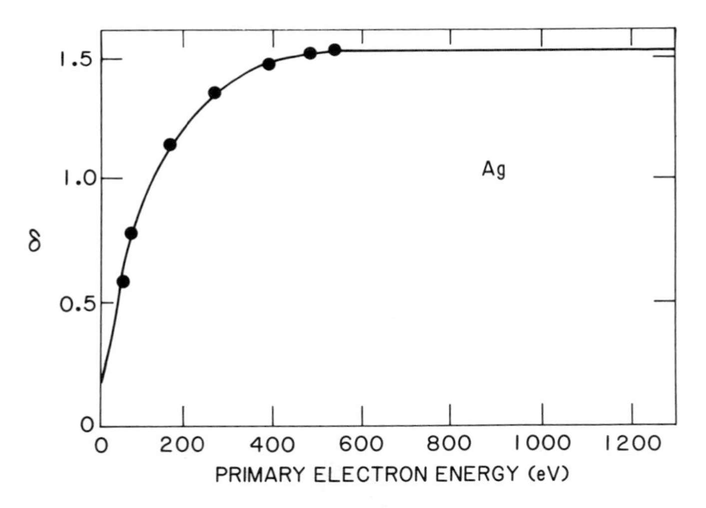

FIGURE 25. Secondary electron emission coefficient from silver as a function of incident electron energy.

been measured accurately. Calculations show the following general behavior (see Fig. 23). If the projectile is more massive than the surface atoms, the projectile will imbed itself in the solid, provided it has enough energy (about 20 eV). If it is less massive, then the projectile will be reflected with about a 30% probability. The reflected atoms have about 30% of their incident energy. The theoretical models are not yet accurate enough to predict the behavior at lower than about 20 eV of energy.

Delayed release is due to diffusion of an implanted atom back to the surface and its subsequent release by thermal desorption, electron-, ion-, or photon-impact desorption, or recombinative desorption.

When ions hit surfaces they can also cause the emission of secondary particles. Sputtered particles are generally of little influence on the plasma behavior; but secondary electrons play a major role. Figure 24 shows the energy-dependent yield of Ar ions and neutrals on Mo. At energies below 1000 eV, potential emission, caused by the deep potential energy well of noble gas ions, gives a constant value for the yield of about 0.1. The yield is about twice larger for He ions and twice lower for Kr ions.

When electrons hit metal surfaces, they too can cause secondaries to be emitted, and their yield can exceed unity. Figure 25 shows the energy-dependent yield for electron bombardment of silver. The yield grows to a maximum at about 400 eV and then slowly decreases.

E. SUMMARY OF PLASMA PARAMETERS IN PRACTICAL UNITS

In Table 1 the expressions for the plasma parameters are converted from cgs (electromagnetic) to practical units. In Section III.G. this was done for the drift velocities. We maintain the same definitions here.

F. INSTABILITIES

The other phases of matter—gases, liquids, and solids—may be readily prepared in states closely approaching thermodynamic equilibrium. Plasmas, in contrast, rarely exist in the laboratory in such states. In the processes of forming, maintaining, and confining plasmas, numerous nonuniformities and anisotropies develop that can result in instabilities quickly connecting the relatively hot plasma to the cooler walls of the

Table 1

Parameter	Symbol	Practical Units	
density	n	cm^{-3}	
ion charge	Z	1 for H^+, 2 for $0^{+2}, \ldots$	
temperature	T	eV	
magnetic field	B	gauss	
ion mass	M	amu	
Plasma Parameter	**Symbol**	**Expression in Practical Units**	
Lengths			
electron gyroradius	r_{Le}	$2.4\ T_e^{1/2}/B$	cm
ion gyroradius	r_{Li}	$102.\ (MT_i)^{1/2}/ZB$	cm
Debye length	λ_D	$743.\ (T_e/n)^{1/2}$	cm
skin depth (transv)	$\delta_\perp$	$5.3 \times 10^5\ n_e^{-1/2}$	cm
Frequencies			
plasma frequency	ω_{pe}	$5.64 \times 10^4\ n_e^{1/2}$	rad/sec
electron cyclotron	ω_{ce}	$1.76 \times 10^7\ B$	rad/sec
ion cyclotron	ω_{ci}	$9.58 \times 10^3\ B$	rad/sec
collision frequency			
ion-ion	ν_{ii}	$5. \times 10^{-7}\ Z^2 n_i/T_i^{3/2}$	/sec
electron-electron	ν_{ee}	$3. \times 10^{-5} n/T_e^{3/2}$	/sec
ion-neutral	ν_{in}	$7. \times 10^{-9}\ n_n(T_i/M)^{1/2}$	/sec
electron-neutral	ν_{en}	$5. \times 10^{-8}\ n_n T_e^{1/2}$	/sec
electron-ion thermalization ($T_i < T_e$)	ν_{ei}	$3. \times 10^{-9}\ Z^2 n_e/MT_e^{3/2}$	/sec
Diffusivities			
Bohm	D_B	$6.3 \times 10^6\ T_e/B$	cm^2/sec
Unmagnetized			
ion-neutral	$D_{\parallel in}$	$3. \times 10^{20}\ (T_i/Mn_n^2)^{1/2}$	cm^2/sec
ion-ion	$D_{\parallel ii}$	$4. \times 10^{18}\ T_i^{5/2}/(Mn_i Z^2)$	cm^2/sec
Magnetized			
ion-neutral	$D_{\perp in}$	$1.5 \times 10^{-4}\ n_n(T_i/M)^{3/2}/(ZB)^2$	cm^2/sec
ion-electron	$D_{\perp ie}$	$5.5 \times 10^{-4}\ n_e/(B^2 T_e^{1/2})$	cm^2/sec
Miscellaneous			
plasma parameter	N_D	$1.7 \times 10^9\ T_e^{3/2}/n_e^{1/2}$	
Spitzer resistivity	$\eta_\parallel$	$0.1\ Z/T_e^{3/2}$	ohm-cm

containment vessels. In some cases, such as the simple z-pinch, instabilities cause the plasma to disappear rapidly; in other cases, such as the abnormal glow, the plasma may exist in steady state with its instabilities.

In plasmas, the definition of equilibrium is broadened to include any situation that is quasi-static, i.e., where the local parameters do not change with time. This need not be a thermal equilibrium—a Maxwellian distribution need not exist. Instabilities, by this criterion, are processes that tend to bring the plasma to a more uniform spatial or velocity distribution on a time scale faster than the Coulomb-collision time.

What drives instabilities are internally generated electric and magnetic fields. The free energy available to make these fields comes from plasma inhomogeneities, both in density and in velocity space. And if these provide unbalanced forces, individual volume elements of the plasma will be convected toward the walls. The question arises, can plasmas be prepared in configurations where these instabilities do not occur or are unimportant? The answer, from experience, is yes. To understand how this is accomplished, we will review the main categories of instabilities and see what promotes and what stabilizes them.

The usual discussion of instabilities is limited to those that exist in the plasma phase alone, with no regard to plasma-wall type instabilities other than how the wall sets up boundary conditions for the electric and magnetic fields. We shall make an effort to remedy this neglect by including information on unipolar arcs. These are a problem not only for what they do to the plasma (quench it), but also for the stresses they place on the power supplies and for the micron-sized droplets they spray around the interior of the containment vessel.

Instabilities cannot be understood using the single-particle equations of motion alone. In these, there is no feedback mechanism to amplify (or damp) the motion of a particle. So we will rely on a pictorial presentation to clarify the physics concepts. The first step is to set up the classification scheme.

As implied earlier, there are two main sources of free energy that may drive instabilities: nonuniform density and temperature profiles; and non-Maxwellian velocity distributions. The first has the clearly apparent tendency to push plasma from a region of higher temperature, density, or pressure to a region of lower. The second manifests itself by generating waves, which predominantly transport energy from one region to another.

Each of these two major categories is further divided into two: an electrostatic and an electromagnetic subdivision. Electrostatic instabilities result from charge accumulation. If $B = 0$, these will usually have a scale size of order λ_D. Electrostatic instabilities along B will behave similarly.

Electromagnetic instabilities result from and reinforce nonuniform current distributions. Their scale lengths and development times are considerably longer than electrostatic instabilities.

Let us now look at two particular instabilities, one from each of the major divisions. These are chosen because many present-day plasma processing devices should be susceptible to them. Yet the devices work well in spite of or possibly because these are present.

The first is the Rayleigh-Taylor instability. This instability occurs in fluid mechanics when a dense fluid is supported over a light fluid in a gravitational field (Fig. 26). A slight perturbation of the initially flat interface between the two fluids results in the heavier one falling through the lighter; and the lighter one rises. The net effect is a conversion of potential energy into kinetic. If this process is rapid, turbulence may set in, eventually turning the directed kinetic energy into heat.

In plasmas, the situation is rather similar. First we shall describe the closest analog and then generalize to a system more appropriate to plasma processing, a mirror machine.

In the closest analog, a plasma with an inverted density gradient replaces the fluids. And a magnetic field is applied transverse to the gravitational field in an effort to prevent the denser plasma from falling. We start with what should give the best confinement of the plasma by the magnetic field, the ideal MHD (magnetohydrodynamic) assumption. This is that the plasma is such a good electrical conductor any magnetic field that permeates the

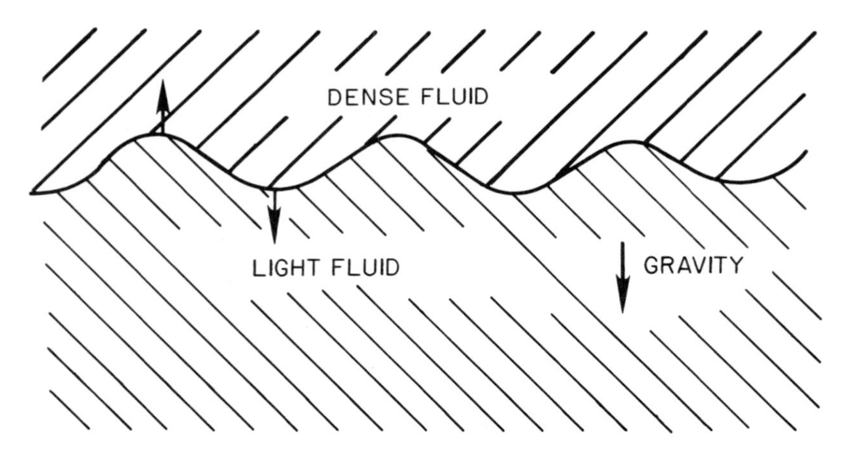

FIGURE 26. Rayleigh-Taylor instability for a dense fluid being supported by a light one in a gravitational field. A perturbation to the fluid interface results in the heavy fluid falling and the light one rising.

plasma is frozen in. In the absence of drift-type motions, charged particles stay on the field lines on which they were born. Will this stop the fall? The picture of what happens when a ripple-shaped perturbation occurs is shown in Fig. 27. The particles of different charge migrate sideways to different sides of the ripples because of the drift due to $\mathbf{F}_g \times \mathbf{B}$, i.e., $\mathbf{v}_g$. This sets up an electric field, which then causes a downward $\mathbf{E} \times \mathbf{B}$ drift of the denser plasma. So the situation is unstable. The magnetic field did not prevent the plasma from falling in the gravitational field.

Of course gravitational forces are unimportant to laboratory plasmas because they are so weak compared to the electric and magnetic ones. So now we go to the mirror machine where a related type of instability, the interchange mode, is predicted and observed to occur.

The gravitational force is replaced by the centrifugal force experienced by the particles as they traverse the curved sections of the magnetic field. When the curvature is outward, as in the central section, we have an exact analogy to the Rayleigh-Taylor problem. The plasma will tend to expand outward there, "falling" into the lower magnetic field region. Nearer to the reflection points, where the B field is strongest, the curvature, hence the centrifugal force, is inward and the situation is stable. Since particles are constantly moving between the two regions, it is not straightforward to say which region dominates. In the simplest mirror arrangement the situation turns out to be unstable. But judicious shaping of the field by adding other coils can yield stability. The two most common solutions are named Ioffe bars and baseball coils and are shown in Fig. 28. These coils create a region of true minimum B near the axis of the mirror machine. Then particle excursions away from the axis are accompanied by a restoring force, which prevents the displacement from growing.

The name interchange mode is given to this type of instability because, from the ideal MHD assumption, one sees the picture that as the plasma

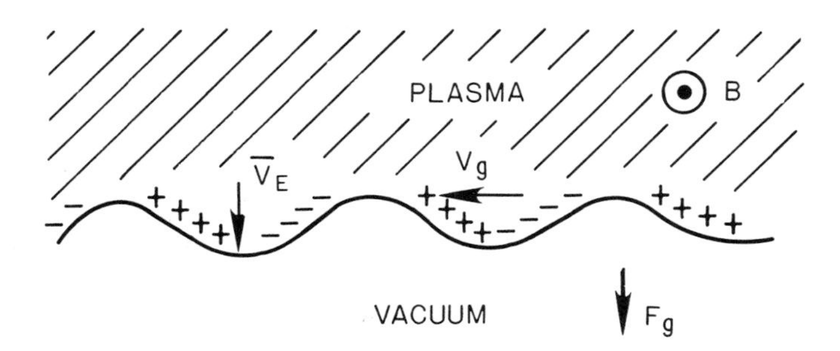

FIGURE 27. Rayleigh-Taylor instability in a magnetized plasma.

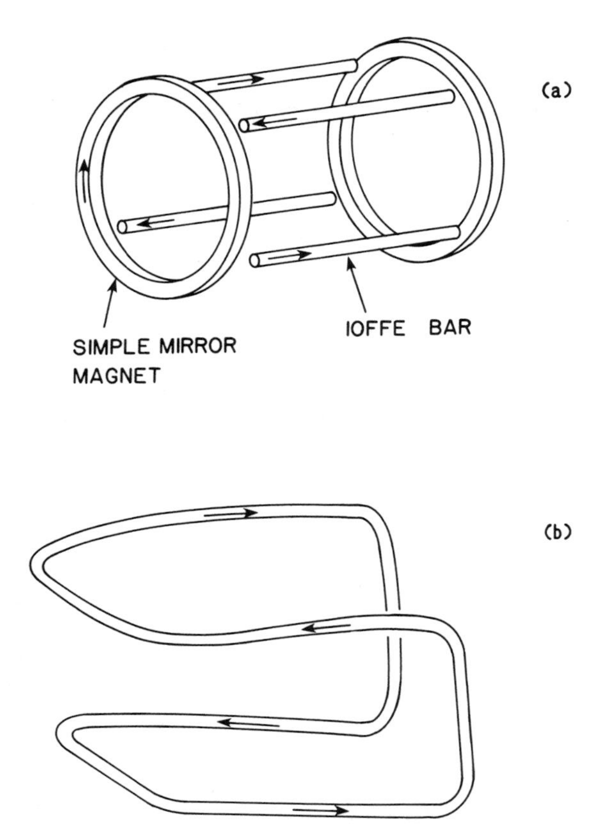

FIGURE 28. a) Stabilization of the simple mirror configuration against Rayleigh-Taylor type instabilities may be obtained by the addition of four Ioffe bars. These cause the net magnetic field to increase with distance from the mirror coil axis. b) Another means to provide stability is to combine the two sets of coils into a single coil with the shape of the stitching on a baseball. The arrows show the direction of current flow.

moves, it carries the magnetic field with it. So, during an instability, one *tube of magnetic flux* may interchange its position with another. This gives yet other views of how to stabilize the plasma. For example, by putting *shear* in the magnetic field (Fig. 29a), one can prevent the flux tubes from sliding past each other. Also by attaching the magnetic field to stationary conductors (Fig. 29b) (remember the plasma is a movable conductor) one

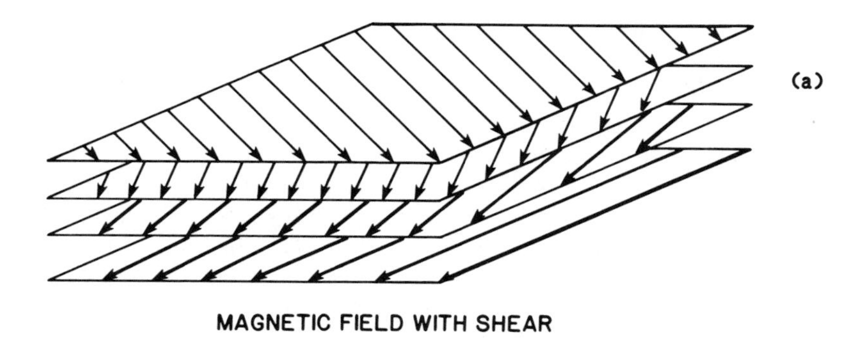

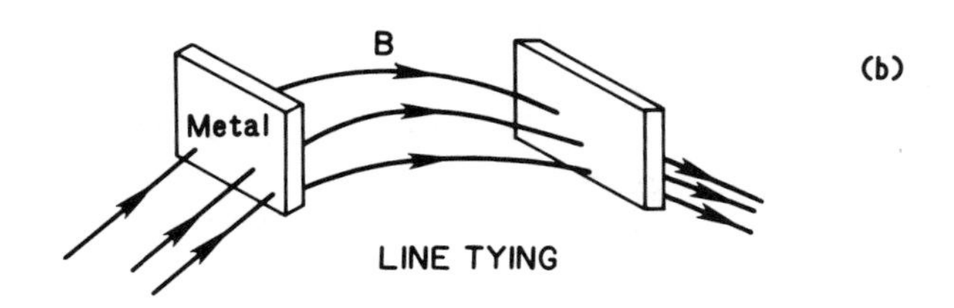

FIGURE 29. Two methods to stabilize interchange modes: a) magnetic shear, and b) line tying.

can inhibit the field from moving. This is called *line-tying*. With the field lines anchored at their ends, the only way plasma motion can carry imbedded field lines is by stretching or bending them. But this frequently requires too much energy. In short, conducting (metal) plates at the ends of the mirrors help to stabilize the plasma.

The instability we just discussed arises from the shape of the confining fields. The configuration caused a spatial deviation from thermodynamic equilibrium. The second type of instability we consider can occur in an infinite and spatially homogeneous plasma. The free energy source is the directed energy of a beam of electrons; the instability is called the two stream. It arises from nonuniformities in velocity space.

In the two-steam instability a beam of particles moves past a counter-streaming beam. Of course one can allow the second beam to be stationary and still get the same results because of translational invariance. For simplicity, we then consider the second stationary and call it the target. This type of situation occurs in most processing plasmas because secondary electrons liberated from the cathode are accelerated to several hundred electron volts by passing through the cathode sheath. They then move through the relatively cold ($T \simeq 5$ eV) background plasma. Classical Coulomb collisions are too weak to transfer the energy from the beam to the target particles. How then is the energy transferred? By instabilities. Consider a perturbation to the streaming electrons, which, for simplicity in analysis, is a single normal mode of motion for this group of particles. One highly likely mode is the plasma oscillation, but with its frequency now Doppler-shifted by the streaming motion. The phase velocity of the perturbation is then

$$v_p = v_{e0} - \omega_{pe}/k \tag{55}$$

where

$$k = \text{the wave number} = 2\pi/\lambda .$$

When this plasma wave is in resonance with a wave that the target particles can carry, (another normal mode), the wave can grow. This is precisely what happens when the target particles are massive ions. Their plasma frequency is $(M/m)^{1/2}$ lower than the electrons'. The resonance takes place when

$$v_p = \omega_{pi}/k . \tag{56}$$

Buneman has described this situation with a musical analogy: the ions are the bow and the electrons the string. To be somewhat more precise, one should consider the bow stationary and the violin moving.

The fastest growing modes have frequencies near the plasma frequency. The directed energy of the beam of electrons creates plasma waves with a level much higher than thermal. This then couples to the background electrons and heats them. So in this case an instability has a very beneficial effect. It withdraws energy from the tenuous stream of energetic electrons and heats the denser background of colder electrons.

We now finish our discussion of instabilities with the unipolar arc. As noted earlier, introducing an electrically floating metal object into a plasma results in the build-up of negative charge on the object and a plasma sheath around it. In this analysis we neglected the possibility of electron emission from the metal. Secondary electrons could be emitted due to electron or ion impacts or even photon impacts. These secondary electrons would be accelerated away from the metal by the sheath. A plasma arc may ignite on the metal surface if a small spot overheats due to the emission of secondary electrons. Then the emission process is enhanced by thermionic emission and the process runs away, i.e., becomes unstable. The arc spot may reemit essentially all the electron current that is collected by the entire metal object. Then the sheath voltage drops and more electron current is collected; and more current may be reemitted through the spot. This is the positive feedback mechanism of the unipolar arc. The cathode is the single solid electrode. The plasma acts as both the other electrode and the electrolyte of this battery (Fig. 30).

When such an arc occurs, the metal object is melted at the arc spot. The metal is explosively released as both vapor and molten droplets of 1 micron typical size. The arc spot on the metal is a few microns in diameter. It is held this small by the combined effects of current channel constriction and thermionic emission. If a magnetic field is present, the arc spot is driven across the surface of the metal by $\mathbf{J} \times \mathbf{B}$ forces. However, the motion is retrograde, that is, in the opposite sense of what is expected from $\mathbf{J} \times \mathbf{B}$. Numerous different mechanisms have been proposed to explain this, but none have yet been proven.

How does one prevent such an arc? There are several methods. First, remove arc initiation sites. These are small embedded insulating occlusions

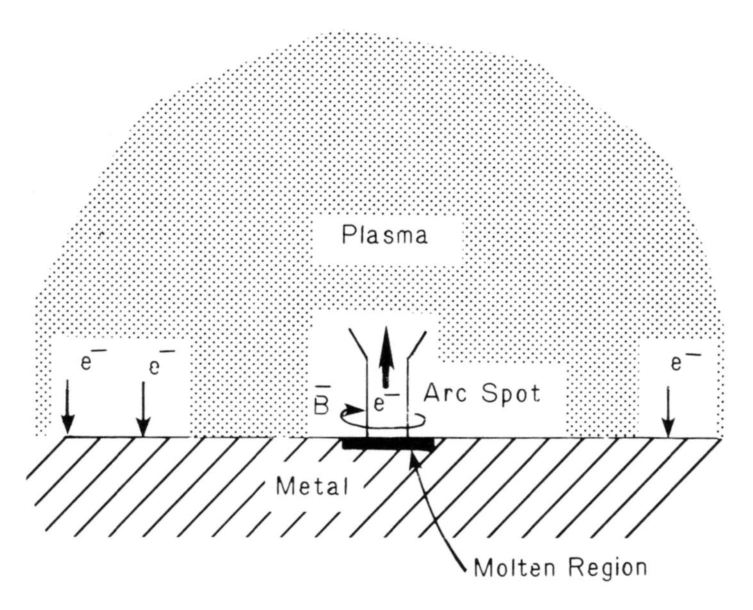

FIGURE 30. Schematic of a unipolar arc. Intense electron emission of all the electrons collected by the entire area of a metal object may occur at one small (1 micron) spot. The metal at that spot melts. The arc will move across the surface of the metal in a direction determined by externally applied magnetic fields interacting with its own fields and currents.

where charge can be built up. Their small size means that strong electric fields will form. Then a spark occurs that ignites the arc. These bits of debris can be removed by standard clean room techniques.

The second method is to avoid metals that form insulating oxides or avoid the formation of oxides. Aluminum is an example of such a problem-prone material. In the case of aluminum, the thin oxide that builds up is a good insulator although it is very thin, 50 Å. Electrons that impact it can raise the potential to a few volts, which then is an electric field of in excess of 10 million volts/cm through the film. The insulating film explodes, releasing many secondary electrons (the Malter effect) and the arc ignites. If aluminum must be used, however, oxygen should be avoided in the plasma chamber. If that too cannot be avoided, then the plasma bombardment of the aluminum must be kept at such a high rate as to keep the aluminum surface clean of oxide.

The third method is to reduce the size of the metal object. This reduces the amount of current that can be reemitted through a potential arc spot.

The final method is to keep the electron temperature low at the metal object. This too reduces the amount of current available to maintain an arc.

G. PLASMA WAVES

The previous section dealt with instabilities. These can be viewed as over-developed waves. Both start out as normal modes of the plasma motion. But instabilities grow to a point wherein they terminate the plasma or, at least, alter its properties greatly. Waves do not change the plasma much. But exactly how much is the reason for their importance to plasma processing equipment.

The most important application of plasma waves is the heating of plasmas. There are other ways to heat plasmas, ways that do not rely on the resonances implicit in wave heating. One method is classical collisional (Ohmic) heating. This was described earlier. Another method, non-resonant rf heating, will be described in Section V and is also a form of Ohmic heating. There are other applications of waves, such as plasma diagnosis, an old and venerable occupation. These are discussed in another chapter. In this section we review some concepts of wave propagation, list the main types of waves in plasmas, and note which ones will be important for heating process plasmas to increase their temperature and possibly their density.

Before embarking on a discussion of the types of waves that may exist in plasmas, we review some elements of general wave terminology. All the waves we consider are small amplitude; they do not perturb the plasma greatly. Linearized equations can then be used to see the changes in plasma properties due to the presence of the waves. The waves are also monochromatic; they are describable by a function $\exp(i(kx - \omega t))$.

The phase velocity of a wave is given by ω/k and the group velocity is given by $d\omega/dk$. If all frequencies (or wavelengths) of a particular type of wave have the same phase velocity, the wave is called dispersionless. A wavepacket composed of a group of such waves will propagate without changing shape.

The equation relating ω to k is called the dispersion relation. In solving the dispersion relation for ω or k, complex values may arise. The real parts of the solution correspond to stationary or propagating waves. The imaginary parts correspond to growing or decaying waves; these result from dissipative processes or resonances in the plasma.

Different types of microscopic mechanisms can result in wave growth or decay. Classical collisions with ions, electrons, or neutrals will cause the decay of a wave. Resonances with other waves, as we saw in our discussion of the two-stream instability, can result in wave growth (or decay). *Landau damping* and *inverse-Landau damping* refer to a related phenomenon wherein fast moving electrons "surf" on the crests or in the valleys of charge density (plasma oscillation) waves. In this manner the charged particles either extract energy from or deposit energy into the waves.

As waves propagate through inhomogeneous media, the wavelength will change in a fashion prescribed by the dispersion relation. If the wavelength grows to infinity, a condition called *cutoff*, the wave is reflected. If the wavelength shrinks to zero, a resonance, absorption, occurs.

Coupling a particular wave to a plasma may be a difficult matter. Electrostatic waves do not propagate in vacuum. Hence any antennae or launchers for these must be located in close proximity to the plasma. And as indicated in the previous paragraph, once a wave is launched into a plasma it need not be absorbed. It can be reflected from the plasma interior or be coupled out of the plasma by an electrode in contact with the plasma.

As in the case of instabilities, the understanding of waves can be greatly simplified by categorizing the various modes. This time electrostatic (longitudinal) and electromagnetic (transverse) form the main categories. The next division is usually made based on the species whose properties dominate the wave motion. That is, the subdivisions are electron and ion waves. Ion waves are lower frequency than electron waves because of the greater mass of the ion. Hybrid waves also occur where the coupled motions of both species must be considered in detail.

A further subdivision is based on the role of the magnetic field. If the wave drives the charged particles across B, cyclotron frequencies naturally occur in the dispersion relation. If the charged particles only have to move along B in response to the wave fields, then the wave is usually the same as in an unmagnetized plasma.

Yet another subdivision is based on whether the plasma is hot or cold. This is a loaded question for processing plasmas, which are relatively cool compared to most other plasmas. One tricky aspect of this question is whether gyroradii are large. Then the massive ions, though rather cold, do dominate the thermally modified behavior of certain waves.

With this basic description of waves, we can now state which subdivisions will be of most use in heating process plasmas. First, it is clear that we want to heat electrons in order that the ionization rate increase. Heating ions would be beneficial if it were desirable to have their energies exceed the electrons' energies. However, the sheath is usually adequate for providing ion energy, mostly to ions impacting surfaces. So we will simply ignore ion waves (such as sound waves or ion-cyclotron waves).

To proceed further we must know whether a magnetic field will be present in the particular device. If the answer is no, then there are only two choices, one from each major category. The electrostatic wave turns out to be our old friend the plasma oscillation; only now it has a thermal part in the dispersion relation. The electromagnetic wave is just a standard light wave. In fact, we derived the dispersion relations for these two waves when we were discussing the skin depth (see Eqns. (23) to (37)). It is difficult to heat the plasma interior with a low-frequency electrostatic wave because its plasma skin depth is only the Debye length. But it is not too difficult to get the low-frequency electromagnetic wave to penetrate deep into the plasma. The question is: will it be absorbed? If so, good. If not immediately, then the wave will rattle around inside the metal containment vessel until it is either absorbed by the walls or plasma or can escape through glass (insulating) windows (or feedthroughs). The exact details have to be worked out for each plasma, taking into account resistivity, collisions, the size of the containment vessel (which determines the lowest frequency, or longest wavelength, allowed), and so forth. But this wave is a good candidate for heating plasmas.

If there is a magnetic field present, there are many other types of waves possible. Of course the electron plasma wave could still be used if the **E** vector was oriented along **B**. Electron cyclotron waves offer an alternative. There are actually three types of electron cyclotron waves, the extraordinary X, the L, and the R. The X propagates perpendicular to B; the L and R parallel. In these cases the frequency of the wave is fixed. For a 1000 g field the required frequency, $\omega/2\pi$, is about 3 GHz. The wave could be launched from a waveguide a distance from the plasma, but wave absorption would only take place in a narrow region, the resonance zone. Depending on field uniformity, this could range from a few millimeters in size to several centimeters.

V. Discharge Initiation

Plasma production and maintenance are two processes bound together by the same basic phenomena: ionization and recombination, plasma transport, and plasma-surface interactions; and one inexorably leads into the others. In this section we will discuss two different techniques for initiating plasmas (dc and rf) and describe the processes responsible for determining the electron temperature achieved in one of them during steady-state operation. In both cases we shall see that the electrical parameters present during discharge initiation are rather different from those of the steady-state phase.

A. DC GLOW

Consider an evacuated glass tube with two internal metal plates separated by a distance *d* (Fig. 31) and biased by an external dc power supply. The electric field between the two plates is approximately uniform, except for the fringing field near the plate edges. If a low pressure gas is introduced into the chamber what might occur? A charged particle, say an electron, may be liberated near the negatively biased electrode, the cathode. The cause of this electron could be the passage of a cosmic ray through the apparatus, the decay of a radionuclide in the metal, photo-electron emission by the sun or fluorescent lights, field emission from the metal plate, or a

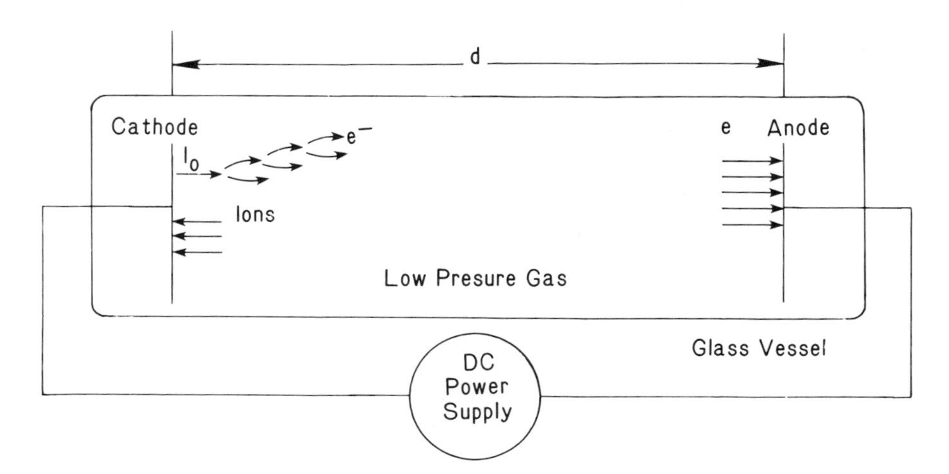

FIGURE 31. Initiation of a glow discharge. An initial current from the cathode is amplified by ionization processes in the low pressure gas. The electrons are collected by the anode. Ions formed in the ionization events are accelerated into the cathode where their impact releases secondary electrons (not shown), restarting the cycle.

host of other processes. The electron now finds itself in the applied electric field and starts to accelerate toward the anode. If the gas pressure is low enough the electron will not suffer any collisions as it proceeds to the anode. It will arrive at the anode, possibly ejecting a secondary electron. However this electron, being born so close to the anode, will simply return there; and there will be no free charges left in our apparatus.

But if the pressure is sufficiently high, the electron will collide with neutral atoms along its path. If the applied electric field is low, the electron will not gain sufficient energy to ionize the atoms. It will simply drift at a speed determined by its mobility. And if the neutral gas is sufficiently electronegative, the electron will attach itself to the atoms. These negative ions will have a substantially smaller mobility.

Let us now consider an applied electric field of such strength that the electron gains enough energy between collisions that it may ionize the next atom it impacts. What happens to the charge pairs created in the volume? The ions drift to the cathode and the electrons to the anode. Secondary electrons are emitted at both. But, as previously noted, only those released from the cathode are useful. The current of electrons toward the anode, $I_e(x)$, grows exponentially with distance from the cathode at a rate determined by the number of ionizations per unit distance, α, the first Townsend coefficient.

$$I_e(x) = (I_0 + \gamma I_i(0)) \exp(\alpha x), \tag{57}$$

where

I_0 = the current of electrons from the cathode due to photo-emission, for example,
$I_i(0)$ = the current of positive ions onto the cathode, and
γ = the ion impact secondary electron emission coefficient.

The current of ions onto the cathode equals only the ions born in the volume. No ions are born during the creation of the photo-current or the secondary electrons from ion impact. Hence,

$$I_i(0) = I_e(d) - I_0 - \gamma I_i(0). \tag{58}$$

Evaluating Eqn. (57) at $x = d$ and substituting in Eqn. (58) for the ion current to the cathode gives

$$I_e(d) = \frac{I_0 \exp(\alpha d)}{\{1 - \gamma(\exp(\alpha d) - 1)\}}. \tag{59}$$

Two cases for Eqn. (59) can be readily seen, based on the value of the denominator: i) For small values of α, γ, and d, the denominator is unity so the only current that flows is that released from the cathode by the photons, but amplified slightly by the small value of $\exp(\alpha d)$; ii) For moderate or large values of α, γ, and d (such that the denominator

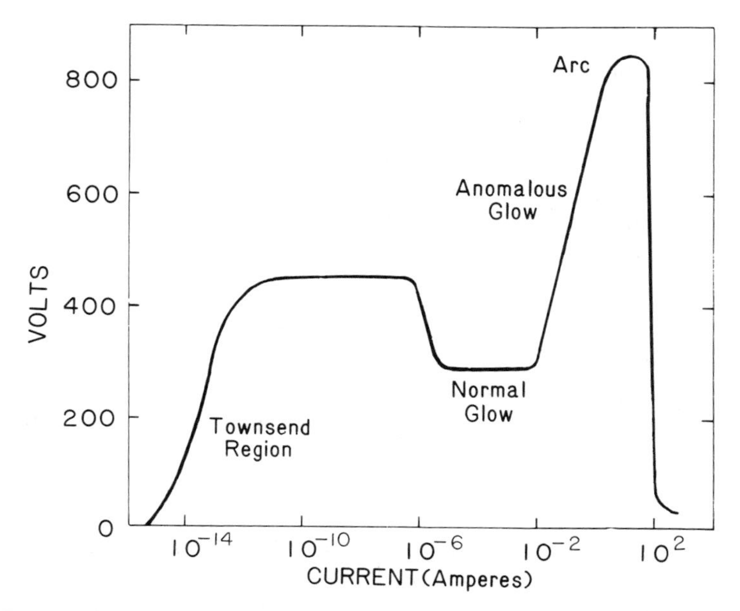

FIGURE 32. Schematic of the I-V characteristics of the apparatus shown in Fig. 31. The metal electrodes are about 100 cm^2 in area and separated by 30 cm. The gas was air at a pressure of 1 T.

precisely equals zero or is negative) the current grows, faster than exponentially, and breakdown occurs. At this point the equation fails because it does not take into account the decrease of neutral atoms in the system because of ionization or the change of potential in the system, and so forth.

The current that flows in such an apparatus is shown versus voltage in Fig. 32. At low voltages, the current grows with the voltage until the ionization probability, related to α, approaches a value large enough to break down the gas. Then the Townsend region occurs, and the voltage is constant as the current rapidly rises.

When the gas is highly ionized, i.e., it is a plasma, the electric field between the two plates has been drastically altered. A sheath exists at the cathode, still promoting the acceleration of ions to the cathode and of secondary electrons into the plasma. The glow does not cover all the cathode. An effort to increase the voltage increases the glowing area and the current. This is termed the normal glow. Note that the voltage required to maintain the normal glow is less than that required to achieve breakdown. This type of constant voltage behavior is useful for high-power voltage regulation applications.

Finally, when the entire cathode is covered with a glow, the I-V characteristics change (see Fig. 32), the current does increase with the voltage, and the abnormal glow has been obtained. This type of discharge is used for many etching processes.

There are several distinct regions of the cathode sheath in a glow discharge. Attached to the cathode itself is the space-charge limited cathode fall region. The ions extracted from the plasma by the cathode fall are limited in current by electrostatic repulsion because the electron density there is less than the ion density. The law describing this was derived by Langmuir and Child for plane parallel geometry and is

$$J_s \left(\text{A/cm}^2\right) = \frac{5.5 \times 10^{-6} V_s^{3/2}}{M^{1/2} d^2}, \tag{60}$$

with V_s in volts, d in cm, and M in amu. Obviously, from the $V^{3/2}$ and d^{-2} dependences, the maximum current that can be drawn increases with the electric field. The $M^{-1/2}$ dependence shows how the particle affects the space charge density near the emitting surface. (For electron emission $M = 1/1836$.) From the current drawn and the value of the cathode fall voltage, the thickness of the region can be calculated. The cathode fall represents the largest potential drop in the plasma. Because ions fall through this drop, they extract most, about 80%, of the energy output from the power supply. The secondary electrons emitted from the cathode extract about 20%, the value of the secondary emission coefficient.

Beyond the cathode fall is the Debye sheath (Fig. 33). Its thickness is about five Debye lengths. This is followed by a presheath region [10]. It is

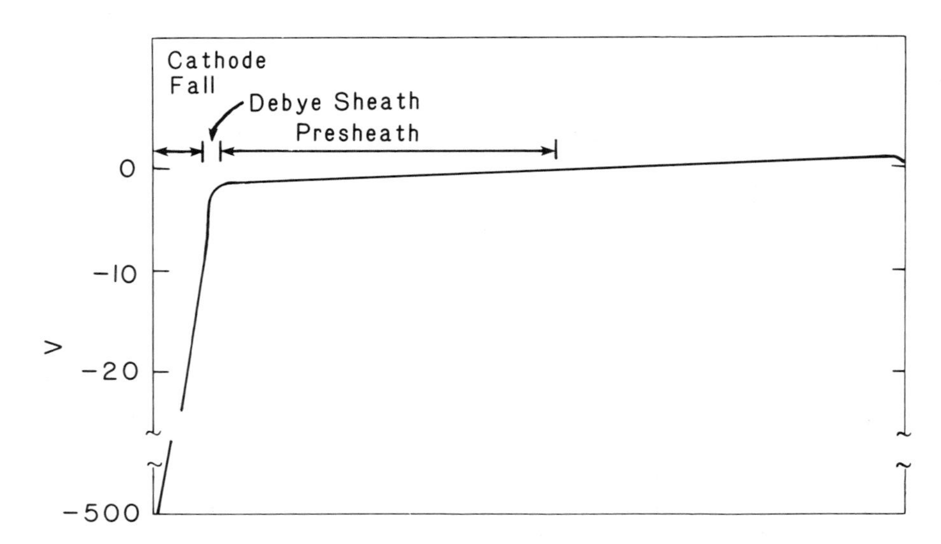

FIGURE 33. Potential as a function of distance for a dc glow discharge.

here that the ions originate which supply the current to the cathode. The continuity condition, called the *Bohm criterion*, relating the flux of ions to the cathode to that from the presheath, requires that the ions leaving the presheath and entering the Debye sheath must have a certain velocity, the ion sound speed. To obtain this velocity, a potential drop of $0.5T_e$ must exist in the presheath. The thickness of the presheath, defined as the distance this potential drop penetrates, depends on the transport mechanism bringing ions into the presheath. The faster the diffusive flow into the presheath from the bulk plasma, the shorter is the presheath. In Section VI we discuss the effects of magnetic fields on the sheath and justify, by a calculation of ionization rates, where the plasma ions are created. Sheaths, for rf and dc situations, are also discussed in the Langmuir probe section of the chapter on diagnostics.

We conclude this section by noting that there is an optimum (minimum) pressure, for a fixed plate separation, to achieve breakdown. If the pressure exceeds this value, electrons gain less energy before a collision and the ionization probability decreases. If the pressure drops below the optimum, then the electrons travel too far before having an ionizing collision and again the voltage required to breakdown increases. A graph of this behavior, called the Paschen relation, is shown in Fig. 34.

B. MICROWAVE BREAKDOWN

If a time-varying electric field were applied to the electrodes, little change in breakdown behavior would be noted until the field reversals occurred

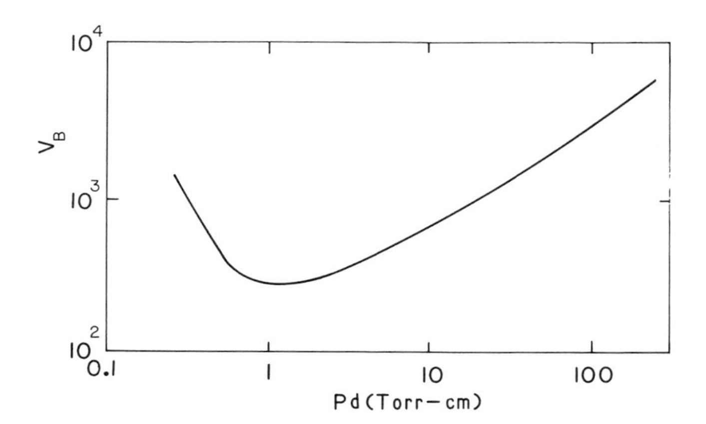

FIGURE 34. Paschen curve, dc breakdown voltage versus pressure *x* plate separation, for hydrogen.

rapidly compared to the particle motions. The slowest particles are the ions. It takes them a few microseconds to traverse the apparatus. Hence, it is not until frequencies of nearly a MHz that an effect is observed. And then it is detrimental; the breakdown voltage rises. The reason is that the rf field interferes with the flow of ions to the cathode, hence fewer secondary electrons are emitted. Once a high enough frequency, about 100 MHz, is applied that the electron losses to the anode via electric-field driven mobility are stemmed, the situation reverses, and a lower breakdown voltage is obtained. At this point the only loss of electrons is by diffusion; and the only loss of electron energy is by collisional heating of the ions and neutrals.

At the walls of the apparatus, the electron losses will build up a sheath. But there is no large cathode fall voltage, only a few T_e. In contrast to the dc glow, it is clear that here most of the applied (rf) electric field energy is used in creating ions, not accelerating them up to high energies. A hybrid apparatus, with rf discharge production and dc electrode bias, is thus a desirable way to separately control plasma density and ion impact energy.

The rf breakdown voltage shows the same general behavior as exhibited by dc discharges, a minimum in the required voltage as a function of pressure. A scale length enters into this problem in the same way that d did in the dc analysis. The scale length is now the excursion distance of the electrons instead of the size of the container. Paschen curves for rf breakdown are shown in Fig. 35.

The effective scale size can be further lengthened by applying a magnetic field. Then the electrons are confined axially. And if the applied frequency

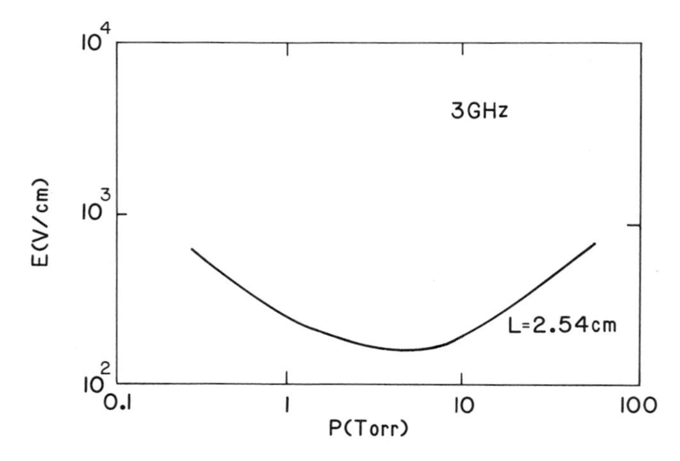

FIGURE 35. Rf Paschen curve for hydrogen in a vessel of 2.54 cm characteristic size (adapted from Ref. 3).

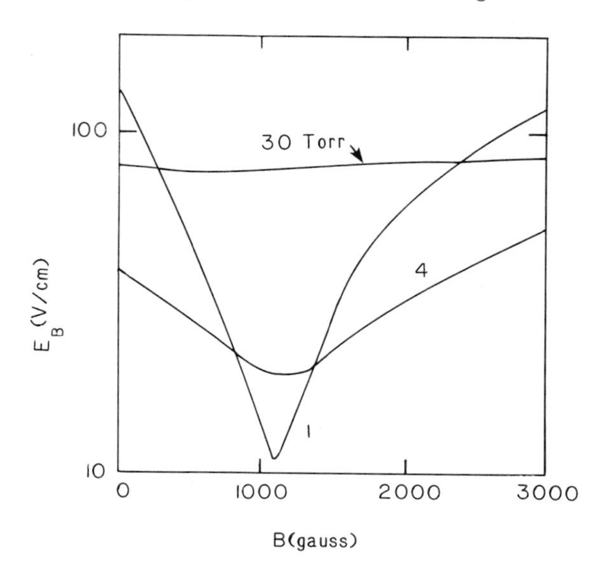

FIGURE 36. The effect of magnetic field on rf breakdown electric field in helium at 3 pressures: 1, 4, and 30 Torr. The applied frequency was 3.1 GHz and the characteristic size was 1.51 cm (adapted from Ref. 3).

matches the electron cyclotron frequency (Fig. 36), then further reduction in required breakdown voltage occurs. This latter improvement is because the electrons now gain energy resonantly from the applied field, instead of "diffusively" as they do when collisional processes are responsible for converting the directed energy into thermal energy.

Let us now examine the electron temperature in a microwave-driven discharge, where the only losses of energy from the electrons are electron-neutral collisions. The force equation is

$$\frac{m\,dv}{dt} + m\nu_{en}v = -qE_0\exp(i\omega t), \tag{61}$$

where ν_{en} is the collision frequency of electrons with neutrals, ω is the frequency of the applied field, and E_0 is its peak amplitude. The solution to this equation is

$$v_f = qE_0\exp\frac{i\omega t}{i\omega m + m\nu_{en}} \cong \frac{qE\nu_{en}}{mv} \tag{62}$$

and the power absorbed by the electrons is

$$P_a = n_n q^2 E_0^2 \frac{\dfrac{\nu_{en}^2}{\nu_{en}^2 + \omega^2}}{4m\nu_{en}}, \tag{63}$$

where $\lambda = v_f/\nu_{en}$ is the mean free path of the electrons and v_f is their thermal velocity due to the electric field causing ohmic heating. The loss of energy by the electrons, assumed to be only to the neutrals, is a fraction $K = 2m/M$ of their thermal energy at each collision. Hence, the energy balance equation is

$$\frac{d\frac{mv_f^2}{2}}{dt} = \frac{qEv_f - Kmv_f^2\nu_{en}}{2} \tag{64}$$

This can be integrated to yield

$$\nu_{en}t = \frac{1}{2}\ln\left\{\frac{1+y}{1-y}\right\} - \tan^{-1} y \tag{65}$$

where $y = v_f/v_f(t = \infty)$. Note that

$$\nu_{en}^{-1} = \left(\frac{1}{2}\right)^{1/4}\left(\frac{m}{e}\right)^2\left(\frac{1}{K}\right)^{3/4}\left(\frac{\lambda}{E}\right)^{1/2} \tag{66}$$

and

$$v_f(t = \infty) = \left(\frac{2}{K}\right)^{1/4}\left(\frac{e}{m}\right)^{1/2}(E\lambda)^{1/2} \sim \left(\frac{E}{P}\right)^{1/2}. \tag{67}$$

And the energy, proportional to $v_f^2(t = \infty) \sim E/p$, is several electron volts. When collisions, or better confinement via magnetic fields, do not form thermal insulation to the wall, it is very difficult to get electron temperatures above ≈ 5 eV because the electron thermal conductivity increases proportional to $T_e^{5/2}$.

Open geometry parallel plates (Fig. 31) do not work well for coupling power at high frequencies; nor do glass walled vessels retain the transmitted (unabsorbed) rf power. The usual way to form a microwave plasma is inside an all-metal vessel fed by a waveguide.

VI. An Application—The Planar Magnetron

In the previous sections we discussed single-particle motions, basic plasma behavior, and discharge initiation. Here we apply our newly gained knowledge to understanding how a particular process-plasma configuration works. We choose the planar magnetron because it has a magnetic field, which is mainly parallel to the electrodes. This generally enhances the particle confinement. Configurations with B normal to the electrodes, as a mirror machine or straight solenoid with end plates, can also get to high plasma densities.

The planar magnetron configuration is more commonly used in deposition than in etching. But related magnetic field shapes have been introduced into commercial etching equipment. Increases in the size of the planar magnetron make it an attractive configuration for both single wafer and batch-etching applications.

A schematic of a symmetrical magnetron is shown in Fig. 37. This magnetron is a figure of revolution. Also common are ones with a race track shape. The poles face each other across a gap of about 2 to 20 cm. For etching applications the larger gap is appropriate for larger wafers. The field strength at the poles is about 2000 g. The field extends away from the poles in the manner shown in the figure.

The surface to be bombarded (called the target) is placed between the poles. It may be flat (as wafers are), angled, bent, or curved. The surface is biased negatively, either by a dc power supply, or by the rectifying properties of the plasma if rf power (typically at 13.5 MHz) is applied to the surface. If the surface is metallic either method is possible; if it is insulating, only rf can be used.

The magnetron is inside a containment vessel, which is evacuated by a set of pumps, typically including roughing, turbomolecular, and cryogenic. The desired species of gases are bled into the vessel at such a rate that the pressure rises to about 0.7×10^{-2} Torr. For this example let us assume that a mixture of equal parts of N_2 and CF_4 are being used. Again, we will only discuss the physics, not the chemistry, of the discharge. So we are mainly interested in the mass of the ions. For our calculations we assume that all ions are 28 amu. The discharge is initiated by the application of a voltage between the anode ring and the cathode. Once formed, the plasma concentrates near the cathode in a region of relatively high field. The main reason for this shall be seen in the ensuing analysis.

Let us assume that the plasma exists with its known, i.e., measured, properties. Then we will see if it is consistent with our understanding. From Fig. 37 we can read and/or calculate the following magnetron parameters:

A = magnetron surface area = 1000 cm^2,
V = volume of plasma = 1000 cm^3 (brightly glowing volume),
B = average magnetic field = 1000 g,
B_M = maximum magnetic field = 2000 g,
R_c = field curvature = 5 cm,
$\nabla B = B/R$ = 200 g/cm,
I = magnetron current = 10 Amp, and
V_B = cathode bias voltage = 500 Volts.

From the pressure measurements we know

$$n_n = 2.5 \times 10^{14} \text{ cm}^{-3}, \qquad \text{and}$$

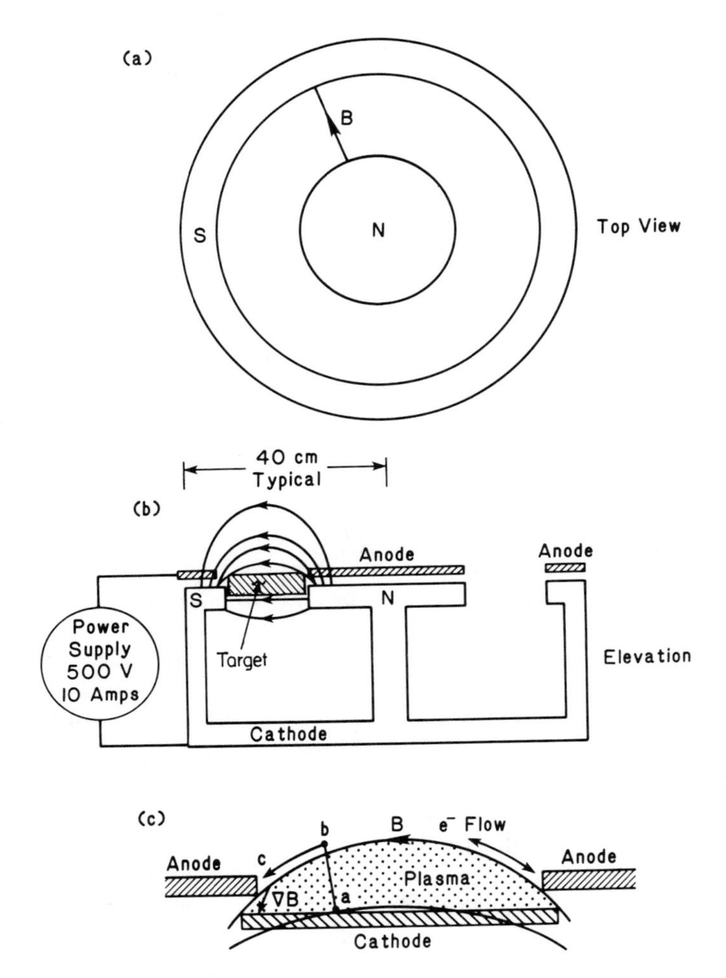

FIGURE 37. Schematic of a circular magnetron. a) Top view showing the magnetic poles. b) Elevation in cross section, showing the placement of the sample surface to be bombarded. The sample may be electrically connected to the poles. Then both would act as the cathode. An anode ring serves to confine the plasma nearer to the sample and away from the containment vessel walls. c) Detail of the sample region showing electron flow along B to the anode and the direction of the magnetic field gradient at one point. Also shown is the line abc, which is an axis in Fig. 38.

from Langmuir probes [15] we find

$$T \cong 3 \text{ eV}, \quad \text{and}$$
$$n_i \cong 2 \times 10^{11} \text{ cm}^{-3}.$$

(Note that these vary with gas pressure, power, and distance from the magnetron.) It is more difficult to get the ion temperature from an experimental measurement. One way is to measure the Doppler shift of radiation. (See Manos and Dylla, Chapter 4.) Let us instead apply our formulae to estimate the ion temperature. The electrons are heated by extracting energy from the electric field. The ions are heated by collisions with the hot electrons. So calculate the ion-electron thermalization time (Section IV.E) getting

$$\nu_{ei}^{-1} = 0.25 \text{ sec.}$$

In contrast, the ion confinement time, found from the ion current to the cathode, is

$$\tau_p = \frac{n_i V}{I \times \text{electrons/coul}} = 3.3 \times 10^{-6} \text{ sec.}$$

We can conclude that the ions in the plasma do not get appreciably heated by collisions with the electrons. They remain near 1/40 eV, room temperature. Of course they will get accelerated in the presheath, in the magnetic sheath, and in the electrostatic sheath.

The first major question we ask is whether the magnetron configuration is stable. From experience, the answer is yes. But what of the Rayleigh-Taylor instability? We predicted that configurations with outward-curving magnetic fields were unstable. We might guess that some stabilizing effects were given by field-line tying to the metallic cathode and anode. Or, more speculatively, we might suggest that the inward electric field balances the outward centrifugal force. But there is actually a clearer reason why no Rayleigh-Taylor instability develops. Let us estimate how long it would take the centrifugal force to move the plasma about $s = 1/2$ cm. Remember that it is the ions that are most important in this instability. The acceleration, a, on the massive ions is $v_{\parallel}^2/R$. So the time to move 1/2 cm is

$$t = \left(\frac{2s}{a}\right)^{1/2} = 5 \times 10^{-5} \text{ sec},$$

which is 10 times longer than the lifetime of particles in the plasma. Thus, the plasma ions are lost before the instability can develop. If the confinement were better or the ions more energetic, Rayleigh-Taylor instabilities could well develop.

The next major question to ask is whether the particle densities and temperatures are consistent with the measured current flow (loss, L, of ions to the cathode) and the creation rate of ions by electron impact ionization. Start with Eqn. (42) and calculate the predicted rate of ionization. This is multiplied by the electron and neutral densities to get the rate per unit volume. Finally, this is multiplied by the plasma volume to get the total predicted source rate, S, which can be compared with the ion current

$$S = R_I n_n n_e V = 6.7 \times 10^{19}/\text{sec} \qquad \text{versus}$$

$$L = I \times (\text{ions/coul}) = 6 \times 10^{19}/\text{sec}.$$

There is only a 10% discrepancy, which is actually too good. The ionization rate could be in error by 50%; secondary electron emission probably changes the true ion current by 20%; the plasma volume could be off by a factor of 2; and the Langmuir probe data could have 20% errors. Each requires scrutiny in an actual experiment. But we accept the agreement as a sign that we have at least done the zeroth order atomic physics part of the problem reasonably well. Now use the summary of plasma parameters in Section IV.E to calculate the other basic plasma parameters.

$r_{Le} = .004$ cm
$r_{Li} = .085$ cm (1/40) eV
$\lambda_D = .003$ cm
$\delta_\perp = 1.2$ cm

$\omega_{pe} = 2.5 \times 10^{10}$ rad/sec
$\omega_{ce} = 1.8 \times 10^{10}$ rad/sec
$\omega_{ci} = 9.6 \times 10^6$ rad/sec
$\nu_{ii} = 2.5 \times 10^6$ /sec
$\nu_{ce} = 1.2 \times 10^6$ /sec
$\nu_{in} = 5.2 \times 10^4$ /sec
$\nu_{en} = 2.2 \times 10^7$ /sec
$\nu_{ei} = 4$ /sec

$D_B = 1.9 \times 10^4$ cm^2/sec
$D_{\|in} = 3.6 \times 10^4$ cm^2/sec
$D_{\|ii} = 71$ cm^2/sec
$D_{\perp in} = 1$ cm^2/sec
$D_{\perp ie} = 65$ cm^2/sec

$N_D = 1.9 \times 10^4$
$\eta_\| = .019$ ohm-cm

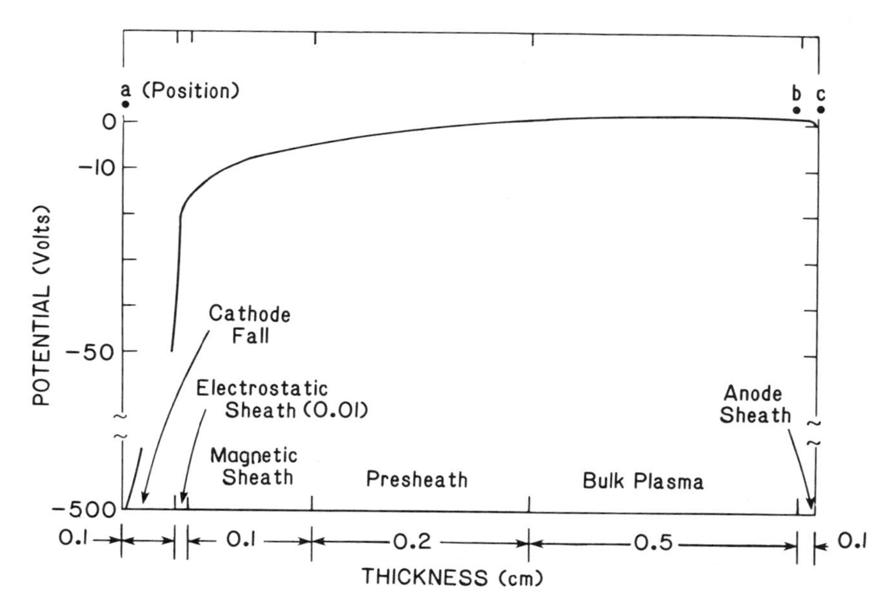

FIGURE 38. Potential along the line abc in Fig. 37. The approximate thicknesses of each region are shown. The potential drop in the anode sheath occurs in the last .01 cm in the B-perpendicular direction and then all along the arc length bc.

We shall now proceed to calculate the particle trajectories. (For experiments describing related measurements, see Ref. [16].) To do this, the electric and magnetic fields inside the plasma are needed. The magnetic field is predominantly the applied one because the current through the plasma, 10 amps, only causes fields < 1 gauss. In contrast, the electric fields are greatly controlled by the plasma. There are both an electrostatic (e.s.) and a magnetic sheath, as shown in Fig. 38. And also a cathode bias voltage zone is present. The electric field for these different regions is shown also in Fig. 38. The strongest field is in the electrostatic sheath, because it is the thinnest. But the field due to the cathode fall is also large. The field in the magnetic sheath is a few percent of that in the electrostatic sheath. With this and the gradients listed earlier, we can now calculate the drift velocities.

Consider the secondary electrons emitted from the cathode. They enter into an electric field of several thousand V/cm. This quickly accelerates them away from the cathode. This occurs before $v_{\nabla B}$ or v_E can develop. So these electrons get up to about 500 eV and pass through the cathode fall and the e.s. sheath into the magnetic sheath. Had we calculated v_E in the e.s. sheath and the cathode fall, these would have corresponded to energies in excess of that to be gained from the electric field ($W = qE \times$ distance). These regions are too short to allow the drift approximation to be valid. Once in the magnetic sheath the drift approximation is valid even for these 500 eV electrons. Their Larmor radii are about 1/6 of the thickness of the magnetic sheath.

Most of the energy the electrons gained in passing through the cathode fall region was in the direction perpendicular to B. So they, of course, are spiraling around B at their cyclotron frequency. These electrons also move along the magnetic field rapidly, at about 10^9 cm/sec. And because of the shape of the magnetic field, they are brought back toward the cathode surface. But most of their energy is, initially at least, perpendicular to B. Combined with the magnetic mirror effect and energy loss to the plasma (via instabilities, for example) it is highly unlikely that the hot electrons will ever again reach the cathode. And the hot electrons are well confined in the plasma—by electrostatic effects parallel to B and by the field itself in the perpendicular direction.

In the magnetic sheath, the hot electrons are also executing drift motions due to the electric and curved B fields. These drift velocities are about 3×10^7 cm/sec, a small fraction of their total velocity. These drifts take the electrons on orbits around the magnetron (Fig. 39). The net pattern of their motions is a zig-zag between the inner and outer poles.

The energy loss processes for these hot electrons include ionization, e-e scattering, e-i scattering, and e-n scattering, as well as instabilities. Using the formulae in section IV.D, these can be evaluated. The time it takes a 500 eV electron to ionize (see Fig. 18) an atom is about 5×10^{-8} sec, during which the electron will drift about 1.5 cm perpendicular to B. In that type of collision the electron loses about 10 eV. Collisions that simply excite the atoms are even more numerous. The net effect is that these electrons lose about 30 eV per cm *of drift circuit* around the magnetron. And in the process they create a new cold (about 1 eV) electron about every cm.

Coulomb collisions with these cold electrons occur at a 10^4 slower rate. And e-i collisions are even less effective at withdrawing the energy from the hot electrons.

No simple formula was given for the energy loss to instabilities. Indeed it is hard to relate, quantitatively, one experiment of this type to another. With that caveat we note that the measured energy loss e-folding distance for a two-stream type instability has been found to be about 10 cm for 500 eV electrons in $\times 100$ less dense plasmas. But even if that slow rate applied here, then the B-parallel velocity of the hot electrons, being so large compared to the drift velocity (10^9 vs 3×10^7 cm/sec), would result in an energy loss rate of 250 eV per cm of circuit (perpendicular to B) around the magnetron.

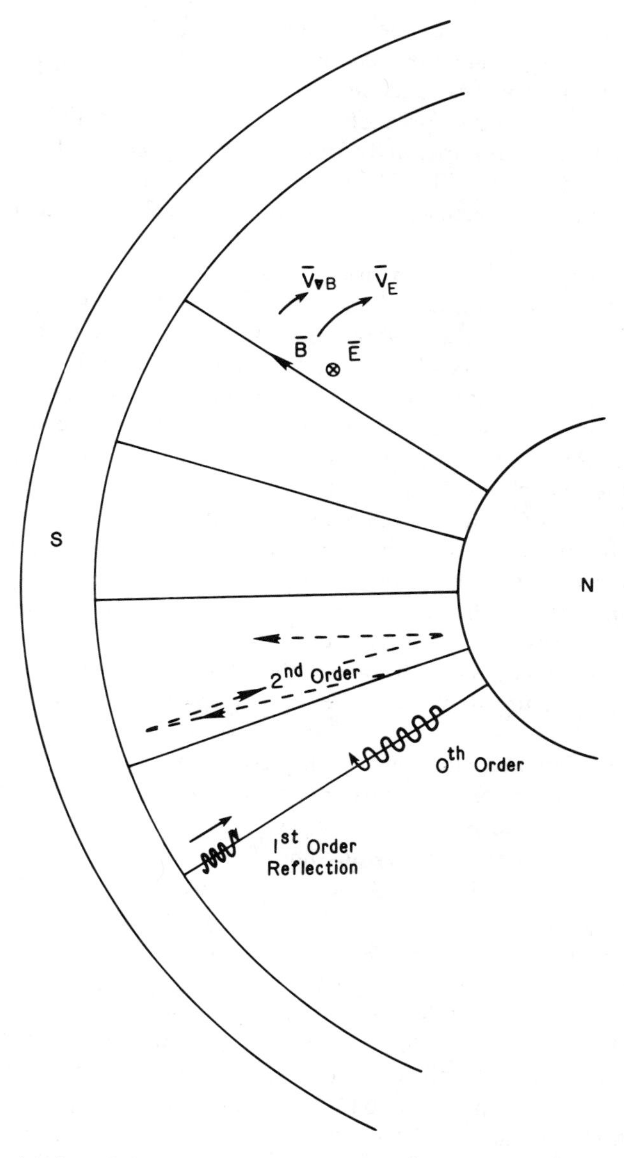

FIGURE 39. Motion of electrons in a magnetron. At 0th order, only the cyclotron motion around B and flow along B are included. The 1st order correction to the motion includes reflection as the electrons enter regions of increasing B and decreasing potential. The 2nd order description includes drift motion due to the crossed E and B fields and the grad B and curved B (not shown) fields. Plasma instabilities and ionization events prevent hot electrons from circulating around the magnetron without energy loss.

And so we conclude that the hot electrons do not circulate around the magnetron, but dissipate their energy mainly (90%) to warming the cold electrons (via instabilities) and also to creating new electrons (10%). It is interesting to note that for each ion that hits the cathode, about 0.2 secondary electrons are liberated. And for each secondary electron entering the magnetic sheath at 500 eV less than 0.5 direct ionizations occur. So these hot electrons replenish less than 10% of the electrons lost by flow to the anode.

The hot electrons have a density, calculated from the ion current, the secondary electron yield, and the sheath voltage of about $4 \times 10^7 \text{ cm}^{-3}$ in the cathode fall region, and about the same in the magnetic sheath (but calculated from the energy loss rate and sheath thickness).

From the e.s. sheath region out to the anode exists the bulk plasma whose electrons have a measured temperature of about 3 eV. Previously we saw that they explain the maintenance of the plasma, i.e., their ionization rate balanced the particle loss rate. In the magnetic sheath, these electrons have a v_E about 10% of their thermal velocity. In the presheath it is 0.1% of v_t. The grad B drift is similarly unimportant. The electrons also have a drift parallel to E, due to their transverse mobility. After traversing about 1 cm perpendicular to the field, they encounter a field line connected to the anode. From there they flow parallel to B, reach the anode, and complete the electrical circuit, which began at the power supply connection to the cathode. This picture shows why the plasma has little extent away from the cathode.

Does the calculated electron mobility correctly explain the electron flow to the anode? There are two parts to this question: flow parallel to B and flow perpendicular to B. Both involve evaluating Eqn. (45) with the mobilities obtained from the diffusivities and the definitions in Eqn. (45). For parallel motion there is no trouble getting agreement. The electrons flow collisionlessly to the anode along B ($\nu_{ei,n} \ll v_t/d$). The parallel resistivity gives good agreement if one assumes an anode drop of 3 eV, and dimensions of 10 cm length and 0.1 cm thickness of the B-parallel region. For perpendicular motion it should come as no surprise to hear that the classical electron-ion $D = 65 \text{ cm}^2/\text{sec}$, and μ_2 are too small by about a factor of 1000. The mobility calculated from Bohm D_B is too small by about a factor of 3, which isn't bad agreement. The perpendicular resistivity of the plasma, including the collisional correction $(1 + \omega_{ce}\tau_m)$ also gives agreement to about a factor of 3.

Let us now consider the ion motion, starting from the region where most ions originate, the positive column. There the ion motion is well described by the drift equations, because the Larmor radius (for 1/40 eV) is small compared to the size of the region. The thermal velocity is small,

4.2×10^4 cm/sec; $v_{\nabla B}$ is smaller $= 10^3$ cm/sec; and v_E is near zero. The ions diffuse into the presheath and are accelerated up to the ion sound speed, about 3×10^5 cm/sec, as they are attracted to the cathode. The drift speeds increase because of the increase in E and W: $v_{\nabla B} = 6 \times 10^4$ cm/sec and $v_E = 3 \times 10^5$ cm/sec. It is here in the presheath that the drift approximation for the ions fails. The ion gyroradii at 3 eV is about 1 cm, equal to the size of the entire plasma thickness.

The ions are further accelerated as they enter the magnetic and e.s. sheaths. Their velocity at the entrance to the e.s. sheath is estimated to be about 3 times the sound speed. The increase above the sound speed is due to the "impedance" magnetic field.

And finally upon reaching the cathode drop, the ions start their final acceleration up to 500 eV. The effects of the magnetic field are small. A detailed calculation should be done to obtain the angle of impact of the ions on the cathode. Based on the fact that the ions can only complete $\cong$ .01 gyroradius during their final acceleration, we estimate that the angle of impact is within a degree of the surface normal.

The flow of the ions is what provides all the current to the cathode. Let us check if it agrees with the measured current. Simply multiply the ion density times the sound speed times the magnetron area,

$$I = (2 \times 10^{11})(3 \times 10^5)(1000)(1.6 \times 10^{-19}\,\text{coul/e}) = 10\ \text{amps},$$

which agrees with the magnetron current. So the most abundant energetic particles are 500 eV ions. They account for about 80% of the power loss. And this goes to the cathode. To reduce the power loss to the cathode, and the possibility of lattice damage there, it would desirable to maintain the ion current (even to increase it) but reduce the ion energy. This could be done by increasing the plasma density and temperature, both of which increase the mobilities and diffusivities. One way to accomplish this might be electron heating at the electron cyclotron frequency. For this magnetic field, the electron cyclotron frequency is close to that used in microwave ovens.

The reader has now seen that the basic operation of a process plasma device can be understood from the fundamental properties of plasmas. The density and its distribution within the chamber, the temperature and energies of the ions, electrons and neutrals, the currents that flow to the different electrodes—all these can be predicted before building an apparatus. And if changes are required in a particular aspect, then the most likely methods to obtain the desired result can be identified by a careful examination of the fundamental laws.

Acknowledgements

I thank the JET project, Oxfordshire, England for its hospitality during the time this manuscript was written, and Dr. D. M. Manos, Dr. S. M. Rossnagel, and Prof. P. C. Stangeby for their comments on the manuscript.

References

[1] F. Llewellyn-Jones, "The Glow Discharge." Methuen and Co., London (1966).
[2] J. L. Vossen and W. Kern, editors, "Thin Film Processes." Academic Press, New York (1978).
[3] A. D. MacDonald, "Microwave Breakdown in Gases." John Wiley and Sons, New York (1966).
[4] "The Collected Works of Irving Langmuir." Pergammon Press, Oxford (1961).
[5] J. M. Meek and J. D. Craggs, "Electrical Breakdown of Gases." Clarendon Press, Oxford (1971).
[6] N. A. Krall and A. W. Trivelpiece, "Principles of Plasma Physics." McGraw-Hill, New York (1973).
[7] L. Spitzer, "Physics of Fully Ionized Gases." (2nd edition). Interscience, New York (1962).
[8] F. F. Chen, "Introduction to Plasma Physics." (2nd edition). Plenum Press, New York (1980).
[9] A. von Engel, "Ionized Gases." Clarendon Press, Oxford (1955).
[10] D. Post and R. Behrisch, editors, "Physics of Plasma-Wall Interactions in Controlled Fusion." Plenum Press, New York (1984).
[11] B. Chapman, "Glow Discharge Processes." John Wiley and Sons, New York (1980).
[12] S. C. Brown, "Introduction to Electrical Discharges in Gases." John Wiley and Sons, New York (1967).
[13] G. Carter and J. S. Colligon, "Ion Impact of Solids." Heinemann, London (1968).
[14] J. M. Lafferty, editor, "Vacuum Arcs." John Wiley and Sons, New York (1980).
[15] S. M. Rossnagel and H. R. Kaufman, Langmuir Probe Characterization of Magnetron Operation. *J. Vac. Sci. and Technol.* **A4**, 1822 (1986).
[16] S. M. Rossnagel and H. R. Kaufman, Induced Drift Currents in Circular Magnetrons. *J. Vac. Sci. and Technol.* **A5**, 88 (1987).

4 Diagnostics of Plasmas for Materials Processing

D. M. Manos and H. F. Dylla
Plasma Physics Laboratory, Princeton University
Princeton, New Jersey

Plasma Etching:
An Introduction

ISBN 0-12-469370-9

I. Introduction

Plasma processing and plasma-assisted processing of electronic materials are essential to the production of present and future generations of large-scale integrated circuits. The advantages of plasma processing are being applied to thin-film processing that involves the deposition or removal of metals, semiconductors, inorganic insulators, and organic films. Increased production rates, more precise production control, and the unique materials properties that can result from the non-thermal chemistry of plasma processing are significant compared to conventional materials processing techniques. Because of the complexity of the physical and chemical environment in a process plasma, a large array of process monitors, historically termed "plasma diagnostics," are required to characterize the plasma, or to properly monitor important control parameters. This paper will review current diagnostics for process plasmas, with the emphasis given to the following techniques: (1) electrostatic probes, (2) surface probes, (3) microwave interferometry, (4) impedance analysis, (5) quantitative plasma mass spectroscopy, (6) emission and absorption spectroscopy, and (7) (laser) fluorescence spectroscopy. Parameters that characterize plasmas are the electron (n_e), ion (n_i), and neutral (n_o) densities, the respective temperatures (energies) of these species, and the magnetic (B) and electric (E) fields. Dc or rf glow discharges, which produce relatively low-temperature, partially-ionized plasmas by applying external electric fields to a gas at low pressure (10^{-3}–10^{-1} Torr), are useful plasma configurations for materials processing. The resulting plasma electron densities are in the range of 10^8–10^{11} cm^{-3} with neutral densities in the range of 10^{13}–10^{15} cm^{-3}. In magnetically enhanced discharge configurations, such as the magnetron, the electron densities can be increased to the low 10^{12} cm^{-3} range. Electron temperatures are in the range of 1–10 eV. Plasma ion temperatures are essentially equal to the temperature of the background neutral gas, except in the vicinity of boundary surfaces where acceleration of ions by sheath potentials or applied voltages can lead to higher ion energies. For process plasma configurations, it is sufficient to know the spatial variation of the applied magnetic field, B. Perturbations to B due to plasma currents and plasma diamagnetism generally can be ignored. In the case of the electric field, knowledge of the spatial and temporal variations of both the applied potential and plasma potential are necessary to fully characterize the plasma.

Several plasma diagnostic techniques, (i.e., emission spectroscopy, Langmuir probes, and total pressure measurements) have a history of application to plasma physics dating back to the early part of this century.

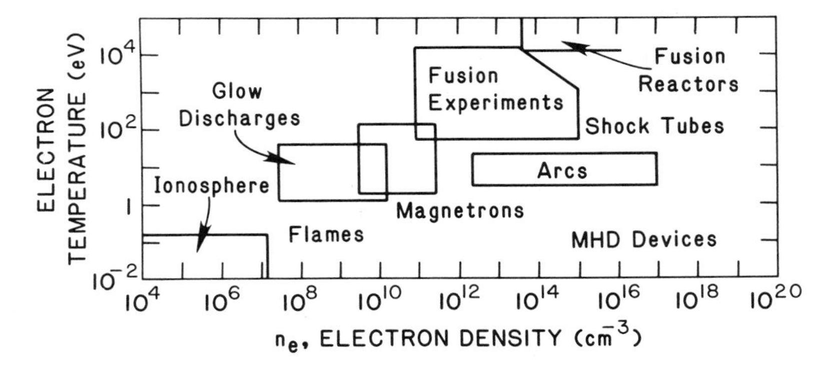

FIGURE 1. Approximate plasma densities and temperatures characteristic of various types of plasmas. Most semiconductor process plasmas fall within the "glow discharge" or "magnetron" regimes.

Since the 1950s, extensive development of plasma diagnostic techniques has resulted from research into high temperature, fully ionized plasmas for the development of magnetic fusion energy. The boundary plasmas in magnetic fusion devices have electron temperatures and densities that are comparable to process plasma parameters. As a result of this similarity, much of the development of diagnostics for fusion device boundary plasmas [1], and the results of plasma-material interaction studies in these devices are directly applicable to process plasma technology [2].

II. Electrostatic (Langmuir) Probes

A. PROBE THEORY

One of the earliest plasma diagnostic techniques was the electrostatic probe, first described and analyzed [3] by Langmuir. Langmuir probes are used for localized measurements of plasma electron densities (n_e) and temperatures (T_e). A Langmuir probe is a conductor that may be immersed in a plasma and biased relative to a second conductor (counter electrode, vacuum vessel wall, etc.) in contact with the plasma. The plasma surrounding the probe develops a boundary layer, termed a "sheath." The theory that yields the necessary relationships to relate n_e and T_e to measured quantities requires an analysis of the sheath. We discuss the essential features of standard probe analysis. We also describe several experimental configurations that are not well known to the process plasma community.

The mathematical theory of a probe in a plasma does not result in closed form general solutions. Instead, it is found that analytical solutions may be derived only for specific, rather narrow regimes. These solutions have been reviewed by a number of authors [4–8]. The complete theory requires the development of the concepts of the plasma sheath and presheath regions and the establishment of the Bohm criterion. Detailed treatment of sheath formation can be found in papers by Emmert and his coworkers [9], Stangeby [10], and Chodura [11].

The regimes of interest divide into two major groups: plasmas with and without magnetic fields. The magnetic field group can be further divided into several cases. The most important subdivision is into cases where the probe is small compared to the ion gyroradius, but large compared to the electron gyroradius. This situation occurs at fairly large values of magnetic fields, such as found in fusion devices or in the high-field regions of plasma magnetron sputtering devices. Use and interpretation of probes in this regime [12–14] is amenable to the simple solutions of the plane probe theory discussed next. When the ion gyroradius and probe size become comparable, ion orbit effects allow the approximate determination of the ion energy by varying the detector location. A simple embodiment of this principle occurs in the Katsumata probe [15]. More complicated ion energy probes, developed by Manos and his coworkers [12, 16], yield information on the velocity distribution of the ions, but require detailed computation of ion orbits. When the probe size is approximately equal to the electron gyroradius, the interpretation of the data is difficult, requiring extensive numerical computation for each individual case.

Plasmas without magnetic fields can be further subdivided into the following regimes: a) small probe (thick sheath) regime, meaning that the probe radius is smaller than the Debye length $\lambda_D = (kTe/4\pi e^2 n_e)^{1/2}$; (b) large probe (thin sheath) regime, meaning that the probe radius is very much larger than the Debye length; and (c) intermediate regime, where the probe is of comparable size to the Debye length. In the small probe regime, it is necessary to include ion orbits; closed form solutions can be derived, however. In the intermediate regime, it is necessary to perform numerical computations to interpret the data. Extensive calculations of this nature were performed and summarized by Laframboise [17], who has given a number of useful graphs for the interpretation of probe characteristics. Thus, even though the theory is most complicated for this case, the actual reduction of data using his results is fairly straightforward. We will return to this special case later. In the large probe regime, ion orbits are not important and the interpretation reduces to the rather simple plane probe approximation first derived by Langmuir.

B. PLANE PROBE

When a Langmuir probe is allowed to electrically float in a plasma, the probe attains a negative potential due to the greater mobility of electrons in the plasma. For cold ions, in the absence of secondary electron emission, the magnitude of this "floating potential," V_f, can be written:

$$V_f = \frac{kT_e}{2e}\ln\left(\frac{m_i}{2\pi m_e}\right), \tag{1}$$

where e and m_e are the electron charge and mass, and m_i is the ion mass. As T_e increases, the ions and electrons arriving at the surface can produce secondary electrons, which will tend to reduce the value of V_f [18, 19]. When the ions have a significant temperature, it is still possible to get a simple expression for V_f [20]:

$$V_f = -\frac{kT_e}{2e}\ln\left[\frac{2\pi m_e}{m_i}\frac{(1 + T_i/T_e)}{(1 - \delta)^2}\right], \tag{2}$$

which includes electron impact secondary electron emission, δ [21]. V_f is the potential that would be measured with respect to the bulk plasma just past the edge of the sheath. The potential at the sheath boundary is a function of the current drawn by the probe [9]. At intermediate values of probe potential, a "hill" forms in the potential, reducing the ion density relative to the bulk plasma value, and increasing the average energy of the collected ions. It is not possible, in practice, to make a direct measurement of the floating potential. Probe potentials are most often referenced to some nominal ground plane such as a metal vacuum vessel wall. The potential at which no current flows through the external circuit is by definition equal to the floating potential. This potential differs from the value given by Eq. (2) by the addition of two terms. The first term represents the potential of the reference relative to the plasma, the second represents the resistive potential drop as the probe current flows through the plasma from the reference electrode. In general, the second term is negligible. The first term is often nearly equal to the floating potential of the probe; thus, it is quite common to observe an apparent floating potential quite near zero when measurement is made relative to the vessel wall.

When a probe in the plasma has its potential (say with respect to the vacuum vessel wall) driven sufficiently negative, it will repel all electrons. The resulting current from the ions is found to be almost constant as the potential is made more negative. This current is called the ion saturation

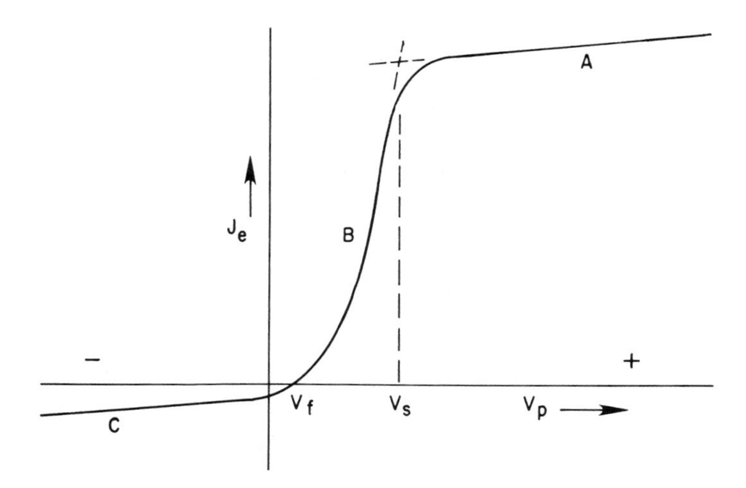

FIGURE 2a. Schematic current-voltage characteristic associated with a Langmuir single probe. The labeled regimes are discussed in the text.

current:

$$j_{sat} = n_e e\left[k\left(\frac{T_e + T_i}{m_i}\right)\right]^{1/2}. \tag{3}$$

The general expression for the total (net) current density to the probe (called the "characteristic") is given by:

$$j = j^+ + j^- = n_e e\left[k\left(\frac{T_e + T_i}{m_i}\right)\right]^{1/2} - \frac{1}{4}n_e e\left(\frac{8kT_e}{\pi m_e}\right)^{1/2} \exp\left(\frac{-eV}{kT}\right), \tag{4}$$

where $V = V_s - V_p$ is the difference between the potential of the plasma and the probe. We emphasize that n_e in the Eqs. (3) and (4) is the plasma electron (or ion) density at the sheath boundary, which will, for plasmas where electrons are the dominant negative charge carrier, be lower than in the bulk by a factor of approximately two. In Fig. 2A, we show a representative plane Langmuir single probe characteristic. The zones of behavior may be interpreted heuristically as follows:

Region C: $V \ll 0$:

This is the ion saturation region. In this region all of the electrons are repelled and all of the ions are collected. The ions are accelerated in the presheath by the applied voltage and the sheath thickness itself is affected slightly by the applied voltage. Thus, there is only a weak dependence on voltage and the formula for "ion saturation" applies.

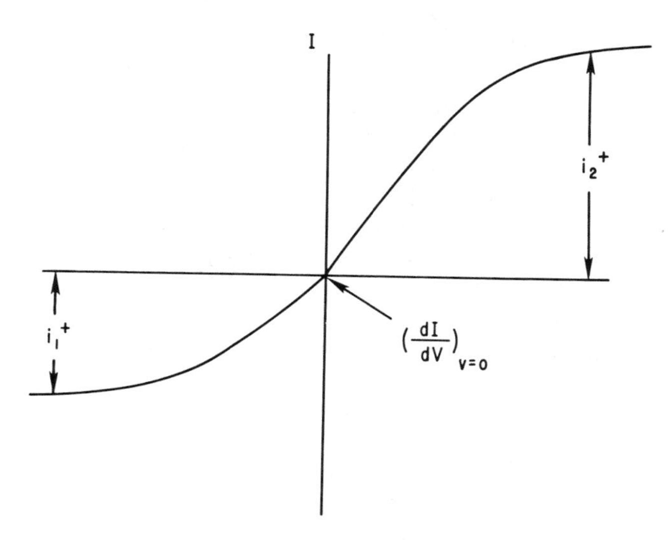

FIGURE 2b. Characteristic for an unequal area Langmuir double probe. The ion saturation levels, i^+, are shown. The electron temperature is easily computed from the derivative, dI/dV, at 0 volts, the electron density follows from T_e and the value of i^+.

Region B: $V_f \sim V < V_s$:

As the voltage increases, the fastest electrons in the plasma begin to strike the probe, and further voltage increase causes increasingly more of the distribution to be admitted. The floating potential, V_f, is achieved when the net current is zero. As V continues to rise, an exponential increase in the electron current is observed, as given by the second term on the right side of Eq. (4). It is from this region that T_e may be computed from the slope of $\ln(j - j_{sat})$ vs. V.

As V rises further, the probe reaches the potential of the plasma, also called the "space potential," V_s. The exponential term approaches one, and the entire electron distribution is admitted.

Region A: $V > V_s$:

Above V_s one again sees only a slow dependence of current density on applied voltage as the presheath attraction slowly reaches further into the bulk plasma. This is the "electron saturation" regime.

C. DOUBLE LANGMUIR PROBE

There are times when a single probe may be difficult to use. There may be no well-defined counter-electrode (ground), as in the case of an electrodeless plasma in a dielectric container. There may be prohibitively large fluctuations in V_s induced by waves or turbulence in the plasma. Such fluctuations lead to noise in the resulting current which, when differentiated, yields large uncertainties in T_e. When the degree of ionization is very low or the plasma is otherwise easily perturbed, the large electron saturation currents drawn by single plane probes are undesirable. In these instances, one may employ a double Langmuir probe. The double probe uses a small counter electrode to supply the return current. The probe power supply system floats at V_f.

The I-V characteristic equation may be developed by treating both electrodes as single probes, requiring that the net external current is zero, and adding the constraint that the system floats, i.e.,

$$i_1 + i_2 = 0 \tag{5}$$

and

$$V_2 = V_1 - V. \tag{6}$$

One arrives at an expression relating the applied potential, the floating potential, and the probe potential:

$$\exp\left(\frac{eV_1}{kT}\right) = \frac{2\exp(eV_f/kT_e)}{1 + \exp(-eV/kT_e)}. \tag{7}$$

This may be substituted into the expression for the current to the probes to yield the general equation of the double probe characteristic:

$$\frac{I + i_1^+}{i_2^+ - I} = \frac{A_1}{A_2}\exp\left(\frac{eV}{kT_e}\right) \tag{8}$$

where

I is the current in the external circuit, (i_1 or $-i_2$),
i_1^+ is the ion saturation current for probe 1, i.e., the current observed when $V_1 \ll V_2$,
i_2^+ is ion saturation for probe 2,
A_1 is the collection area of probe 1,
A_2 is the collection area of probe 2.

A representative double probe characteristic is shown in Fig. 2b. Differentiating I(V) we find

$$\left(\frac{dI}{dV}\right)_{V=0} = \frac{e}{kT_e}\frac{(i_1^+)(i_2^+)}{(i_1^+ + i_2^+)}. \tag{9}$$

We note that in the most commonly employed special case of equal probe areas, $A_1 = A_2$ and $i_1^+ = i_2^+$; the characteristic may be written:

$$i_1 = i_+ \tanh\left(\frac{eV}{2kT_e}\right) \tag{10}$$

and differentiating yields

$$\frac{kT_e}{e} = \frac{i_+}{2(dI/dV)_{V=0}}. \tag{11}$$

Using Eq. (11), T_e is computed quite easily from the slope of I vs. V at $V = 0$. As for the single probe, n_e is derived from the ion saturation value, the calculated T_e, and the assumed, or independently measured, T_i.

One additional note of caution attends the use of double probes. The probes must be far enough apart so that one probe is not residing in the sheath of the other. The thickness of the sheath is not an exact quantity. Most of the potential drop usually comes within five Debye lengths of the surface, however, some field gradient associated with the presheath extends out as far as 100 λ_D. A separation of at least 10–20 λ_D is desirable. If there are gradients in the plasma density or temperature, the probes must be situated close enough to each other to sample similar conditions. (The analysis assumes identical conditions.) It may be advisable to vary the separation between the probes to test the sensitivity of the measurement to this parameter.

D. TRIPLE PROBE

Another style of Langmuir probe requires a third probe, close to the double probe, which is used to independently measure the floating potential. This configuration is called a triple probe [22, 23]. From the double probe analysis:

$$\exp\left(\frac{eV_1}{kT_e}\right) = \frac{2\exp(eV_f/kT_e)}{[1 + \exp - (eV/kT_e)]}, \tag{12}$$

where $V = V_1 - V_2$. If the applied potential is chosen so that $V \gg kT_e$, then from this we obtain

$$\frac{eV_1}{kT_e} = \frac{eV_f}{kT_e} + \ln 2, \text{ or} \tag{13}$$

$$\frac{kT_e}{e} = \frac{1}{\ln 2}(V_1 - V_f). \tag{14}$$

A measurement of the voltage between the positive double probe electrode and the floating probe yields T_e directly. A detailed comparison of double

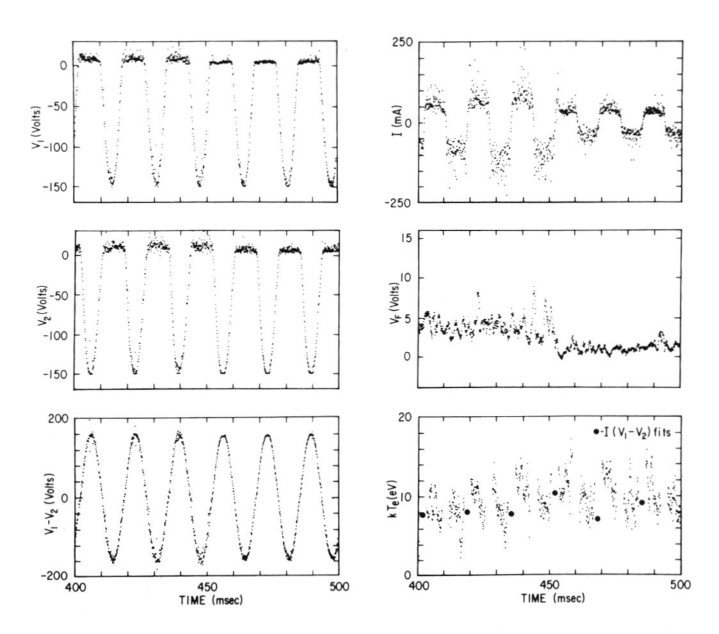

FIGURE 3. Comparison of equal area, planar double and triple probe behavior. Bottom left: Applied voltage between probes. Middle left: Voltage of one of the probes relative to the vacuum vessel wall. Top left: Voltage of the second probe relative to the vacuum vessel wall. Top right: Current through the probe circuit (note an important plasma transient caused an abrupt density decay just after 450 msec). Middle right: Independently measured floating potential on third electrode. Bottom right: Small dots are electron temperature from the triple probe analysis, which is obtained quasi-continuously $[T_e \propto (V_1 - V_f)]$. Large dots are T_e extracted from numerical fits of Eq. (10) of the text to each of the double probe characteristics resulting from the voltage sweeps (six full excursions, only the ascending voltage sweep analysis is displayed) (from Ref. 24 with permission).

and triple probes in magnetic plasmas has been given by Budny and Manos [24].

Figure 3 shows results from that comparison. Figure 4 shows a rough schematic of the apparatus used to make the measurements. The lower right frame of Fig. 3 indicates that the temperature derived from the triple probe compares very well to that extracted from the double probe. The frames at the upper left show how plasmas redistribute applied potentials. The probe bias was applied to the two equal area electrodes of the double probe using a floating power supply. The resulting potentials on the electrodes as measured relative to the large metal vacuum vessel wall are quite different. The division of the applied sine wave (bottom left frame) into symmetric components is clearly evident. Note that neither electrode ever achieves

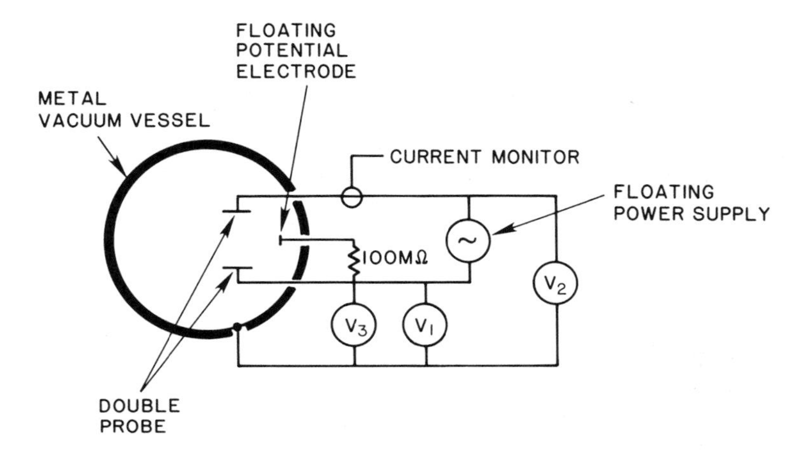

FIGURE 4. The experimental arrangement used to compare double and triple probe behavior of Fig. 3. V_3 is taken to be equal to the floating potential, V_f.

large positive potentials relative to the vessel wall. In this plasma, where no negative ions exist, the higher mobility of the electrons forces the plasma to be the most positive potential in the system. Since the plasma is reasonably well connected to the large surface area of the vacuum vessel wall, the wall can deviate only slightly from its floating potential, without large currents resulting. Thus, even at the maximum of their voltage excursions, the electrodes are still slightly below the plasma potential. We note that this behavior serves only as a rough guide for understanding the potentials on rf driven electrodes when neither is grounded. As shown in the work of Flamm *et al.* [25], there are important frequency dependent effects, and the analogy to probe behavior is further weakened because power electrodes are the source of the plasma, and thus, cannot be considered to be non-perturbing as this style of probe is assumed to be.

E. CYLINDRICAL PROBES (LAFRAMBOISE)

The conditions of many process plasmas, densities of 10^9 to 10^{10} cm^{-3} and electron temperatures of less than 10 eV, place them in the regime of analysis considered by Laframboise [17] for cylindrical probes. Large planar probes are difficult to employ in such discharges because they represent far too large a perturbation to the system. For a probe to be truly planar in such systems, it needs to be approximately 1 cm in size. When driven into electron collection, a probe of this size would draw off an appreciable fraction of the plasma source current, driving the system far away from its original conditions. In such systems the use of a small cylindrical probe is necessary. Steinbruchel has given a compact summary of the relevant parts of Laframboise's work, which we follow in this brief treatment [26].

Laframboise's expression for the ion current drawn by a cylindrical probe is:

$$I_i = en_i A\left(\frac{kT_e}{2\pi m_i}\right)^{1/2} i_1, \tag{15}$$

where n_i is the ion density, A is the probe area, i_1 is a dimensionless parameter related to the ion current density, which depends on the applied potential $X = e(Vp - Vs)/kT_e$, and on the parameter r_p/λ_D (probe size/Debye length), and also depends very weakly on ion temperature. One finds that if r_p/λ_D is less than about 3 and T_i/T_e is very small, that to within a few percent $i_1^2 = 1.27(-X)$. Thus, the square of the probe current will be a linear function of the probe potential whose slope depends on ion density according to:

$$\frac{-\Delta(I_i^2)}{\Delta v_p} = 0.20e^3\frac{n_i^2A^2}{m_i}. \tag{16}$$

The plasma potential is found from the equation

$$V_s = V_p(I_p = 0) - \frac{kT_eX_f}{e}, \tag{17}$$

where X_f, the floating potential, is found by solving the equation

$$1.1\left(\frac{m_e}{m_i}\right)^{1/2}(-X_f)^{1/2} = \exp(X_f). \tag{18}$$

There are several ways to proceed with the analysis for cylindrical double probes, some involving iteration [27] using the plots given by Laframboise, others such as that of Sonin [28] avoiding iteration. The method used by Steinbruchel involves plotting I_p^2 against V_p at large negative V_p, where $I_p = I_i$. One determines n_i from the slope, as described above, and extrapolates back into the transition region using the value of n_i to obtain $I_e(V_p)$ from the difference between I_p and I_i. A plot of $\ln(I_e)$ against V_p yields T_e from the following expression for the electron current:

$$I_e = en_i A\left(\frac{kT_e}{2\pi m_e}\right)^{1/2} \exp(X). \tag{19}$$

Analyses of probe data by these techniques have been performed for both etching and deposition plasma, examples of which can be found in Refs. 29 and 30.

F. OTHER LANGMUIR-LIKE PROBES

Other electrical probes include the hot electron emitting probe [31]. By operating the probe at an elevated temperature, it will emit electrons whenever the probe potential equals or exceeds the plasma potential, and hence the plasma potential can be measured directly. Such probes are often physically delicate and until recently have been generally of use only in relatively low density plasmas. However, they can be constructed to survive in denser plasmas [32].

Another probe useful for examining the electron energy distribution function is the retarding field analyzer [33] shown schematically in Fig. 5. The retarding field is swept while the current of charged particles is collected in a manner similar to a Langmuir probe. The resulting characteristic is differentiated to provide the distribution function of electron energy. This analyzer avoids the sophisticated orbit effects of the cylindrical probe, while limiting the current drawn by separating the ions and electrons prior

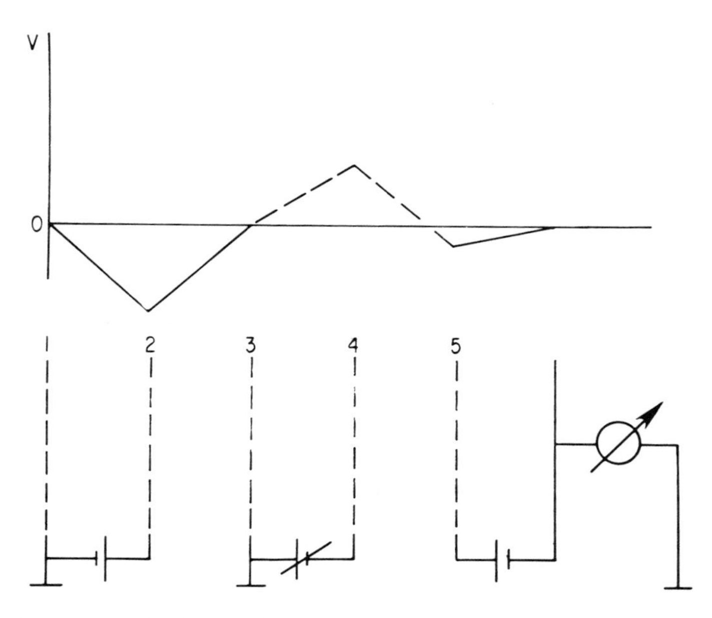

FIGURE 5. Schematic of a retarding field analyzer biased for ion collection. The electrons are separated from the ions by the negative bias of grid 2. The ion velocity distribution is analyzed by sweeping the retarding field between grids 3 and 4. Grid 5 is held slightly negative to suppress secondary electron emission from the collector at the right (from Ref. 6 with permission).

to their collection. The difficulties associated with the use of retarding field analyzers have been reviewed in Ref. 6.

G. POWER FLOW MEASUREMENTS

One way of obtaining information about T_i, at least in principle, is to independently measure the power (P) to a Langmuir probe. This may be done with a calorimeter probe to monitor the rate of rise of the temperature of a thermally isolated collector [34]. The power flow to the probe is approximately given by:

$$P \cong mC\frac{dT}{dt}, \tag{20}$$

where m is the collector mass, and C is the specific heat of the collector element.

A representative heat flux probe [35] is shown in Fig. 6. According to the elementary theory of probe sheaths, the power, P, arriving at a probe at the

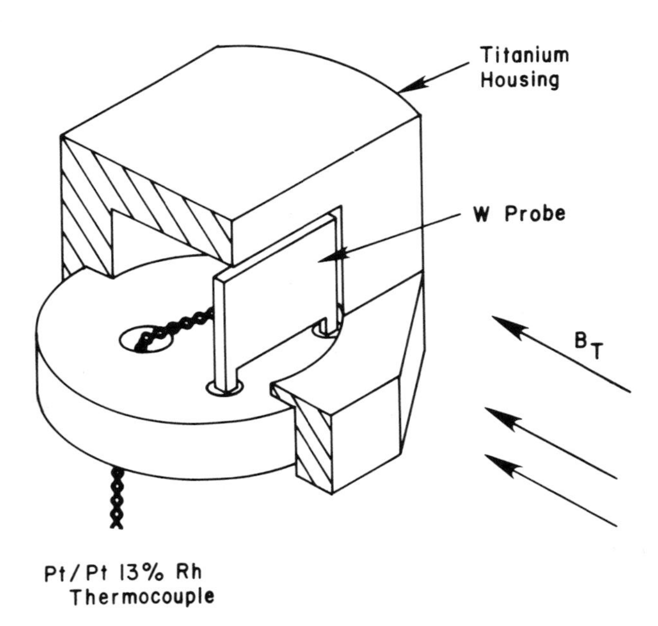

FIGURE 6. Schematic of a calorimeter probe for use in a magnetized system. The materials shown were chosen for maximal performance at intermediate heat fluxes (100–1000 watts/cm^2). For process plasma work, material choice may be dominated by plasma purity, for example, in a carbon coating plasma, most of the probe could be graphite (from Ref. 35 with permission).

floating potential is:

$$P = I_s \frac{kT_e}{e}\left[2\left(\frac{T_i}{T_e}\right) + 2(1-\delta)^{-1} + \frac{eV_f}{kT_e}\right], \tag{21}$$

from which can be obtained:

$$\frac{T_i}{T_e} = \frac{eP}{2I_s kT_e} - \frac{eV_f}{2kT_e} - \frac{1}{1-\delta}. \tag{22}$$

We also have from Eq. (2) that

$$\frac{eV_f}{kT_e} = -\frac{1}{2}\ln\left[2\pi\left(\frac{m_e}{m_i}\right)\frac{1+(T_i/T_e)}{(1-\delta)^2}\right]. \tag{23}$$

From Eqs. (22) and (23), T_i and V_f can be obtained by iterative substitution, and hence T_i obtained independently of T_e. In practice there are a number of uncertainties that lead to large errors in T_i, and at present it is not a very reliable technique. For example, if the electron energy distribution has an unexpected high energy component, this will give an unexpectedly high value to P, resulting in a spuriously high value for T_i. Uncertainties in the secondary electron emission coefficient or in the ion energy reflection coefficient can also produce substantial errors in the calculated value of T_i. Thus, the use of the measured power to the probe to calculate T_i may yield results that are wrong by factors of ~ 2. The technique is nevertheless useful for measuring material surface temperatures to estimate the effects of evaporation, chemical sputtering, thermal desorption, etc.

H. TIME DEPENDENT EFFECTS: ALTERNATING BIAS AND PLASMA OSCILLATIONS

In this section we consider the effects of oscillations in plasma parameters on the behavior of a dc-biased probe and the inverse problem of an oscillating probe bias in a steady plasma. So far we have assumed that the applied potential, V_p, varies slowly compared to the time required for the plasma to equilibrate to such changes, i.e., $dV_p/dt \ll V_p/\omega_{ci}$, where ω_{ci} is the ion plasma frequency, $\omega_{ci} = (4\pi n_i Z^2 e^2/m_i)^{1/2}$. Conventional second derivative [36] or second harmonic [8] probes, driven by alternating voltages, operate well within this regime. We will not consider these here; the interested reader should see Ref. 8 for a thorough discussion of their use. As the frequency of the applied voltage increases, inertia prevents the ions from equilibrating.

The plasma reacts to changing probe bias by readjusting the thickness of the sheath and the surface charge on the probe and the sheath boundary.

On time scales shorter than the inverse of the ion plasma frequency, only the electrons can move to cause this readjustment. Oliver *et al.* [37] have performed experiments using bias frequencies above the ion plasma frequency but below the electron plasma frequency, $\omega_{ce} = \omega_{ci}(m_i/m_e)^{1/2}$ to measure the effective resistance and capacitance of the probe sheath. From these measured values they inferred the plasma electron density and temperature using Bohm's expression [4] for the capacitance, together with a model developed by Crawford and Grard [38]. Agreement between this method and conventional dc probe methods was within 20% for both n_e and T_e as long as the probes are kept clean. The presence of resistive thin films on the surface of the probe [39] can lead to large errors in such measurements.

At frequencies near ω_{ce}, a resonance occurs in the current drawn by the probe. The diagnostic use of this resonance was pioneered by Takayama [40] and Ichikawa and Ikegami [41], who first derived the theory predicting it. Their theory gives a good approximation for the resonant peak height [42] and resonance width:

$$\delta j_r = j_o \frac{1}{\sqrt{2}} \frac{\omega_R}{\nu} \frac{\lambda_D}{L} \frac{e\,\delta V}{kT_e} \left(\frac{eV}{kT_e}\right)^{1/2} I_1\left(\frac{e\,\delta V}{kT_e}\right) \tag{24}$$

$$\Delta\omega_{1/2} = 2\nu, \tag{25}$$

where ω_R is the resonance frequency, j_o is the current in the absence of the ac field, δV is the amplitude of the applied ac voltage, L is the distance over which the voltage is applied, V is the dc voltage difference between the plasma and the probe, $I_1(Z)$ is the modified Bessel function of the first order, and ν is the electron-neutral collision frequency. The theory erroneously predicts that $\omega_R = \omega_{pe}$, the electron plasma frequency. Mayer [43] and Harp [44] showed that the resonance actually occurs at a frequency below ω_{pe}, given by:

$$\omega_R^2 = \frac{\omega_{pe}^2}{[1 + (R/s)]}, \tag{26}$$

where R is the probe radius and s is the total sheath thickness (usually taken to be $5\lambda_D$). The density of the plasma may be extracted from the above expression. The method is especially useful for low density plasmas, $n_e = 10^6 - 10^7\,\text{cm}^{-3}$ [45].

The problem of a conventional, low-frequency Langmuir probe in an oscillating plasma is quite similar to that of the ac probe. Even dc glow discharge plasmas are subject to high frequency oscillations, which interact with biased probes [46, 47]. The amplitude spectra of such "noise" has been studied to provide a measure of the electron temperature and has been found to agree reasonably well with temperatures obtained from dc probes using Langmuir's methods described previously [48]. The communication between a probe and a plasma occurs exclusively through the sheath, which, up until now, we have treated as having only a resistance and capacitance. As stated earlier, the sheath exists because of the different mobilities of positive and negative charge carriers and so also acts as a rectifier. Indeed, the characteristic shown in Fig. 1 is very similar to that of a diode. When an alternating voltage is applied across the sheath, either by using an ac bias on the probe keeping the plasma space potential fixed, or by holding the probe voltage constant and oscillating the space potential, the sheath will rectify this voltage and a dc current will result. This current distorts the pure dc characteristic of Eq. (4).

One may attempt to eliminate the dc distortion experimentally. This can be done fairly well when the oscillating component is well-localized at a single frequency, ω. One can make the impedance of the probe very high at that frequency by introducing a parallel LC filter (rf trap) tuned to resonate at ω. The resulting rectified current is then small. Traps work well in 13.56 MHz rf discharges, particularly when the plasma is in contact with a large area of grounded conducting wall. In such systems, the small plasma-wall impedance causes most of the rf current to flow to the wall, making the rf probe current small even before the trap is added [50]. Alternatively, one may account for the existence of fluctuations in the theoretical derivation of the characteristic. Because of the symmetry of sheath bias alluded to before, the analysis is similar to that of ac and resonance probes.

Garscadden and Emelius [49] have given a theoretical derivation of the characteristic that includes the effects of fluctuations on Langmuir probes. They considered the effects of independent fluctuations in electron density, plasma (space) potential, and electron temperature, on the retarding portion of the characteristic. The treatment was extended by Crawford [51] who considered the simultaneous variations of these quantities, including all cross-terms up to second order. In the case where the fluctuation is restricted to a single frequency, the more complicated expressions of Crawford reduce to those of Garscadden and Emelius. Both treatments are simplified considerably by incorporating the experimental observation that temperature fluctuations are often small compared to fluctuations in density or space potential. Experimental tests have concentrated on the variation of space potential [52] although limited variation of density was also done [51]. In the electron retarding region, the ac rectification causes a constant increment to the current, whose magnitude depends only on the magnitude of the fluctuation in the space potential. This means that temperature [extracted from $d \ln i/dV_p(\mathrm{dc})$] is unaffected by the presence of such fluctuations. As fluctuations drive the probe beyond the retarding region into electron saturation, the "knee" in the curve (at $\approx V_s$ in Fig. 1) is found to become progressively rounder; breakaway from the linear regime of $\ln i$ vs. V occurs at lower values of V as the peak-to-peak fluctuating voltage, $\delta\tilde{V}$, increases. This leads to errors in estimating the density for cylindrical probes when $\delta\tilde{V}$ is equal to or greater than T_e. One finds that the floating potential shifts [52] by an amount ΔV_f given by:

$$\exp\left(\frac{e\,\Delta V_f}{kT_e}\right) = I_o\left(\frac{e\,\delta\tilde{V}}{kT_e}\right), \tag{27}$$

where I_o is the modified Bessel function of zero order.

Note that when such fluctuations are large, considerable advantage is gained by using double probes (or triple probes) instead of single probes. As long as the wavelength of the fluctuations is large compared to the distance between the probe tips, and as long as the impedance of the double probe lead wires and measuring circuit is balanced, the oscillating components and the resulting rectified dc currents will cancel.

I. OTHER COMPLICATIONS IN THE USE OF PROBES

There are numerous complications that arise in the use of electrostatic probes. Probes are intrinsically perturbing to the distributions they measure. Probes collect particles in a portion of phase space and reflect others, thereby modifying distributions. The existence of a sheath modifies the plasma density and electric field at its boundary [55]. The electric field, in turn, modifies the ion distribution at the probe location. Simple estimates of the degree to which a probe intrinsically perturbs the plasma are possible [56]. If the probe collector dimensions become comparable to the mean free path for velocity changing collisions, the probe will distort the distribution function of the collected species. If the net probe current is greater than a few percent of the volume plasma source rate, the probe is even more seriously perturbative. Changes in the surface conditions of an electrical probe may lead to varying secondary electron emission, ion particle and energy reflection coefficients, or contact potentials, rendering the analysis quite difficult. Overheating of the probe can drive it into thermionic emission where the above theory does not apply. A more thorough analysis of such difficulties has been given in Ref. 8. We note that, in spite of the problems, probe measurements are in reasonable agreement with other methods like microwave interferometry [53, 54].

J. SURFACE PROBES[†]

A novel extension of probe techniques is the use of specifically designed probes and probe materials to collect plasma particles for later surface analysis [55–60]. This class of plasma probes, called surface (or retention) probes, provides plasma species identification, and flux and energy distributions of both major and minor plasma species. Surface probes were developed originally for diagnosing the edge plasmas of magnetic fusion devices and have also been used for determining the species and energy distributions of high power ion and neutral beam sources developed for plasma heating in magnetic fusion devices [61, 62]. In general, surface probes are useful for characterizing any configuration of low temperature plasma or ion source. Surface probes, like electrostatic probes, can tolerate only limited amounts of power, depending on their material properties. This results [63] in limits to the maximum product of plasma density, temperature, and sampling time. In addition, surface probes can measure ion energy distributions down to a minimum energy beyond which reflection coefficients of incident plasma particles are not well characterized (typically, ~ 20 eV).

K. BASIC CONFIGURATION OF A SURFACE PROBE

The design of a surface probe depends on the plasma configuration. Typically, a surface probe (Fig. 7) consists of a collector surface fabricated from a clean, well-characterized material. Alternatively, the collector material can be deposited as a coating on a carrier probe. The collector material should have the following properties: (1) high purity, (2) absence of elements expected in the plasma, and (3) known sputtering and implantation properties. Suitable materials, which have been successfully used for collector surfaces, include: silicon (both single crystal and amorphous), carbon (usually various forms of graphite), and beryllium. The probe collector surface usually is shielded with a metallic shield with apertures that both limit the solid angle of exposure to plasma and provide unexposed probe surface behind the shield for background measurements. By shuttering the aperture, or by moving the collector beneath the aperture, time resolved measurements of plasma fluxes can be made. The increase in surface concentration of a particular species after plasma exposure (Δn_s) is

[†] This section has been adapted from Ref. 76 (with permission).

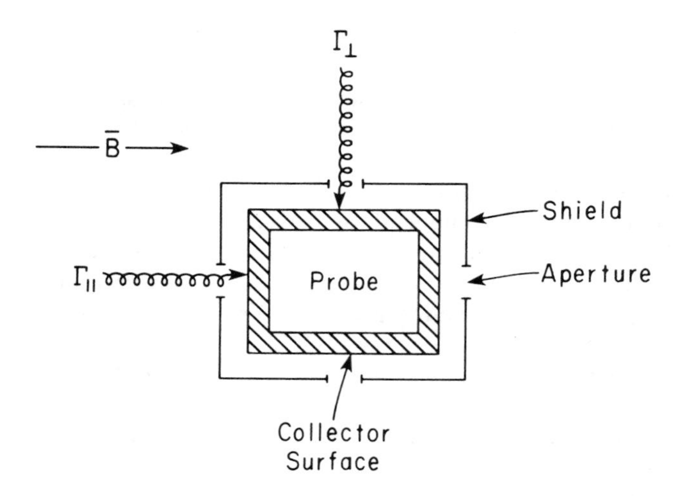

FIGURE 7. Schematic diagram of a surface probe immersed in a plasma (from Ref. 76 with permission).

related to the incident flux of this species (Γ_i) by the following:

$$\Delta n_s = g\Gamma_i \Delta t, \tag{28}$$

where Δt is the exposure time, and g is a constant of proportionality that depends on the probe shield-geometry, the probe-plasma electrostatics, and the reflection/retention properties of that species on the collector surface. Quantitative surface analysis of Δn_s gives a measure of the plasma flux Γ_i if the reflection and retention properties of that species (i) on the collector surface are known. The simplest cases involve sampling plasma species for which the sticking coefficient is high and the surface and bulk diffusion coefficients are low for the collector material. Most moderately energetic, medium and high atomic number ions on the light substrates mentioned as possible collectors (C, Si and Be) fall into this category. Calibrations are necessary to characterize the constant of proportionality (g) in Eq. (28), if the reflection and retention properties of the plasma species are not well known. Examples of this latter case include the use of C and Si collectors to characterize hydrogenic plasmas, where calibrations and model calculations in the literature [63–65] allow quantitative flux and energy distribution measurements over a wide range of incident energy.

The sensitivity of surface probe measurements is limited by the sensitivity of the surface analysis technique used for analysis, and by the exposure

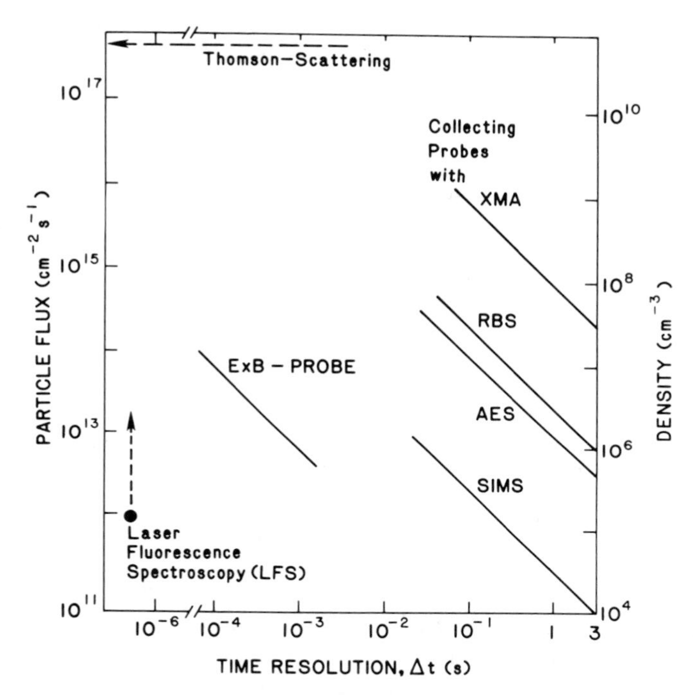

FIGURE 8. Plasma fluxes and densities that can be sampled with surface probes as a function of the desired time resolution of the measurement. The sensitivity of surface probes with four different analysis techniques is shown: XMA, X-ray microanalysis, RBS, Rutherford back scattering, AES, Auger electron spectroscopy, and SIMS, secondary ion mass spectrometry (adapted from Ref. 58 with permission).

time, Δt (Fig. 8). Using Auger electron spectroscopy (AES), particle fluxes of most elements (except H) can be detected above 10^{14} cm^{-2} s^{-1} for 100 ms sampling times. For 10 eV carbon ions, this minimum detectable flux corresponds to a minimum detectable plasma ion density of $n_i < 10^8\ cm^{-3}$, or of the order of 0.1–1 ppm of the particle density in a typical glow discharge plasma (at 10–100 mtorr). Such a lower limit of detectability is respectably close to the detection limits of laser induced fluorescence spectroscopy (LIF), although the time resolution of LIF is much better ($\lesssim 1\ \mu s$). The dynamic range and accuracy of flux measurements by surface probes can be improved by incorporating other surface analysis techniques. For example, the minimum detectable flux can be decreased to $\sim 10^{12}$ cm^{-2} s^{-1} for $\Delta t = 100$ ms by using secondary ion mass spectrometry

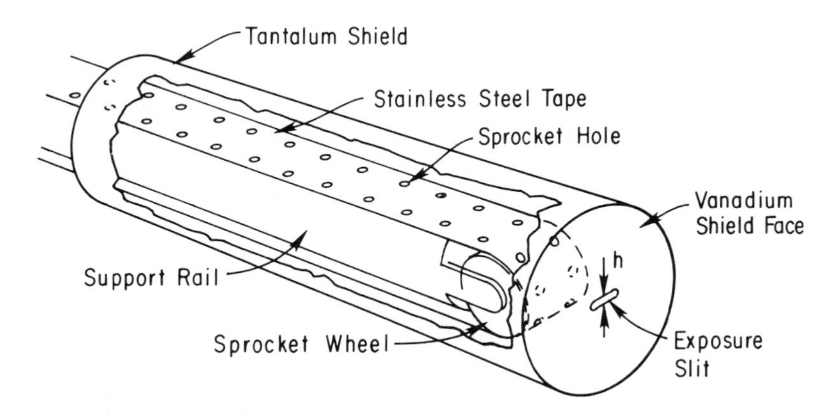

FIGURE 9. One type of surface probe used for time-resolved sampling of plasma fluxes in the edge plasma of magnetic fusion devices. This probe was coupled with an in-situ Auger electron spectrometer (from Ref. 76 with permission).

(SIMS); and the accuracy of measurements at higher flux rates ($> 10^{15}$ $cm^{-2}\ s^{-1}$) can be improved by using Rutherford ion backscattering (RBS), or X-ray microanalysis (XMA).

The usual precautions and limitations with respect to the size and placement of electrostatic probes are applicable to surface probes; i.e., the probe size should be small if a non-perturbing plasma measurement is desired. A large size probe or arbitrary placement of probes in a given plasma configuration is usually prohibited by power flux considerations. Surface probes can provide spatially resolved information on plasma fluxes by using any of three approaches: (1) probes with small collector areas can be moved to different regions of interest within the plasma volume; (2) an array of collector surfaces can be placed over an interesting area of a plasma device (such as the substrate plane of a sputter-deposition or plasma etching device); or (3) a spatially resolved surface analysis technique can be used to resolve the deposition on a single larger-area collector surface.

Figures 9 and 10 show examples of surface probes originally designed for the investigation of the boundary plasmas in magnetic fusion devices, that provide both time-resolved and space-resolved measurements of plasma fluxes. The collector surface of the probe shown in Fig. 9 [66, 67] consists of a continuous loop of stainless steel tape, exposed at one end to plasma fluxes through a small (1 × 3 mm) exposure slit. The other end of the tape loop is coupled to an ultrahigh vacuum chamber containing evaporation sources and an Auger electron spectrometer for preparing and analyzing the

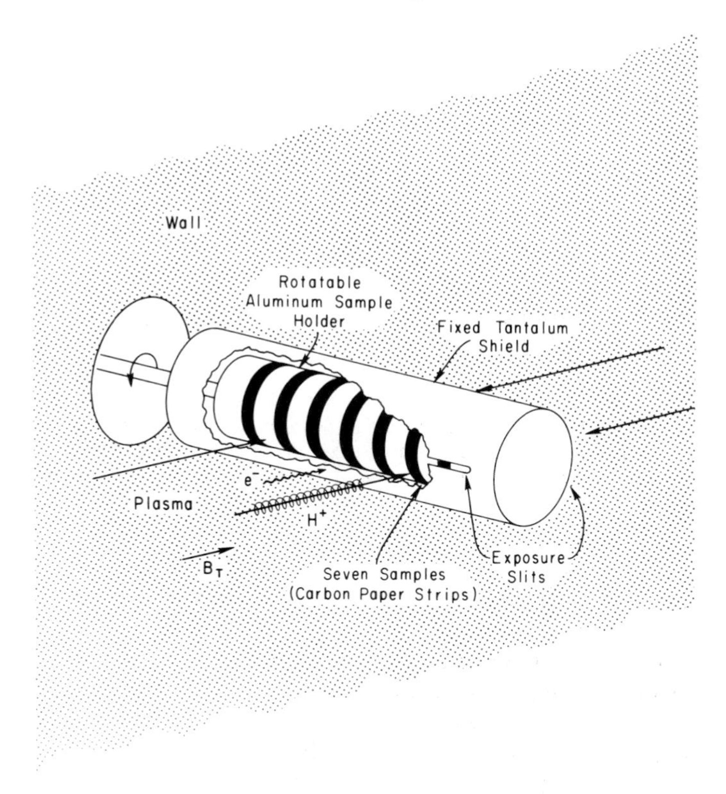

FIGURE 10. Second type of surface probe used in fusion device edge plasmas. This probe provided time and space resolved plasma flux and density measurements. The collector samples (carbon strips) are removed for ex-situ surface analysis (from Ref. 6 with permission).

surface composition of the tape. Spatial resolution is provided by translating the probe in one dimension, and time resolution is provided by rotating the tape loop past the exposure slit. An example of spatially resolved plasma flux measurements using this probe is shown in Fig. 11.

For the probe design shown in Fig. 10 [57], time-resolved flux measurements are obtained by rotation of the collector surface beneath the exposure slit. However, spatially resolved flux measurements are obtained in this design by the incorporation of an array of seven (carbon) collector strips distributed beneath an elongated exposure slit. After plasma exposure, the probe is removed from the plasma chamber and the carbon strips are removed for surface analysis in an external system.

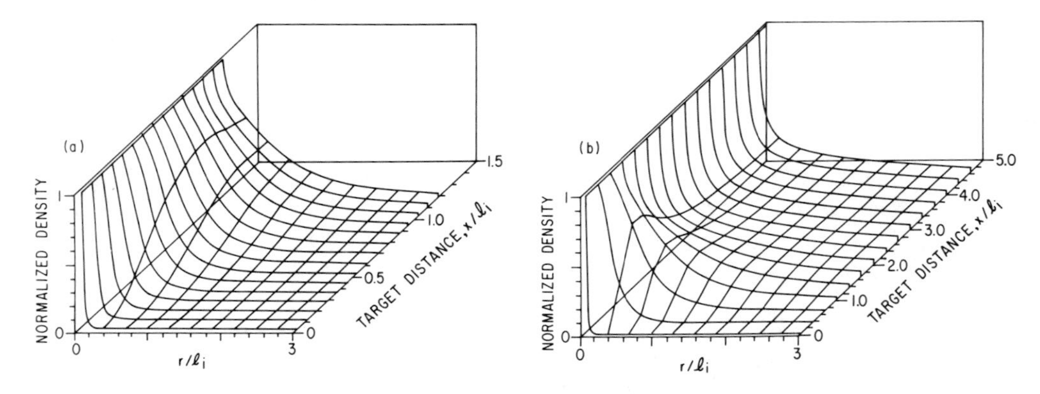

FIGURE 11. Calculated lateral distribution of deposited surface density on a surface probe in a magnetized plasma as a function of the aperture dimensions, r, (in units normalized to the ion Larmor radius, l_i), and as a function of the aperture to target distance (from Ref. 57 with permission).

L. PLASMA FLUX AND ENERGY ANALYSIS

Relating the inferred plasma flux from a deposition measurement on a surface probe, to the local plasma ion density, n_i, ion velocity distribution, $f(v_i)$, or, in the case of a Maxwellian distribution, the ion temperature, T_i, is not a generally solvable problem. However, with certain simplifying assumptions about the probe-plasma electrostatics and with the additional information provided by spatially resolving lateral or depth distributions of the implanted plasma species on collector surfaces, measurements of plasma ion densities and temperatures can be obtained as discussed below. The functional relationship between the incident plasma flux and the local plasma ion density and velocity distribution is complex. At a minimum, this relationship:

$$\Gamma_i = f(n_i, v_i) \tag{29}$$

depends on: (1) the electrostatic potential of the probe with respect to the plasma potential; (2) the probe size relative to other characteristic lengths in the plasma (i.e., ion mean-free-paths, Debye length, Larmor radii, etc.); and the plasma collisionality [57]. With respect to the probe-plasma electrostatics there are two simple cases that lead to analytic expressions relating the incident flux to the local plasma ion density. These two cases are taken directly from the analogous situation with electrostatic probes [57, 10] that are either (1) biased negatively, at the ion saturation potential or (2) biased at the plasma potential. When biased to a potential less than the floating potential, the probe will sample an ion flux proportional to the ion saturation current, (Eq. 3),

$$\Gamma_i = n_i v_s = \left[\frac{k(T_e + T_i)}{m}\right]^{1/2}, \tag{30}$$

where v_s = ion sound velocity. When biased at the floating potential, the probe is electrically floated and thus no net current flows. An ambipolar ion flux

$$\Gamma_i = D_a \nabla n_i \tag{31}$$

is sampled by the probe, which is proportional to the plasma ambipolar diffusion coefficient, D_a, and the gradient in the local plasma ion density, ∇n_i. Biasing at the floating potential has the added advantage of minimizing the power flow to the probe while still sampling a large ion flux [57].

Magnetic fields complicate the relationship between the flux to surface probes and the local plasma parameters. For typical applications in magnetostatic process plasma configurations, the applied magnetic fields produce Larmor radii that are much larger than the probe dimensions, and thus, the probe ion collection efficiency is not significantly affected. The collection efficiency in the high field case has been treated by a number of authors [27, 57, 58, 66, 69]. There are two limiting cases for analyzing the ion flux from a magnetized plasma incident upon a surface probe: (1) a collector surface perpendicular to the applied magnetic field (and negatively biased) will sample a streaming flux proportional to the ion saturation flux; and (2) a collector surface parallel to the applied magnetic field will sample a diffusive flux proportional to the cross-field diffusion coefficient and gradient in the local plasma ion density.

In either of the above cases, or for an arbitrary placement of a collector surface with respect to the magnetic field direction, the deposition pattern on a surface probe will depend on the magnetic field when the dimensions of the probe aperture are comparable to the Larmor radius. This property has been exploited in a number of applications of surface probes in magnetic fusion devices [70–72]. Since the Larmor radius of an ion is given by:

$$r_L = \frac{(kT_i m_i)^{1/2}}{zB}, \tag{32}$$

the transmission probability of an ion through an aperture of diameter d

(or past a step of depth d, where $d \simeq r_L$) and the corresponding deposition profile on the collector surface are well-defined functions for a given ion temperature, T_i, and charge state, z. Figure 11 shows calculations performed by Cohen [57] of deposition profiles as a function of the aperture dimensions and the shield to collector surface distance. Measurements of lateral deposition profiles have been used in a number of studies to infer the average ion energy and charge state [70–72]. However, in these applications the energy resolution of the measurements tends to be poor because the mixture of charge states that are often present in fusion device boundary plasmas broadens the deposition profile. Deconvolution of the deposition profile into individual charge state components is difficult without an independent measurement of the charge state distribution.

A more successful technique for obtaining ion energy distributions with a surface probe is the use of depth distributions of the deposited species. When an energetic ion is incident upon a surface, the penetration depth, or "range," is dependent on the incident energy of the ion. Due to the importance of ion implantation ranges in solids to fundamental studies of ion-surface interactions and implantation technologies, numerous Monte-Carlo-type calculations have been performed yielding the first moment (mean range) and second moment (straggling parameter) of ion implantation depth profiles [73–75]. In general, the range increases with increasing ion energy and measurements of implantation depth profiles can be used to infer incident ion energies if the ion-retention properties of the solid are known. An implanted ion can be trapped in a solid by a variety of mechanisms: the formation of a stable chemical compound between the incident ion and a host lattice atom; solubility of the incident ion within vacancies, interstitial positions, grain boundaries; or ion damage induced defects in the lattice. The implantation profile will be stable in a solid if the diffusivity of the incident ion in the host solid is small and if the peak concentrations are below saturation values.

The elemental materials, C, Si, and Be are most often used for collector materials in surface probes because the retention properties are known over a wide range of energy for a variety of incident ions, and the retention properties satisfy many of the above criteria. Since the atomic number of the collector elements is relatively low, the mean range of incident ions will be larger compared to the range in high atomic number target materials. This property is important for detection of low energy ions (< 100 eV), since at low incident energies the probability of remission from shallow layers of the solid (i.e., particle reflection) is high. In addition, C and Si are useful target materials because of the availability of high purity target material and their well-characterized sputtering properties. Figure 12 shows examples of the type of calculations and measurements that are necessary

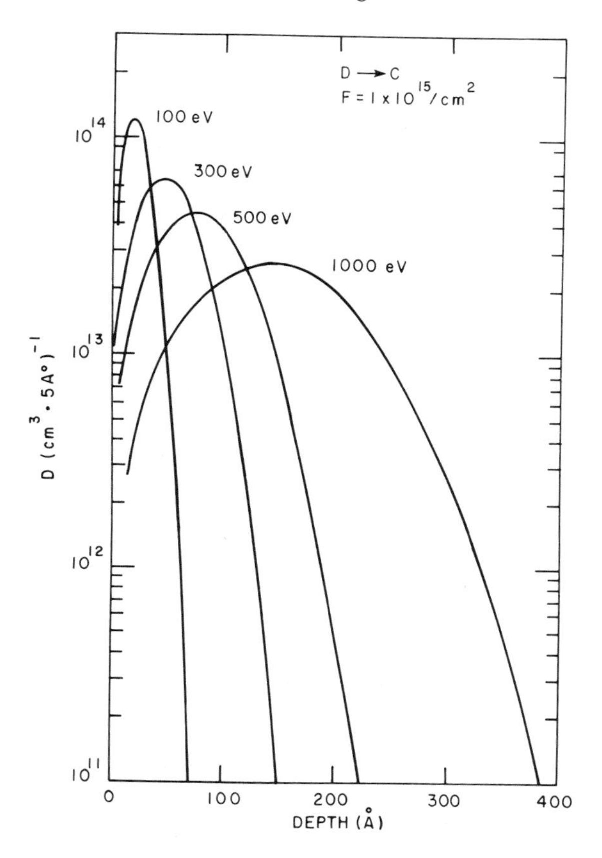

FIGURE 12a. Calculated implantation depth profiles of deuterium incident upon a carbon target as a function of the hydrogen energy (from Ref. 63 with permission).

to characterize a collector material for flux and ion energy distribution measurements [63]: the calculated implantation depth profiles (Fig. 12A) and retention of deuterium incident upon a carbon target as a function of the deuterium energy (Fig. 12B). Both the depth distribution and retention calculations can be used to calibrate incident ion energy measurements with carbon collector probes [63–65]. In addition, the retention curves shown in Fig. 12B can be used to calibrate the total incident fluence since the saturation retention is dependent on the incident ion energy. Plotted with the calculated retention curves in Fig. 12B are measurements of the retention of incident deuterium vs. incident fluence and a as function of energy of an incident (monoenergetic) beam. There is reasonable agreement be-

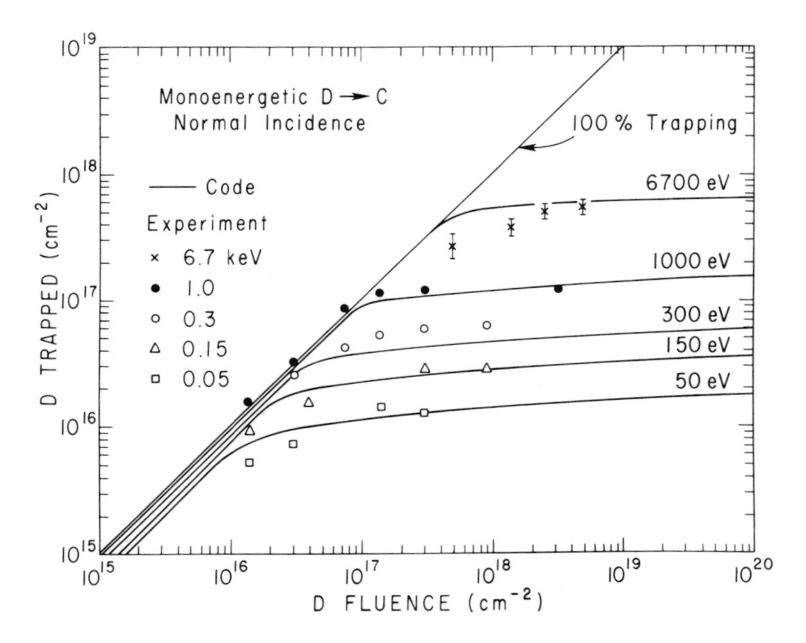

FIGURE 12b. Retention of deuterium incident upon a carbon target as a function of the deuterium energy (from Ref. 63 with permission).

tween the measured and calculated retention vs. fluence. Since the shape of the retention curve is dependent on both the incident energy and fluence, a measurement of the retention in a surface probe at several increasing values of exposure time can be used to obtain both the incident energy and fluence by matching the measured retention curves with the calibration curves. Retention calibration curves for surface probe measurements have been published for hydrogen isotopes incident on C and Si targets for both monoenergetic and Maxwellian incident ion energy distributions [63–65].

M. APPLICATION OF SURFACE PROBES TO GLOW DISCHARGES

We describe a recent application [76] of surface probes for flux and energy distribution measurements in a hydrogen glow discharge [77, 78]. This problem is interesting because of the importance of hydrogen glow discharges for conditioning magnetic fusion devices [77, 82–85] and particle accelerators [75–77] and for plasma-etching oxides from semiconductor

(Si, Ge) surfaces [86] without causing the ion-impact damage to the substrate associated with high mass (O_2, Ar) etchant gases.

The measurement used an array of Si surface probes positioned in a glow discharge chamber, attached to the vessel wall behind a movable shutter, to characterize the plasma ion flux from a nominal 400 V dc glow discharge over a range of discharge pressures (5–90 mTorr). After exposure, the Si targets were removed from the vacuum vessel and the depth distributions of implanted hydrogen were measured in a separate system, which incorporated a high resolution SIMS instrument [87].

Figure 13 shows the results of SIMS depth profile analysis on the Si surface probes. The resulting values of hydrogen retention span the range $(5.4–7.3) \times 10^{16}\ \mathrm{cm}^{-2}$ for the probe exposures, which is in good agreement with the total incident fluence, $6.4 \times 10^{16}\ \mathrm{cm}^{-2}$ calculated from the total ion current, assuming a uniform flux to the cathode surface. Thus, the Si probes produce a reasonably accurate measurement of the cathode fluence, and since in this particular example, the discharge current was constant, the flux is also obtained.

Estimation of the mean energy of the cathode ion flux from the data of Fig. 13 is less accurate than the flux measurement, due to the relatively shallow implantation depths. With the assumption that the range parameters for incident hydrogen are not too different for the thin passivation oxide layer on the Si target and bulk Si, then the peak in the measured implantation profile (or mean range) is approximately 40 (± 5) Å. Comparing this value with the calculated range parameters [88] shown in Fig. 14, yields a value of 200 eV $\simeq V_a/2$ for the mean energy of the incident hydrogen ions. This result is consistent with the more sophisticated mass spectrometer measurements of the cathodic ion energy distribution, which show the primary ion in the discharge is H_2^+ [89]. Thus, each proton impacts at a peak energy of $V_a/2$ after the molecular ion dissociates.

In summarizing surface probes, the techniques can provide useful information on plasma flux and energy distributions. A key advantage of surface probe techniques is the ease of qualitative plasma species analysis. Elemental identification is usually unambiguous and highly sensitive using any of a variety of standard surface analysis techniques. An important advantage of surface probes is the high spatial resolution, which can provide useful information on flux distributions across substrate planes. Energy distributions of plasma species can be obtained from measurements of the depth distributions of implanted species, and in the case of magnetized plasmas, lateral distribution measurements on surface probes can also provide energy distributions. If the retention properties of the surface probe collector material are known for the expected plasma species, then reasonably accurate plasma flux measurements can be obtained from surface density

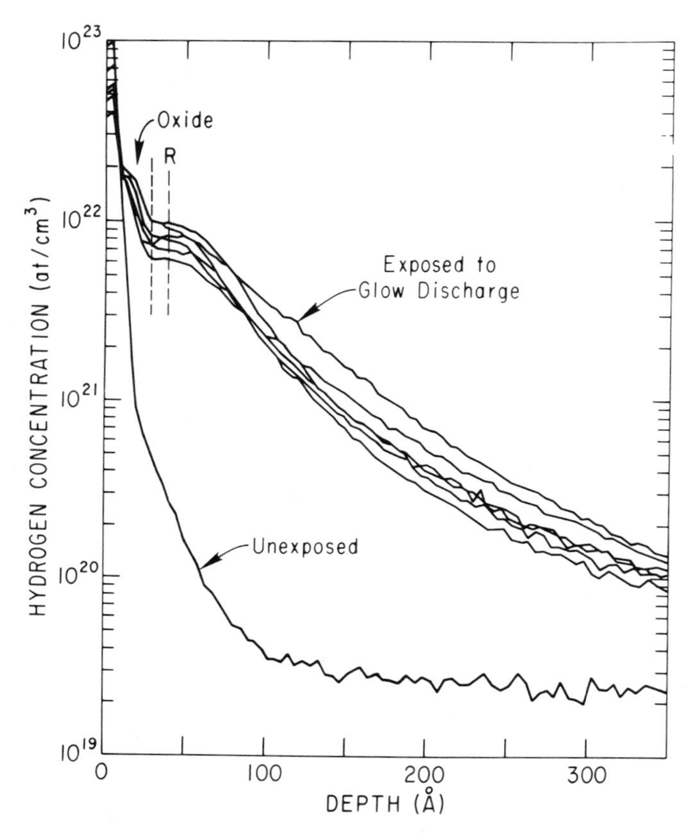

FIGURE 13. SIMS depth profile from an array of Si surface probes exposed to 15 minutes of a 0.1 A hydrogen glow discharge. Samples were exposed at discharge pressures of 5, 10, 20, 30, 60, and 90 mTorr, and one sample was shielded from direct plasma exposure (from Ref. 76 with permission).

measurements on the probe. In general, it is difficult to relate the measured surface density and inferred incident flux to the local plasma ion density in the vicinity of the probe. However, with properly designed probes and shielding apertures, operation of surface probes in an analogous fashion to electrostatic probes biased at ion saturation or at the floating potential, yields quantitative measurements of local plasma ion densities. Finally, an important caveat with the use of surface probes is the possible complication of additional surface processes that compete with the plasma species deposition. Processes such as evaporation, erosion, redeposition, surface and bulk diffusion of implanted species can compete with the initial

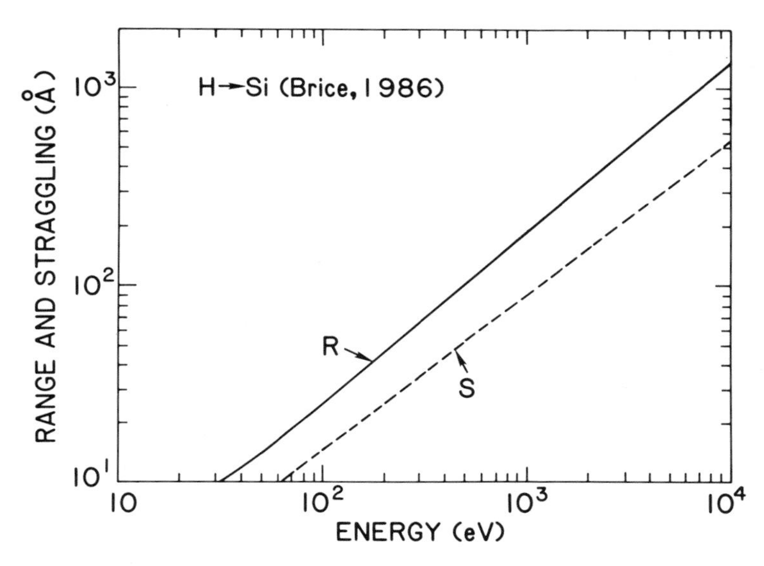

FIGURE 14. Calculated mean range (R) and straggling (S) of low energy deuterium ions incident upon a silicon target. (Figure supplied by D. Brice, Ref. 88.)

deposition of plasma species on surface probe collector surfaces and seriously complicate the interpretation of the change in surface composition.

III. Microwave Interferometry

One of the limitations with the use of electrostatic probes for providing plasma density measurements is the invasive nature of the technique. As discussed previously, an electrostatic probe can seriously effect the local plasma conditions, and conversely, the plasma can affect the physical condition of the probe (i.e., overheat, coat with a dielectric, etc.) leading to changes in the probe characteristics. An alternative plasma diagnostic technique for providing electron density measurements, which is non-invasive, is microwave interferometry. The technique involves the use of microwaves over the range of 1–200 GHz depending on the plasma density range of interest. Line-averaged plasma densities from 10^{10}–10^{14} cm^{-3} can be determined with high accuracy (better than $\pm 2\%$) and with high time resolution (< 1 ms). Although the measurement provides only the line average density, and no information about the density profile, the interferometer technique can be useful for providing real time density measurements and is ideally suited for process control applications. Microwave interferometry is used routinely in magnetic fusion devices [90] both as a diagnostic measurement and as a control element for providing feedback information for plasma density control systems [91]. The development and most applications of the technique are described in the fusion literature [92, 93]. To date, there are few examples of applications of microwave interferometry to process plasma systems.

The technique is applied most usefully to higher density process plasma systems such as magnetron configurations. A scheme for providing microwave density measurements for a magnetron is provided in the conclusion of this section. We also discuss briefly an alternative resonant cavity scheme that can be used for plasma density measurements when the interferometric configurations are inappropriate.

The principle of microwave plasma density measurements is based on the fact that the presence of plasma in the path of an electromagnetic wave will change the index of refraction, which is a function of the plasma electron density. By arranging the microwave transmission through the plasma device in one leg of an interferometer configuration, the net change in index of refraction is manifest as a phase change in the interferometer output. This phase change is proportional to the line-integrated plasma density over the microwave path length through the plasma.

These principles can be illustrated by considering the ideal situation of the transmission of a plane electromagnetic wave through a fully ionized plasma. The wave equation (in one dimension) for electromagnetic waves with magnetic field B, electric field E, and current density J, is given by:

$$\frac{d^2E}{dx^2} - \frac{4\pi}{c^2}\frac{\partial J}{\partial t} - \frac{1}{c^2}\frac{\partial^2 E}{\partial t^2} = 0. \tag{33}$$

The current density can be written as:

$$J = n_e e v, \tag{34}$$

where v is the electron velocity. For the case of a fully ionized plasma the equation of motion for the electrons is:

$$m\frac{dv}{dt} + m\nu v = -eE - \frac{e(v \times B)}{c}, \tag{35}$$

where m is the electron mass and ν $(= \nu_{ei})$ is the electron-ion collision frequency. Using the given equation of motion to solve the wave equation (Eq. 33) and seeking plane wave solutions of the form $E = E_o \exp[i(k\nu - \omega t)]$ yields an expression for the index of refraction, n, known as a dispersion relation:

$$n^2 = \frac{k^2c^2}{\omega^2} = 1 - \frac{\omega_p^2}{\omega^2} \frac{1}{[1 - (i\nu/\omega) - (i\omega_c/\omega)]}, \tag{36a}$$

where substitutions have been made for the two characteristic frequencies: the plasma frequency, $\omega_p = (4\pi n_e e^2/m)^{1/2}$; and the cyclotron frequency, $\omega_c = eB/mc$.

For many applications (including process plasmas) the following approximations can be made.

$$\frac{\nu}{\omega} \ll 1 \text{ and } \frac{\omega_c}{\omega} \ll 1,$$

which reduces the dispersion relation to the simple result:

$$n^2 = 1 - \frac{\omega_p^2}{\omega^2}. \tag{36b}$$

At $\omega = \omega_p$, the index of refraction, n, approaches zero and the wave will not propagate. This is termed the cut-off frequency and there is a corresponding critical electron density, n_c, where wave propagation is cut off equal to:

$$n_c = \frac{\omega^2 m}{4\pi e^2}. \tag{37}$$

A graphical representation of the dispersion relation given in Eq. (36a) is shown in Fig. 14, where the real and imaginary parts of the index of refraction are plotted as a function of the normalized plasma density, n_e/n_c. Figure 14 shows for a given microwave frequency, ω, there is a range of plasma density where the index of refraction is linearly proportional to the plasma density. This regime is given approximately by $n_e/n_c \leq 0.3$, and in this range plasma density measurements by microwave interferometry are most straightforward.

A. MICROWAVE INTERFEROMETER CONFIGURATIONS

Figure 15 shows the simplest microwave interferometer configuration for measuring plasma electron densities. This configuration is based on the Mach-Zehnder interferometer, which is well known in classical optics [94]. Waves from a single microwave source are split into two optical paths, one of which traverses the plasma volume. Both paths recombine at a single detector. The reference path is first adjusted for zero phase difference with

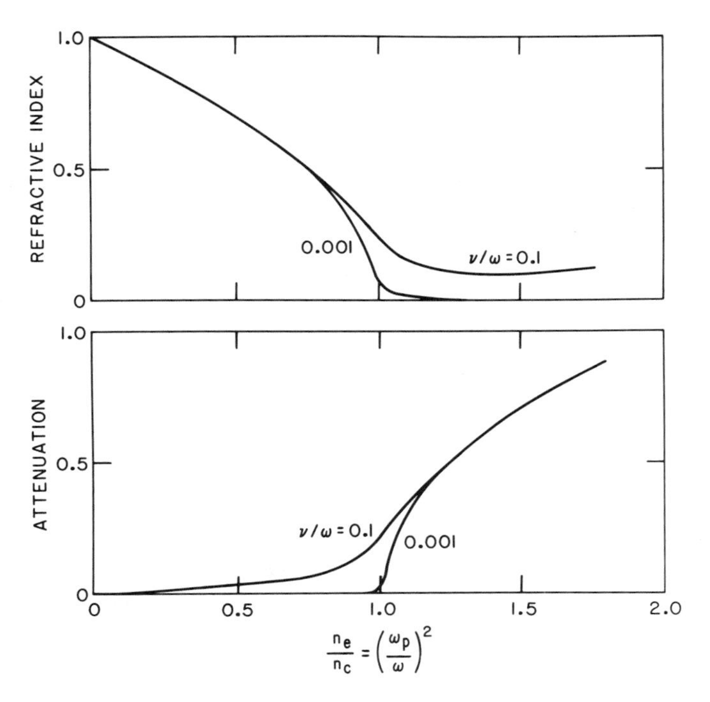

FIGURE 15. Dispersion relation for a plane wave in a plasma showing the variation of the real and imaginary components of the index of refraction as a function of frequency (adapted from Ref. 90 with permission).

the plasma path by adjusting the phase shifter in the absence of plasma. The detector will then output a sinusoidal variation in response to changes in plasma density. The simple Mach-Zehnder interferometer is not the configuration of choice for continuously monitoring the quasi-steady state plasma density of a process plasma, because frequency drifts in the microwave source will cause phase changes. A more sophisticated configuration, the "zebra-stripe" interferometer [90] is shown in Fig. 16. This configuration has several advantages over the Mach-Zehnder configuration: it is less sensitive to source phase changes; insensitive to source amplitude variations; has a greater dynamic range; and outputs a signal proportional to the line-averaged density (as opposed to density changes). The zebra-stripe interferometer produces a repeating interference pattern as the source frequency is modulated (typically at ~ 1 MHz). The unusual name of this type of interferometer comes from the appearance of the output as the fringe pattern is displayed on an oscilloscope (see inset in Fig. 16). Recent

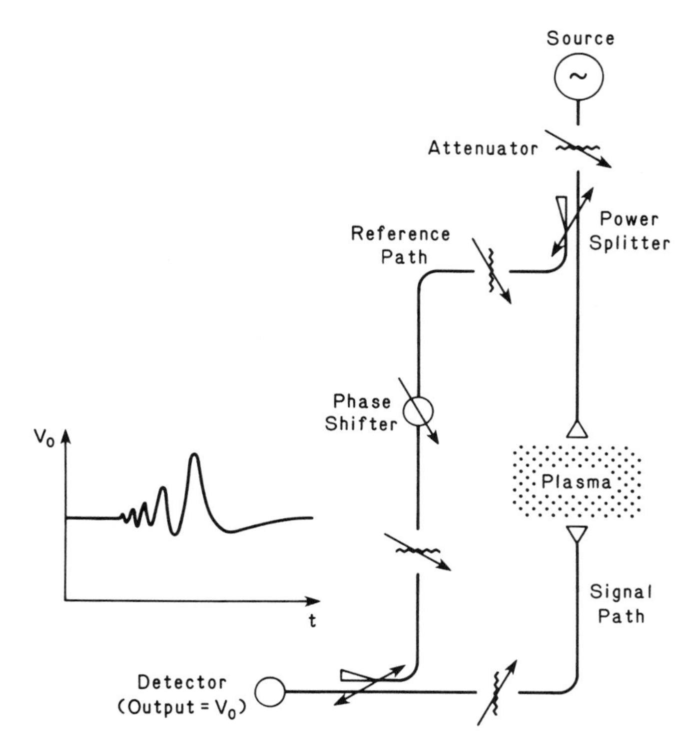

FIGURE 16. Schematic diagram for microwave interferometer configured for plasma density measurements. The inset signal output voltage (V_0) shows a representative output responding to changes in plasma density within the microwave beam path (adapted from Ref. 90 with permission).

versions of the zebra stripe interferometer have replaced the fringe display with phase counting electronics, which output a voltage proportional to the plasma density.

In either of the interferometer configurations described, a line-averaged electron density, n_e is determined equal to

$$\bar{n}_e = \frac{1}{l}\int_0^l n_e(x)\,dx,$$

where l is the path length of the microwave in the plasma. We can evaluate the dynamic range and sensitivity of microwave density measurements by estimating the magnitude of the phase shifts as a result of the presence of plasma in the signal path of the interferometer. Continuing the analysis with the use of one-dimensional plane waves traversing the plasma, the phase shift $\Delta\phi$ is given by

$$\Delta\phi = -\frac{2\pi}{\lambda}\int(n-1)\,dx = \int\left[1-\left(1-\frac{n_e}{n_c}\right)^{1/2}\right]\frac{2\pi}{\lambda}\,dx, \qquad (38)$$

where we have substituted Eq. (36b) for the refractive index. Equation (38) has two implicit assumptions with regard to the plasma: (1) the plasma properties vary slowly over the microwave wavelength (the so-called "adiabatic limit"); and (2) the plasma is non-dissipative, i.e., ω is far from any resonant frequency.

For the range of plasma densities where the index of refraction is proportional to the plasma density, i.e., $n_e \ll n_c$. Equation (38) can be approximated by:

$$\Delta\phi \simeq \frac{\pi}{\lambda n_c}\int n_e(x)\,dx = \frac{2\pi e^2}{mc\omega}\int n_e(x)\,dx. \qquad (39)$$

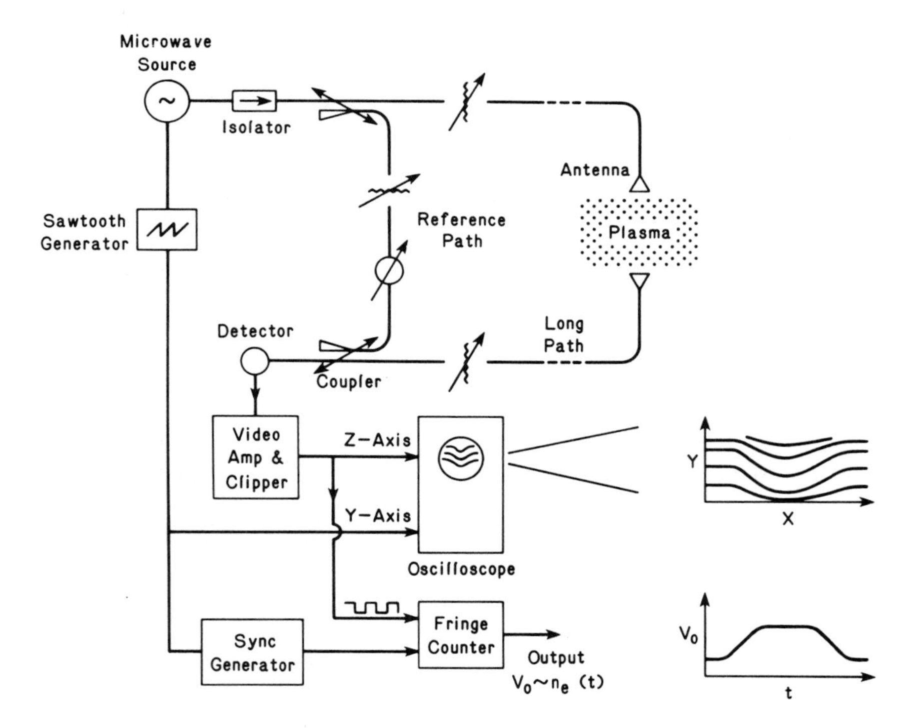

FIGURE 17. Schematic diagram of the "zebra-stripe" interferometer. The inset graphs show a representative oscilloscope output in which the stripe density is proportional to plasma density, and the output of a phase counting circuit, which generates a voltage proportional to the line-averaged plasma density (adapted from Ref. 90 with permission).

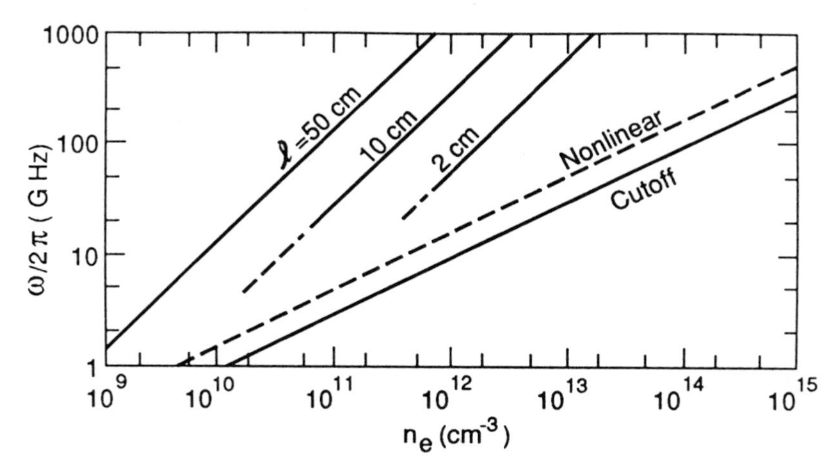

FIGURE 18. Accessible range of plasma density measurements using microwave interferometry as a function or the source wavelength (λ) and plasma path length (l), assuming a phase resolution of 18°. With care, phase resolution of 1–2° is possible (adapted from Ref. 90 with permission).

Therefore, for a path length, l, in the plasma, the line-averaged electron density, $\bar{n}_e$, is equal to:

$$\bar{n}_e \simeq \frac{1}{l}\int_o^l n_e(x)\,dx = \frac{mc\omega\,\Delta\phi}{2\pi e^2 l}. \tag{40}$$

In practical units, the above expression is:

$$\bar{n}_e(\text{cm}^{-3}) = 118\,\frac{\omega/2\pi(\text{Hz})\,\Delta\phi(\text{rad})}{l(\text{cm})} \tag{41}$$

As previously stated, the maximum density suitable for density measurements is $n_e(\text{max}) \leq n_c/3$, because of the non-linear and dissipative effects that become important near the cutoff density. The minimum density is directly proportional to the minimum phase measurement [via Eq. (40)], which depends on the detector noise level and stability of both the source and detector. Figure 18 shows the range of accessible plasma electron density measurements using microwaves as a function of the microwave frequency and plasma path length, assuming a very conservative phase resolution of $\pi/10 = 18°$.

In practice, choosing one of the standardized microwave frequency bands for the design of a microwave interferometer increases the availability of microwave hardware such as sources, detectors, and waveguides. Table 1 lists some of the useful frequencies for microwave density measurements

Table 1 Useful Frequencies for Interferometric Measurements of Plasma Densities

Frequency ($\omega/2\pi$)	Wavelength (λ)	Band Designation	Cut-off Density $n_c(\mathrm{cm}^{-3})$
18–26.5 GHz	1.2 cm	K	5×10^{13}
8.2–12.4 GHz	3 cm	X	1×10^{12}
2.6–4.0 GHz	10 cm	S	1×10^{11}
1.1–1.7 GHz	20 cm	L	2.9×10^{10}

along with the band designation and cutoff density. The table shows that higher frequencies provide access to higher density plasmas; however, for a given plasma density, use of higher frequency microwaves requires a smaller phase shift measurement per unit path length [90].

For the interferometer configurations shown in Figs. 16 and 17, the microwaves are transmitted through the plasma using a pair of microwave horn antennae. Several criteria must be met with regard to the microwave optics in order to optimize interferometer operation. The antennae must be separated a distance, which is large compared to the diffraction limit (A^2/λ) of the antennae aperature (A), in order to maintain phase coherence. Conversely, the separation distance cannot be too large or the signal will be degraded by refraction in the plasma (since $n < 1$), or the phase shift detection may be compromised by microwave ray paths which do not pass through the plasma. Satisfying all the above criteria is not always possible. In particular, for low density ($< 10^9$ cm^{-3}), small diameter (< 1 cm) plasmas it is impossible to satisfy the optics requirements for the interferometer configuration. There are examples in the literature of microwave density measurements in this regime where the plasma chamber is considered as a resonant cavity whose frequency is dependent on the presence of plasma [2].

There are a few examples of microwave density measurements for process plasma configurations [2, 53, 54, 95]. Figure 19 shows two possible configurations for density measurements with the typical "race-track" magnetron used for plasma sputtering or etching. The microwave beam is transmitted through one of the long legs of the race track, either with opposing horn antennae or with isolated probes. In the latter configuration, the plasma chamber is treated as a waveguide. For either configuration, if 10 GHz microwaves are used and the path length is 40 cm, the typical magnetron plasma density, $\bar{n}_e \simeq 10^{11}$ cm^{-3}, would correspond to a phase change of ~ 0.5 fringe.

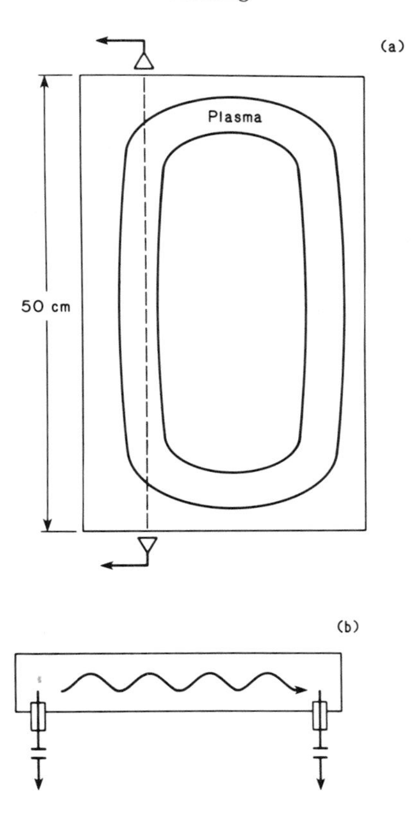

FIGURE 19. Possible configurations for microwave density measurements in a plasma magnetron using external microwave horns (a) or internal probes and (b) for launching the microwaves through the plasma column.

IV. Impedance Analysis

Another tool for the analysis of glow discharges is the study of the current-voltage characteristics of the plasma column. Most of the early fundamental studies of dc glow discharges were attempts to deduce such characteristics from the underlying surface and gas phase ionization mechanisms [96, 97]. The resulting (dynamic) resistance,

$$R(t) = dV/dI, \tag{42}$$

represents the convolution of numerous underlying parameters. For the rf driven plasma case, discussed here, there is an analogous measure in the dynamic (complex) impedance,

$$Z(t) = dV(t)/dI(t). \tag{43}$$

Before discussing the methods for measurement of this parameter, we digress briefly to describe the electrical characteristics of rf discharges. The simple electrical analysis considered next results from the existence of the sheath in a contained discharge.

The electrical characteristics of a plasma system can be modeled in many different ways. Koenig and Maissel [98] have represented rf discharges for insulator sputtering as an impedance network, shown in Fig. 20. Their analysis of this network assumed that the rf current through the sheath was carried exclusively by displacement current so that the voltages of the nodes across the sheaths couple capacitively. Conduction by ions is small because of their large mass and low mobility. Conduction by electrons is small because of their greatly reduced density in the sheath. When the two electrodes have unequal area, the analysis proceeds by assuming: that their ion current densities are equal, that the ions undergo no collisions in the

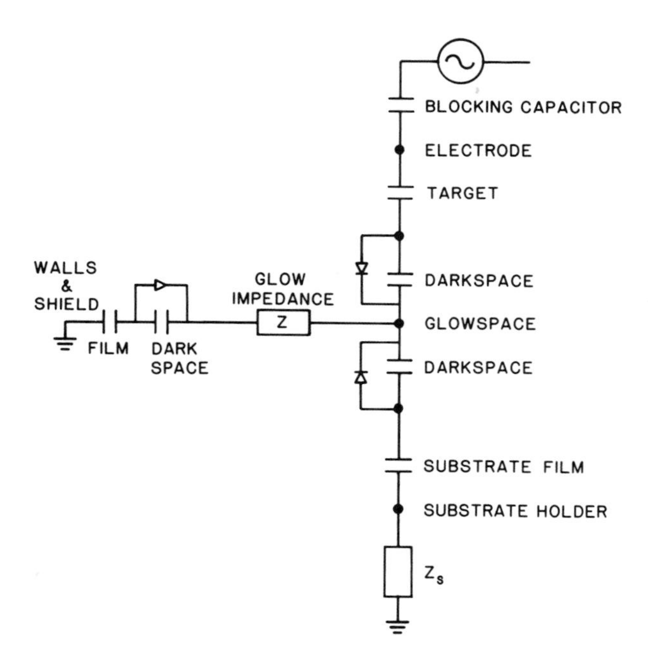

FIGURE 20. One possible schematic representation of an rf-driven diode sputtering system. The "dark spaces" in the figure are the "sheaths" described in the text and result from low visible excitation of the background gas in these zones (from Ref. 98 with permission).

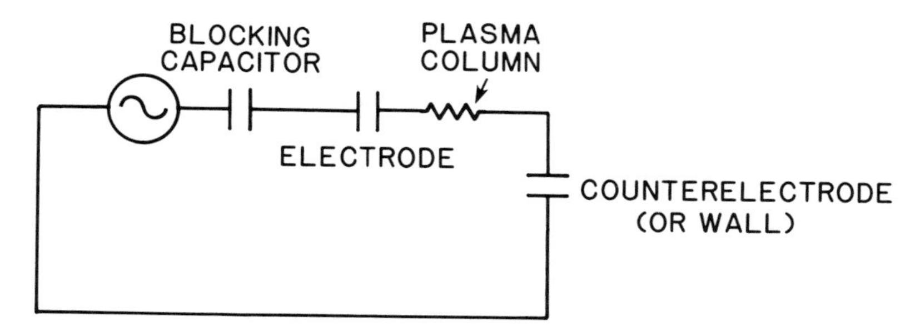

FIGURE 21. Simplified representation of an rf-driven diode reactor.

sheath (space charge limited flow), and that the capacitance of the sheath is proportional to the electrode area and inversely proportional to the sheath thickness. With these assumptions the applied rf voltage is divided capacitively by the network. The sheaths are also assumed to be ideal rectifiers, so that the resulting rf potentials are converted to dc bias voltages on the plates. This model predicts that the ratio of the plate voltages is inversely proportional to the fourth power of the ratio of their areas. The assumptions in the above model are incorrect for most process plasma configurations [99], and experimental evidence [100] shows that the voltage ratio depends on a smaller power (near unity) of the inverse area ratio. The same result is obtained if one substantially simplifies the picture of the discharge to that shown in Fig. 21, considering only the capacitance associated with the powered electrode(s) and the walls, and assuming that the plasma column is entirely resistive. With such a simple model one can write:

$$V_{\mathrm{dc}} = V_{\mathrm{rf}}(C_t - C_w)/(C_t + C_w), \tag{44}$$

where C_t and C_w are time-averaged capacitances of the sheath at the powered electrode and at the wall (or counter-electrode). At low pressure (20 mTorr) and at low power, the experimental agreement with this expression is fairly good [101].

This analysis establishes that the smaller electrode is always at a lower potential than the larger electrode when the negative charge carriers have a higher mobility than the positive charge carriers. We note in passing, that this is quite independent of which electrode is powered and which is grounded [102], and indeed, is true even if both electrodes are powered. In real systems, stray capacitances, electrode end-effects, electrode surface conditions, and power dissipation in the sheaths are not calculable a priori; so in order to determine the resulting potentials, it is necessary to measure them directly. Power dissipation in the sheath can be represented as a

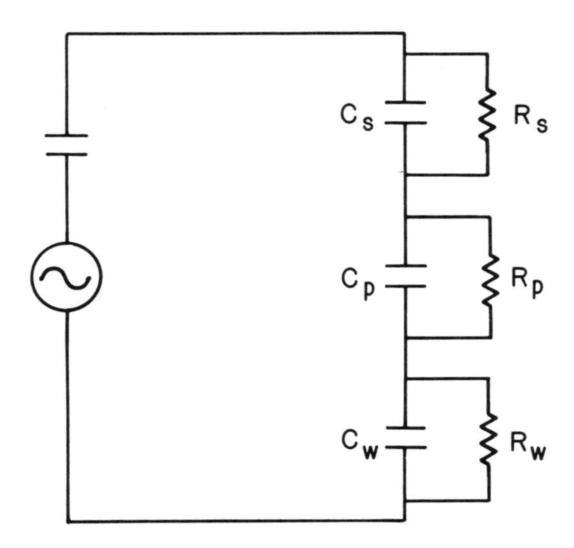

FIGURE 22. An alternate representation of the plasma regions (from Ref. 103 with permission).

sheath resistance in parallel with the sheath capacitance [103], shown in Fig. 22. One finds that at low frequency ($\leq$ 1 MHz) this circuit behaves like three resistors in series, while at high frequency (1–100 MHz) the sheaths are mainly capacitive and are best analyzed as above. It is also found that the sheath fields at low frequency are substantially larger than at high frequency, for equal excitation voltages.

The most direct measure of the impedance comes from measuring the voltages and currents on the electrodes, as done by a number of researchers [104–108]. Currents are measured with Rogowski coils or with an alternate form of transformer. Rf voltages are measured using calibrated capacitive divider inputs to fast analog or digital oscilloscopes. At higher frequencies, considerable attention must be paid to the measurement circuit design and implementation to avoid introducing spurious inductance or capacitance. Figure 23 shows the apparatus used by Hebner and Verdeyen [108]. Note that an rf filter is used for the measurement of the dc bias, which develops at the electrode. The time dependent values of V and I recorded by these workers are shown in Fig. 24. The power delivered to the plasma is computed by multiplying these traces as shown in the bottom portion of the figure. Note that the power peaks in phase with the current and that the

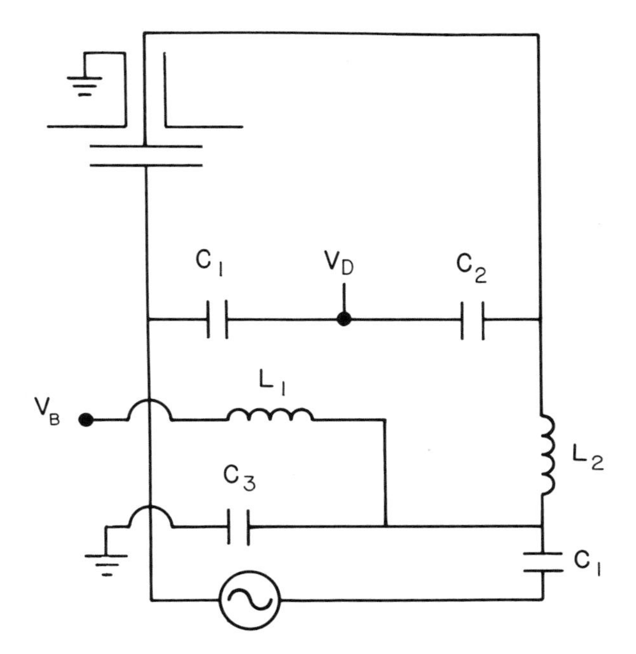

FIGURE 23. Schematic of a discharge circuit used for impedance studies. The values employed were $C_1 = 0.02\ \mu F$, $C_2 = 30$ pF, $C_3 = 200$ pF for 2.6 MHz operation, 700 pF for 1.0 MHz operation, $L_1 = 200$ mH, $L_2 = 28\ \mu H$ for 2.6 MHz operation, 400 μH for 1.0 MHz operation. The dc bias voltage is measured at the node labeled V_B on the left, and rf voltage is measured at the node labelled V_D. Current is measured with a coil probe (not shown) (from Ref. 108 with permission).

power is negative for a portion of the rf cycle because the discharge is reactive.

Measuring the dc bias allows the average energy of the bombarding ions to be estimated. More accurate estimation of the bombardment energy would require measuring the potential on the electrode relative to that in the plasma column (rather than to ground). In practice, the actual energy of bombarding species will also depend upon charge exchange collisions and, under some conditions, on electron impact ionization occurring in the sheath. Sampling mass spectrometry with energy analysis, discussed later, is a more direct method for determining the ion bombardment energy. The resistive (R) and reactive (ωC_s) components of the impedance can be directly computed form the $I(t)$ and $V(t)$ waveforms. Within the framework of very simple models [107], these can be related to fundamental

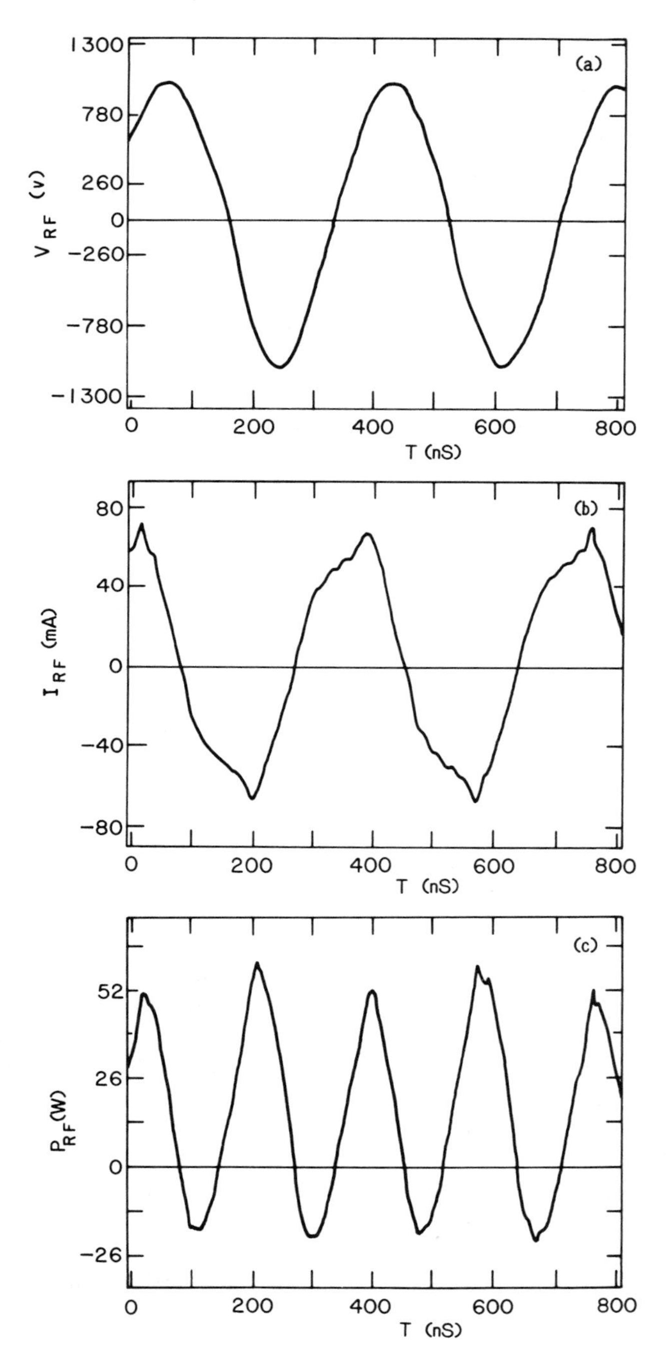

FIGURE 24. The voltage (a) and current (b) measured in the apparatus of Fig. 23. Power (c) is obtained by multiplying the upper traces (from Ref. 108 with permission).

parameters of interest as follows:

$$\frac{E}{p} = \frac{IR}{\sqrt{2}\,pd} \tag{45}$$

$$V_s = \frac{I}{\omega C_s} \tag{46}$$

$$n_e = \frac{d}{Ae\mu_e R}, \tag{47}$$

where d is the electrode spacing, A is electrode area, p is the pressure, μ_e is the mobility, and V_s and C_s are the sheath voltage and capacitance.

When the direct measurement of voltages and currents is difficult, there are alternatives. One may use a matching network to measure the plasma impedance. When a matching network is tuned so that the generator and the load are optimally coupled, the combined impedance of the rf source and matching network is the complex conjugate of the impedance of the plasma. The details of this method have been reviewed in a paper by Norstrom [109], which shows how to construct a variety of such matching networks. Plasma impedance is estimated by tuning the network for maximum forward power into the plasma, and calculating the discharge impedance from the known values of the components in the matching network and the known impedance of the associated cabling and connectors. An alternative method is to fix the impedance of the generator at 50 ohms (purely resistive). Once tuned, the reactance of the matching network is equal to the conjugate of the reactance of the plasma. It is often difficult to make measurements of the complex impedance with sufficient accuracy, and these techniques ignore the losses that may occur in the lines or in the matching network. Compensation for such losses can be performed by the construction of dummy loads (designed to simulate the plasma) [110], for which it is easy to measure dissipation in the matching network.

Mlynko and Hess have described a technique using an operating impedance bridge [111]. A schematic of their apparatus is shown in Fig. 25. The bridge samples forward and reverse voltages along the transmission line. Resistances and reactances are added to the bridge until a null occurs, indicating that the resistance and reactance are identical to that of the load, which consists of the plasma, the chamber, and the connections. This technique has been used [111] to measure the resistance and capacitance of nitrogen discharges as functions of rf power.

In addition to providing an estimate of electron density and sheath voltages, impedance measurements are useful in production environments as a gross measure of process quality and reproducibility. Because discharge impedance rapidly responds to changes in fundamental parameters like

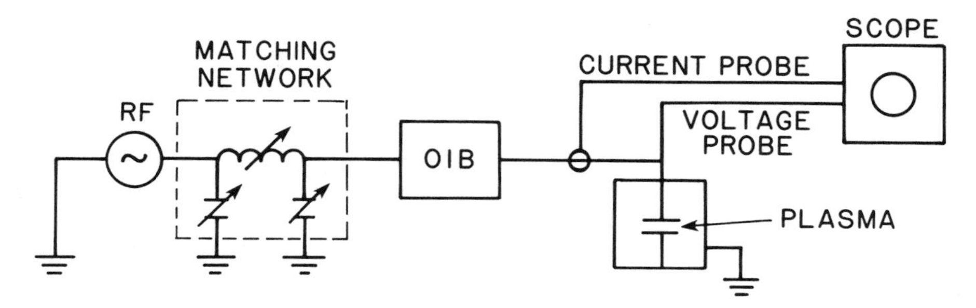

FIGURE 25. A schematic of an apparatus using an operating impedance bridge (OIB) for the measurement of plasma resistance and capacitance (from Ref. 111 with permission).

electron density, surface conditions, species mix, and so on, it is useful as an endpoint monitor. For this purpose, the simplest method is to fix the rf power supplied to the electrode, P_d. The impedance, $Z_d = E_d^2/P_d$, is monitored by simply measuring the voltage. A comparison of this method with an optical emission endpoint for Al reactive ion etching is shown in Fig. 26 [112].

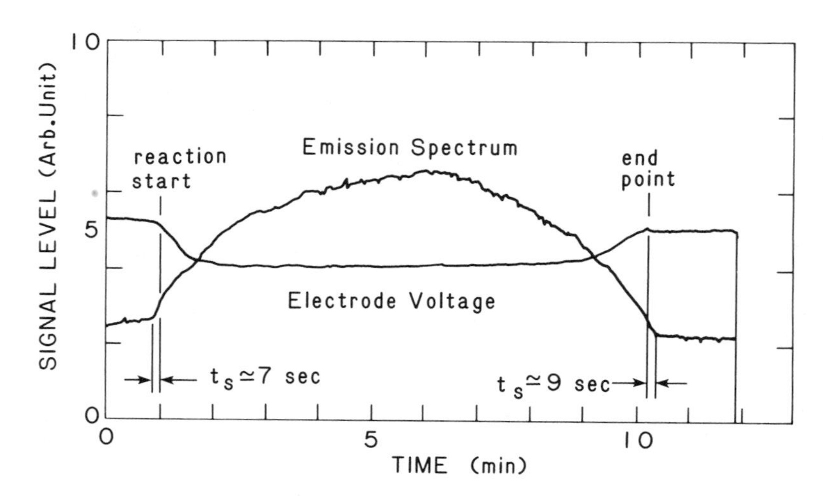

FIGURE 26. Comparison of an Al etch endpoint determined by a change of plasma impedance, represented by the electrode voltage at constant power (see text) to the change in optical emission at 261.6 nm (from Ref. 112 with permission).

V. Mass Spectrometry of Plasmas†

Mass spectrometric analysis of the particle emission from a process plasma is one of the more commonly applied plasma diagnostic techniques [113–115]. The relative simplicity of the mass spectra of the gaseous components in a plasma, in comparison to the optical emission spectra [116–118], makes mass spectrometry useful for quantifying plasma composition and impurity concentrations for the study of gas-phase plasma reactions and plasma-surface interactions.

One can distinguish two general schemes for mass spectrometric analysis of plasmas: flux analysis and partial pressure analysis. Flux analysis involves sampling the plasma directly by coupling the line-of-sight emission of plasma particles through a small aperture into the ion optics of the mass spectrometer. This method is best suited for plasma species and energy analysis. In comparison, partial pressure analysis allows simpler vacuum connections between the mass spectrometer and plasma chamber, and generally less stringent differential pumping requirements. In addition to its experimental simplicity, it is the method of choice for monitoring glow discharge treatments of vacuum vessels [77] or plasma processing of large area materials, because the measured signals are representative of the integrated effects of the relevant plasma-surface interactions.

Reviews of plasma mass spectrometry have been given by Drawin [119], Dylla [114], and Vasile and Dylla [115]. The primary features of plasma flux analysis by mass spectrometry and partial pressure analysis are discussed next.

A. PLASMA FLUX ANALYSIS

First, we consider the methodology of plasma flux analysis. Figure 27 shows a schematic diagram of the typical apparatus required for plasma flux analysis, which includes four stages: (1) the particle extraction/collimation optics, (2) the energy analyzer, (3) the mass analyzer, and (4) the ion detector. The most complicated component in the system, in terms of its design and effect on signal collection and interpretation, is the extraction optics. The simplest extraction system employs an aperture that separates the plasma vessel from a separately pumped chamber containing the energy and mass spectrometers. The resulting flux of neutral particles from the plasma (background atomic and molecular neutrals and plasma charge-

† This section has been adapted from Ref. 114 (with permission) and an expanded version appears in Ref. 115.

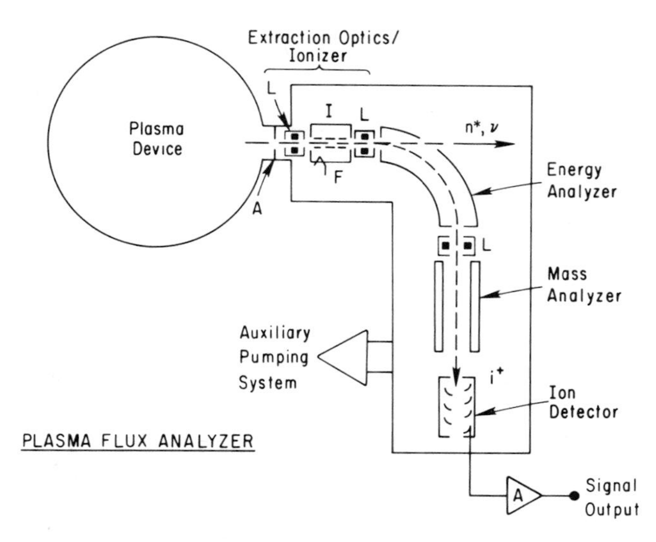

FIGURE 27. Apparatus for plasma flux analysis (from Ref. 114 with permission).

exchange neutrals) will be collimated by the aperture according to the appropriate kinetic gas laws. Once collimated by the sampling aperture, the flux of neutrals can be ionized to allow energy and mass analysis by electrostatic and/or magnetic deflection. For particles with thermal or near thermal energies (0.02–10 eV), the ionization is usually performed with an electron bombardment ionizer [120].

Flux analysis of plasma ions is not as straightforward as the analysis of neutrals because of the electrostatics of the interaction of the sampling aperture with the plasma. The nature of the charged particle flux through the aperture depends on the plasma conditions in the vicinity of the aperture, the geometry and electrostatic potential of the aperture. This problem is not solvable in general; however, a few special cases of practical importance are well-treated in the literature. Drawin [119] discusses three configurations for extracting plasma ions from an electrostatic plasma (i.e., glow discharge), that depend on the relative magnitude of the ion mean-free-paths, λ_i, the Debye shielding length, λ_D, and the extraction aperture diameter, D. The simplest case is collisionless extraction through a plasma sheath, which occurs when $\lambda_i \gg D, \lambda_D$. In this case, if the extraction aperture is biased negatively (or is simply an aperture in the cathode if the plasma is a dc glow discharge), the sampled flux is proportional to the ion saturation current, Eq. (3). The case of collision dominated extraction

through a plasma sheath, $\lambda_i < \lambda_D$, D, is more complicated. Various collision processes between plasma particles can lead to the development of an extracted ion flux with a very different ion population from the unperturbed plasma far from the vicinity of the aperture. A third configuration is extraction by ambipolar effusion, which results when the extraction aperture is biased at the floating potential, allowing an ambipolar current $j \sim D_{\mathrm{amb}} \nabla n$ to effuse through the aperture. In certain cases, such as the analysis of the effusive flux from the positive column of glow discharge [121], this method can lead to a relationship between the extracted and plasma ion flux that is similar to the ion saturation method.

Magnetized plasmas complicate the design and analysis of extraction optics for plasma flux analysis and, in addition, can cause practical problems with the design of the downstream instrumentation because of the necessity of magnetic shielding or the complication of moving sensitive apparatus to far-field regions. The sampling of ion fluxes from magnetized plasmas entails additional geometric constraints. Because of the confinement of ions along magnetic field lines, the ion flux effusing through an aperture will vary with the geometry of the aperture relative to the magnetic field. If the aperture axis is parallel to the magnetic field, the streaming flux will be sampled; whereas, if the aperture axis is perpendicular to the magnetic field, the cross-field diffusive flux will be sampled. Both fluxes can be related to the local plasma ion density, but the analysis is complicated. This problem has been considered in the design of probes for space plasma research [69], and more recently [76] in the design of surface probes for sampling the boundary of fusion devices, as discussed.

Flux analysis has been applied to plasma sputtering and etching devices primarily for plasma species analysis. The technique is capable of identifying ionized and neutral species in the vicinity of the substrate plane of sputter-deposition devices such that the chemical composition of the sputter deposited film can be monitored in real-time [123]. The line-of-sight sampling of plasma flux analysis is useful for analyzing reactive or transient species in plasma-etching systems [124–127], and has proved helpful for identifying reaction mechanisms in such devices [128, 129]. Similarly, flux analysis has been used for identifying the ion species in rf [130] and high current plasma ion sources [131, 132].

These measurements illustrate the primary advantage of the flux analysis technique for mass spectrometry of plasmas. With proper design of the sampling aperture and extraction optics, plasma ions and neutrals that are representative of the plasma composition in the vicinity of the aperture can be sampled such that the extracted species is unaffected by subsequent gas-phase and surface collisions. The primary disadvantages of the flux analysis techniques are basically practical concerns: (1) the technique requires careful design of the sampling and extraction optics, which cannot always be made nonperturbing to the plasma device; (2) the differential-pumping requirements to maintain collision-free conditions may be nontrivial, especially for sampling the higher pressure process plasma; and (3) for applications with magnetic plasma systems, the magnetic shielding requirements for the spectrometers can be severe.

B. PARTIAL PRESSURE ANALYSIS

The alternative experimental scheme for mass spectrometry of plasmas, which employs the same or somewhat simpler instrumentation as required for flux analysis, is partial pressure analysis. Figure 28 shows a schematic diagram of the typical experimental arrangement for partial pressure analysis in which a quadrupole mass spectrometer (QMS) is indicated for the partial pressure measurement. Depending on the operating pressure regime of the plasma device, there is either a high conductance or low conductance connection tube between the plasma device and the QMS chamber, which is usually differentially pumped.

The instrumentation for partial pressure analysis can be made simpler than that for flux analysis, because the extraction optics and energy analyzing stages are unnecessary. Usually, plasma particles are allowed to thermalize before reaching the QMS chamber by repeated collisions on the connection tube wall. Thus, for the typical electron-bombardment ionizer of

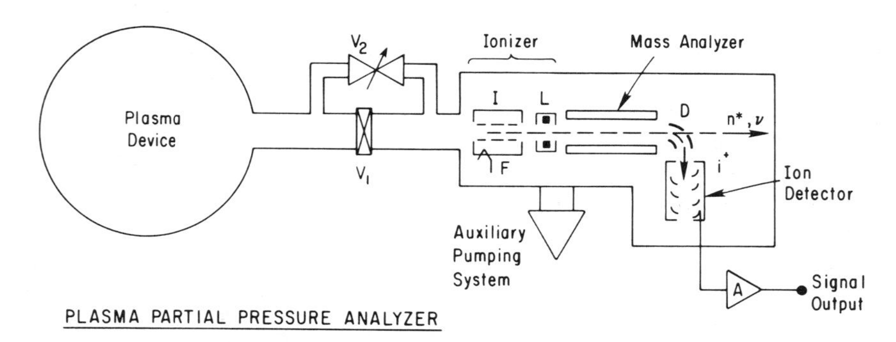

FIGURE 28. The use of a QMS for plasma partial pressure analysis. The transition piece shows a high and low conductance path; the number and conductance of such paths is selected to match the various operating pressures in the plasma device (from Ref. 114 with permission).

a QMS, the mass analyzed ion current, which is proportional to the local gas density, can be calibrated in terms of a partial pressure equilibrated to the temperature of the QMS chamber walls.

Partial pressure analysis is useful because the technique samples volatile reaction products resulting from plasma-material interactions summed over the entire surface area of the plasma chamber. In contrast, the flux analysis technique samples only the reaction products that are representative of plasma-material interactions in the vicinity of the sampling aperture. This is particularly useful when mass spectrometry is used as a diagnostic or monitor of: (1) plasma-etching [128] of large area samples; and (2) discharge cleaning [77] of large vacuum vessels. There are a number of problems, however, concerned with interface of a partial pressure analyzer to a plasma device, which have been poorly addressed in the literature and in commercial documentation, leading to confusion and erroneous data interpretation. These problems include: (1) the necessary trade-off of pressure reduction versus dynamic range when applying a high vacuum instrument to sample low vacuum ($10^{-3} - 1$ Torr) processes, (2) high background problems induced by the primary process gas, and (3) system response and system calibration. We refer the reader to the previously referenced reviews [114, 115] for a detailed discussion of these problems. We will expand on the latter topic here, because of the general lack of absolutely calibrated mass spectrometer data that is evident in the literature.

The experimental scheme for partial pressure analysis diagrammed in Fig. 28, which shows the QMS separated from the plasma chamber by a differentially pumped connection tube, results in the dynamic response of a pressure change in the QMS chamber, $\Delta P_Q(t)$ being different from the pressure change in the plasma device, $\Delta P_D(t)$. Figure 29 shows the experimental scheme of Fig. 28 redrawn with the important kinetic factors in the system identified: the plasma chamber with volume V_D, is pumped with speed S_p, and may have one or more gas input sources Q; the QMS chamber with volume V_Q, is pumped with a separate pump with speed S_Q, and is connected to V_D through a tube with conductance C_T. The quantities S_p, S_Q, and C_T can be a function of the gas temperature, T, and the pressure, P, mass, M, and viscosity of the gas depending on the flow regime and type of vacuum pump.

For proper calibration of a partial pressure analyzer, the relationship between $\Delta P_D(t)$ and $\Delta P_Q(t)$ needs to be evaluated. First, we consider the simplified case of non-reacting gases, i.e., gas-surface interactions within the connection tube and the QMS are ignored, so that only the kinetic relationship between $\Delta P_D(t)$ and $\Delta P_Q(t)$ need be considered. For static measure-

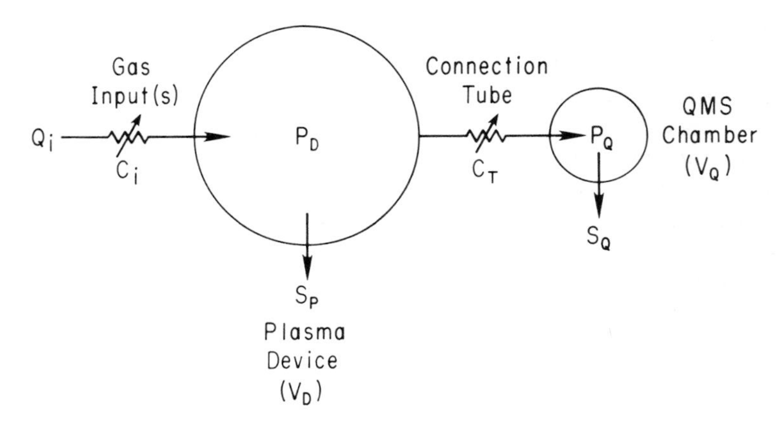

FIGURE 29. Schematic of the apparatus of Fig. 28 showing the equivalent paths for analysis of the flows and pressure (from Ref. 114 with permission).

ments, ΔP_D and ΔP_Q are related simply by a multiplicative throttling factor, ρ, which is the pressure reduction ratio.

$$\Delta P_D = \rho(\Delta P_Q) \tag{48}$$

The throttling factor, ρ, is inversely proportional to the net series conductance (C_T) of the connection tube and any intervening apertures or flow restrictions, and proportional to the differential pumping speed, S_Q.

For dynamic measurements, a differential equation describing the time-response of the system must be solved:

$$P_D(t) = \left(1 + \frac{S_Q}{C_T}\right) P_Q(t) + \frac{V_Q}{C_T} \frac{d[P_Q(t)]}{dt} - \left(1 + \frac{S_Q}{C_T}\right) P_B, \tag{49}$$

where the last term represents an explicit background pressure P_B correction in the QMS. Equation (49) usually has to be solved numerically [133].

A common occurrence in the process plasma literature is the presentation of partial pressure data in an uncalibrated format, i.e., the signal output of a QMS is displayed as raw data (the electron multiplier or Faraday cup output current), or in arbitrary intensity units. Without, as a minimum, a calibration that encompasses the entire system, the interpretation of QMS data can be very misleading. Reconsider the simplified vacuum system schematic shown in Fig. 29. A system calibration determines the functional relationship between P_D and P_Q for one or more gases over a pressure range of interest. In addition, the relationship between P_Q and the measured output current of the QMS must be determined.

In certain situations the calibration procedure can be quite complicated. For example, if the pumping speed of the plasma chamber or QMS chamber pump is not a well-defined function of gas species and pressure; or if the calibration must span a pressure regime which causes the conductance of the connection tube to overlap more than one kinetic flow regime; or if the gas species of interest has significant physical or chemical reactions with the vessel wall materials; then the calibration procedure is not straightforward.

There are several examples in the literature of experimental arrangements for partial pressure analysis that are difficult to calibrate, unstable with respect to maintaining a calibration, or capable of being calibrated only over a limited range of parameters. The use of an ion pump for the differential pump on the QMS chamber [134] is a bad choice because of the large variation of pumping speed with gas species, and the tendency of the pumping speed to decrease with time and total throughput. The closed-cycle cryopumps operating at 20 K have the same disadvantages and also should be avoided for this application [135]. The optimum type of pump for this application is either a turbo-pump or a diffusion pump, because the pumping speed for both types of pumps can be characterized as a function of the mass of the gas and the gas pressure; for all practical purposes, the speed does not vary with time or total throughput. This property greatly simplifies the calibration of a QMS system over a wide range in mass and pressure. A second common pitfall, which complicates calibration procedures, is the use of needle-type throttling valves in the connection tube to accomplish the pressure reduction. The mechanical hysteresis that is inherent in the design of most valves of this type makes repeatability difficult. A better choice is the use of a series of well-defined apertures.

No optimal calibration procedure exists for the partial pressure analysis scheme shown in Fig. 28. However, a number of relatively simple approaches can be taken that yield satisfactory results. If the process to be monitored is a static process or otherwise slowly varying with respect to the characteristic vacuum time constants, then a static calibration will suffice. In this case a predetermined gas mixture can be introduced to V_D; the amount of introduced gas is quantified either by means of a calibrated leak (Q_i) or by comparison with a calibrated total pressure gauge on the volume V_D; and finally, the response of the QMS as a function of mass is recorded. For dynamic measurements, single component gases must be introduced individually, and the system differential equation, Eq. (49), can be obtained numerically from the measured pumping transients (V_D/S_D, V_Q/S_Q) and the form of $dP_Q(t)/dt$. Calibration procedures for the partial analysis are described in detail in Refs. 114 and 115.

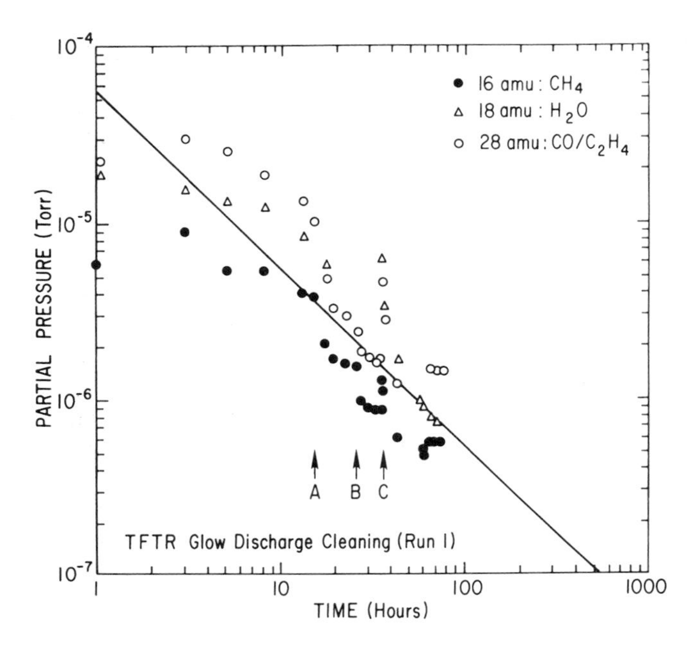

FIGURE 30. The time evolution of various gaseous components evolved during hydrogen glow discharge cleaning of the large stainless steel chamber of the TFTR magnetic fusion device. The effectiveness of the procedure in removing tenacious surface contaminants is clearly evident (from Ref. 84 with permission).

We conclude this section with an example of a quantitative application of partial pressure analysis in a plasma system. Partial pressure analysis has been the primary technique for monitoring and optimizing the hydrogen discharge cleaning techniques [77], which have been developed over the last decade to condition the large vacuum vessels of contemporary fusion devices. Figure 30 shows an example of this type of experiment: the time evolution of the primary impurity gases produced during the hydrogen glow discharge cleaning of the 86 m^3 stainless steel vacuum vessel of the TFTR device [84]. Since the instrument was absolutely calibrated, the data of Fig. 30 were used to calculate the total amount of carbon (0.6 g) and oxygen (1.2 g) that was removed from the vessel as a result of the hydrogen plasma interaction with the vacuum vessel surfaces.

VI. Emission Spectroscopy

Emission spectroscopy is one of the oldest of plasma diagnostics, early workers having noticed that the colors of their discharges were dependent on the choice of fill gas and the pressure in the tube. Developments in spectroscopic equipment paralleled the study of low pressure gas discharges, when it was realized that, along with flames and electric arcs, gas discharges were bright, easily controlled ways of exciting light emission from particular chemical species. The fact that the spectra emitted by excited chemical species are practically unique, makes spectroscopy extremely useful as an indicator of the composition of a gas mixture. The ability to relate the intensity of emitted or absorbed light to the concentration of the active species has made spectroscopy invaluable to the study of physics and chemistry in this century.

A. QUALITATIVE DESCRIPTION

The atomic and molecular physics that governs the interaction of light and matter is too complicated to review in detail. Comprehensive treatments of atomic and molecular spectroscopy have been given by Herzberg [136–138]. For readers who may be unfamiliar with the principles and the notation of optical spectroscopy, this section briefly describes the fundamental features of the physics. More advanced readers may wish to skip directly to the quantitative use of optical spectroscopy.

The solutions to the quantum mechanical problem of the motion of an electron in the field of a positively charged nucleus yield discrete, quantized levels of energy. This central field problem is quantized in each of its spatial (spherical) coordinates, r, θ, ϕ. The levels associated with r, θ, ϕ are indexed by quantum numbers, n, l, m_l, respectively. There is an additional degree of freedom associated with the spin, or intrinsic angular momentum of the electron, giving an additional quantum number, s (or m_s). The radial (or principal) quantum number, n, is by far the most important in determining the magnitude of the energy of the electronic state. The other spatial quantum numbers, l and m_l, are associated with the angular momentum of the electron. Quantization means that the electron can be found only in states having particular (integral or half-integral) values of these numbers; n may only be an integer ranging from 1 to ∞, l may have integral values from 0 to $n - 1$, m_l may have any integral value from $-l$ to $+l$, and s can be $+1/2$ or $-1/2$. Figure 31 shows a diagram of the resulting energy states for the hydrogen atom [136]. Only the n quantization is shown because the states associated with various values of l and m_l are too close in energy to resolve on this diagram. Optical transitions can be imagined as occurring when an electron "jumps" from one level to another in the atom. When the energy of the initial state is higher than that of the final state, a photon is emitted to carry away the energy difference. When the final state is of higher energy, the missing energy is supplied by the absorption of a

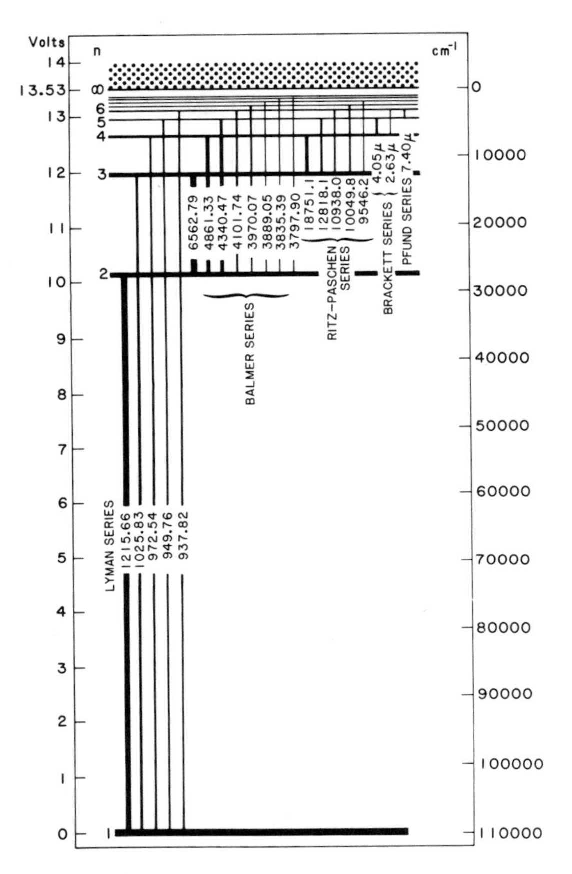

FIGURE 31. Energy levels of the H atom. The principle quantum number, n, is shown on the left. Energies are given in electron volts (left) and cm^{-1} (right). Various transitions between these levels are identified, the numbers shown on the lines identifying these transitions are the emission (or absorption) wavelengths in angstroms (redrawn from Ref. 136 with permission).

photon. Knowing the energy of the two states involved allows the use of Planck's relation to determine the frequency of the absorbed or emitted photon:

$$\nu = (1/hc)(E_f - E_i), \tag{50}$$

from which we may calculate the wavelength by $\lambda = c/\nu$. The wavelengths of a number of the well-known emission lines of hydrogen are indicated on the lines connecting the states in Fig. 31.

The uniqueness of the overall pattern of the absorption or emission spectra of atoms arises from the fact that the energy levels of different atoms are different. There are, however, numerous overlaps between particular transitions of different atoms, and experimentally, it is often necessary to work at high spectral resolution, or to monitor a number of different emissions from the species of interest to separate such overlaps.

There are a large number of bound states known for all atoms. In fact, each ionization charge state of each atom has a large number of bound states. Moore [139] has given extensive tabulations of these states. For example, these tables list 64 bound states for the neutral fluorine atom. If one naively presumed that each such state could connect pairwise to all of the others to give radiative transitions, one would expect 2018 lines to result in the neutral fluorine (FI) spectrum. There are, however, selection rules associated with the angular momentum and spin quantum numbers, which permit only specific changes in the quantum numbers to occur in a radiative transition. These rules are:

$$\begin{aligned}
\Delta n &= 0, 1, 2, \ldots \text{ (unrestricted)} \\
\Delta l &= + \text{ or } - 1 \text{ for one optical electron} \\
\Delta L &= \Delta \Sigma l_i = 0, \pm 1, \text{ for all optical electrons} \\
\Delta S &= 0 \text{ and} \\
\Delta J &= 0, \pm 1 \text{ where } J = L + S \text{ is the total angular momentum.}
\end{aligned} \tag{51}$$

These selection rules greatly reduce the allowed combinations of atomic levels that can be connected by radiative transitions. For example, in the case of neutral fluorine considered above, only 33 prominent lines have been observed. This reduction is fortunate, since if each ionization state of each atomic species in a plasma were to radiate without restriction, thousands or tens of thousands of lines would result. The resulting apparent "continuum" would not be useful for spectroscopic identification at all.

The physics of radiative transitions in molecules is complicated by the motion of the atomic nuclei relative to the center of mass of the molecule. In polyatomic molecules the internal motion is quite complicated. We restrict our discussion to diatomic molecules that are considerably easier to describe. In a diatomic molecule, the two nuclei can vibrate along the molecular axis, and they can rotate around the center of mass. The lowest order solution of this simplest case of vibratory motion is obtained by assuming that the molecule is a simple harmonic oscillator: two point masses connected by a massless spring, oscillating with a small amplitude. Rotation is represented to lowest order as a rigid rotor: two point masses connected by a massless rigid rod, rotating about the center of mass. Within this approximation, one finds that the vibrational states are quantized, with energy given by

$$E_v = k\nu_o(v + \tfrac{1}{2}) \qquad \text{where } v = 0, 1, 2, \dots, \tag{52}$$

and the rotational states are also quantized, with energy given by

$$E_R = B_o N(N + 1) \qquad \text{where } N = 0, 1, 2, \dots . \tag{53}$$

In real molecules, these vibrational and rotational motions are coupled, and the expressions are modified by the addition of higher order corrections.

The problem of the electron motion in molecules is solved by assuming that the nuclei are stationary, fixed at their equilibrium positions (Born-Oppenheimer approximation). Even this simplified problem is more difficult than the central field problem of the atom. Solution sets for the molecule are built up by appropriate superposition of the wavefunctions obtained by solving the equations of the separated atoms making up the molecules. This molecular orbital problem looks similar to the atomic case, but unlike the atomic case, a strong electric field exists along the internuclear axis. The result is that the projections of the electron angular momenta along the internuclear axis, rather than the momenta themselves, are the quantities most easily described. These quantities are found to be quantized in a manner analogous to the quantization for the atomic electrons. The quantum number L is replaced by Λ, quantizing the projection of the orbital angular momentum onto the internuclear axis. The total spin quantum number, S, is replaced by its projection, Σ, and the total angular momentum, $J = L + S$, is replaced by its projection $\Omega = \Lambda + \Sigma$. To provide a compact representation of states having various values of these quantum numbers, spectroscopists developed term symbols, such as $^2\Sigma$ or $^2\pi$, which are merely a shorthand listing of the relevant quantum numbers for a diatomic molecule. For example, in the term $^2\Sigma$, $\Lambda = 0$, and the spin, Σ, is 1/2. A full explanation of the rules by which these term symbols are formed can be found in Herzberg [137]. Analogous to the atomic case, selection rules apply to the electronic angular momentum projections, as well as to the angular momentum of nuclear rotation ($\Delta N = 0, \pm 1$).

What finally results for molecules is a set of energy levels shown schematically in Fig. 32. Each of the electronic terms is analogous to the level designated by the principle quantum number, n, in Fig. 31. These levels may also be further split (not shown) into multiple values of Ω, the total electronic angular momentum. The nuclear motion gives rise to the v, J energy levels shown superimposed on the electronic state. The ordering of the energy splittings is schematically correct; rotational states in the same term are separated by small differences in energy. Vibrational states are

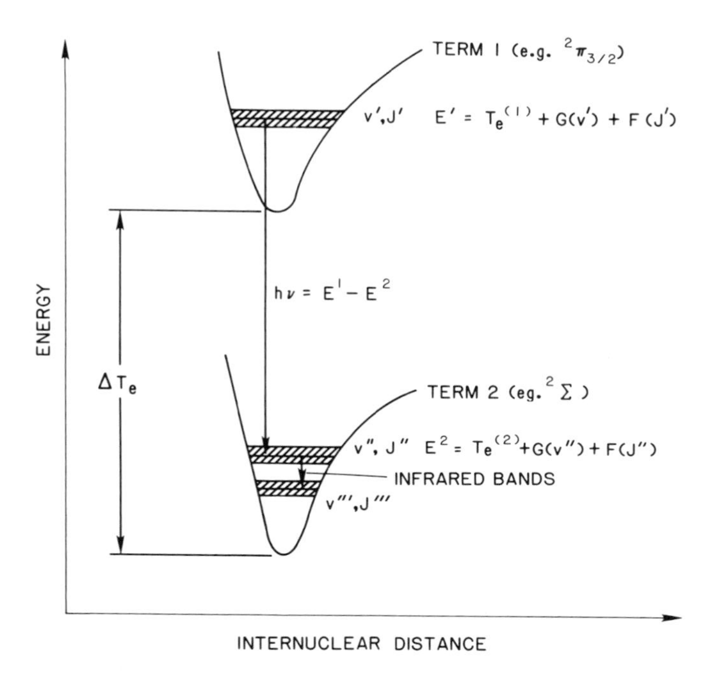

FIGURE 32. Schematic diagram of the intermolecular potential functions associated with two states of a diatomic molecule. Pure rotation spectra, resulting from transitions within a fixed V' are not shown.

separated by larger differences. Thus, for molecules with a permanent dipole moment, where such transitions are allowed, spectra arising from transitions within an electronic term (keeping the vibrational level also unchanged—so-called pure rotation spectra), occur at microwave frequencies. Spectra between vibrational levels ($\Delta V = \pm 1$) of the same electronic state generally occur in the infrared, while transitions between different electronic terms generally occur in the ultraviolet and visible region.

Because the total number of resulting states in molecules is so large, the number of radiative transitions allowed is also very large, generally in the hundreds or thousands for diatomic molecules with moderate amounts of excitation. When the excitation of vibrational and rotational levels is very high, the spectrum resulting from transitions between two electronic terms can appear to be nearly continuous over quite a broad range of wavelengths. Figure 33 [140] shows such an apparently continuous spectrum for AlO, which was formed with a high degree of excitation of nuclear motion by the reaction Al + O_3. In general, plasma excitation of nuclear motion is

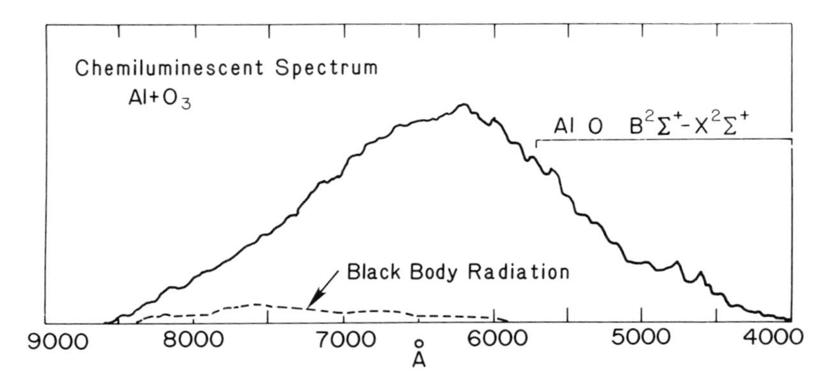

FIGURE 33. The spectrum of AlO formed from reaction of Al with ozone. The very high degree of nuclear excitation in this chemiluminescent system results in the apparent near-continuum in this low resolution spectrum. At higher resolution, the three broad bumps between 4250 and 4850 Å can be resolved into approximately 60 individual vibrational bands (from Ref. 140 with permission).

substantially lower than this, so the transitions tend to cluster into bands at distinct wavelengths associated with various amounts of change in vibrational energy. Thus one may use modest spectral resolution to measure the intensity of these vibrational bands in a manner similar to atomic line spectra. Quantitative calculation of the number density of the molecular state from the photon intensity is more complicated than for atomic states, but it is still possible. It is often desirable to monitor only relative changes in concentration as conditions are varied. For this purpose, vibrational band intensity is nearly equivalent to atomic line intensity.

B. QUANTITATIVE ANALYSIS USING EMISSION

As mentioned, there are many transitions possible for even simple atomic systems; each transition contains information about the state from which it originates. In most plasma applications, not including plasmas to produce lasers and other light sources, the desired processes, such as sputtering or enhanced chemical reactivity (see Flamm, this volume), tend to be mostly insensitive to the specific atomic states of the ions or neutrals involved. The problem in such systems is to find a representative measure for the total density (or flux) of the active species of interest. The most obvious candidate for such a representative state is the ground electronic state, since it is usually the most highly populated. The problem is then reduced to the relationship between the emission intensity and the density of the ground state.

Figure 34 shows a schematic of a three-level atomic system. Various optical processes, including emission, absorption, and fluorescence are shown. For the transition $n_2 \to n_1$, the intensity can be written:

$$I = g_2 A_{21} n_2 h\nu_{12}, \tag{54}$$

where A_{21} is the Einstein probability for spontaneous emission, n_2 is the density of the emitting state, g_2 is the state multiplicity, equal to $2J + 1$, where J is the total angular momentum quantum number described above, and ν_{12} is the frequency of the emitted light. For the absorption process $n_1 \to n_2$, the rate of absorption of photons (i.e., photons/s) can be written:

$$\frac{d\phi}{dt} = B_{12}\rho_\nu, \tag{55}$$

where B_{12} is the Einstein coefficient for absorption, and ρ_ν is the density of the photons. Fluorescence, shown in Fig. 34, is seen to be an absorption followed by emission. There is another radiation process, stimulated emission, in which a photon interacts with an excited species causing it to emit a second photon of the same wavelength. The rate of such emission is written in terms of the Einstein probability for stimulated emission, B_{21}. Einstein

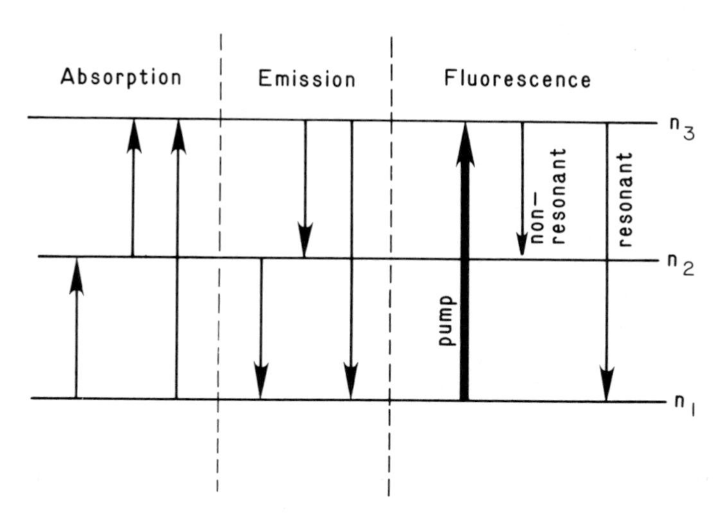

FIGURE 34. Schematic of some possible optical processes in a three-level system.

derived the following relationships between these fundamental rate coefficients:

$$g_2 B_{21} = g_1 B_{12} \tag{56}$$

and

$$A_{21} = 8\pi h c \nu^3 B_{12}. \tag{57}$$

Clearly, knowledge of any one of the coefficients can be used to compute the other two. The values of these transition probabilities (or the equivalent quantities called Ladenburg oscillator strengths) have been tabulated [141, 142] for a very large number of species.

By measuring the absolute emission intensity and knowing the Einstein coefficient, it is possible to compute the excited state density. One might assume a Boltzmann distribution of electronic states to compute the total density of the species, knowing the energy and density of this single state. However, plasma systems cannot generally be assumed to be in equilibrium and such a procedure is not acceptable. Instead, it is necessary to make kinetic models of the relationship between the total density and the density of the upper state. The usual assumptions are: that the excited state is formed solely by electron impact excitation, that this excitation occurs in a single step from the ground state and that emission must occur once the state is excited. In this case, one can write the intensity of the emission as:

$$I(\lambda_{ij}) = N P_i A_{ij} T(\lambda_{ij}),$$

where N is the ground state density, A_{ij} is the Einstein emission probability, $T(\lambda_{ij})$ is a correction factor to account for the apparatus collection efficiency at the transition wavelength, λ_{ij}, and P_i is the electron impact excitation function

$$P_i = \int n_e(\mathbf{x}, \mathbf{v}) \sigma(\mathbf{x}, \mathbf{v}) \, d\mathbf{x} \, d\mathbf{v} \tag{58}$$

representing the probability of exciting the state i by electron impact on the ground state. In the above expression, σ is the cross section for excitation, and $n_e(x, v)$ is the electron density distribution, both strong functions of electron velocity. A number of very serious problems arise with the attempt to use these expressions. Rarely are cross sections or the electron velocity distributions sufficiently well known to allow accurate computation of the ground state density. Also, while it is possible to calibrate collection optics to give the absolute photon production rate, it is extremely difficult. The assumption that only electron impact on the ground state contributes to the excitation may be quite wrong; the emitting state may be achieved by a multi-step sequence of electron impact processes passing through lower non-radiating states. In this case, collisional quenching of the lower states will enter the formation kinetics of the emitting state. Other processes contributing to the formation of the emitter may include metastable electronic energy transfer [143], radiative or collisional cascade [144] from higher states, and chemiluminescence [145]. Finally, if the pressure is high enough, collisional quenching of the emitting state may compete with radiation and the above kinetic model would require modification.

Use of emission spectroscopy for even relative density measurements, therefore, requires some other method of calibration. Usually one compares the observed intensity to an independent measurement of the concentration of the species. One way to achieve such an independent measure is by gas-phase titration reactions, the principles of which have been reviewed by Fontijn and his coworkers [146, 147]. Another method is to use absorption spectroscopic methods, which we consider next, to calibrate the emission measurements. Finally, actinometry has been used to circumvent the unknown excitation kinetics.

Actinometry involves the addition of a known amount of a gas phase species, such as argon, to the discharge. One attempts to find an emission in this species that originates from a state whose energy is near that of the excited state in the species of interest [148, 149]. The assumptions are made that the excited state is formed only by single step electron excitation from the ground state, that the velocity dependence of the cross section for this excitation is similar to that of the species of interest, that collisional quenching is unimportant, and that the emission probabilities are truly constant (electric field quenching is unimportant). Under these circumstances, the variation of the electron energy distribution function with features of the discharge like pressure, power, etc. can be compensated by taking the ratio of emission of the species of interest, Z, to that of the inert gas, M:

$$\frac{I(\lambda_Z)}{I(\lambda_M)} = \frac{N_z}{N_M}\frac{P_z}{P_M}\frac{A_z}{A_M}\frac{T(\lambda_z)}{T(\lambda_M)}. \tag{59}$$

If the wavelength of the two lines is similar, the apparatus function also cancels and the emission ratio is related only to density and Einstein probability ratios. Figure 35, from Coburn and Chen [149], shows how such a ratio gives a more plausible result for using emission to obtain the dependence of F atom density on feedgas pressure. The apparent successes of actinometry in providing such zero-order correction have led to its widespread use as a plasma tool. The assumptions made are serious, however, and for many or most process plasmas are probably not valid. Flamm and Donnely [25] have shown that for chlorine discharges, frequency-dependent effects on the spatial variation of the electron energy distribution function can lead to serious errors in the method unless viewing area and rf phase are properly accounted for. Walkup and coworkers [150]

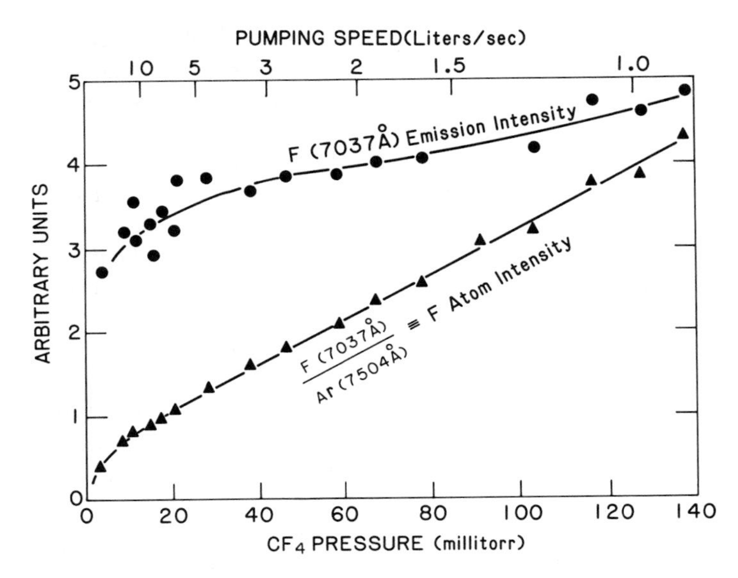

FIGURE 35. Emission intensity of neutral *F* atoms (top) as pressure is varied in a CF_4 discharge compared to the ratio of this emission to the emission from trace amounts of Argon added to provide an actinometric reference (from Ref. 149 with permission).

have shown that the contribution from dissociative excitation in O_2 discharges leads to errors in Ar actinometry for O atom density determination. Gottscho [151] has shown that actinometry can be wrong by an order of magnitude in N_2 discharges. In principle, these difficulties also may be overcome if proper care is used.

Emission spectroscopy, especially when corrected by actinometric methods or calibrated against secondary density measurements, is a very useful tool for determining endpoints in plasma etching or deposition. With the addition of a window and comparatively inexpensive optical components outside of the process chamber, a non-invasive, real-time process control tool can be implemented. Quantitative measurement, even of the relative concentrations, is not necessary for this purpose. The emission from a key species in a process can be monitored as a function of time and the achievement of a particular intensity, perhaps in combination with a particular value of its time derivative, can be defined as an acceptable point at which to discontinue the process. In practice, such a determination may be made entirely empirically, as shown in Fig. 36 taken from Oshima [152]. Such endpoints have been described for resist stripping [153, 154], etching of Si [149], Si_3N_4 [155], and other films. Table 2 [156, 157] shows monitored species and wavelengths commonly used for endpoint detection.

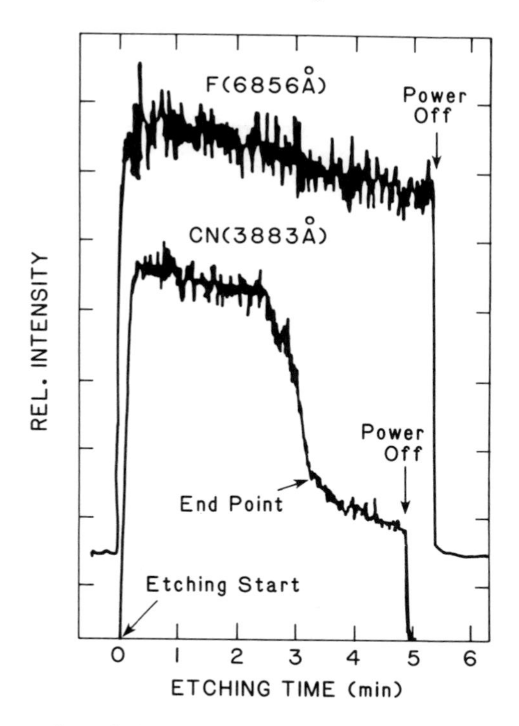

FIGURE 36. Comparison of *F* atom emission and *CN* molecular emission intensities during etching of Si_3N_4 over Si. The inability to use an *F* atom endpoint is clearly demonstrated, along with the rather good endpoint capability provided by the *CN* emission (from Ref. 152 with permission).

Emission measurements can be used to extract a variety of temperatures in gas discharges. If a sequence of atomic levels having known emission probabilities can be found for an inert gas species like He or Ar, it is possible to extract an "internal electronic" temperature by plotting $\ln(gf/I\lambda^3)_k$ vs. E_k. Such internal states, particularly members of a common spin-orbit multiplet, are often in equilibrium with the external electron population, thus providing a measure of T_e [158]. From diatomic molecular band spectra, it is possible to extract population distributions of vibrational and rotational levels, from which vibrational or rotational temperatures may be computed [159]. Rotational temperatures extracted this way are reasonably good measures of the translational temperature of the neutral gas; however, vibrational temperatures in low temperature reacting gases have little or no meaning except to chemical dynamicists [160].

Table 2 Common Endpoints [157]

Material Etched	Species Monitored	Wavelength (nm)
Al	AlCl	261.4
	Al	396.2
GaAs	Ga	287.4
	As	278.0
Si, poly-Si	F	703.8
	SiF	777.0
SiN, Si_3N_4	F	703.8
	CN	387.1
	N_2*	337.1
SiO, SiO_2	CO	483.5
	CO	519.5
Resist	CO	483.5
	CO	519.5

*The authors of this chapter advise against the use of N_2 due to the prospect of large error caused by even small air leaks.

C. ABSORPTION AND FLUORESCENCE

As mentioned earlier, absorption and fluorescence both have their origin on the ground state level; neither of these processes depends on the electron energy distribution. Thus, it is possible to obtain a direct measure of the ground state species density from these methods. The fact that both processes depend on external light sources also makes the absolute calibration of the optical system very much easier. Full understanding of absorption depends on the details of the lineshapes of both the external light source and the absorbed species. The full treatment is beyond the scope of this chapter, but is described in the superb book by Mitchell and Zemansky [161]. Details regarding specific light sources for such measurements have been summarized by Braun and Carrington [162, 163]. In most process plasmas, where the dominant contribution to the line shape is Doppler broadening, one may write the absorbance:

$$A = \ln\left(\frac{I}{I_o}\right) = \frac{kl}{(1 + \alpha^2)^{1/2}} - \frac{(kl)^2}{2(1 + 2\alpha^2)^{1/2}} + \cdots + (-1)^{n-1}\frac{(kl)^n}{n!(1 + n\alpha^2)^{1/2}}, \quad (60)$$

where k, the absorption coefficient, is given by

$$k = N_j\left[\frac{2}{\Delta\nu_o}\left(\frac{\ln 2}{\pi}\right)^{1/2}\frac{\pi e^2}{m_e c}f_{ji}\right] \tag{61}$$

and l = path length of the light through the absorber, α = linewidth of source/linewidth of absorber, $\Delta\nu_o$ = Doppler width of absorber, f_{ij} is the oscillator strength, N_j the absorber number density, m_e, e, and c are the mass and charge of the electron, and the speed of light. Using a hollow cathode excitation lamp to study absorption in a process plasma, α can be made very close to unity. By choosing conditions so that $N_j f_{ji} l$ is small, only the first term in the series needs to be retained. In this so-called "Beer's law" regime, the absorbance is a linear function of absorber density. Felder and Fontijn [147] have given a detailed discussion of the effects of temperature and pressure on deviations from Beer's law. The main effect occurs through α; it is found that for higher temperature the linear regime persists to higher values of Nfl. Pressure effects are ignorable for pressures less than 10 Torr. The presence of blended (overlapping) hyperfine lines complicates absorption spectroscopy somewhat; a thorough treatment is found in Tellinghausen and Clyne [164]. In the absence of such complications, it is relatively easy to measure densities to within 20–30% when the oscillator strengths are known; better accuracy is achievable by calibrating with gas phase titration. Using synchronous detection of a chopped light source, the lower limit of sensitivity is about 10^9 cm^{-3}. It is possible to extend the useful range of hollow cathode lamps up to 10^{13} cm^{-3}. To do this, one may use a variety of lines, each covering one or two orders of magnitude in concentration, and one may also vary the path length. Concentrations higher than 10^{13} cm^{-3} require more sophisticated methods using continuum lamp sources. Lasers are also quite useful for absorption studies, particularly in the infrared. We defer discussion of this application to the next section.

VII. Fluorescence

The equations of fluorescence are derived by combining the rate for absorption to the excited level with the rate for emission to a lower level. Lineshape matching is similar to absorption, so hollow cathode lamps turn out to be very suitable for fluorescence work in process plasma systems. Lasers have recently gained in popularity for absorption and fluorescence measurements in plasmas. This discussion assumes the use of lasers, but the reader should understand that for low light-intensity work, hollow cathode lamps may serve as well.

Laser induced fluorescence (LIF) is routinely employed as a spectroscopic method in fundamental chemical and physical research. Excellent reviews of such applications exist in the literature [165–170]. A natural outcome of this widespread employment as a research method is its development as a diagnostic tool, first for routine measurement in analytical laboratories, and ultimately for process evaluation. Such applications have occurred; LIF has become a routine diagnostic in applied plasma physics, the majority of the development effort coming from the interest in plasma-material interactions in magnetic fusion devices. Reviews of this work have been given by Hintz [171] and Bogen and Hintz [172]. In this brief overview we rely heavily on these references and recommend that the interested reader see them for more detail.

The fundamental equations governing fluorescence intensity [161] have been presented, so we review them only briefly here.

For a resonant fluorescent system, ignoring collisional quenching, we can write the rate of change of the upper population level as

$$\frac{dn_2}{dt} = \frac{I_l}{c}[B_{12}n_1 - B_{21}n_2] - A_{21}n_2, \tag{62}$$

where I_l is the pump source intensity. We make the simplification that the total density, N, can be approximated by the sum of the two levels above. This approximation will be fairly good since n_1 is usually the ground state and excited states have only a small population from electron impact excitation compared to the photon excitation of n_2 from the pump source. Substituting $n_1 = N - n_2$ above, and integrating, we arrive at an expression for $n_2(t)$. This yields $n_{\phi(t)}$ the photon production rate as a function of time, which can be integrated over time to give the total photon yield:

$$N_\phi = \frac{N}{k}\frac{B_{12}}{c}I_l\left[A_{21}\,\Delta t - \frac{A_{21}}{k} + 1 + \left(\frac{A_{21}}{k} - 1\right)\exp(-k\,\Delta t)\right], \tag{63}$$

where N is the total atom density, A_{21} and B_{12} are the Einstein coefficients for spontaneous emission and absorption, respectively. I_l is the pump intensity, which has a temporal duration, Δt. We have defined the parameter $k = I_l/c[B_{12} + B_{21}] + A_{21} = A_{21} + S$, where S is the saturation parameter. In general, one is interested in obtaining absolute (or relative) measurements of the total density, N, as process parameters are varied. For this purpose, it is useful to consider the following limiting regimes:

Case 1: Weak pumping

The laser intensity is small enough so that

$$A_{21} \gg I_l/c(B_{12} + B_{21});$$

therefore, $k \approx A_{21}$ and Eq. (63) becomes $N_\phi = NB_{12}I_l\,\Delta t/c$.

In this limit the integrated fluorescence intensity is proportional to the laser intensity and a knowledge of B_{12} is required to extract N from N_ϕ.

Case 2: Strong pumping
The laser intensity is large enough so that

$$A_{21} \ll I_l/c[B_{12} + B_{21}];$$

therefore, $k \approx S = I_l/c[B_{12} + B_{21}] = I_l/cB_{21}(1 + g_2/g_1)$ (where g_1 and g_2 are level degeneracies) and Eq. (63) becomes

$$N_\phi = N\left(\frac{g_2}{g_1 + g_2}\right)(A_{21}\,\Delta t + 1). \tag{64}$$

In this limit, the integrated fluorescence intensity is *independent* of laser intensity (hence this regime is called "saturated"). In general, the Einstein coefficient must still be known. However, it is sometimes possible to use a sufficiently short laser pulse that $A_{21}\,\Delta t \ll 1$, so Eq. (64) becomes

$$N_\phi = N\left(\frac{g_2}{g_1 + g_2}\right), \tag{65}$$

and the species density is simply the total photon count multiplied by a constant. When possible, experimental tests of the dependence of N_ϕ on laser power and pulse length are advisable to determine the regime.

Similar expressions may be derived for multilevel systems by properly accounting for the branching ratios of the allowed transitions from the upper level to the various lower levels. An important consequence is that the saturation parameter is smaller by the factor A_{21} divided by the sum of the A's over all allowed transitions from the excited state. As a consequence, a more intense laser is required to achieve the saturated regime.

We note that these equations are quite approximate, in general, since the specific lineshape of the absorbing medium has been ignored and the lineshape of the laser (light source) has been assumed continuous [172, 164]. The laser output has been assumed to be a square wave in time, a condition that is never actually satisfied, though the error is often inconsequential for saturated conditions.

Calibration of the system is best performed by comparison to alternative techniques for measuring the emitter density. Numerous possibilities exist whenever a relatively steady set of reproducible conditions may be achieved in the process. For example, Manos and Fontijn [144] have measured metal atom densities in a fast flow reactor by simultaneously monitoring absorption and fluorescence and comparing the results to densities inferred by a gas phase chemical titration. In some transient processes it is not possible to provide alternative measurements. In these cases, absolute calibration of the system collection optics is required to obtain N_ϕ. This may be done by

Rayleigh scattering [173] from a known density of argon introduced into the system. Bogen and coworkers [174] have also used sputtering sources of controlled geometry to provide a density of Mo atoms calculable from known sputtering rates.

A. ACCESSIBLE SPECIES

The species accessible by laser fluorescence measurements are limited at present by the wavelength tunability and the power of available lasers. Neutral metal atoms are species of great interest to many process systems including sputter sources, evaporative coaters, combustion systems, etc. These species have numerous allowed transitions, which lie between the near ultraviolet and the near infrared, a region where pumped dye lasers of adequate power (10–500 mJ) and proper pulse length (10–1000 ns) are readily available. Figure 37 shows some of the metal atom lines that are accessible for study using tunable dye lasers, operating on various dye media. The majority of metals are most easily studied in the near ultraviolet using frequency doubled output from such lasers.

The narrow linewidth and precise tunability of such lasers makes it possible to extract velocity distributions by Doppler profile measurements in cases where there is adequate metal atom density. Figure 38a shows the Fe I density in front of a stainless steel surface sputtered by hydrogen ions in the boundary plasma of the TEXTOR fusion device [175]. The time resolution of the measurement is determined by the repetition rate of the laser. Note the ease of measurement in spite of the fact that the average

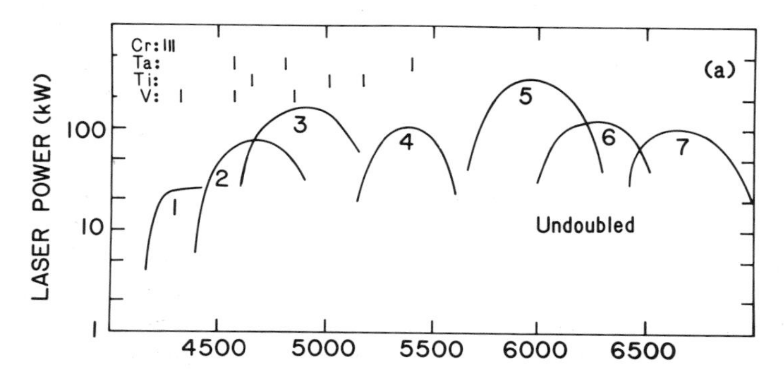

FIGURE 37a. Tunable dye laser powers and wavelengths in the visible available using dyes (numbered) such as coumarins (1–4), rhodamines (5–6), and cresyl violet (7). Overlay shows visible lines of selected metals, which are accessible to study at these wavelengths.

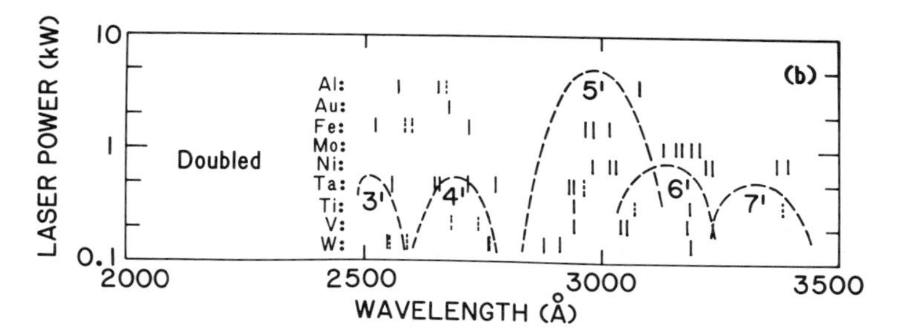

FIGURE 37b. Shows possible metal atom studies using the same dye lasers as in (a) but adding a KDP doubler. (Figure supplied by P. Bogen and E. Hinze, KFA/Jülich, FRG.)

density is $\approx 10^9\ cm^{-3}$, a low value for most diagnostic techniques. This is more than three orders of magnitude beyond the limit of detection for Fe, hence velocity profiles for sputtered Fe, such as that shown in Fig. 38b, are also measurable [176]. Measurements of Ti I densities in the DIII fusion device [177], have been made at even lower densities of $< 10^6\ cm^{-3}$. Limits of detectability vary widely with the choice of metal atom, resonance line, and system conditions. In a few systems [178, 179], 10^4 atoms cm^{-3} may be

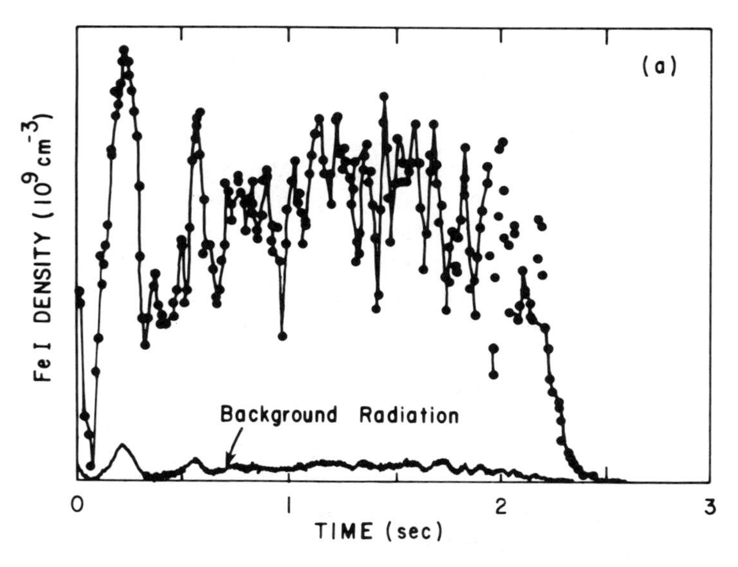

FIGURE 38a. Laser fluorescence measurement of the time dependence of Fe I density from sputtering of a steel target by an H plasma (from Ref. 175 with permission).

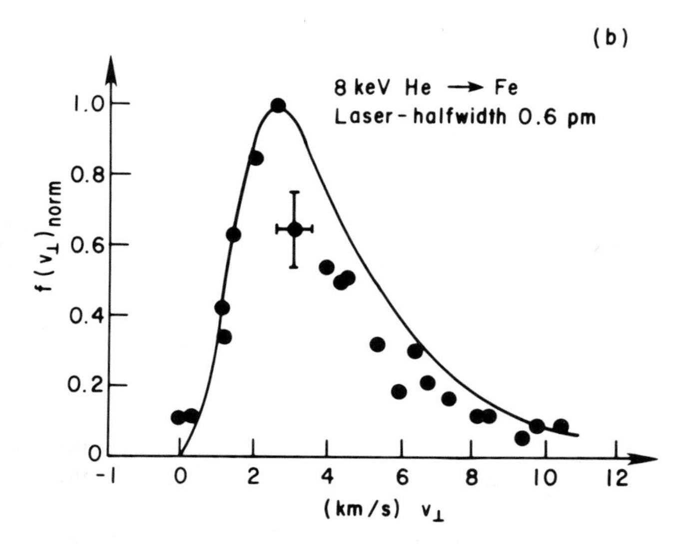

FIGURE 38b. Perpendicular component of the velocity distribution of the Fe atoms sputtered in (a) (from Ref. 176 with permission).

done with moderate care, although, in general, 10^5–10^6 cm^{-3} is a more useful lower bound even under favorable conditions.

Low ionization states of light atoms, and higher ionization states of metals, require far u.v. or vacuum u.v. lasers, which are now under development. Bogen and coworkers [180] have successfully measured H-atom densities of 10^7–10^9 cm^{-3} using a frequency tripled dye laser tuned to the Lyman alpha line (1216 Å). Neutral diatomic molecules can be studied in the visible [181]. Molecules of process importance which have been studied by LIF include AlO [182], LiH [183], NaH [183], OH [184], CCl [128], and others too numerous to review here (see Ref. 170 and references therein).

Various infrared lasers exist [185]. The long radiative lifetimes in the infrared make fluorescence studies difficult except under special conditions at low pressures; these lasers are therefore much better suited for absorption studies. Figure 39 [186] shows some of the molecular species that may be studied using various salt lasers. Numerous species present in process plasmas have transitions in the 3–30 μ region of the spectrum where these lasers are designed to operate. Most of the attention to these systems has come from combustion and environmental studies laboratories. Work on etching and deposition systems using such lasers is increasing. For example, Wormhoudt, Stanton, and Silver [187, 188] have reported semiconductor

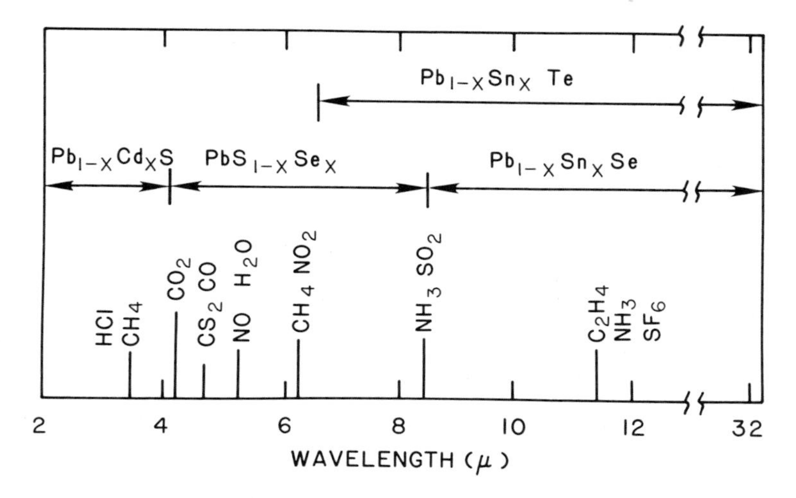

FIGURE 39. Infrared wavelength ranges over which various lead salt diode lasers can be tuned. Selected molecular species having strong absorption lines in these ranges are shown (from Ref. 186 with permission).

laser spectroscopic studies of CF_2, SiF_2, SiH_2, and other species of interest for semiconductor processing.

Clearly, laser optical spectroscopies are useful diagnostic techniques, which have the potential for wide application in process plasma characterization and monitoring. With the recent availability of "off-the-shelf" tunable lasers, much development along these lines may be expected in the near future.

VIII. Summary

In this chapter we have attempted to outline the fundamentals of some of the more useful diagnostics for process research and development. We have not discussed every possible method for examining processes, and we ask the reader's indulgence if we have omitted his or her favorite tool in our discussion. We hope to have given some hints of where diagnostic progress can be made to advance the art of production of commercial products using plasmas.

The application of plasma diagnostic tools to production equipment is proceeding rapidly. Numerous types of endpoint monitor and quality assurance tools are now offered as standard or optional features on most etch equipment. In order to make use of the full potential of automated and computer assisted methods on the manufacturing floor, much additional progress in developing specific diagnostics will have to be made.

References

[1] For example, see the Proceedings of the International Conferences of Plasma Surface Interactions in Controlled Fusion Devices published in *J. Nuc. Mater.*, Vols. **53** (1974); **76 / 77** (1980); **111 / 112** (1982); **128 / 129** (1984); **145 / 147** (1987).

[2] Course Notes: *Production and Diagnosis of Process Plasmas*, (Princeton Scientific Consultants, Princeton, New Jersey, 1982).

[3] I. Langmuir and H. M. Mott-Smith, *Gen. Elec. Rev.* **26**, 731 (1923), **27**, 449, 583, 616, 726, 810 (1924). Also see I. Langmuir and K. T. Compton, *Rev. Mod. Phys.* **3**, 191 (1931).

[4] D. Bohm, E. H. S. Burhop, and H. S. W. Massey in "Characteristics of Electric Discharges in Magnetic Fields." H. Guthrie and R. K. Wakerling eds., McGraw-Hill, New York (1949) Chapter 2.

[5] P. M. Chung, L. Talbot, and K. J. Touryan, *AIAA Journal* **12**, 9133 (1974).

[6] D. M. Manos and G. M. McCracken, in "*Plasma Wall Interactions in Controlled Fusion*." D. E. Post and R. Behrisch, eds. Plenum Press, New York (1986).

[7] R. M. Clements, *J. Vac. Sci. Technol.* **15**, 193 (1978).

[8] J. D. Swift and M. J. R. Schwar, "*Electrical Probes for Plasma Diagnostics*." Iliffe, London (1970).

[9] G. A. Emmert, R. M. Wieland, T. Mense, and J. N. Davidson, *Phys. Fluids* **23**, 803 (1980).

[10] P. Stangeby, in "*Plasma Wall Interactions in Controlled Fusion Devices*." D. Post and R. Behrisch, eds. Plenum Press, New York (1986) Chapter 3.

[11] R. Chodura, in "*Plasma Wall Interactions in Controlled Fusion Devices*." D. Post and R. Behrisch, eds. Plenum Press, New York (1986) Chapter 4.

[12] D. M. Manos, *J. Vac. Sci. Technol.* **A3**, 1059 (1985).

[13] S. M. Rossnagel and H. R. Kaufman, *J. Vac. Sci. Technol.* **A4**, 1822 (1986).

[14] M. B. Hopkins and W. G. Graham, *Rev. Sci. Instrum.* **57**, 2210 (1986).

[15] I. Katsumata, *Jap. J. Appl. Phys.* **6**, 123 (1967).

[16] D. M. Manos, R. Budny, S. Kilpatrick, and D. Stangeby, *Rev. Sci. Instrum.* **57**, 2107 (1986).

[17] J. G. Laframboise, Univ. of Toronto, Inst. for Aerospace Studies, Report No. 100 (1966).

[18] G. D. Hobbs and J. A. Wesson, *Plasma Phys.* **9**, 85 (1967).

[19] P. J. Harbour and M. F. A. Harrison, *J. Nucl. Mater.* **76 / 77**, 513 (1978).

[20] P. C. Stangeby, *Phys. Fluids* **27**, 682 (1984).

[21] P. J. Harbour, Culham Laboratory Report No. CLM Preprint-535 (1978) unpublished, (available from Culham Lab. on request).

[22] M. Kamitsuma, S. L. Chen, and J. S. Chang, *J. Phys.* D**10**, 1065 (1977).

[23] S. L. Chen and T. Sekiguchi, *J. Appl. Phys.* **36**, 2363 (1965).

[24] R. Budny and D. Manos, *J. Nucl. Mater.* **121**, 41 (1984).

[25] D. L. Flamm and V. M. Donnelly, *J. Appl. Phys.* **59**, 1052 (1986).

[26] Ch. Steinbrüchel, *J. Electrochem. Soc., Sol. State Sci. and Technol.* **130**, 648 (1983).

[27] J. G. Laframboise and J. Rubinstein, *Phys. Fluids* **19**, 1900 (1976), see Ref. 6a for a brief description of the method.

[28] A. A. Sonin, Univ. of Toronto, Inst. for Aerospace Studies, Report No. 109 (1966).
[29] Ch. Steinbrüchel, B. J. Curtis, H. W. Lehmann, and R. Widner, *IEEE Trans. on Plasma Science* **PS-14**, 137 (1986).
[30] R. E. Hurley, *Vacuum* **34**, 351 (1984).
[31] M. H. Cho, C. Chan, N. Hershkowitz, and T. Intrator, *Rev. Sci. Instrum.* **55**, 631 (1984).
[32] N. Hershkowitz, B. Nelson, J. Pew, and D. Gates, *Rev. Sci. Instrum.* **54**, 29 (1983).
[33] J. A. Simpson, *Rev. Sci. Instrum.* **32**, 1283 (1961).
[34] D. M. Manos, R. V. Budny, and S. A. Cohen, *J. Vac. Sci. Technol.* **A1**, 845 (1983).
[35] S. K. Erents and P. C. Stangeby, *J. Nucl. Mater.* **111 & 112**, 165 (1982).
[36] S. L. F. Richards, G. J. Lloyd, and R. P. Jones, *Journal of Physics E: Scientific Instruments* **5**, 595 (1972).
[37] B. M. Oliver, R. M. Clements, and P. R. Smy, *J. Appl. Phys.* **41**, 2117 (1970).
[38] F. W. Crawford and R. Grard, *J. Appl. Phys.* **37**, 180 (1966).
[39] T. A. Anderson, *Phil. Mag.* **38**, 179 (1947).
[40] K. Takayama, H. Ikegami, and S. Miyasaki, *Phys. Rev. Lett.* **5**, 238 (1960).
[41] Y. H. Ichikawa and H. Ikegami, *Prog. Theor. Phy.* **28**, 315 (1962).
[42] H. Ikegami and K. Takayama, *Resonance Probe*, Inst. Plasma Phys. Nagoya University Report No. IPP-10, March 1963.
[43] H. M. Mayer, *Proc VI Int. Conf. Ion Phenom in Gases*, July 1963, Paris, 1964, Vol. **4**, p. 129.
[44] R. S. Harp, *Appl. Phys. Lett.* **4**, 186 (1964).
[45] R. B. Cairns, *Proc. Phys. Soc.*, London **82**, 243 (1963).
[46] C. Singh, *Proc. Phys. Soc.*, London **74**, 42 (1959).
[47] A. R. Galbraith and A. Van der Ziel, *Physica* **29**, 1115 (1963).
[48] K. S. Knol, *Philips Res. Rep.* **6**, 288 (1951).
[49] A. Garscadden and K. G. Emelius, *Proc. Phys. Soc.*, London **79**, 535 (1962).
[50] E. Eser, R. E. Ogilvie, and K. A. Taylor, *Thin Solid Films* **68**, 381 (1980).
[51] F. W. Crawford, *J. Appl. Phys.* **34**, 1897 (1963).
[52] A. Boschi and F. Magistrelli, *Il Nuovo Cimento* **29**, 487 (1963).
[53] M. Sicha, J. Gajdusek, and S. Veprek, *Brit. J. Appl. Phys.* **17**, 1511 (1966).
[54] J. C. Hosea, UHF Plasma Interaction: The Two Stream Instability of a Current Carrying Plasma. Ph.D. Thesis, Stanford University, Aug. 1966, University Microfilms, Ann Arbor.
[55] P. C. Stangeby, *J. Nucl. Mater.* **121**, 36 (1984).
[56] J. F. Waymouth, *Phys. Fluids* **7**, 1843 (1984).
[57] S. A. Cohen, *J. Nucl. Mater.* **76 / 77**, 68 (1978).
[58] P. Staib, *J. Nucl. Mater.* **111 & 112**, 109 (1982).
[59] R. A. Zuhr, J. B. Roberto, and B. R. Appleton, *Nucl. Sci. Appl.* **1**, 617 (1984).
[60] E. Taglauer, *J. Nucl. Mater.* **128 / 129**, 141 (1984).
[61] C. W. Magee, *J. Vac. Sci. Technol.* **A1**, 901 (1983).
[62] R. A. Langley and C. W. Magee, *J. Nucl. Mater.* **93 / 94**, 390 (1980).
[63] S. A. Cohen and G. M. McCracken, *J. Nucl. Mater.* **84**, 157 (1979).
[64] G. Staudenmaier, J. Roth, R. Behrisch, J. Bohdanski, W. Eckstein *et al.*, *N. Nucl. Mater.* **84**, 149 (1979).
[65] W. R. Wampler, D. K. Brice, and C. W. Magee, *J. Nucl. Mater.* **102**, 304 (1981).
[66] H. F. Dylla and S. A. Cohen, *J. Nucl. Mater.* **63**, 487 (1976).
[67] S. M. Rossnagel, S. A. Cohen, H. F. Dylla, and P. Staib, *J. Vac. Sci. Technol.* **17**, 301 (1980).
[68] J. Sanmartin, *Phys. Fluids* **13**, 103 (1970).

[69] R. Krawec, NASA Technical Note TND-5747 (1980).
[70] G. Staudenmaier, P. Staib, and W. Poschenrieder, *J. Nucl. Mater.* **93 / 94**, 121 (1980).
[71] P. Staib, H. F. Dylla, and S. M. Rossnagel, *J. Nucl. Mater.* **93 / 94**, 166 (1980).
[72] S. K. Erents, E. S. Hotston, G. M. McCracken, C. J. Sofield, and J. Shea, *J. Nucl. Mater.* **93 / 93**, 115 (1980).
[73] M. T. Robinson and I. M. Torrens, *Phys. Rev.* **B9**, 5008 (1974).
[74] J. P. Biersack and L. G. Haggmark, *Nucl. Instr. Meth.* **174**, 257 (1980).
[75] J. F. Ziegler, ed. "*Handbook of Stopping Cross-Sections for Energetic Ions in All Elements*." Pergammon Press, New York (1980).
[76] H. F. Dylla, *Proc. 6th Symp. on Plasma Processing*, San Diego, **87–6**, p. 18, Oct. 1986; (*Electrochem Society*, Pennington, NJ, 1987).
[77] H. F. Dylla, *J. Nucl. Mater.* **93 / 94**, 61 (1980).
[78] H. F. Dylla, J. L. Cecchi, and M. Ulrickson, *J. Vac. Sci. Technol.* **18**, 1111 (1981).
[79] A. Mathewson, *Vacuum* **24**, 505 (1974).
[80] R. Calder, A. Grillot, F. LeNormand, and A. Mathewson, in *Proc. 7th Intern. Vac. Cong.*, Vienna, 1977, p. 231.
[81] H. C. Hseuh, T. S. Chou, and C. A. Christianson, *J. Vac. Sci. Technol.* **A3**, 518 (1985).
[82] F. Waelbroeck, J. Winter, and P. Wienhold, *J. Vac. Sci. Technol.* **A2**, 1521 (1984).
[83] H. F. Dylla, S. A. Cohen, S. M. Rossnagel, G. M. McCracken, and P. Staib, *J. Vac. Sci. Technol.* **17**, 286 (1980).
[84] H. F. Dylla, W. R. Blanchard, R. B. Krawchuk, R. J. Hawryluk, and D. K. Owens, *J. Vac. Sci. Technol.* **A2**, 1188 (1984).
[85] J. Winter *et al.*, *J. Nucl. Mater.* **128 / 129**, 841 (1984).
[86] R. P. H. Chang, C. C. Chang, and S. Darack, *J. Vac. Sci. Technol.* **20**, 45 (1982).
[87] C. W. Magee, W. L. Harrington, and R. E. Honig, *Rev. Sci. Instr.* **49**, 477 (1978).
[88] Calculated implantation parameters courtesy of D. K. Brice, Sandia National Laboratories, Albuquerque, New Mexico using the TRIM 86 code.
[89] A. G. Mathewson, A. Grillot, S. Hazeltine, K. Lee Li, A. Foakes, and H. Störi, in *Proc. 8th Intern. Vacuum Congress*, Cannes, 1980, Vol. **II**, p. 395.
[90] M. Heald and H. Wharton, "*Plasma Diagnostics with Microwaves*." Kreiger Publishing Co., New York (1978); First Edition, Wiley, New York (1965).
[91] M. E. Thompson *et al.*, *J. Vac. Sci. Technol.* **A4**, 317 (1986).
[92] P. C. Efthimion *et al.*, *Rev. Sci. Instrum.* **56**, 908 (1985).
[93] J. A. Fessey, C. W. Gowers, C. A. J. Hugenholtz, and K. Slavin, *J. Phys.* **E20**, 169 (1987).
[94] F. Jenkins and H. White, "*Fundamentals of Optics*." McGraw Hill, New York (1976).
[95] K. E. Greenberg, G. A. Hebner, J. T. Verdeyen *et al.*, *Appl. Phys. Lett.* **44**, 299 (1984).
[96] J. H. Ingold in "*Gaseous Electronics*," Vol. I. M. N. Hirsch and H. J. Oskam, eds. Academic Press, New York (1978) p. 19.
[97] R. N. Franklin, "*Plasma Phenomena in Gas Discharges*." Clarendon Press, Oxford (1976) Chapter 4.
[98] H. R. Koenig and L. I. Maissel, *IBM Journal of Research and Dev.* **14**, 168 (1970).
[99] B. Chapman, "*Glow Discharge Processes*." Wiley Interscience, New York (1980) p. 159.
[100] J. W. Coburn and E. Kay, *J. Appl. Phys.* **43**, 4965 (1972).
[101] K. Kohler, J. W. Coburn, D. E. Horne, and E. Kay, *J. Vac. Sci. Technol.* **A3**, 638 (1985).
[102] K. Wellerdieck and K. Hofler, in *Proc. Symp. Atomic and Surface Physics*, Maria Alm/Salzburg, Jan. 29 to Feb. 4, 1984, F. Howorka, W. Lindinger, and T. D. Mark, eds. Univ. Innsbruck Press (1984) p. 325.
[103] C. B. Zarowin, *J. Vac. Sci. Technol.* **A2**, 1537 (1984).

[104] D. E. Ibbotson, D. L. Flamm, and V. M. Donnelly, *J. Appl. Phys.* **54**, 5974 (1983).
[105] O. A. Popov and V. A. Godyak, *J. Appl. Phys.* **57**, 53 (1985).
[106] V. M. Donnelly, D. L. Flamm, N. C. Dautremont-Smith, and D. J. Werder, *J. Appl. Phys.* **55**, 242 (1984).
[107] H. H. Sawin, A. Yokozeki, A. J. Owens, and K. D. Allen, *MRS Res. Soc. Symp. Proc.* **38**, 235 (1985).
[108] G. A. Hebner and J. T. Verdeyen, *IEEE Trans. Plasma Sci.* **PS-14**, 132 (1979).
[109] H. Norstrom, *Vacuum* **29**, 341 (1979).
[110] S. E. Savas, D. E. Horne, and R. W. Sadowksi, *Rev. Sci. Instrum.* **57**, 1248 (1986).
[111] W. E. Mlynko and D. W. Hess, *J. Vac. Sci. Technol.* **A3**, 499 (1985).
[112] K. Ukai and K. Hanazawa, *J. Vac. Sci. Technol.* **16**, 385 (1979).
[113] D. L. Flamm, *Sol. State Technol.* **4**, 109 (1979).
[114] H. F. Dylla, *Proc. 9th Intern. Vacuum Congress*, Madrid, 1983 (A.S.E.V.A., Madrid 1983).
[115] M. J. Vasile and H. F. Dylla, in "Plasma Diagnostics," Vol. 1. O. Auciello and D. L. Flamm, eds. Academic Press, New York, in press.
[116] H. R. Griem, "*Plasma Spectroscopy*." McGraw Hill, New York (1964).
[117] R. W. B. Pearse and A. G. Gaydon, "*The Identification of Molecular Spectra*." J. Wiley & Sons, New York (1963).
[118] J. E. Greene, *J. Vac. Sci. Technol.* **15**, 1718 (1978).
[119] H. W. Drawin in "*Plasma Diagnostics*." W. Lochte-Holgreven, ed. North-Holland, Amsterdam (1968) pp. 777–841.
[120] W. E. Austin, A. E. Holme, and J. H. Leck in "*Quadrupole Mass Spectrometry and its Applications*." P. H. Dawson, ed. Elsevier Scientific, Amsterdam, New York (1976) pp. 121–152.
[121] M. Pahl, *Z. Naturforsch* **12a**, 632 (1957).
[122] P. Staib, *J. Nucl. Mater.* **93 & 94**, 351 (1980).
[123] J. W. Coburn, E. Taglauer, and E. Kay, *J. Appl. Phys.* **45**, 1779 (1974).
[124] J. W. Coburn and E. Kay, *Sol. State Technol.* **22**, 117 (1979).
[125] R. H. Bruce, *J. Electrochem. Soc.* **81**, 243 (1981).
[126] A. J. Purdes *et al.*, *J. Vac. Sci. Technol.* **14**, 98 (1977).
[127] K. M. Eisele and D. Hoffman, *Proc. 8th Intern. Vacuum Congress*, Cannes, FR, Sept. 1980 (LeVide, Paris, 1980).
[128] D. L. Flamm and V. M. Donnelly, *Plasma Chem. and Plasma Proc.* **1**, 317 (1981).
[129] D. L. Flamm, V. M. Donnelly, and D. E. Ibbotson, *J. Vac. Sci. Technol.* **81**, 23 (1983).
[130] I. Ishikawa, T. Iino, and S. Suganomata, *Jap. J. Appl. Phys.* **17**, 937 (1978).
[131] A. Wekhof, R. R. Smith, and S. S. Medley, *Nucl. Technol./Fusion* **3**, 462 (1983).
[132] H. R. Kaufman, J. J. Cuomo, and J. M. E. Harper, *J. Vac. Sci. Technol.* **21**, 725 (1982).
[133] W. Poschenrieder, *J. Vac. Sci. Technol.* **A5**, 2265 (1987).
[134] B. A. Raby, *J. Vac. Sci. Technol.* **15**, 205 (1978).
[135] B. F. T. Bolker *et al.*, *J. Vac. Sci. Technol.* **18**, 328 (1981).
[136] G. Herzberg, "*Atomic Spectra and Atomic Structure*." Dover, New York (1944).
[137] G. Herzberg, "*Molecular Spectra and Molecular Structures: Spectra of Diatomic Molecules*," Vol. I. Van Nostrand, Princeton, New Jersey (1950).
[138] G. Herzberg, "*Molecular Spectra and Molecular Structure: Electronic Spectra and Electronic Structure of Polyatomic Molecules*," Vol. III. Van Nostrand, Princeton, New Jersey (1966).
[139] C. Moore, "*Atomic Energy Levels*," Vol. I–III. National Bureau of Standards Circular 467, U.S. Govt. Printing Office, (1971).
[140] J. Gole and R. N. Zare, *J. Chem. Phys.* **57**, 5331 (1972).

[141] W. L. Wiese, M. W. Smith, and B. M. Glennon, "*Atomic Transitions Probabilities*," Vol. I. NSRDS-NBS-4, U.S. Govt. Printing Office, (May 1966).
[142] W. L. Wiese, M. W. Smith, and B. M. Miles, "*Atomic Transition Probabilities*," Vol. II. NSRDS-NBS-22, U.S. Govt. Printing Office, (Oct. 1969).
[143] M. F. Golde and B. A. Thrush, in "*Advances in Atomic and Molecular Physics*," Vol. II. 361 (1975).
[144] D. Manos and A. Fontijn, *J. Chem. Phys.* **72**, 416 (1980).
[145] V. M. Donnelly and D. L. Flamm, *J. Appl. Phys.* **51**, 5273 (1980).
[146] A. Fontijn, in "*Chemiluminescence and Bioluminescence*." M. J. Cormier, ed. Plenum Press, New York (19..).
[147] W. Felder and A. Fontijn, in "Reactive Intermediates in the Gas Phase." D. Setzer, ed. Academic Press, New York (1979).
[148] J. Coburn and M. Chen, *J. Appl. Phys.* **51**, 3134 (1980).
[149] J. Coburn and M. Chen, *J. Vac. Sci. Technol.* **18**, 353 (1981).
[150] R. Walkup, K. Saenger, and G. S. Selwyn, *MRS Symp. Proc.* **38**, 69 (1985).
[151] R. A. Gottscho, G. P. Davis, and R. H. Burton, *Plasma Chemistry and Plasma Processing* **3**, 193 (1983).
[152] M. Oshima, *Jap. J. Appl. Phys.* **20**, 683 (1981).
[153] E. O. Degenkolb and J. E. Griffiths, *Appl. Spectro.* **31**, 40 (1977).
[154] E. O. Degenkolb, C. J. Mogab, M. R. Goldrick, and J. E. Griffiths, *Appl. Spectro.* **5**, 520 (1976).
[155] W. R. Harshbarger, R. A. Porter, T. A. Miller, and P. Norton, *Appl. Specto.* **31**, 201 (1977).
[156] J. G. Shabushnig, P. R. Demko, and R. N. Savage, *MRS Symp. Proc.* **38**, 77 (1985).
[157] J. G. Shabushnig and P. R. Demko, *American Laboratory* (Aug. 1984) p. 60.
[158] I. M. Podgornyi, "*Topics in Plasma Diagnostics*." Plenum Press, New York-London (1971) Chap. 3.
[159] G. Heinrich, H. Vickel, M. Mazurkienicz, and R. Anni, *Spectrochem. Acta.* **33B**, 635 (1978).
[160] D. M. Manos and J. M. Parson, *J. Chem. Phys.* **69**, 231 (1978).
[161] A. C. G. Mitchell and M. W. Zemansky, "Resonance Radiation and Excited Atoms." University Press, Cambridge (1934; reprinted 1961).
[162] W. Braun and T. Carrington, *J. Quant. Spectrosc. Radiat. Transfer* **9**, 1133 (1969).
[163] W. Braun and T. Carrington, *Line Emission Sources for Concentration Measurements and Photochemistry* NBS Tech Note 476, March 1969, U.S. Govt. Printing Office.
[164] J. Tellinghausen and M. A. A. Clyne, *J. Chem. Soc. Far. Trans.* **72**, 783 (1976).
[165] H. Walther, ed., "Topics in Applied Physics," Vol. 2. *Laser Spectroscopy*, Springer-Verlag, New York (1976).
[166] C. B. Moore, ed., "*Chemical and Biological Applications of Lasers*," Vol. 1. Academic Press, New York (1974).
[167] C. B. Moore, *Ann. Rev. of Phys. Chem.* **22**, 387 (1971).
[168] K. Shimada, "Topics in Applied Physics **13**, *High Resolution Laser Spectroscopy*." Springer-Verlag, New York (1976).
[169] J. L. Kinsey, *Ann. Rev. of Phys. Chem.* **28**, 349 (1977).
[170] R. N. Zare, *Science*, **226**, 298 (1984).
[171] E. Hintz, *J. Nucl. Mater.* **93 & 94**, 86 (1980).
[172] R. Bogen and E. Hintz, in "*Physics of Plasma-Wall Interactions in Controlled Fusion*." D. E. Post and R. Behrisch, eds. Plenum Press, New York (1986) p. 211.
[173] W. Lochte-Holtgreven, ed., "*Plasma Diagnostics*." North-Holland Publication, Amsterdam (1986).

[174] P. Bogen, B. Schweer, H. Ringler, and W. Ott, *J. Nucl. Mater.* **111 & 112**, 67 (1982).
[175] H. L. Bay and B. Schweer, private communication.
[176] B. Schweer and H. L. Bay, *Appl. Phys.* **A29**, 53 (1982).
[177] C. H. Muller, III, D. R. Eames, and K. M. Burrell, *J. Vac. Sci. Technol.* **A1**, 822 (1983).
[178] P. J. Dagdigian and R. N. Zare, *J. Chem. Phys.* **61**, 2464 (1974).
[179] P. Gohil and D. D. Burgess, *Plasma Physics* **25**, 1149 (1983).
[180] P. Bogen, R. W. Dreyfus, H. Langer, and Y. T. Lie, *J. Nucl. Mater.* **111 & 112**, 75 (1982).
[181] A. Dienes in "*Laser Applications to Optics and Spectroscopy.*" Jacobs *et al.*, eds. Addison-Wesley, New York (1975) Chapter 2, p. 53.
[182] P. J. Dagdigian, H. W. Cruse, and R. N. Zare, *J. Chem. Phys.* **62**, 1824 (1975).
[183] P. J. Dagdigian, *J. Chem. Phys.* **64**, 2609 (1976).
[184] P. Hogan and D. D. Davis, *Chem. Phys. Lett.* **29**, 555 (1974).
[185] J. F. Butler, K. W. Nill, and A. W. Mantz, *SPIE*, Vol. **158**, *Laser Spectroscopy* (1978).
[186] R. S. Eng., J. F. Butler, and K. J. Linden, *Opt. Eng.* **19**, 945 (1980).
[187] J. D. Wormhoudt, A. C. Stanton, and J. A. Silver, *MRS Res. Soc. Symp. Proc.* **38**, 91 (1985).
[188] J. D. Wormhoudt, A. C. Stanton, and J. A. Silver, *Proc. SPIE* **452**, 88 (1984).

5 Plasma Etch Equipment and Technology

Alan R. Reinberg
Perkin-Elmer Corporation
Norwalk, Connecticut

ISBN 0-12-469370-9

I. Introduction

Dry etching as currently practiced is part art and part science. While the selection of appropriate hardware and chemistry is most often based on experience, details of a particular process are usually a matter of trial, error, and physical intuition. Since selection of an optimum system is difficult and often not unique, methods like Response Surface Analysis (RSM) [1] or Simplex [2] may be used to plan process development. Many available variations in equipment, gas combinations, operating conditions and desired results require that limits be put on the range of process parameters that can reasonably be explored.

A. HARDWARE

Hardware choices range from manual, large batch, high pressure, chemical, barrels; to computer-controlled, single-wafer, low pressure, ion-intensive, complex etchers. The best choice will depend on many factors including material to be patterned, whether it is to be used in development or production, required geometry control and, not insignificantly, the available budget.

B. CHEMISTRY

The hardware choice depends on the chemistry choice. Equipment for handling relatively inert gases such as CF_4 may be completely inadequate for corrosives like BCl_3 or even for O_2. Care must be taken to insure that gas handling apparatus and vacuum systems are consistent with the application. Much of contemporary etch technology uses gases that are potentially hazardous, so safety is a principle concern [3]. Not surprisingly, there are no universal etching methods that satisfy all needs. Although general principles can be established, etching is an application specific process. What is suitable for one device or facility may be unsuitable for another. Process acceptability will vary with device, device level, facility, time and management philosophy.

C. SYSTEM APPROACH

A successful etch process will depend on previous process steps, may effect subsequent process steps and may change if either of them is modified. A successful pattern transfer is the result of considering the constraints on the total system. Prior factors that influence an etch result include: the substrate material, the structure and composition of the previously defined layers, the underlying topography, the masking material composition, the defining lithographic method, and the pre-etch treatment (after lithography). Post-etch factors that may influence the etching choices are: resist removal method, next layer deposition and, the final judgement, device performance.

D. COST EFFECTIVENESS

The cost of the etch process is inconsequential if it is the only manner in which a device can be manufactured. Most often, financial considerations of the equipment cost, throughput, and the operating expenses are critical. There are many ways in which the cost of a piece of equipment can be rationalized. Return-on-investment calculations are important for larger manufacturing operations, where several pieces of identical costly hardware are needed. At the time of this writing (1987), the cost of production equipment ranges up to approximately one million dollars. Factors that have been shown to be important for economic optimization include: original equipment cost, cost of needed accessories, installation cost, floor-space requirement, labor costs for operation, throughput in wafer area/time, chemical and gas consumption and cost, and maintenance cost.

II. Classification of Etch Equipment

A. BULK OR SURFACE LOADING

Surface loaded reactors, as the name implies, place one side of the material being processed in direct contact with a surface, wall, electrode or other part of the reactor. In bulk loaded reactors, material is stacked so that only a small portion of each wafer is supported, with all surfaces in contact with the gas or vacuum space. The back (inactive) side of the material may be partially protected by loading wafers back to back. Surface loading usually results in better thermal contact providing better wafer cooling.

B. FREQUENCY

The exciting source of an electrical discharge is said to be high or low frequency depending on its relationship to the collision rate between the particles in the gas. This subject is discussed in more detail in other chapters [4]. In this section we outline the characteristics of the different regimes insofar as they alter etching characteristics and equipment requirements. The excitation source supplies reactive atoms, ions and radicals and, electric fields that control the motion of charged particles.

1. High Frequency 13.56 MHz / 27 MHz

The common radio frequency designated for experimental devices is 13.56 MHz and its low harmonics [5]. They are frequently used when an alternating field is needed for excitation without the use of conducting electrodes in contact with the discharge. Generators operating in this range readily couple into a gas chamber through a dielectric wall such as a glass or quartz tube without the need for excessively high applied voltage.

2. Low Frequency 10 KHz–1 MHz

If voltages are sufficiently high, "electrodeless" excitation down to very low frequency is possible. At sufficiently low frequency, ions are able to follow the changes in electric field so that discharges behave similarly to dc discharges with alternating cathodes and anodes. Selection of an operating frequency depends on many factors. An important consideration has been the availability of suitable generators. Commercial solid state devices are obtainable with power outputs up to several kilowatts. Matching circuits are needed to adapt the impedence of the source to that of the plasma [5]. At high frequency ordinary L-C circuits are most common. Ferrite core transformers are useful up to about 0.5 MHz. The frequency band between this and the standard high frequency rf band presents special problems.

C. MICROWAVE

Microwaves are a convenient source of excitation for several applications. The electric fields associated with a microwave discharge system are very complex so that they are used principally as free radical generators, or as a source of ions in conjunction with an acceleration system. The usual frequency is 2.45 GHz, identical to that used in microwave ovens.

Most microwave discharge systems are non-resonant, using a wave guide with the discharge tube passing through it, and an appropriate matching device such as a stub tuner. Discharge volumes are usually small and require adequate cooling if high powers are used. Special techniques are needed to get large volume discharges [6].

D. DC (BEAMS)

DC discharges are limited to those cases where there are conducting electrodes, free of insulating films, in contact with the plasma. An externally supported dc bias can be applied to conducting material in conjunction with a high frequency discharge [7].

E. OPERATING PRESSURE

Classification of pressure into high, medium and low is convenient to differentiate among the various types of pumping systems and to illustrate differences in the plasma and neutral gas background.

The usual gas kinetic conditions apply. Lower pressure means decreasing (neutral) particle density, longer mean-free-path, fewer collisions, etc. We call pressures below about 20 Pa low, since above this, back scattering of sputtered material becomes significant. “High pressure etching” usually refers to considerably higher pressure, about 100 Pa and above. Most etching is performed at pressures less than 1000 Pa. The various pressure regimes, their characteristics and the methods employed to establish them are outlined in Table 1.

F. ELECTRODE ARRANGEMENT

Electrodes couple energy from the power supply into the plasma. They must be matched to the mechanism by which this power is transferred into the gas. Three often-used methods are: Direct coupling, in which there is

Table 1 Pressure Ranges Used in Dry Etching and Important Gas Parameters.

Pressure	Density	Mean Free Path	Particle Flux to Surface	Pumping Method
	(m^{-3})	(nm)	($m^{-2} - s^{-1}$)	
High; > 150 mTorr (20 Pa)	4.90×10^{21}– 1.64×10^{23}	$3.47–10^{-4}$– 1.04×10^{-5}	4.81×10^{23}– 1.61×10^{25}	Mechanical Pumps, Blowers
Low; 5–50 mTorr (0.67–6.7 Pa)	1.64×10^{20}– 1.64×10^{21}	1.04×10^{-2}– 1.04×10^{-3}	1.61×10^{22}– 1.61×10^{23}	Blowers Diffusion Pumps Turbo Pumps
Very Low; .5–5 mTorr (0.067–0.67Pa)	1.64×10^{19}– 1.64×10^{20}	1.04×10^{-1}– 1.04×10^{-2}	1.61×10^{21}– 1.61×10^{22}	Diffusion Pumps Cryopumps Turbo Pumps
Ultra Low; 0.1 mTorr	$< 3.19 \times 10^{18}$	> 0.533	$< 3.13 \times 10^{20}$	Diffusion Pump Cryopumps Turbo Pumps

free exchange of charge between the plasma and a conducting surface; capacitive coupling, where there is a solid dielectric barrier between the electrode and the plasma; and inductive coupling, where a changing magnetic field induces an alternating voltage that sustains the discharge. Often, coupling consists of a combination of these methods, even when it is not intentional. True inductive coupling, although possible, is difficult to achieve at 13.56 MHz. The common electrode forms are shown in Fig. 1.

G. BARRELS

Most barrel systems, including simple rectangular metal boxes, use capacitive coupling. Sometimes barrels are driven by winding a coil around the perimeter, which appears to be inductive coupling. Unless special precautions are taken, coupling with such a coil is mainly capacitive between the winding and the plasma. The magnetic field of such an "inductor" has very little effect on the discharge except at very high power density (coil currents).

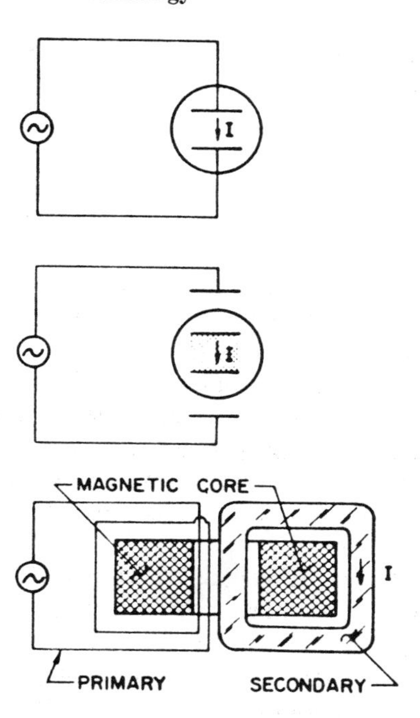

FIGURE 1. Electrode arrangements, direct, capacitive, inductive. Methods of coupling a power supply into a plasma. Direct coupling requires that the electrode be free of insulating films and may cause undesirable contamination from sputtering. Capacitive coupling dominates at high frequency. Inductive coupling can be enhanced at low frequency and at high power.

In the case of shielded barrels or "tunnels," a perforated metal cylinder, which acts as the shield, may also be a counter electrode to a set of capacitive plates on the outside of the tube.

H. PLASMA ETCHERS "ANODE LOADED"

Since an electrode, which is capacitively coupled to an rf generator, will develop a negative potential with respect to the plasma [8], it is common to refer to it as a cathode. Although the opposite electrode is the anode, it is not generally positively charged with respect to the plasma but will also assume some negative potential. The magnitude of these potentials depend in a complex manner on many parameters, including the relative electrode areas, [9] the area of nearby grounds, and the pressure and nature of the gas. Nevertheless, etch systems in which the wafer rests on a grounded electrode, i.e., one not directly attached to the generator, are often referred to as "anode loaded."

I. REACTIVE ION ETCHERS "CATHODE LOADED"

Cathode loaded reactors are those in which wafers are placed on an electrode, which is capacitively coupled to a generator, usually operating at 13.56 MHz. They are analogous to rf sputtering systems with the wafers serving as the target [10]. It is common to refer to these as reactive ion etchers or RIE reactors. The terminology is meaningful only under specialized conditions, including appropriate area ratio of "cathode" to other surfaces, anodes or grounds. Operating pressure and power also need to be chosen so that the desired negative bias develops on the wafer support plate.

The terminology RIE is unfortunate. Ions present in the discharge may well not be "reactive," particularly if inert gas diluents are present. It also leads to the false impression that anisotropic etching, which is related to the directionality of ions, reactive or otherwise, is only achievable for cathode loaded systems. As in the anode loaded system, the cathode may come in various shapes and sizes. Hexagonal right prisms known as Hex reactors have been particularly popular and are used in many manufacturing operations.

J. OUTSIDE, DOWNSTREAM, AFTERGLOW

Wafers are most frequently located directly in the plasma, separated from it only by the natural boundaries of the sheath, where they are bombarded by all of the energetic species of the discharge. These include positive and negative ions, electrons, photons from decaying excited atoms or radicals, photons created by electron bombardment of surfaces, short-lived radicals, and hot and metastable atoms. It is possible to protect the wafers from some of these particles by appropriate shields or by removing the wafers from the region where the plasma is created. In the latter case, longer lived radicals and atoms may be transported to the wafers by forced convection or diffusion. When the transport mechanism is primarily through convection created by fast flowing vacuum pumping, it is termed a down-stream reactor. The absence of ions, electrons or other directed species results in chemical reactions providing primarily isotropic etching. Microwave excitation is a popular choice for downstream systems. Figure 2 portrays a

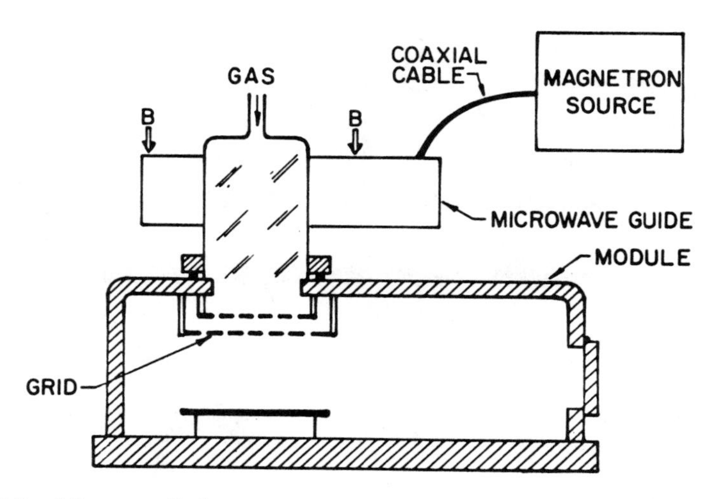

FIGURE 2. Microwave discharge attached to etching system. High power density in tube requires adequate cooling.

generic microwave etch system, which includes grids for retarding or accelerating ions.

K. LOAD SIZE

Production reactors are classified, primarily by the number of wafers that are processed in a single etching step. Load size is frequently limited by factors such as uniformity, rather than by the dimensions of the chamber.

L. SINGLE WAFER SYSTEMS

The main attribute of a single wafer system is that it presents an identical environment to every wafer. Usually a single wafer etching machine is designed so that the number of wafers processed per hour is independent of the wafer size. Scaling of a process to a larger wafer size may be as simple as scaling all of the known process parameters (gas flow and power). It may also require changes in less obvious parameters such as electrode gap and pressure. Single wafer reactors must operate at 10 to 20 times the etch rate of batch reactors in order to have similar throughput.

Except for simple laboratory type reactors, all single wafer systems (sometime referred to as one-at-a-time systems) have sufficient automation

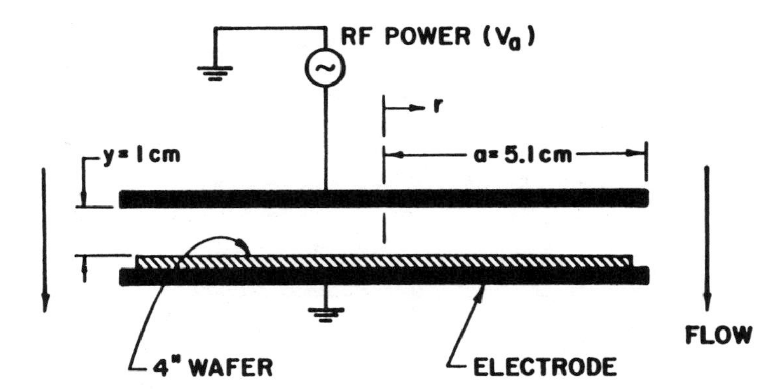

FIGURE 3. Scale drawing of single wafer etcher configured for 100 mm wafers. The small electrode gap is needed to increase ion energy for oxide etching or other ion intensive processes.

to remove wafers from a cassette or line, feed them into the processing unit and return them to a receiver. There are many variations in construction, with major differences in the wafer handling mechanism. Air tracks, where wafers are moved on an air cushion, "rubber-bands," which move wafers on stretched rubber-like bands driven by pulleys, and mechanical pick-and-place are all used. Most systems come complete with gas handling hardware, gauging, rf generators and sufficient automation to operate without attendance once wafers have been placed in the input. There are many variations of chamber configuration with major differences in electrode structure.

1. Planar Plasma

Most single wafer systems on the market consist of variations of the planar plasma configuration. Dimensions of a typical single wafer etch chamber used for 4″ wafers are shown in Fig. 3. High etch rates, needed for production single wafer etch systems, usually require substantially higher power density than batch systems. Special techniques are necessary to handle the resultant thermal and electrical loads. Removal of heat from wafers are discussed next. Enhanced backside cooling may require hardware features like wafer clamping and special gas control features to permit adding high pressure gas behind the wafer—so-called "helium chucks." In all cases, coolant fluids need to be circulated to the electrodes in sufficient quantity to remove the heat created by the discharge. It is desirable to control the temperatures of all electrodes because they may affect process performance, so refrigerated cooling systems are common. Electrode temperatures from 0° to 90°C may be needed for different processes. For aluminum etch systems, it is also desirable to heat other portions of the reactor and its exhaust system to prevent condensation of low vapor pressure etch products.

Proper selection of electrode material and other structural parts in contact with the plasma is important to prevent sputtering of contaminants onto wafers. Trace amounts of ferrous metals can substantially degrade the performance of most ICs. For this reason electrode materials such as silicon, graphite, SiO_2, silicon carbide or various polymers are preferred over metals. Most commercial systems make use of special coatings, which, though they sputter, are usually much less harmful than metals to device performance.

2. "RIE"

Some commercial single wafer etching systems connect the high side of the generator to the electrode supporting the wafer and the "ground" connection (low side) to the chamber or opposite electrode. A push-pull, or balanced connection system in which each electrode is attached to one side of the generator is used in at least one commercial system. An earthed ground connection is made to the chamber walls, which are isolated from the electrodes. Many single wafer systems operate at a pressure greater than what is normally considered the RIE regime in batch systems. The primary problem encountered in this mode of operation is to achieve sufficient etch rate. Some commercial systems attempt to further promote RIE conditions by providing an appropriate electrode area ratio similar to lower pressure systems. These systems also employ power densities similar to other single wafer systems, which are much greater than batch RIE.

3. Magnetrons

Magnetic fields are used to enhance ionization by confining a discharge. Arrangements similar to those used for magnetron sputtering have been incorporated into etching hardware. In dynamic magnetron apparatus, a racetrack-like region of enhanced ionization is mechanically scanned over the wafer surface [11, 12]. A static magnetic design that creates an intense discharge zone without the need for mechanical motion is shown in Fig. 4. A belt of ionization surrounds the central core. These are two examples of

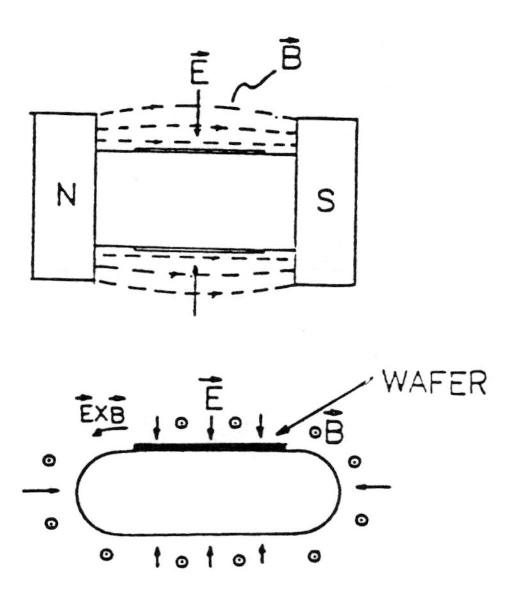

FIGURE 4. Static magnetic enhancement discharge system. This type of system is related to post magnetrons used for sputtering. High rates are achievable at low pressure in this system.

commercially available etch systems that afford high enough etch rates at low pressure to compete as single wafer systems. The two systems apply the magnetic field in the vicinity of the wafer. In an alternate approach, magnets are arranged around the periphery of the chamber to reduce electron diffusion losses similar to the arrangement used in broad beam ion sources. Other arrangements include interactions between magnetic and microwave fields in beam-like machines [12].

M. TRIODES

Triodes, as their name implies, are three element discharge systems. The purpose of the third element is to create two zones with distinct discharge properties. One such arrangement is shown in Fig. 5. In this system, a grounded grid separates the two regions forming what are, in effect, two planar etch chambers. The wafer is placed in one chamber configured like an RIE system. This chamber is downstream of, and shielded from, the other portion. There are two separate power supplies, which can operate at different power levels and which need not be at the same frequency. By adjusting the power level of each part, the amount of chemical etching from

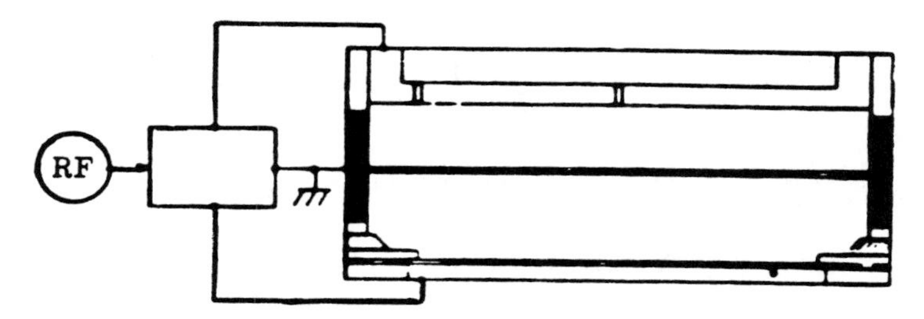

FIGURE 5. Triode discharge system. Other three terminal arrangements are possible. A metal wall can also be used to supply additional discharge power at an alternate frequency.

radicals produced in the upper section can be varied with respect to the ion-enhanced and presumably anisotropic portion generated in the bottom chamber.

N. BATCH AND MINI-BATCH

There are two types of mini-batch reactors, parallel processing single wafer systems and small batch systems usually with provision for four to six wafers. The former may share some common subsystems such as gas supplies, but tend to be relatively independent. The latter type may have only a single set of critical parts, gas system, pumps, power supplies, electronics, etc.

The capacity of batch systems varies considerably with reactor type and wafer size. Bulk, barrel-type reactors may hold several hundred wafers or equivalent material. Surface loaded reactors range from a few wafers (mini-batch) to large parallel plates or hex-type systems that accommodate from 10 to 50 wafers depending on wafer diameter. Usually the size of the reactor batch is matched to that of other process equipment. Twenty-five wafers is a common cassette size for wafer handling. It is desirable to maintain cassette identity during processing. If it is necessary to process a fraction of a load in a batch, it may be necessary to supply dummy wafers to balance the reactor operation.

O. BARRELS

A barrel is any chamber in which material is placed and exposed to the reaction products of an active discharge without any attempt to specify the

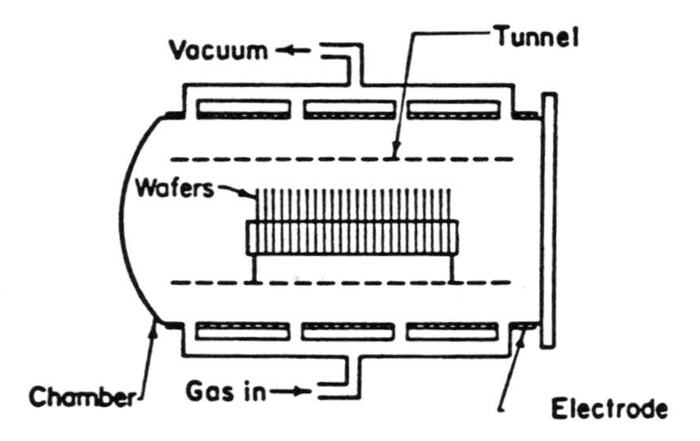

FIGURE 6. Tunnel etch system. The tunnel or shield can be grounded or floated. Grounding provides more positive control of the plasma location. At low pressure material in the center can still receive some bombardment by charged species.

directions of the exciting fields. Barrels are available in many different sizes and shapes. Metal containers several feet in diameter are used for printed circuit board desmearing [13]. The principle use is photoresist stripping and non-critical etching. Other applications are: cleaning of organic residue, and surface modifications such as changing adhesion and wetability [14].

The usual form of excitation for barrels is a 13.56 MHz rf discharge and may use either capacitor plates or a wrapped coil (still capacitive). If low frequency is used, care must be taken to prevent sputtering of electrode material at the higher voltages needed to maintain the discharge. Typical maximum power inputs are about 50 Watts/liter of chamber volume. As shown in Fig. 6, chemical effects can be enhanced over ion induced effects by using perforated metal shields to confine the discharge to the region between the outer wall and the shield. The process material sees primarily long-lived radicals. Lowering the pressure, however, permits diffusion of charged particles into the inner zone.

Most barrel systems do not provide for uniform passage of gases over the (wafer) surfaces and rely mostly on diffusion to obtain uniformity. Because the production rate of chemical species is limited, loading effects are common. Most process rates increase with temperature, so thermal control is important for reproducibility. Some systems come equipped with special programmed preheat cycles. Uniformity is a complex function of load size, pressure, gas flow and of arrangement of the material in the chamber. Chambers with shields (sometimes referred to as tunnels) usually produce more uniform processing than those in which the material is actually in contact with the plasma. Processing proceeds from the periphery inward with a constantly increasing rate with time. This is due to a combination of decreased load and wafer heating. Several attempts have been made to calculate optimum wafer spacing to minimize non-uniformity [15].

Barrel systems remain the most cost effective system for processing material. They are useful when control of feature size is not critical.

P. PLANAR REACTORS

Planar reactor load size is directly related to the diameter of the plate holding the wafers. Small reactors, 6 to 12 inches in diameter find application in development and special purpose production where throughput is not critical. Typical production systems may be up to three feet in diameter and can hold a large number of wafers. The useful number of wafers, however, that can be treated at one time depends on details of the process performance, particularly uniformity. Methods to achieve uniformity in planar reactors and in other systems can be divided into two major categories: hardware and process.

Hardware variations are usually determined by the equipment vendor, although minor modifications by the user can lead to significant improvements. Some hardware characteristics that may be candidates for modification are: materials in contact with the plasma, electrode design and construction, gas distribution system (this may be part of the electrode design), and gas metering system (primarily size change). Less frequently, the user may modify the plasma power supply and end point monitoring system to satisfy special needs.

Both plasma and "RIE" versions of batch planar systems are available. Besides electrode connection, major differences between the two are in the gas, pumping and control systems.

Q. HEXODES

Hexodes are RIE systems having six sided cathodes on which wafers are placed. Both manual and automatic versions have been very successful. Wafer loading is done in two steps. The first, is placing wafers on special rectangular pallettes. One places wafers on special rectangular pallettes using a special mechanical loader/unloader. The six pallettes are then placed on the central core either by hand or by a robot. A considerable amount of expertise has developed in using this etcher, which is, however, not available in the open literature. Systems have been developed to handle all of the currently used production size wafers. As with other batch systems, wafer number decreases as wafer size increases unless the physical dimensions of the system are increased. A typical batch size for a hex system is 24 100 mm wafers.

III. Process Control

A. GAS CONTROLS

Important characteristics of the gas control systems used in any etcher are the number, type and available gas flow range. Some very early etchers used proprietary, pre-mixed gas compositions that were available only from the equipment vendor or their licensee. Modern etch technology requires that users be able to adjust compositions for their special requirements so that flexibility has become important. It is not uncommon for a single etch step to use four different gases in a special mix to obtain a needed result. It is not clear whether the complex mixtures used are actually necessary. In some instances it is possible to get similar results with fewer components by varying other system parameters. The creation of special etch compositions is likely to remain as a primary method for tailoring etch processes to special device fabrication needs. Since a real process may use several sequential steps, control of many different gases may be needed. For process equipment dedicated to a particular process four gases may be sufficient. It is useful, however, to have a readily implementable option by which either more gases can be added or controllers may be shared among more than one gas. Careful consideration should be given in the latter instance to the compatability of the gases that use shared controllers.

Most contemporary etch equipment uses thermal conductivity-type mass flow controllers (MFCs) for gas monitoring. These are available to cover a wide range of gases. Special forms of MFCs are made for liquid sources. Most high vapor pressure liquids can be used as sources for etching. They may either be pumped directly or used with a carrier gas to entrain vapor. MFCs have the advantage of continuous adjustment and programmability, making it easy to incorporate them in multistep processes under computer control.

Rotameters are a convenient method for measuring gas flows when programming is not important. They consist of precision bore glass or plastic tubing containing a float or ball. Different gas flow ranges are obtained by changing the tube diameter and/or the float material. A range of about 3X is available by changing the ball material from glass to a metal such as Hastalloy. Rotameters are simple and reliable and can be used in many applications.

Fixed flow needle valves or critical flow orfices are an alternate type of gas flow controllers that can be used in special applications. They are inexpensive but lack flexibility.

1. Gas Flow Considerations

Gas flow is commonly quoted in units of "standard cubic centimeters per minute" (sccm), which is 1 cm^3 of gas at $P = 1$ atm, $T = 25°C$. Normally, larger quantities of gas are used in etchers that operate at high pressure than in RIE systems. This is partly because very large pumps would be required to use large gas flows in the low pressure RIE regime. Conversely, it would be difficult to control low gas flows in the higher pressure plasma mode. Outgassing and other sources of impurities also become a significant fraction of the total gas flow for very low flows at high pressure.

The following relationships between gas flow and vacuum parameters are important in scaling:

$$T_r = PV/Q$$
$$Sp = Q/P$$

where,

T_r is the residence time of the gas in the chamber in seconds,
P is chamber pressure in Torr,
Sp is pumping speed in liters/second, and
Q is gas throughput in Torr-liters/second (1 Tl/s = 79 sccm),
V is the chamber volume in liters including connecting pipes.

Typical values are:

Large Batch Reactor	*Small Single Wafer Reactor*
Q = 20 sccm (0.25 Tl/s)	Q = 200 sccm
P = 0.01 Torr	P = 0.01 Torr
V = 100 liters	V = 1.0 liters
T_r = 4 seconds	T_r = 0.4 seconds

Gas input is usually low enough that laminar flow results. The criterion for laminar flow is that the Reynolds number, N_{Re}, is less than approximately 1000. N_{Re} is a dimensionless quantity, which is the ratio between the dynamic and kinematic viscosities [16] given, for a round tube by:

$$N_{Re} \approx \frac{[\text{Flow (cm}^3\text{/min at STP)}] \times [\text{Molecular Weight}] \times 10^{-6}}{\mu \times d},$$

where μ = viscosity in poise and, d = tube diameter in cm.

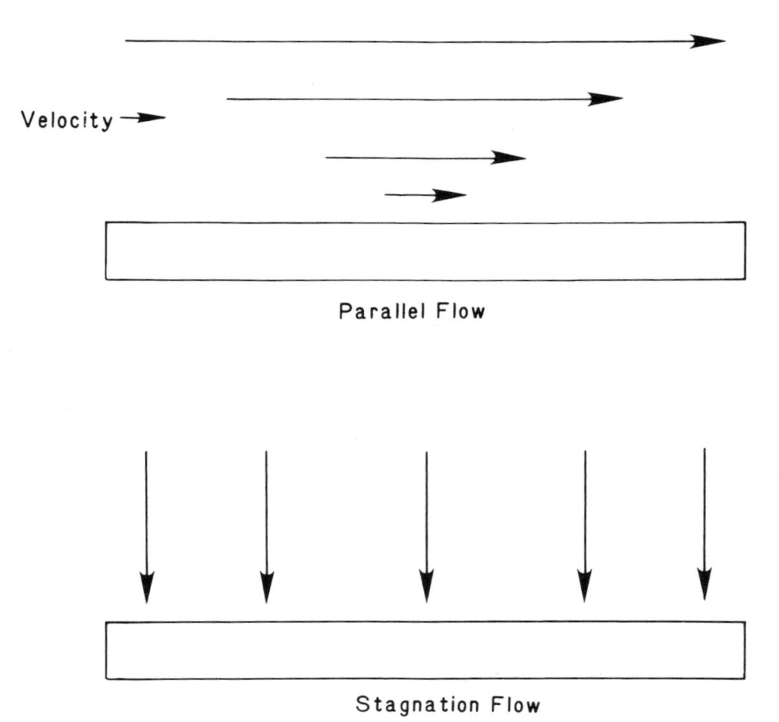

FIGURE 7. Common flow patterns used in plasma etchers. Gas expansion from nozzles, shower heads or porous media usually result in complex flow situations with parallel flow favored. Results often attributed to flow patterns are second order in that they primarily affect the plasma currents. Hollow cathoding from shower head holes is an example of this effect.

Gas flow is modified by the presence of diffusion. A relative measure of importance is obtained by comparing the flow velocity in the discharge region with the diffusion velocity. Most plasma systems have complex flow patterns, which are difficult to analyze. Gas flow may be either parallel or normal (stagnation flow) to a surface as depicted in Fig. 7. An approximate value for the flow velocity can be obtained by dividing the volumetric flow in the plasma zone by an appropriate area. For a parallel plate reactor utilizing stagnation flow this is the area of the plates. For parallel flow it is the cross-sectional area normal to the plates and varies as a function of position. The diffusion velocity may be approximated by dividing the diffusion coefficient by the density gradient scale length. For the reactor shown in Fig. 3, assuming a normal flow of 100 sccm at, $v \approx 20$ cm/sec, the self diffusion coefficient for CF_4 is about 40 cm^2/sec at 0.76 Torr, so if the density gradient scale length is ≈ 1 cm, the effects of flow and diffusion are comparable.

B. PARTICULATE CONTROL

The need for particulate control is becoming the most important issue in production plasma equipment. Requirements of less than .01 particles/cm^2, 0.5 μ in diameter or larger are desired goals. Performance is usually much poorer. Several sources of contamination associated with plasma machines are: (1) generation from hardware movement, valves opening and closing, wafer transport and other mechanical parts; (2) process generated hardware problems from corrosion and material degradation; (3) etch product coating and flaking; (4) improper operation, pumpout too fast, dirty gas lines, etc.

The first two sources are best reduced through proper system design and engineering. Avoidance of rubbing metal parts, and correct selection of elastomers (fluorine-based material are good). The last two sources can be reduced by proper operation and maintenance. Several selective etching processes exploit the balance between polymer formation and etching to achieve selectivity. If these are operated too close to the "polymer point," films may deposit on portions of the hardware and provide a source of flaking material. If it is necessary to operate under these conditions, then frequent cleaning of the hardware will be necessary. Cleaning may be either with different plasma conditions, or as is most common, by opening of the system and wiping with appropriate solvents.

Point-of-use filters in gas lines are common in production equipment, as are controlled pump-out and purges. Studies indicate that purge times are more important in the elimination of particulate generation than pump-out times. Gas flows should be controlled so that turbulence (high Reynolds numbers) is avoided. The size and placement of the gas inlets can be a significant factor in controlling turbulence. Some advanced production equipment is capable of programmed pump-out and purges to reduce particulate generation from this source.

C. THERMAL CONTROL

Most of the power input to a discharge eventually ends up as heat that needs to be dissapated. That portion of the heat, which deposits onto the wafer(s), is most critical. In some processes it is necessary to assure that sufficient heat is supplied to hardware to control deposition of undesirable products. Standard electric heating elements are most common, but care is needed in the selection to assure long life when exposed to the plasma environment. Most resists are organic polymers with limited thermal stability. Excess heat can result in changes of dimension from resist flow and in severe cases complete destruction of the pattern from resist burning. Also, because chemical effects are thermally activated, process control is reduced if the temperature is allowed to rise too high.

Circulating liquids and refrigeration units are the major means of heat removal from hardware components. A single wafer etch system for large wafers may input 1–2 KW to the discharge, dictating the need for substantial cooling capacity.

Cooling of wafers in a plasma is significantly more complex than hardware temperature control. The predominant method of heat transfer between a wafer and its support structure is by gas conduction. The heat transfer coefficient between the wafer and its support depends on the pressure, the gap between the wafer and the nearest cooled surface and the type of gas. Figure 8 illustrates the dependence on pressure and gap for

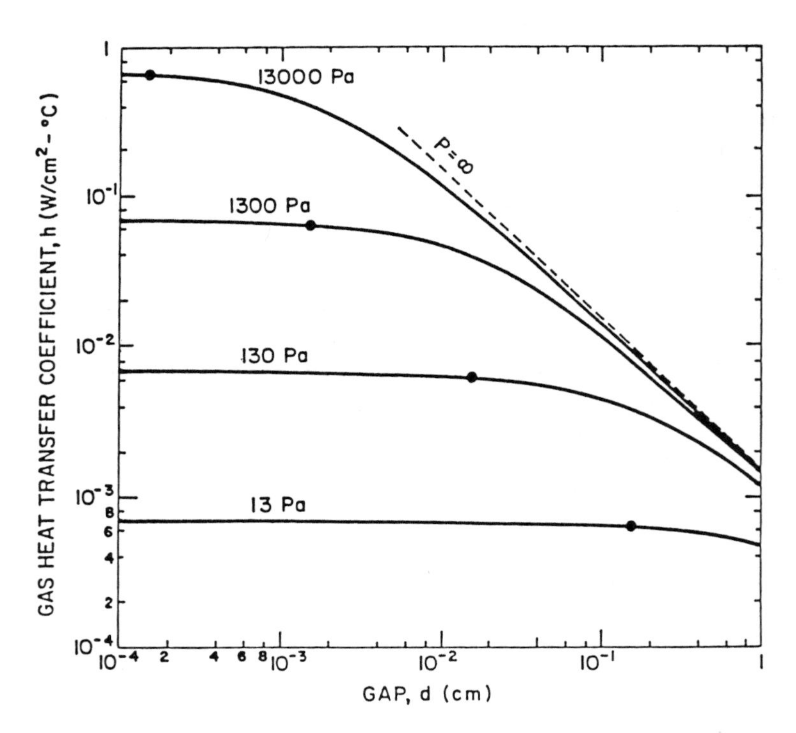

FIGURE 8. Heat transfer coefficient vs. wafer-platform gap for various pressures of helium. The dots on the curves depict where the gas mean-free-path is equal to the gap.

helium. Helium is commonly used as a background gas (diluent) because of its high heat transfer coefficient. To improve the heat transfer further, wafers may be clamped to the substrate table and high pressure helium introduced behind the wafer. A back pressure of several Torr is adequate for most applications. Care must be taken so that as small a portion of the wafer as possible is obscured by the clamps. A pressure differential of a few Torr can deform the center of a wafer by several thousandths of an inch, which needs to be compensated for by proper design of the table shape. With properly designed backside cooling, a wafer with resist can accommodate several watts/cm^2 without overheating.

D. AUTOMATION

Contemporary production etching equipment is often highly automated. Most systems are able to automatically process a wafer through all of the steps necessary for etching without the need of operator intervention. These include automatic loading of the wafer from a cassette to the etch chamber, pump down, gas selection and set-up, pressure setting and automatic stabilization, rf turn-on and turn-off as determined from automatic detection of end point or by time (if appropriate) and, purging and return of the wafer to the output station. In addition, most systems are able to store several process recipes and sequence them in a predetermined manner, usually under control of a small micro computer.

In the ultimate level of automation, the entire system is under direction of a host computer, which keeps track of wafers by a code number on the wafer and automatically assigns the correct process step without operator intervention. Also, the system performs self-diagnosis and takes action when it is not in peak operating condition.

IV. Etching Methods—Films

In the following sections we review the materials important for integrated circuit manufacture, the forms in which they are often encountered, common etching requirements and the means by which they may be approached [17].

A. SILICON DIOXIDE

In the manufacture of semiconductor devices SiO_2 is the most frequently etched material. Even in the case of non-silicon devices it is very frequently used for its properties as an insulator, and because it is readily patterned with dry etching. As the properties of SiO_2 were essential to the construction of silicon integrated circuits, so too, the ability to selectively and anisotropically etch it are crucial to the production of high density structures.

SiO_2 is produced by several different physical or chemical methods:

Thermal conversion of silicon
- Single crystal
- Polysilicon doped/undoped

Chemical Vapor Deposition (CVD)
- Atmospheric pressure
- Low pressure (LPCVD)
- Plasma enhanced (PECVD)
- TEOS CVD (tetraethylorthosilicate)

Deposited
- Sputtered from quartz
- Electron beam evaporation
- Spin-on-glass

Anodization.

Material may be undoped, doped with phosphorous (PSG), boron, or boron and phosphorous (BPSG). Variations in density are common between deposited films and those made by thermal conversion of silicon. Sputtered films tend to be more like thermally oxidized silicon. Etch rates are similar for most materials with the major exception of the phosphorous doped glasses. These can have etch rates that are considerably faster than undoped material. The actual rate depends on the doping concentration.

Typical Applications

Thickness range encountered and type are:

Gates: 10–100 nm, thermally converted crystal Si

Interlevel insulation:	200–2500 nm, deposited, converted polysilicon, combinations
Protection, overcoats:	1000–3000 nm, deposited, PECVD, CVD

Typical underlayers include single crystal silicon, polysilicon, silicides, refractory metals, aluminum, silicon nitride, other forms of SiO_2 (CVD over thermal) and organics (used in multi-level resist structures or organic interlevel insulation).

SiO_2 has the widest range of etching requirements because it has the widest range of uses in device manufacture. In addition to the performance parameters discussed next, there are often other special conditions that arise. Radiation damage, metal contamination from reactor components, and particulate generation are three common areas of concern. It is likely that radiation damage is related more to voltage and ion energy than to etch rate itself. There is also some evidence that details, such as how the power cycle is ended may be important [18].

Contamination from reactor components was common in early versions of planar oxide etch systems, plasma or RIE. Most metal contamination, particularly iron, is traceable to sputtering of stainless or aluminum vacuum parts. Evidence of metal sputtering may be quite gross as in the case of aluminum. It usually appears as very non-uniform spotty etching with a great many unetched microscopic regions or "grass."

Heavy metal contamination may not be as obvious, showing up only later as device failure or degraded performance. Excess junction leakage is a well-known result of iron contamination in sensitive bipolar devices. This type of contamination is now well recognized and most equipment manufacturers make an effort to coat sensitive areas with special materials such as Teflon®, material with low sputter rates such as aluminum oxide or with material that is not damaging even if sputtered, such as silicon or silicon oxide coatings. It may be necessary to periodically replace parts or renew coatings if the etching process attacks them. Small holes and channels are particularly susceptible.

B. RATE, UNIFORMITY, SELECTIVITY AND ANISOTROPY

The most often quoted criteria used in describing the result of an etching procedure are the rate, uniformity of rate, selectivity to underlying films and resists and the anisotropy or dimensional control. In most instances, the number of wafers processed per hour is the most important factor. It is a combination of rate and overhead time due to wafer transport, gas stabilization, pumping times, etc.

Uniformity of rate is frequently a critical issue requiring experiments to assess correctly. It is typically easier to obtain good uniformity on wafers that are smaller than the reactor's maximum capacity. Non-uniformity can result from poor gas flow, non-uniform electric fields or currents, and from so called "edge effects." Boundary conditions resulting from the presence or absence of reactor walls and change of material, including changes from wafer to support structures, affect the plasma uniformity and resulting etch process. For planar reactors it is advantageous if the wafer is as indistinguishable as possible from the supporting plate.

Rates are unimportant for thinner films as long as radiation damage is avoided. Generally, higher rates involve higher power and/or voltage, which induce damage. Typically, rates are adjusted to provide between 40–60 wafers/hour for a one-micron thick film including wafer handling and other overhead time.

The need for uniformity varies widely with the particular process. For example, when opening windows through a protective overcoat to aluminum bonding pads, uniformity is not critical. This is because the geometry control is not important and because the underlying layer is not strongly influenced by overetching. Small contact etching to silicon, planarization, thin oxide removal and other oxide definition layers may require very tight control of uniformity. Etch rates, which vary less than $\pm 5\%$ over more than 98% of the wafer area, are often requested [19].

Two selectivity values are significant, to the underlying film and to photoresist. Selectivity is usually expressed as a ratio of etch rates. For selective etching of SiO_2 over silicon, values of 4 : 1 to more than 25 : 1 are cited with 15 : 1 as a reasonable compromise. Higher values carry with them the potential for greater polymer production and result in particulate generation from supporting hardware. Reported selectivity is often based on a single etch step, whereas most production requires multistep processing. A better way to define selectivity is to specify the maximum amount of underlayer that can be removed, including any required overetching needed to allow for variations in material thickness and system non-uniformity.

Consider for example, a film that has a maximum thickness of 1 μ and a minimum thickness of 0.5 μ and an etch process with an inherent selectivity of 10 : 1. If a 20% overetch is used, the silicon under the thinner layer will be etched to a depth of 0.07 μ, while that under the thicker layer will be etched only 0.02 μ.

Selectivity to photoresist is needed to assure accurate transfer of critical dimensions. Vertical resist profiles provide a greater margin for linewidth control even with resist erosion. Selectivity of 3–5 : 1 are nominally acceptable for optical resists roughly 1 μ thick. If resist profiles are not steep, higher selectivity is necessary for accurate dimension and profile control.

Resist selectivity is influenced by resist type, pre-treatment and state of cross-linking, temperature and temperature ramping during processing, ion bombardment energy, and chemical attack from the species created in the plasma from the etch gas. High rate single wafer etchers have the propensity to produce higher temperature than batch systems. Because they operate at high pressure, however, this is offset by improved heat transfer mechanisms.

V. Etching Techniques

Etch rates of most materials increase with power input. Oxide etching is also dependent on volume power density (greater for smaller electrode gaps in planar systems). Figure 9 depicts the etch rate variation with power (for constant gap) in a reactor, where the power is known to be well coupled to

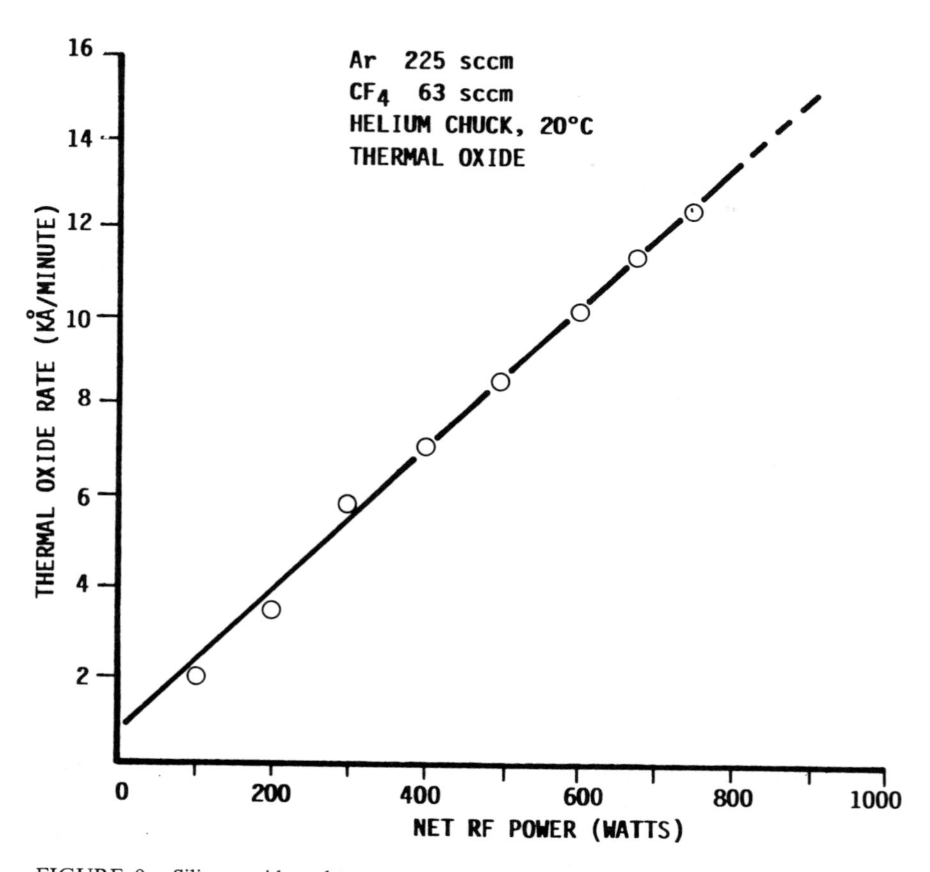

FIGURE 9. Silicon oxide etch rate vs. power. In this system care was taken to insure that most of the power input was restricted to the volume between the wafer and the upper electrode. Current leakage to other surfaces is a major contributor to rate limitation and to non-uniformity.

the wafer. A common difficulty in interpreting the effects of power increases is the uncertainty in the actual plasma density. Increasing wafer diameter and power also produces higher electrode voltage.

Besides power input to the plasma, oxide etch rates depend on the ion flux to the surface. As a result, the rate is a function of the coupling of the wafer to a system electrode. Current to the wafer surface is dependent on the capacitances between the wafer and the electrode and between the wafer and the plasma boundary. It is particularly evident at lower frequency where the capacitive reactance is high.

Using the relation between current, frequency f, and voltage,

$$J = 2\pi fCV$$

it can be shown that

$$R_0/R = 1 + C_s/C_w,$$

where R_0 is the etch rate for perfect coupling (as might be achieved by painting an electrode directly on the back of the wafer) and R is the etch rate for some finite coupling.

C_s and C_w are the capacitances between the wafer and the plasma body (sheath capacitance), and between the wafer and the electrode surface (coupling capacitance) respectively. This dependence is illustrated in Fig. 10. One cause of non-uniformity is due to the change in coupling for different points on the wafer. A bowed wafer, for example, resting on a flat surface will have a slower etch rate where the wafer is further from the table. The differences may be masked by a low wafer resistance, which can short-circuit the effect.

Process variations that improve uniformity will often cause variations in other etch characteristics. In oxide etching, changes that improve uniformity can adversely affect selectivity and rate.

One particularly useful way of adjusting uniformity is total gas flow. The rate increases with total gas flow gradually flattening out before finally showing a slow decrease. The initial increase is the result of some reactant species limitation at low flows. The saturation region is due to the onset of some other rate limiting process and the final decline is due to a transport limitation where gas phase reactants are removed before getting to the surface. By operating in the flat portion of the curve, variations in etch rate due to flow differences over the wafer may be reduced.

A general relation existing among selectivity, uniformity and rate is that factors increasing selectivity generally decrease rate and uniformity, i.e., less selective gas compositions are faster and more uniform. Performance is improved by using a two step process. The first step, which is relatively non-selective, is terminated just before the underlying layer (silicon or

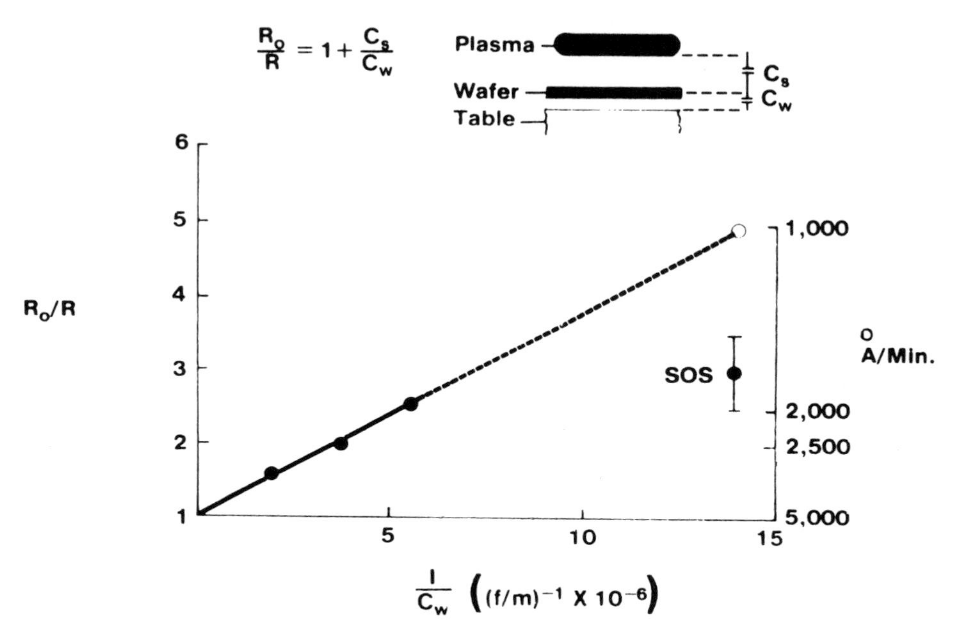

FIGURE 10. Oxide etch rate variation with coupling to an electrode. Capacitive coupling between the wafer and its support electrode can limit rate and effect uniformity. The effect is more pronounced at low frequency where the capacitive reactance is high.

polysilicon) is exposed. Etching is then completed with a high selectivity process. This is usually adjusted by a change in gas composition as described next.

A. TAPERING AND PLANARIZATION

Several situations arise in oxide etching where anisotropic profiles are not wanted. To facilitate coverage of the next layer metal into deep contact windows, for example, it is advantageous to provide a gently sloped edge instead of a vertical wall. If conformal metal coating could be achieved, then vertical holes for contacts would be more desirable as a means to increase packing density. Until this becomes possible, however, some form of window taper process is needed. Isotropic oxide etching, although

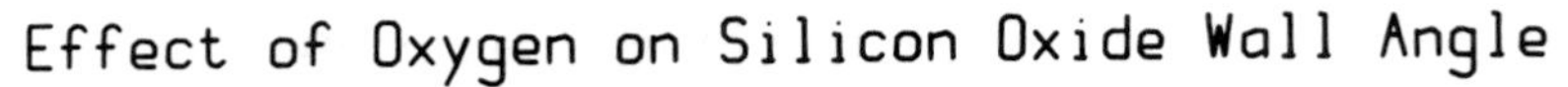

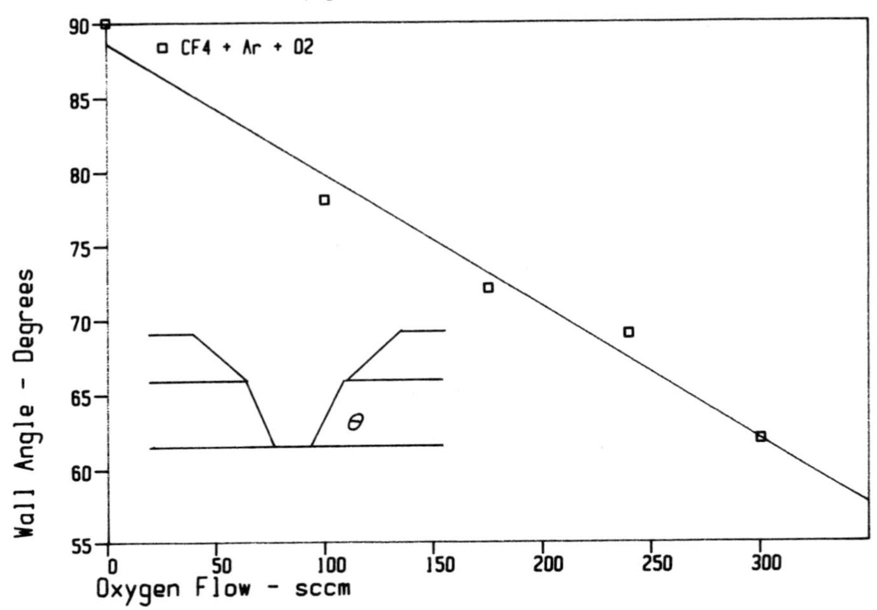

FIGURE 11. Controlled resist erosion for tapering features. Effective isotropic etching is difficult to achieve for oxide etching. A combination anisotropic/isotropic etch can be used to obtain a tapered edge.

possible in downstream configurations, is inadequate for several reasons, including low rates (for one at a time processing) and non-selectivity to silicon. The required fluorine rich processes generally etch silicon many times faster than oxide.

Ion enhanced SiO_2 etching is inherently anisotropic, so some alternate mechanism is needed to produce a taper. The method most commonly employed is controlled resist failure at the edge of the feature, as illustrated in Fig. 11. If the resist erosion is also anisotropic, the slope of the feature depends on the relative rates of resist: oxide and the initial slope of the resist as shown. Usually the problem is that of increasing the resist erosion rate sufficiently to achieve the needed profile. It is difficult to achieve the needed resist profile without compromising dimensional control. Thermal flow of the resist is commonly used for this purpose.

Resist erosion rate is increased by addition of oxygen to the etch gases. Typical change of resist and oxide etch rate upon oxygen addition to an

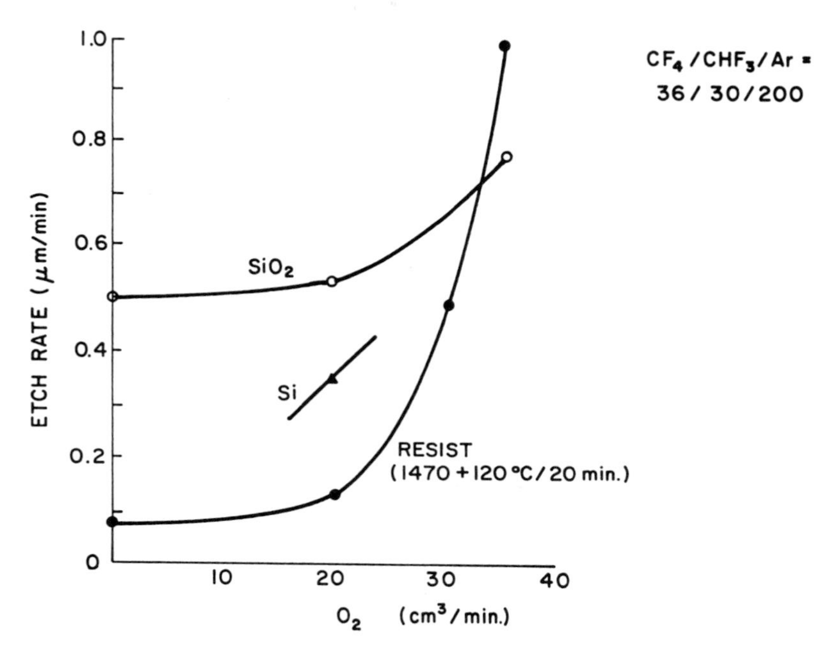

FIGURE 12. Effect of oxygen addition to oxide etch gases. Oxygen increases resist erosion and decreases selectivity to silicon. High concentrations of O_2 are needed to taper vertical resist walls as produced by steppers.

oxide-selective etch recipe is shown in Fig. 12. If sufficient oxygen is added it is also possible to create a tapered feature with initially vertical resist profiles. This is a desirable feature for stepper produced patterns. Oxygen addition, however, has the undesirable effect of increasing the silicon etch rate so that the selectivity is reduced. With sufficient control, a multistep process may be used to remove the oxygen prior to breaking through to the underlying silicon (or polysilicon). In some end-point detection schemes oxide etching is detected by monitoring the intensity of a CO emission band. In this case it may also be necessary to allow sufficient time for the added oxygen to be removed from the system to permit the end-point to function properly.

Planarization etching is becoming increasingly important for the production of state-of-the-art devices [20]. As already discussed, vertical steps, although needed for dimensional control, produce severe topography that make next level coverage difficult. They also contribute to significant height variations on the wafer that can exceed depth-of-field restrictions in lithography. Therefore, it is becoming more important to provide a planarized or

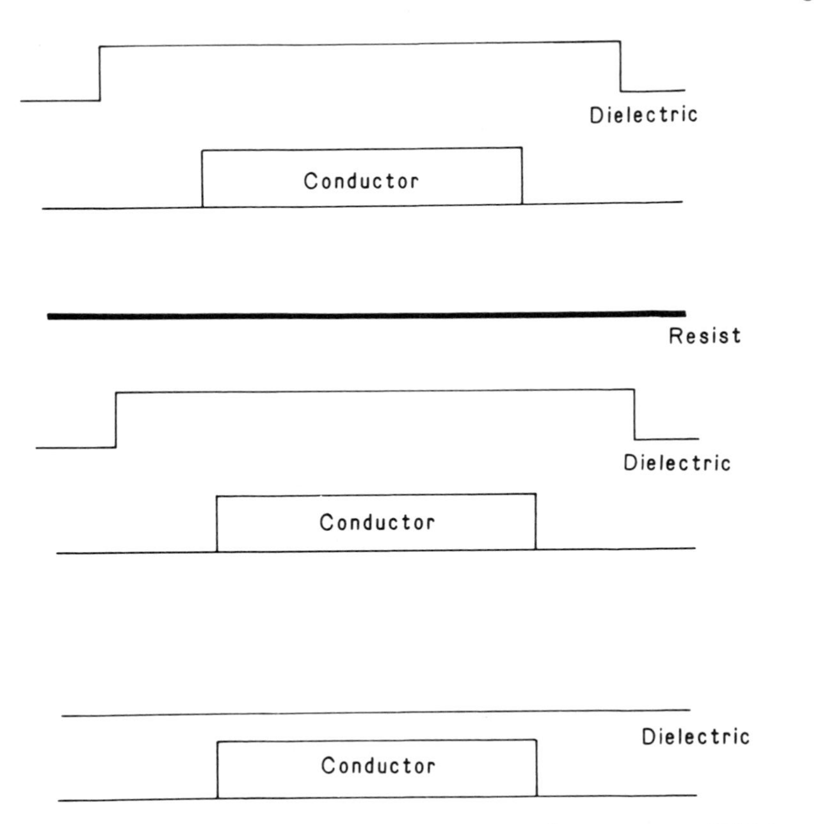

FIGURE 13. Planarizing etch back process. Photoresist is the common sacrificial layer. Other materials can be used to adjust the removal rates of material. Ion-intensive processes are less prone to micro-loading effects than chemical processes as result from the addition of oxygen.

level surface between successive layers. One method of achieving this is through a planarizing etch back. This process is illustrated in Fig. 13.

After anisotropic patterning, a sacrificial layer (often photoresist) is applied to the surface in some manner, such as spinning that reduces the surface height variations. The process is adjusted to cause the sacrificial layer and the underlying film, principally oxide (or other dielectric such as PECVD silicon nitride) to etch at equal rates. In this manner a point is reached in the etching where the surface is fairly flat (see Fig. 13). Process adjustment is commonly done by adding oxygen to the etch gases.

Special attention needs to be given to device structures that have large, high, flat areas of underlying oxide covered by a thin (sacrificial) resist layer and also deeper narrow grooves. When the large areas are etched through the top resist layer, the system becomes "unloaded" [21] with respect to resist etching, the exact amount depending on the amount of oxygen present. Consequently, the etch rate of the resist increases dramatically. The relative rates are no longer 1 : 1, the resist in the deep shallow grooves is rapidly etched away and the planarizing effect ceases. This can be avoided in part by using a more ion-intensive process to attack the resist without oxygen. Ion induced processes exhibit less tendency toward loading than chemical etching. To achieve equal resist and oxide rates is not easy, especially if the underlying film is a fast etching oxide such as heavily phosphorous doped glass. Photoresist is often used for the levelling layer but other materials with higher etch rates can be employed.

B. ETCHING CHEMISTRY REVIEW

CF_4 and $CF_4 + O_2$ are used for isotropic, non-critical, non-selective etching. An example where this would be a satisfactory batch process is etching bond pads to a non-reactive surface such as aluminum [22]. For selective etching in reactors used for oxide etching, a combination of gases that result in fluorine deficient discharge products are employed. Table 2 lists some compositions reported useful for oxide etching. Various mixtures are possible using either CF_4 or C_2F_6 and a hydrogen containing gas such as H_2, CHF_3, or heavier fluorocarbon. The hydrocarbon component of resist erosion products also contributes to the reaction. Electrode material as

Table 2 Silicon Oxide Etch Gases

Etch Gas	Addant	Remarks
CF_4	O_2	non-selective to silicon
	H_2	selective
	CHF_3	selective
CHF_3		polymerizes except at low concentration
	O_2, CO_2	oxidant controls selectivity
C_2F_6		concentration controls polymerization/selectivity
C_3F_8		similar to C_2F_6
NF_3		non-selective etches silicon very fast
SF_6		similar to NF_3

Table 3

Factors that Tend to Increase Polymerization and Selectivity	Factors that Tend to Decrease Polymerization and Selectivity
Decreased temperature	Increased temperature
Hydrogen concentration	Oxygen concentration
Concentration of monomer (CHF_3)	Dilution with inert gas
Increased pressure	Lower pressure
Lower power	Higher power
Increased electrode coupling	Decoupling from electrodes

electrodes can also participate in adjusting the equilibrium fluorine content of the gas through catalytic surface recombination. Graphite or organic polymer coated electrodes are known to behave in this manner.

Selectivity is determined primarily by the polymerizing potential of the gas composition. If a sufficiently high concentration of hydrogen containing gas or other polymer forming material is present, film deposition (polymerization) will occur instead of etching. The composition where this happens is often referred to as the polymer point. The polymer point will vary with temperature (decreasing), position in the reactor (variable), and the nature of the surface. It is in fact the surface variation that gives rise to the selectivity. Any factor that has a tendency to increase polymerization rate will increase selectivity. Factors believed to control polymerization are listed in Table 3.

C. SILICON NITRIDE

Silicon Nitride is prepared in three different forms, high temperature CVD (often referred to as pyrolytic material), low temperature plasma enhanced CVD (PECVD), and physically deposited by sputtering, which may be reactive in nature, i.e., from silicon in a nitrogen discharge. Pyrolytic silicon nitride often approaches the stoichiometric compound Si_3N_4, while the other forms frequently vary from this composition. It is also possible for pyrolytic composition to vary, particularly if the deposition system has oxygen as an impurity. As the etching characteristics, particularly rate and selectivity, depend on its composition, the method of production needs to be considered carefully, particularly if the results appear to be outside the expected range. PECVD material, in particular, will vary widely depending

on the deposition conditions, containing considerable quantities of hydrogen as well as oxygen and carbon impurities.

Silicon nitride is used as a diffusion mask, gate material, etch mask interlevel insulator, and passivation overcoat. In all but the latter two applications it is usually present as a thin film, under 100 nm thick.

Silicon oxide is the most common underlayer and in state-of-the-art devices may be about 10 nm thick. In the other applications, underlayers may be any of the common materials used in circuit production. It is not common to use silicon nitride (particularly pyrolytic forms) directly on single crystal silicon.

1. Etching Requirements

Silicon nitride is considered to be an easy material to etch in most applications. In many cases films are quite thin so that etch rate is usually not a significant factor. Even thicker films do not usually present difficulties, because they are often prepared by the lower temperature methods, which have lower density and higher etch rates. Anisotropy is important for very fine geometries and/or thicker films. It may become particularly bothersome in cases where thin oxides are the underlayer. Even more difficult problems may arise if nitride is deposited directly on single crystal silicon and damage to underlying material must be kept to a minimum. Nitride films that have been used for some step of device manufacture are often removed prior to further processing without additional masking. This is referred to as a "blanket etch."

2. Etching Methods

Most forms of silicon nitride can be etched in either barrel or planar plasma or RIE systems. It is most cost-effective to use barrel etchers for non-critical nitride etching. If anisotropic nitride etching is needed, then a system that provides the requisite ion bombardment is usually necessary. If selectivity, particularly with respect to underlying silicon oxide is important, then a two-step process can be used. The first step makes use of high ion energy by using high power or reduced pressure. The second step consists of a lower power, or better yet, downstream process. If silicon nitride is a middle layer of a multilayer film consisting of various types of other material, it may be difficult to maintain anisotropy in the layer without compromising some other aspect of the process.

Table 4 Relative etch rates of silicon compounds of nitrogen and oxygen with Fluorine etchants.

Material	Normalized Etch Rate
Thermal SiO_2	1.0
CVD SiO_2	2.0
CVD SiO_2 Phosphorus doped	2.0–3.0
SiOxNy	2.0–3.0
Si_3N_4 (stoichiometric)	3.0
SixNy (x/y > 3/4)	3.0–10.0
Silicon—single crystal	10.0

3. Etching Chemistry Review

Fluorine etchants are the most common for etching silicon nitride. CF_4 and $CF_4 + O_2$ are the most generally employed etchants. Nitrides of nominal composition Si_3N_4 will usually etch at a rate between that of single crystal Si and SiO_2. Table 4 summarizes the relative rates of silicon oxides, nitrides and silicon as observed in barrel-type systems. NF_3 and SF_6, with and without O_2, etch nitrides significantly faster than CF_4. SiF_4 has been

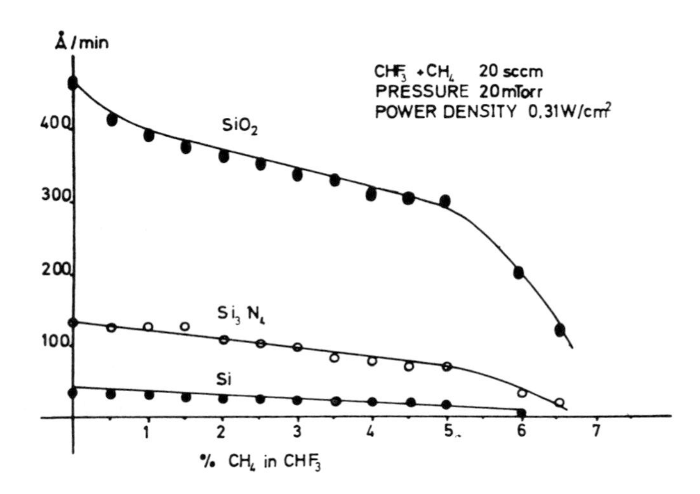

FIGURE 14. Selective etching of silicon nitride with CH_4/CHF_3. Selective etching is obtained at high CH_4 concentrations. As this is near the "polymer point" care must be taken to guard against contamination from particulates and excess polymerization. H. Norstrom *et al. Vacuum* **32** 737, 1982.

reported [23] to etch silicon nitride faster than both single crystal silicon and silicon oxide in a barrel etcher.

Selective etching of silicon nitride over silicon can be achieved with etchants similar to those used for selective etching of oxide to silicon. The selectivity will be less than that for oxide. Figure 14 shows data for the system CH_4 and CHF_3. Very high selectivity was achieved for high CH_4 concentrations near the polymer point. Other oxide etch systems may be expected to behave in a similar manner.

Selective etching of silicon nitride over silicon oxide presents a different set of problems. The starting place for this reaction is the selective etching of silicon over oxide. As discussed next, selective (and anisotropic) silicon etching is principally carried out with etch gases containing halogens other than fluorine, both in combination with fluorine and alone. Sanders [24], has shown that small additions of CF_3Br to CF_4—O_2 mixtures suppress the etch rate of SiO_2 substantially more than that of the nitride increasing the selectivity of nitride over oxide by a factor of 10 or more.

D. SINGLE CRYSTAL SILICON

Numerous applications exist for etching features into single crystal silicon. Isotropic etching is used to make islands. Trenches of various depths are used in the manufacture of MOS devices. Shallow trenches (0.5 μ to 1.5 μ) are used to make flat structures (planarization) and to aid in reducing the effects of lateral diffusion during processing. Deeper trenches (1.5 μ to 10 μ) are used to create structures that become capacitors and isolation regions in integrated circuits. Special structures such as microchannels and gratings have also been made with dry etching. In some of these, consideration can be given to an alternate etching method using orientation dependent wet etchants like KOH or ethylene diamine/pyrocatechol [25]. In cases where fairly shallow etching is needed, or etching is performed totally chemically as in a downstream system, photoresist is an adequate mask. For deeper etching, such as for isolation trenches, a silicon oxide mask may be necessary. However, the etch characteristics may depend dramatically on the presence or absence of the photoresist polymer.

1. Chlorine and Fluorine Etchants

Except at very low pressure, etchants containing only fluorine will etch silicon with at least some isotropic component. Selective oxide etchants usually result in anisotropic silicon profiles as well, but at reduced rates

[26]. Addition of even small concentrations of other halogens significantly reduce the isotropic component of most fluorine-based etchants and increase the selectivity to silicon oxide. For pure chlorine systems such as Cl_2 or HCl, the etch rate of SiO_2 can be so low that the presence of thin native oxides completely prevent the silicon etch from starting. This is particularly evident in cases where the ion bombardment is very low. Besides gas composition, reactor materials and the presence of organic compounds can influence anisotropy and selectivity of silicon etching. Carbon containing compounds increase the etch rate of SiO_2, while at the same time increasing the anisotropy presumably reducing the fluorine to carbon ratio [30]. The carbon can come from the etch gas, photoresist or reactor components such as carbon electrodes [27]. It has been reported [28] that single crystal silicon exhibits orientation dependent etch rates with CCl_4—O_2 in a shielded barrel reactor. The series CCl_xF_{4-x} produces profiles of increasing anisotropy in silicon as x increases from 0 to 4 [29]. $SiCl_4$ when used together with Cl_2, [30] or O_2 [31] have been reported to produce good results in etching single crystal silicon. Chlorine etchants and combinations like $CHCl_3$, which provide sidewall passivation, may also be used to advantage. It is likely that temperature of the substrate is an important parameter in determining the profile resulting from chlorine etching of silicon [32, 33].

Silicon etch rates with chlorine are very sensitive to ion energy, which is reflected by sensitivity to power and frequency [34]. Applying a dc bias to a wafer [35] produces etch rates in excess of 10 microns/minute with a trend for small areas to etch faster than large ones [36]. Usually the higher rates are accompanied by lower selectivities.

Etching deep trenches into silicon is one of the more challenging processes needed for state-of-the-art devices. Required features may be 1 μ wide and 6–15 μ deep. Special applications such as micro-machined silicon parts can be several times deeper. The desired attributes include straight or positively tapered walls with rounded bottom corners and smooth floors. Silicon oxide is the material of choice for masking. Even so, care needs to be taken to assure survival of the mask for very deep structures. For example, a 10 μ trench will need 2 μ of SiO_2 if the selectivity is 5 : 1. Although very high selectivity of silicon to SiO_2 is possible with chlorine based etchants, this is compromised by the need for high ion energy to achieve anisotropy. For this reason, it may be desirable to include some additional sidewall passivation component to the etch gas. Temperature has already been mentioned as important to the control of line shape. Lower frequencies may be advantageous in increasing rate and anisotropy for deep trenches. Figure 15 illustrates several mechanisms that can contribute to poor line shape control.

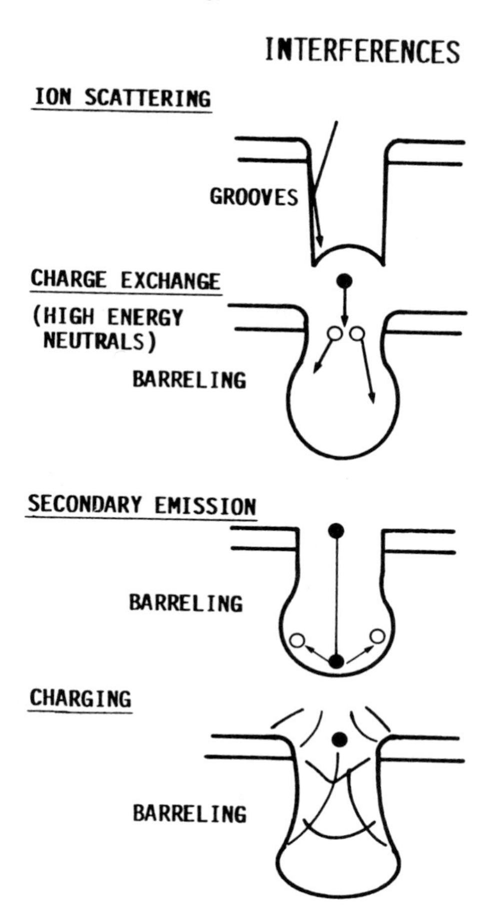

FIGURE 15. Interference effects in deep trench etching. Grooving is more likely at low pressures and is common in sputter etching. In addition to these postulated mechanisms, material variation is a common source of poor profiles.

E. POLYSILICON

Polycrystalline silicon (Polysilicon) is used primarily for gate electrode material and interconnects. It is also used as an oxidizable source for insulation or, when doped with oxygen, as a passivation layer. It is most often applied to wafers by low pressure chemical vapor deposition (LPCVD). Other methods are sputtering, PECVD and by bombardment and damage of single crystal silicon. Doping with phosphorous, arsenic, boron or hydrogen in amorphous PECVD films is done during and after growth by diffusion or ion implantation. The resulting material can have many microstructural variations, profoundly influencing the etch characteristics.

1. Etching Requirements

Since polysilicon films are usually thin, etch rate is not often a problem. Dimensional control and maintaining the integrity of the underlying film are the most important factors. Final feature size is a combination of anisotropy, resist erosion and uniformity. SEM analysis is the most common form of linewidth verification, though linewidth measuring tools that use resistance measurement of test structures may also be needed for a careful evaluation.

Undercut requirements are often stated as "zero bias," meaning that no detectable undercut is observable in SEM photographs. Simultaneously, loss of underlying gate oxides must be kept to a minimum requiring selectivity of $\approx 30:1$ or greater depending on how much underlying oxide loss can be tolerated. Usually the acceptable amount of oxide loss of a 150-Angstrom thick gate is less than 50 Angstroms. The amount of underlayer loss is determined by the amount of necessary over etching, relative initial thickness of the layers, and the selectivity according to:

$$\%\ \text{Underlayer loss} = \frac{\text{Poly thickness}}{\text{Oxide thickness}} \times \frac{\%\ \text{Over-etch}}{\text{Selectivity}}.$$

For polysilicon 0.3 μ thick over 150 Å gate oxide requiring 50% over-etch, a selectivity (based on a single process step) of 30 : 1 is needed. The amount of over-etch necessary is determined by the uniformity of the process and the height and angle of the steps, if any, over which the films are deposited [37]. Over-etching is required whenever a material deposited over a step is anisotropically etched. Two factors can be identified that contribute to the need for over-etching are illustrated in Fig. 16. First, there is a greater thickness of the material at the edge by an amount equal to the step height. Failure to etch for a sufficiently long time to remove this extra height leaves a residue along the corner of the feature called a "stringer." The second factor is directly related to the nature of anisotropic etching, which results from the directionality of plasma particles. Depending on details of the original film deposition process, the material covering the step will display some curvature over the corner. As a result, there is a finite zone along the corner where the etch rate changes from a high value perpendicular to the wafer to a low value parallel to the wafer. The variation will depend on

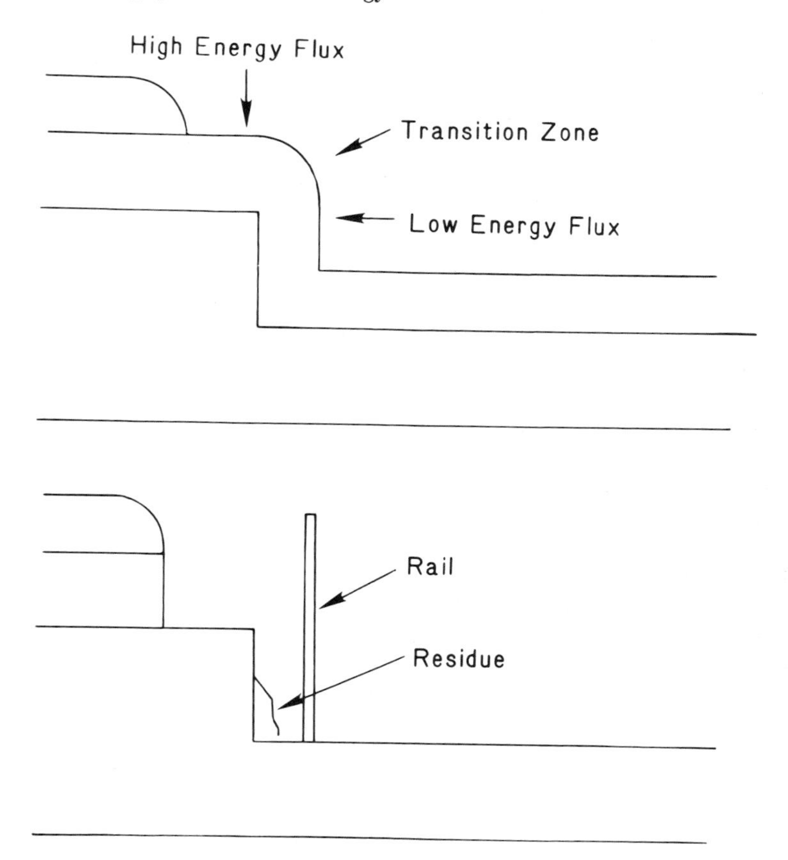

FIGURE 16. Two types of residues resulting from etching over vertical steps. Rails, or fences, are commonly encountered when attempting very anisotropic etching of material, which is deposited over vertical steps. Sidewall passivation from feed gas or from etch products also contributes to the problem. This problem is a good argument for planarization.

the ratio of anisotropic to isotropic component in the etch as well as the details of the film profile over the corner. The result is that material displaced from, and parallel to, the edge of the underlying vertical step will etch at a very reduced rate from the horizontal film surface, giving rise to a residue called a "fence" or "rail." Fences are different from stringers in that they are displaced from the edge of the feature by approximately the original film thickness. These residues are very difficult to eliminate and are best avoided by sloped features in combination with some isotropic component during etching.

Isotropic etching of polysilicon with fluorine etchants results by eliminating ion bombardment in barrel or downstream etchers or in various planar etchers at low power. With sufficient uniformity, isotropic etches may be adequate for devices whose minimum polysilicon dimension is $> 2.5\ \mu$. Isotropic processes can also be useful when silicon is deposited over topography, which includes reentrant profiles or "bat-caves." All other conditions require ion bombardment and some form of chlorinated etch. In order to minimize etching of underlying thin oxide gates, a multistep process is often required. In the first step, process conditions are selected to enhance anisotropy. These include increased ion energy or power, low pressure and the presence of passivation components in the etchant. The second step concentrates on lowering the oxide etch rate by lowering power, increasing pressure and removing oxide reducing components of the gas mixture.

Polysilicon etch properties depend on doping type and level, the amount of open area (pattern sensitivity) and the concentration of chlorine in the etch gas. With little or no chlorine, doped and undoped material etch very similarly [38]. With increasing chlorine the variation becomes significant. Heavily phosphorous doped polysilicon etches up to 15 times faster than undoped with pure chlorine. Similarly, the variation with resist coverage increases with increasing chlorine. The etch rate variation of various doped polysilicons in a mixture of $CF_4 + CFCl_3 + O_2$ [39], have been reported. An approximately 25% increase in etch rate when the donor concentration increased from 10^{18} to 3×10^{20} was observed. For boron-doped polysilicon the etch rate decreased by about 15% when the active carrier concentration increased from 10^{19} to 2×10^{20}.

In addition to etch rate increases, anisotropy can also change with carrier type and concentration. An anisotropic process at high resistivity can undercut severely with increased phosphorous doping. Also, poor lineshape, i.e., reentrant profiles are also associated with higher chlorine content and high doping. A process that is satisfactory at a low n-type concentration may no longer suffice at higher levels. Experience indicates that pure chlorine-based processes are likely to lead to more profile variation with doping level than fluorine–chlorine mixtures.

There are numerous single, binary, and tertiary mixtures of gases in use for etching polysilicon. Many of them are mixtures of compounds of the form $C_wH_xF_yCl_z$ such as C_2F_5Cl. These are often mixed with SF_6. Other common mixtures include the addition of oxygen and/or hydrogen containing gases. Binary mixtures typified by $SF_6 + Cl_2$ present a particularly interesting behavior with respect to anisotropy. Although not usually capable of "zero bias" etching, these mixtures exhibit a small but finite undercut, often of the order of $0.1\ \mu$ or less that increases very slowly, if at all with over-etch time [40].

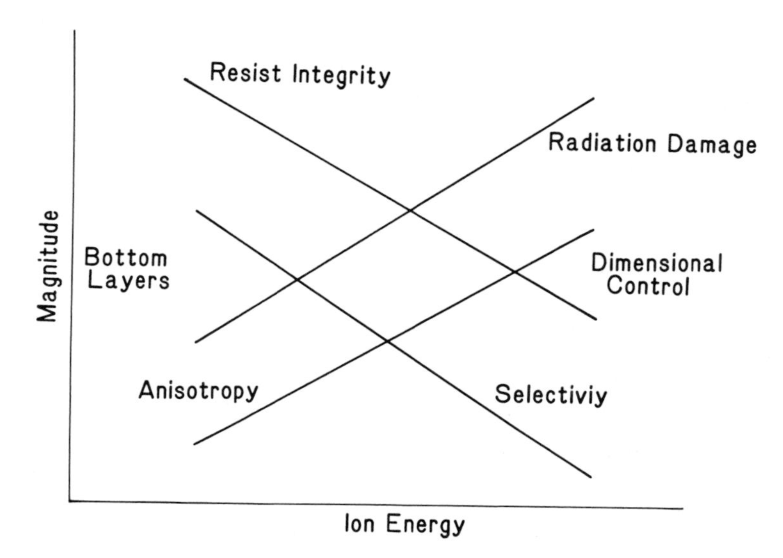

FIGURE 17. Opposing influences of ion energy on pattern transfer. Ion energy aids in anisotropic etching but can enhance resist erosion and damage and reduce selectivity. Vertical resist profiles help in controlling dimension but can cause other problems by acting as nucleating sites for sidewall passivants and etch products that are difficult to remove.

Most poly etch gases show a dramatic increase in selectivity to oxide with decreasing power. Below a certain power level, which will vary from system to system, there is virtually no oxide etching. In some cases, deposition is observed similar to polymerization in selective oxide etching.

While ion energy increases anisotropy, it has several negative effects important to polysilicon etching, illustrated in Fig. 17. Resist integrity, together with anisotropy, diminish with increasing ion energy.

	Chlorine	Passivant	Fluorine	Power	Pressure
RATE	↑	↓	↑	↑	↑
SELECTIVITY	↓	↑	↓	↑	○
ANISOTROPY	↑	↓	↑	↓	↑
UNIFORMITY	↑	↓	↑	↓	○

FIGURE 18. Etching trend chart example for polysilicon etching trend charts are helpful guides but are not as efficient as response charts, which may actually require less data.

The large number of variables involved in polysilicon etching can make selection of an appropriate process formidable. Some guidance in the form of a trend chart is presented in Fig. 18. Headings are generic, e.g., "Cl_2" represents any chlorine containing gas, "passivant," one that is believed to provide side wall protection, and "fluorine" any fluorinated gas.

F. SILICIDES AND REFRACTORY METALS

Refractory metals and silicides are important adjuncts and replacement materials for polysilicon for fine line geometry. The major advantage is lower resistivity than doped polysilicon. Four metals, which together with their silicides are currently used by various device manufacturers, are titanium, tantalum, tungsten and molybdenum.

With the exception of titanium, the pure refractory metals form fluorides with high vapor pressure at low temperature. Titanium chloride is a stable liquid with a boiling point of 136°C. Molybdenum and tungsten also form highly volatile oxychlorides. As with polysilicon, it is difficult to achieve highly anisotropic etching with pure fluorine etch gases, although high rate chemical etching is possible [41]. The metals and silicides are frequently used as part of a multilayer structure including doped polysilicon. Achieving good dimension control with a simple, single step process is unlikely. It is made even more difficult by the frequent requirement of stopping on very thin underlying oxide [42]. There are many variations in film composition and layer structure. Problems can arise from non-uniform films, interlevel doping, interdiffusion, variation in etch rate and anisotropy between layers and subsequent overhanging structures, redeposition from low volatility components, difficult to etch interfacial layers and poor selectivity to resist and underlying oxide.

Many of the chlorine-based etchants that are effective for anisotropic polysilicon etching can be modified for use with silicides. Oxygen can be added to enhance the production of oxychlorides. Oxygen can have a negative effect, however, on dimensional control by enhancing resist erosion.

G. ALUMINUM AND ALUMINUM ALLOYS

Aluminum is used almost universally as the metal of choice in integrated circuit production for connection between the device and external circuits. Aluminum is usually deposited in thickness ranging from 1 to 2 μ to provide contact to underlying single crystal silicon, gate material in MOS devices and to bonding pads to which wires or other connecting devices are attached. Devices may have one or more levels of metal separated by an insulator.

Although the majority of aluminum is deposited by evaporation or sputtering, there is considerable interest in aluminum deposition by chemical vapor deposition (CVD) as a means to improve conformality and step coverage. Evaporation can be by means of electron beam heating, from a heated filament, or from an externally heated crucible. The properties of the deposited material will depend on the method and deposition conditions such as temperature, rate, pressure, etc. Sputtering is the more widely used technique, however.

Aluminum can be doped with impurities. Silicon is sometimes added in concentrations ranging from 1% to 4%. Copper is added at 0.5% to 4% for electromigration resistance, though at 4% average doping, local concentrations in excess of 15% can occur as a result of segregation and deposition conditions. Often silicon and copper are added together to produce a common alloy consisting of 1% silicon and 4% copper. Difficulties associated with corrosion and with etching heavily copper-doped aluminum have prompted the search for other ways of preventing electromigration. Crystal morphology [43] and layered structures, incorporating titanium, have been proposed.

Doped or undoped SiO_2 is the most common underlayer when aluminum is used alone. When used as part of a multilayer metallization a thin layer of titanium, titanium/tungsten or other barrier metal are the most common underlying materials. In some instances a thin layer of these metals may also be used on the top of the aluminum.

Aluminum etching needs are similar to those of other materials with several additional caveats. Since films are relatively thick while horizontal dimensions are near those of the smallest feature size, anisotropy is important. The same considerations apply to stringer formation as for polysilicon. In some instances dry etching may be followed by a short wet etch to insure stringer removal.

Selectivity of 15 : 1 over SiO_2 is often acceptable, while 4 : 1 or better over the masking resist is desired. This oxide selectivity is routinely achieved while resist selectivity of 2 : 1 or less is common. In spite of high resist loss, removal of remaining resist can be troublesome. This is particularly true for negative-type photoresists, which contain a high proportion of solvents.

Aluminum etching is particularly susceptible to formation of residues, which are a source of particulate contamination. All surfaces in the immediate vicinity of the wafer need to be held at a sufficiently high temperature to prevent low vapor pressure material such as $AlCl_3$ etch product from collecting. A temperature between 70 and 100°C is adequate.

Safety, equipment reliability [44], and preventive maintenance are particularly important in aluminum etching. In the presence of moisture, chlorine residues are readily converted into highly corrosive hydrochloric acid, which rapidly attacks aluminum. This can also lower the reliability of the etched product. Techniques for addressing this requirement are discussed next.

Aluminum is etched in machine types where material is in contact with the plasma. Low and high pressure systems are used as are Reactive Beam systems [45]. Aluminum etching proceeds in the following steps:

1. *Preparation:* Special resist treatment may be required if the hardware/chemistry combination has poor resist selectivity. PRIST [46] or deep UV [47] and high temperature bakes may be required. Removal of water vapor from the resist and vacuum chamber are needed for reproducible results. Single wafer systems for aluminum etching usually have load-locks that keep the process chamber from exposure to atmospheric gases. Non-load-locked batch systems may need extended vacuum baking to reach acceptable limits of contamination. The process chemistry can be useful for reducing contamination as discussed in Step 4.

2. *Aluminum Oxide Removal:* Aluminum oxide is present in varying amounts on the surface of all aluminum films. Its properties depend on deposition method, environment prior to etching and pretreatment. Aluminum oxide etches significantly slower than aluminum and protects the metal from attack by chemical agents including chlorine, the usual aluminum etch agent. It can be removed by sputtering or ion bombardment and reaction with a reducing agent such as carbon, boron, or silicon atoms. The time required to etch through the oxide film is sometimes referred to as the initiation or induction time. Water vapor has a deleterious effect on the removal of the oxide layer. $SiCl_4$ or BCl_3 etchants aid in removal of residual water vapor markedly reducing the induction time [48].

3. *Etching the Metal:* Aluminum etching requires a chlorine-containing etch gas. Other gases often include some type of passivant and diluent used to enhance resist survival. The metal etching step can be fast. Rates in excess of 1 μ/minute are not unusual. Anisotropy requires careful balance of ion bombardment and chemistry. Increasing the concentration of chlorine increases rate but usually causes significant isotropic etching.

4. *Residue Removal:* In the case of aluminum containing impurities, it is sometime necessary to include a special step to remove residual material resulting from insufficient etch rate of one of the components. If the alloy

contains silicon, for example, a short CF_4 etch may be needed to remove residual silicon nodules. It is not clear why this happens. The relative etch rate of each of the constituents taken separately indicate that, at the levels of $\approx 1\%$, there should be sufficient etch rate to remove the silicon. Nonetheless, frequently material is left on the surface. Segregation at grain boundaries may account for some increased concentrations leading to the effect.

5. *Passivation:* This step is necessary to remove residual chlorine. Chlorine absorbs in resists and adsorbs on surfaces. Removal of the resist either in-plasma or down-stream of a standard oxygen discharge is usually effective. If it is desired to keep resist on the wafer for inspection purposes, a pure CF_4 discharge can be used instead. This process is believed to replace the chlorine with fluorine, which will not attack the metal. Another method of displacing chlorine is heating in flowing purge gas like nitrogen.

6. *Resist Removal:* Dry resist removal is most common and is most often a result of Step 5. Sometimes, however, additional cleanup processing may be needed to remove all vestiges of resist. Negative resist, for example, may become so tenacious that ordinary methods are unsuccessful. In this case it is best to consider some changes in the process that prevent the resist hardening rather than attempting to find a successful removal procedure.

Aluminum etching relies on chlorine to produce volatile $AlCl_3$ and ultimately $(AlCl_3)_2$. The overall reaction is

$$2Al + 3Cl_2 \longrightarrow Al_2Cl_6.$$

This reaction proceeds with pure Cl_2 even in the absence of a discharge once the protecting oxide layer is removed. The chemical reaction does not require ion bombardment and hence is inherently isotropic. Anisotropy is produced by limiting the amount of chlorine, providing high ion energy and most importantly by the presence of a sidewall protection mechanism. Material that protects or passivates the sidewall and prevents attack by chlorine may be derived from the input etch gas or as an etch reaction product probably involving the resist [49]. Chlorine may be derived from a gas containing the other component needed for reduction of the oxide and/or gettering of residual water vapor. Some common etch gases of this type are:

$$BCl_3, CCl_4, SiCl_4, HCl, PCl_3.$$

Also, $AlCl_3$ can etch aluminum. A possible reaction path that proceeds with the participation of $SiCl_4$ is,

$$2Al + AlCl_3 \longrightarrow 3AlCl$$

$$AlCl + SiCl_4 \longrightarrow AlCl_3 + SiCl_2.$$

Aluminum containing more than trace amounts of copper is one of the more difficult materials to etch with plasmas. The main reason is that copper does not form halogen compounds with high vapor pressure at operating temperatures. Copper and aluminum form several intermetallic compounds, the most important being $CuAl_2$, which can precipitate. A nominally 4% copper-containing metal can contain regions of high concentration that make etching very difficult. Highly anisotropic etches are more susceptible than others.

Methods of increasing the copper removal rate include increasing temperature and increasing ion energy, since sputtering is important in removing copper. Some of these may also be removed by a short dip in weak acid solution after etching.

Whereas many chlorine etchants may be used for pure aluminum, those containing carbon appear to produce more residues than carbon free etchants. $SiCl_4$ and BCl_3 are better for this purpose than CCl_4 and related compounds.

Copper aluminum alloys are even more prone to corrosion after dry etching than pure metal. In part, this is due to the large electronegative potential difference between the two metals; which readily leads to galvanic corrosion. Some form of wet cleanup after etching Cu/Al is highly recommended.

VI. Other Materials

A. III–V'S

III–V etching, (most commonly GaAs) is basically chlorine etching. Very high etch rates are achievable. With the exception of the base material, many of the materials used in GaAs device construction are similar to those used for silicon. Deposited oxides and nitrides can be etched with the same techniques used for silicon devices. Since GaAs forms no volatile fluorides, it presents an excellent etch stop for etching with CF_4 or other fluorine gases.

A process unique to GaAs microwave devices is the etching of connecting holes completely through the substrate. Various chlorine-containing etchants such as $SiCl_4$ and Cl_2 can be used to etch 100 μ or larger holes through the substrate [50].

B. METALS

Along with aluminum and the transition metals discussed in connection with silicides, a number of other metals of interest are amenable to plasma etching. Production of a volatile metal compound by reaction with a discharge species is the key issue. Two frequently encountered candidate materials used in IC manufacture are chromium and gold. Both of these are attacked by oxygen or oxygen atoms, particularly at elevated temperature. When these materials are present, care needs to be taken when using oxygen discharges for cleaning organics. Various additions, such as hydrogen to the gas can surpress the formation of volatile oxides without substantially reducing the cleaning capability.

Chromium is a particularly interesting example of a material where the pure chloride is not volatile but which possesses a highly volatile oxychloride (M.P. $-96.5°$, B.P. 117°C). Thus the etch rate of chromium increases rapidly with the addition of oxygen to chlorine or chlorine containing gas such as CCl_4 or $CHCl_3$ [51]. Mixtures near 1 : 1 are typical. Of course, the addition of oxygen reduces the survival of photoresist. Care must be taken to limit the ion energy incident on the surface in order to keep the resist erosion manageable.

Gold requires chlorine etchants and special procedures to produce ion assisted chemical etching. It has been reported [52] that etching only procedes if the interior surface (opposite electrode) is coated with gold. Temperature and ion energy are important factors in obtaining a reproducible process.

VII. Discovering and Characterizing Processes

Because process optimization may involve five or more independent variables, some systematic procedure needs to be used for development. Experimental design procedures and response surface analysis [1] are valuable tools to explore unknown regions of parameter space, to obtain initial information on important variables and to refine the information so that optimum conditions can be obtained. Factorial designs and central composite models [53] are the most common design forms and are useful when optimization is limited to three variables. Gas composition, pressure and power are three commonly used factors. More sophisticated methods,

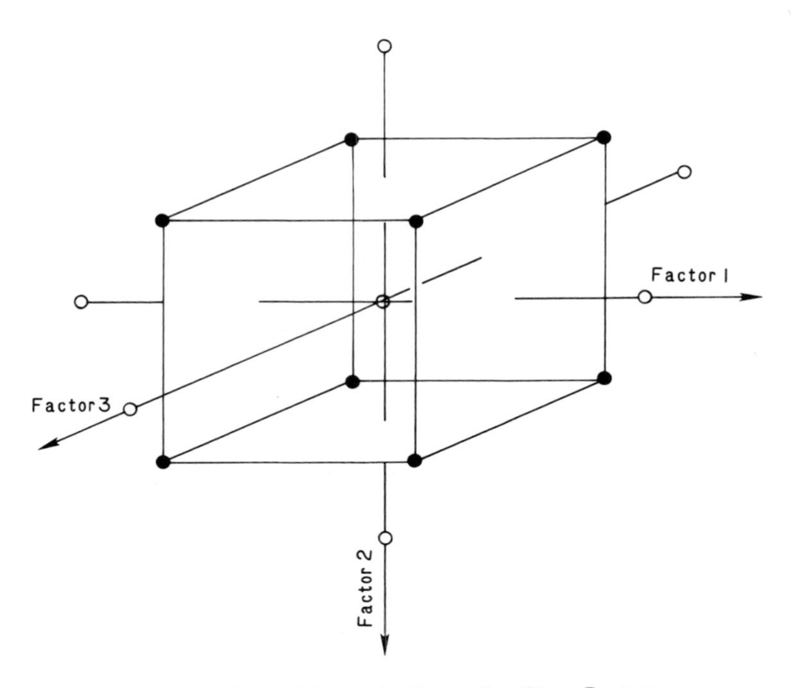

FIGURE 19. Central composite model in three variables. One model of data selection for response surface analysis. Analysis of this type of data is readily accomplished with a personal computer.

orthogonal designs, for example [54], are readily adapted to cases where more variables need to be considered. Figure 19 displays data points used for a central composite model in three variables. Data so obtained is sufficient to describe the result (say etch rate) over the entire space represented according to a complete second degree equation including interactions of the form,

$$\text{Response} = A_0 + A_1X + A_2Y + A_3Z + A_4XY + A_5XZ + A_6YZ + A_7X^2 + A_8Y^2 + A_9Z^2.$$

A complete array, including replication to determine experimental error would require 17 experiments. A minimum of 11 experiments are sufficient to solve for the coefficients in the equation.

Some of the common effects and the variables that are useful in controlling them are listed in Table 5. These relations are useful in setting up an appropriate experimental array.

Table 5 Factors and Common Dependent Functions in Plasma Etching

Dependent Variable	Independent Variable
Rate Selectivity Anisotropy	Composition/Dilution
Rate Uniformity	Flow/Dilution
Rate Uniformity	Loading
Rate Anisotropy Selectivity Uniformity	Pressure

References

[1] George E. P. Box, William G. Hunter, and J. Stuart Hunter, "Statistics for Experimenters," John Wiley & Sons, Inc. New York (1978).
[2] S. N. Deming and S. L. Morgan, *Analytical Chemistry* **45**, 278A, March 1973.
[3] See Chapter 7 by G. K. Herb in this volume.
[4] See Chapters 3 and 4 in this volume.
[5] R. B. McDowell, *Solid State Technol.* Feb. 1969, p. 23.
[6] R. G. Bosisio, C. F. Weissfloch, and M. R. Wertheimer, "The Large Volume Microwave Plasma Generator," *Journal of Microwave Power* (1972).
[7] R. Bruce and A. Reinberg, *J. Electrochem. Soc.* **129**, 393 (1982).
[8] H. S. Butler and G. S. Kino, *Physics of Fluids* **6**, 1346 (1963).
[9] H. R. Koenig and L. I. Maissel, *IBM J. Res. Dev.* **14**, 168 (1970).
[10] J. A. Thornton, Chapter 5 in "Deposition Technologies for Films and Coatings," R. Bunshah, editor, Noyes Publications, Park Ridge, New Jersey, 1982.
[11] Y. Horiike, H. Okano, T. Yamazaki, and H. Horie, *Jpn. J. Appl. Phys.* **20**, L817 (1981).
[12] K. Suzuki, S. Okudaira, N. Sakudo, and I. Kanomata, *Jpn. J. Appl. Phys.* **16**, 1979 (1977).
[13] H. Markstein, *Electronic Packaging and Production*, pp. 66–68 (1980).
[14] F. Robb, B. Svechovsky, J. Helbert, and N. Saha, in *Extended Abstracts of the Fall Meeting*, October, San Diego, Calif. Abstract #287, Electrochemical Society, Princeton, New Jersey (1986).
[15] R. C. Alkire and D. J. Economou, *J. Electrochem. Soc.* **132**, 648 (1985).
[16] S. Dushman, J. Lafferty, editor, "Scientific Foundations of Vacuum Technique," 2nd edition, John Wiley & Sons, Inc., New York, (1962).

[17] An excellent compilation of etch gas compositions is given in "The Design of Plasma Etchants," D. L. Flamm and V. M. Donnelly, *Plasma Chemistry and Plasma Processing* **1**, 4, p. 317 (1981).

[18] T. Watanabe, Y. Yoshida, and M. Shibagaki, in *Extended Abstracts of the Fall Meeting*, October, Washington D.C., Abstract #204, Electrochemical Society, Princeton, New Jersey (1983).

[19] Specifications on uniformity vary from user to user and vendor to vendor. They are often stated as $\pm N\%$ $K\sigma/\langle R\rangle$ where K is 1, 2 or 3, σ is the standard deviation of measurements taken over a significant number of points and $\langle R\rangle$ is the average rate.

[20] A. C. Adams and C. D. Capio, *J. Electrochem. Soc.* **128**, 423 (1981).

[21] C. J. Mogab, J. *Electrochem. Soc.* **124**, 1262 (1977).

[22] Although an etch process may appear to not etch some materials, one cannot conclude that there is no effect at all. In particular, non-etching may be symptomatic of the fact that deposition is occurring. In the case of aluminum bond pads, fluorcarbon deposits may interfere with successful wire bonding or lead to premature failure of bonds.

[23] H. Boyd and M. S. Tang, *Solid State Technology* **22** (4) 133, April (1979).

[24] F. Sanders and J. Dielman, H. Peters and H. Sanders *J. Electrochem. Soc.* **129**, 2559 (1982).

[25] D. L. Kendall, "Vertical Etching of Silicon at Very High Aspect Ratios," in *Ann. Rev. Maer. Sci.* **9**, 373–403, Annual Reviews Inc., Palo Alto, California (1979).

[26] N. C. Us, R. W. Sadowski, and J. W. Coburn, *Plasma Chemistry and Plasma Processing*, 1, **6**, 1 (1986).

[27] W. W. Yao and R. H. Bruce, in *Extended Abstracts of the Fall Meeting*, October, Denver, CO. Abstract #268. Electrochemical Society, Princeton, New Jersey (1981).

[28] H. Kinoshit and K. Jinno, *Jpn. J. Appl. Phys.* **16**, 2 (1977).

[29] Hirata *et al. IEEE Trans on Electron Devices* **ED-28**, 11, 1323.

[30] K. O. Park and F. C. Rock, in *Extended Abstracts of the Spring Meeting*, May, San Francisco, CA. Abstract #155. Electrochemical Society, Princeton, New Jersey (1983).

[31] S. M. Cabral, D. D. Rathman, and N. P. Economou, in *Extended Abstracts of the Fall Meeting*, San Francisco, CA. Abstract #153. Electrochemical Society, Princeton, New Jersey (1983).

[32] G. C. Schwartz and P. M. Schaible *J. Vac. Science. Technol.* **16**, 2 (1979).

[33] R. Carlile, O. Palusinski, M. S. Madi, V. Liang, G. Djaja, and B. Chapman, in *Extended Abstracts of the Fall Meeting*, October, San Diego, CA. Abstract #268. Electrochemical Society, Princeton, New Jersey (1981).

[34] R. Bruce, D. L. Flamm, V. M. Donnelly, and B. S. Duncan, in *Extended Abstracts of the Electrochemical Society Fall Meeting*, October, St. Louis, MO. Abstract #243, Electrochemical Society, Princeton, New Jersey (1981).

[35] V. Ho and T. Sugano, *Thin Solid Films*, **95**, 315 (1982).

[36] R. Bruce and A. Reinberg *J. Electrochem. Soc.* **129**, 393 (1982).

[37] K. Wang, T. Holloway, R. Pinizzotto, Z. Sobczak, W. Hunter, and A. Tasch Jr. *IEEE Trans. on Electron Devices* **ED-29**, 547 (1982).

[38] C. J. Mogab and H. J. Levinstein *J. Vac. Sci Technol.* **17**, 72 (1980).

[39] L. Baldi and D. Beardo *J. Appl. Phys.* **57**, 2223 (1986).

[40] M. Mieth and A. Barker *J. Vac. Sci. Technol*, A, **1**. 629 (1983).

[41] T. P. Chow and A. J. Steckl, in *Technical Digest*, 1980 Intl. Elect. Dev. Meeting. p. 149, Washington D.C., The Inst. of Electrical & Electronics Eng. Inc. Piscataway, N.J. (1980).

[42] R. J. Saia and B. Gorowitz, *Electrochem. Soc. Extended Abstracts*, **86-2**, 475.

[43] S. Vaidya and A. K. Sinha, *Thin Solid Films*, **75**, 253 (1981) reference on bamboo aluminum.

[44] See Chapter 7 by G. K. Herb, this volume.
[45] See Chapter 6 by J. Harper, this volume.
[46] W. H-L. Ma, in *Technical Digest*, 1980 International Electron Devices Meeting, Washington D.C., The Inst. of Electrical & Electronics Eng. Inc. Piscataway, N.J. (1980).
[47] R. Allen, M. Foster and Y. Yen, *J. Electrochem. Soc.* **129**, 1379 (1982).
[48] K. Tokunaga, F. C. Redeker, D. A. Danner, and D. W. Hess, *J. Electrochem. Soc.* **128**, 851 (1981).
[49] M. Oda and K. Hirata, *Jpn. J. Appl. Phys.* **19**, L405 (1980).
[50] C. B. Cooper III, M. E. Day, C. Yuen, and M. Salimian, *Electrochem. Soc. Extended Abstracts*, **86-2**, Abstr. #288.
[51] B. J. Curtis, H. R. Brunner, and M. Ebnoether, *J. Electrochem. Soc.* **130**, 2242 (1983).
[52] C. B. Zarowin in *Extended Abstracts of the Electrochemical Society Fall Meeting*, October, Pittsburgh, PA. Abstract #195. Electrochemical Society, Princeton, New Jersey (1978).
[53] G. R. Bryce and D. R. Collette, *Semiconductor International* **71**, (1984).
[54] D. W. Jillie, P. Freiberger, T. Blaisdell, and J. Mutani, in *Extended Abstracts of the Electrochemical Society Fall Meeting*, October, San Diego, CA. Abstract #289. Electrochemical Society, Princeton, New Jersey (1981).

6 Ion Beam Etching

James M. E. Harper
IBM Thomas J. Watson Research Center
Yorktown Heights, New York

I. Introduction

Ion beam etching is a versatile process for pattern delineation and materials modification. The low pressure, line-of-sight nature of beam techniques provides a flexibility of directional bombardment not available in other plasma processes. Also, the ability to separately control ion energy and flux gives quantitative yield data, which are applicable to other ion-surface applications. Reviews by C. M. Melliar-Smith and C. J. Mogab [1], R. E. Lee [2] and by B. A. Heath and T. M. Mayer [3] describe the principles of ion milling (inert gas) and reactive ion beam etching, and give numerous examples. This chapter is not intended to be a complete review of the topic

Plasma Etching:
An Introduction

ISBN 0-12-469370-9

of ion beam etching. Here, we focus on the characteristics of ion beam source operation and beam measurement, which must be understood to make full use of the quantitative control offered by beam techniques. We include a discussion of the limits of fine line resolution. Some aspects of operation with reactive gases are described, with examples of the ion-surface chemical interactions which provide increased selectivity and etch rate. Finally, ion beam etching in combination with surface layer growth or thin film deposition is briefly addressed, to point out the significant effects of ion bombardment in deposition processes.

II. Broad-Beam Ion Sources

A. KAUFMAN ION SOURCE DESIGN AND OPERATION

The multi-aperture electron bombardment ion source was invented in 1961 by H. R. Kaufman as an ion thruster for space propulsion, and generates collimated, well-characterized ion beams from a wide choice of gas species [4]. A detailed description of the fundamentals of ion source operation is given by Kaufman [5], and a typical configuration is shown in Fig. 1. A broad ion beam system contains two plasma regions. Ions are generated in the discharge plasma by electron bombardment of the operating gas. They

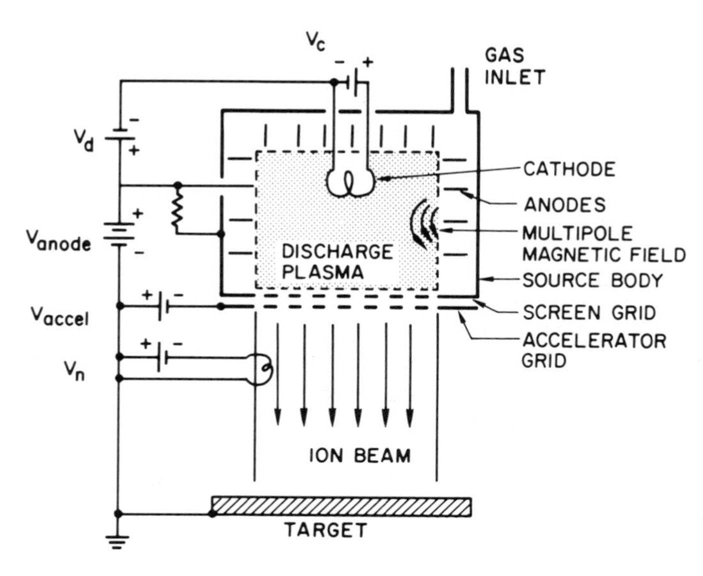

FIGURE 1. Broad-beam multiaperture ion beam source.

are then accelerated through the extraction grid region and travel through the beam plasma region.

Kaufman sources produce several mA/cm^2 ion flux at 500–1000 eV, and are available up to about 30 cm diameter, generating total beam currents exceeding one ampere. At lower energy, the flux is limited to lower values, as discussed later. Measurement and control of the ion energy and flux from a Kaufman source are straightforward, making this type of broad-beam ion source very suitable for quantitative beam processing over large areas. The development of these sources is reviewed by Kaufman *et al.* [6] and their application to sputtering processes is reviewed by Harper *et al.* [7]. In addition to etching by sputtering and reactive beam processes, Kaufman sources are used to synthesize thin surface compound layers, and to bombard thin films during deposition to modify their properties. For example, ion bombardment of a multicomponent film during deposition will change the composition according to the ratios of sputtering yields of the component materials. Microstructure, stress, density and other film properties also show marked response to ion bombardment during deposition, and ion beam techniques give quantitative information on the basic processes occurring during growth [7].

Together with microwave ion sources, described later, Kaufman sources have the following attributes, which enable highly quantitative etching

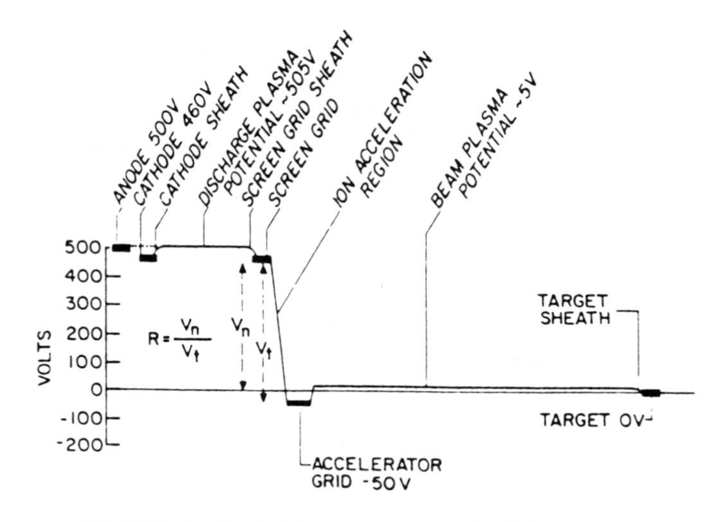

FIGURE 2. Typical electrical potentials for 500 eV ion beam.

processes to be performed:

1. independent control of ion energy and current density,
2. isolation of samples from ion generation process,
3. isolation of samples from high electric fields,
3. control of angle of ion impact,
4. narrow ion energy spread,
5. low background pressure for line-of-sight operation.

To describe the operation of an ion source, we use the example of a multiaperture electron bombardment ion source of several cm beam diameter, which generates an argon ion beam of typically 0.5–1.0 mA/cm^2 at 500 eV. Typical electrical potentials from the source to the target are given in Fig. 2.

1. Discharge Plasma Region

Ionization. Ions are generated in the discharge chamber by electron bombardment of neutral gas atoms. Since the cross section for electron bombardment ionization peaks at an electron energy of 70 eV for argon, it is not necessary to use highly energetic electrons for ionization. Electrons are emitted by a hot filament (typically Ta or W) and accelerated by a potential difference V_d between the cathode and anode. This discharge voltage must be higher than the gas ionization potential (15.8 eV for argon), and is typically operated at several times this value, about 40–50 V, in order to establish a glow discharge.

Magnetic Field. To obtain a large number of ions per discharge electron, a magnetic field is applied such that electrons cannot pass directly from the cathode to the anode, but follow helical paths between collisions, greatly increasing their path length and ionization efficiency. The cyclotron radius of a 100 eV electron in a field of 0.01 Tesla (100 Gauss) is 3.2 mm, so relatively low field strengths are adequate. By using multipole permanent magnet designs, non-cylindrical configurations are made, such as long rectangular beam shapes [6]. A discussion of magnetic field distributions for optimum ionization is given by Robinson [8].

Discharge Characteristics. To maintain a glow discharge in the ion source, three conditions must be present:

1. sufficient gas pressure in the discharge chamber,
2. a source of electrons (hot filament),
3. accelerating voltage for electrons (discharge voltage V_d).

To ignite the discharge, the gas pressure is increased to about 1 mTorr in the discharge chamber, the filament is heated and the discharge voltage applied. The pressure in the sample region is substantially lower, typically 10^{-5} to 10^{-4} Torr, depending on the conductance of the grids and the pumping speed of the vacuum pump. The discharge plasma establishes itself between the cathode, anode, chamber walls and screen grid (Fig. 1). The potential of the discharge plasma is determined by the loss rate of electrons and ions from the plasma, and is typically a few volts positive of the anode. The screen grid and chamber walls are at cathode potential, creating at each of these surfaces a cathode dark space or sheath. Electrons from the cathode are accelerated across the cathode dark space to an energy equivalent to the discharge voltage. These primary electrons bombard gas atoms and create both positive ions and low energy secondary electrons, which form a Maxwellian distribution with typical energies of several electron volts (eV). Ions are also accelerated across the cathode sheath by the discharge voltage. To minimize sputtering inside the ion source, the discharge voltage is kept no higher than necessary to maintain a stable discharge. This also minimizes multiple ionization (5).

Typical discharge parameters are:

argon pressure = 1 mTorr,
ion density = electron density = $10^{10}/\text{cm}^3$,
ion energy in plasma = 0.05 eV,
primary electron energy = 40 eV,
Maxwellian electron energy = 3 eV,
plasma potential = about 5 V above anode potential,
discharge voltage = 40 V,
discharge current = 0.2–2.0 A,
mean free path for electron impact ionization = about 0.8 cm,
mean free path for Ar-Ar collisions = 5 cm,
Debye length = about 0.1 mm.

The Debye length is the shielding distance in the discharge plasma, determined by the electron density. The main control on ion density in the source (and thereby in the beam) is the cathode heating current.

2. *Ion Extraction Region*

To extract an ion beam from the discharge plasma (Fig. 2), we raise the anode to a positive voltage above ground. The cathode, chamber walls and screen grid are also increased to a positive voltage, which is lower than the anode by the magnitude of the discharge voltage. Raising the anode potential increases the discharge plasma potential to nearly the same value. Thus any ion leaving the discharge plasma and striking the grounded target surface arrives with an energy determined by the anode potential. The anode potential is therefore the primary control over ion beam energy. The ion energy distribution is narrowly peaked around the anode potential, with a spread of 5–10 eV.

Referring to Fig. 2, to generate an ion beam of 500 eV, a voltage of 500 V is applied to the anode. The plasma potential is then about 505 V. The accelerator grid is held at a negative potential of about -50 V relative to ground. Ions are first accelerated by the 555 V potential difference between the plasma and the accelerator grid, then decelerate to about 500 eV to enter the beam plasma, and eventually to strike the target, which is held at ground potential.

Obtaining Maximum Ion Flux. We are often interested in maximizing the ion current density at a given ion energy, in order to minimize the etching time. To understand the constraints on current density, we first analyze the source of ions, which is the discharge plasma sheath at the screen grid of a dual grid source (Fig. 3a). The single grid configuration (Fig. 3b) is described later. Since the discharge plasma has a screening distance (Debye length) of less than 1 mm, ions moving in the plasma only feel the effect of external electric fields when they reach the edge of the plasma and enter the sheath region.

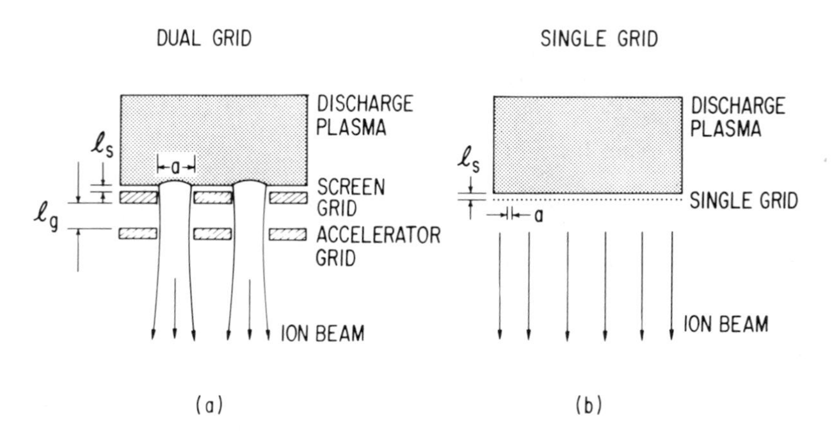

FIGURE 3. Extraction grid region for (a) dual grid and (b) single grid extraction system.

Ions that cross the sheath at the screen grid either strike the screen grid at low energy or are accelerated through the screen grid apertures by the electric field between the screen and accelerator grids. These two grids are typically spaced about 1–2 mm apart and have an equal number of aligned apertures of typically 2 mm diameter, with about 50% open area.

The maximum ion current density that can be extracted from the plasma is space charge limited, and given approximately by:

$$j = K(q/m)^{1/2} V_t^{3/2} / l_g^2,$$

where j is the maximum current density, V_t is the total potential difference between the grids (Fig. 2), K is a constant depending on grid geometry, q/m is the charge to mass ratio of the ion, and l_g is the separation of the grids, which determines the distance over which the ions are accelerated [4]. This equation is a statement of Child's law of space charge limited current flow and is the most important relationship in understanding the performance of the ion extraction region. The ratio $j/V_t^{3/2}$ is called the perveance. The maximum perveance is fixed for a given grid system and ion species, hence the maximum current density is proportional to $V_t^{3/2}$. An example of this relationship is shown in Fig. 4 for a source of 3.0 cm beam diameter. Values of current density lower than the space charge limit are easily achieved by operating at lower discharge current. Higher maximum perveance can be achieved by decreasing the grid separation. Spacing the grids too close, however, causes problems with mechanical stiffness, alignment, and electrical breakdown. An optimum is reached with grid spacings

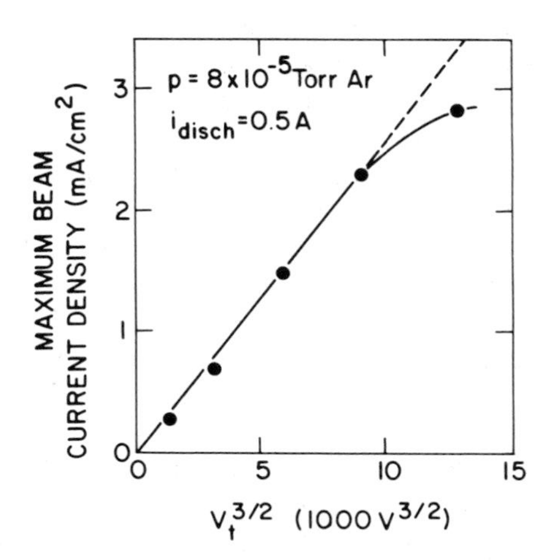

FIGURE 4. Maximum ion current density as a function of $V_t^{3/2}$ showing space charge limit on ion flux.

in the range 1–2 mm for ion energies up to 1 keV. Current densities over 0.1 mA per aperture at 1000 eV are achieved [6], giving current densities of several mA/cm^2 at the grids, while maintaining low beam divergence half-angles of 5–7 degrees. Molybdenum and graphite are the most commonly used grid materials because of their stiffness and thermal stability.

Beam Divergence. Beam divergence is determined by the ion trajectories in the extraction region. At very low perveance the plasma sheath at the screen grid apertures is concave, allowing ion trajectories to cross over, and leading to high divergence angles and impingement of the beam on the accelerator grid. This direct impingement shows up as current in the accelerator grid power supply circuit (Fig. 2). By increasing the perveance to the normal operating regime, the converging effect of the plasma sheath is decreased and is counteracted by the defocusing effect of the accelerate-decelerate sequence, giving a minimum beam divergence of 5–10° and almost no direct impingement on the accelerator grid. Under these conditions there is negligible erosion of the accelerator grid by direct impingement. Increasing the perveance further causes more divergent trajectories between the grids and again leads to high beam divergence and direct impingement. The range of normal operation is quite broad, however, and a wide range of perveance can be successfully achieved without significant impingement on the accelerator grid.

Effects of Accelerator Grid Potential. Increasing the accelerator grid potential V_{acc} (Figs. 1, 2) to more negative values enables slightly higher ion currents to be extracted due to increasing V_t. Since the accelerate-decelerate sequence of the extraction grid region acts as a defocusing lens, however, increasing V_{acc} causes increased divergence. To minimize divergence, V_{acc} must be kept as near ground potential as possible. The ratio $R = V_n/V_t$ (Fig. 2) characterizes the potential distribution. Small R gives increased beam divergence, and R close to unity gives minimum divergence. Thus beam divergence can be manipulated by varying V_{acc}. The accelerator grid serves a second purpose. It provides a negative potential barrier to prevent electrons in the beam plasma from backstreaming into the positive discharge plasma and registering as a false high value of beam current. Preventing backstreaming sets an upper limit on R of about 0.9 [5, 6].

Single Grid Extraction System. As shown in Fig. 4, the maximum current density obtainable from a dual grid extraction system is limited by space charge to being proportional to $V_t^{3/2}$. This seriously limits the current density available at low ion energies. For example, a grid system producing 1 mA/cm^2 at $V_t = 500$ V gives only 0.03 mA/cm^2 at $V_t = 50$ V. The single grid extraction system (Fig. 3b) (6) circumvents the space charge limit of a dual grid system and produces current densities of about 1.0 mA/cm^2 at ion energies as low as 20 eV. Ion energy spread is similar to that observed with the dual grid system. The improvement in low energy current density is achieved by using the plasma sheath as the accelerating distance for the ions, instead of the separation distance of a two-grid system. A drawback is the direct impingement of ions on the grid, which limits operation to below about 200 eV.

3. Beam Plasma Region

Beam Plasma Characteristics. In the beam region the ion current density is typically high enough that electrostatic repulsion between ions alone (space charge) would sharply fan out the beam once it left the extraction region. This space charge repulsion is prevented by the presence of an equal number of electrons and positive ions in the beam plasma. The electrons screen the electrostatic repulsion between ions and allow much higher current densities to propagate than are possible with a pure positive ion beam. The presence of a high density of electrons classifies the beam region as a plasma and distinguishes it from the type of beam typically used in sputter etching for analytical purposes, which is a positive ion beam with few electrons. Electrons in the beam plasma have low energy (several eV) and come from either an external neutralizer (Fig. 1) or are secondary electrons produced by ion impact on the target. These electrons do not directly recombine with the ions in the beam, however, because the cross section for this process is very small. The beam plasma potential is a few volts positive with respect to ground (Fig. 2), forming a potential well for electrons.

Neutralization. If the ion beam strikes a conducting grounded surface, enough secondary electrons are generated to balance the space charge of the beam, and external neutralization is not necessary. If the ion beam strikes an insulating surface, however, with no external source of electrons, the insulator surface charges positively almost to the equivalent of the beam energy. This charging may cause sparking, voltage breakdown, or deflection of the ion beam. The beam may be deflected until it strikes a grounded surface, which provides a large enough flux of secondary electrons to balance the beam current flow. The beam then no longer reaches the intended target, and sputter-contamination and non-reproducible bombardment profiles will result. The remedy is to supply electrons from an external neutralizer, either a filament immersed in the beam, or a small secondary discharge coupled to the main beam by a plasma bridge [6]. For most etching applications it is advisable to operate a neutralizer to avoid surface charging problems. The exact degree of electron emission by the neutralizer is not critical, however, since slightly positive target voltages will attract a large flux of electrons from the low energy tail of the Maxwellian energy distribution of electrons in the beam.

Measurement of Beam Properties. The ion current density is most easily measured by an open probe of the type shown in Fig. 5. The probe surface is biased to about −20 V to repel the low energy electrons in the beam, and allow only positive ions to be measured. Secondary electrons leaving the probe surface add to the total current density reading, however, causing an error typically around several percent for stainless steel or refractory metal probes with ions of several hundred eV. If necessary, this small error can be removed using a screened probe.

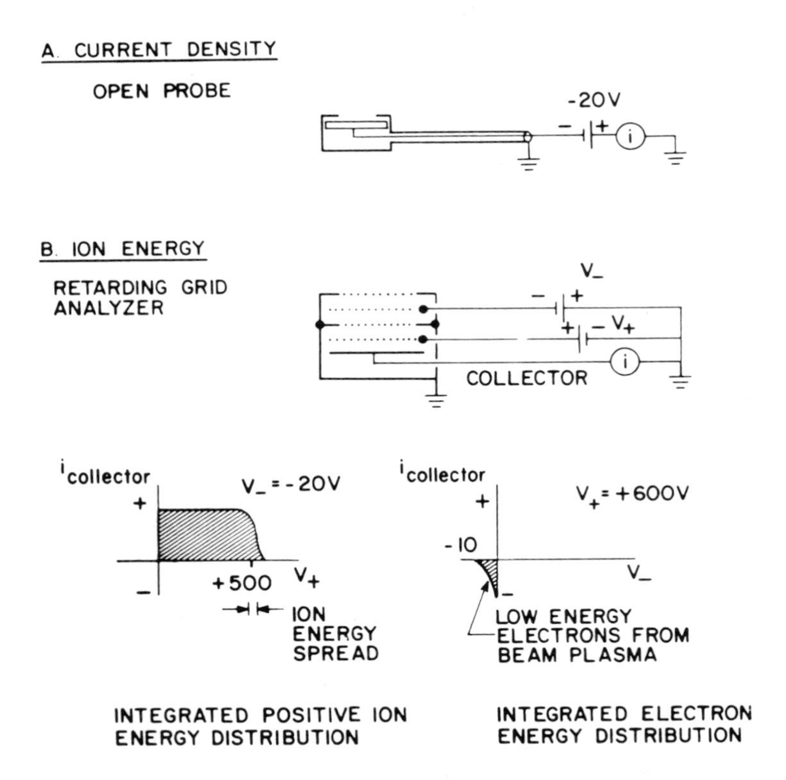

FIGURE 5. Beam measurement probes for (a) ion current density and (b) ion energy.

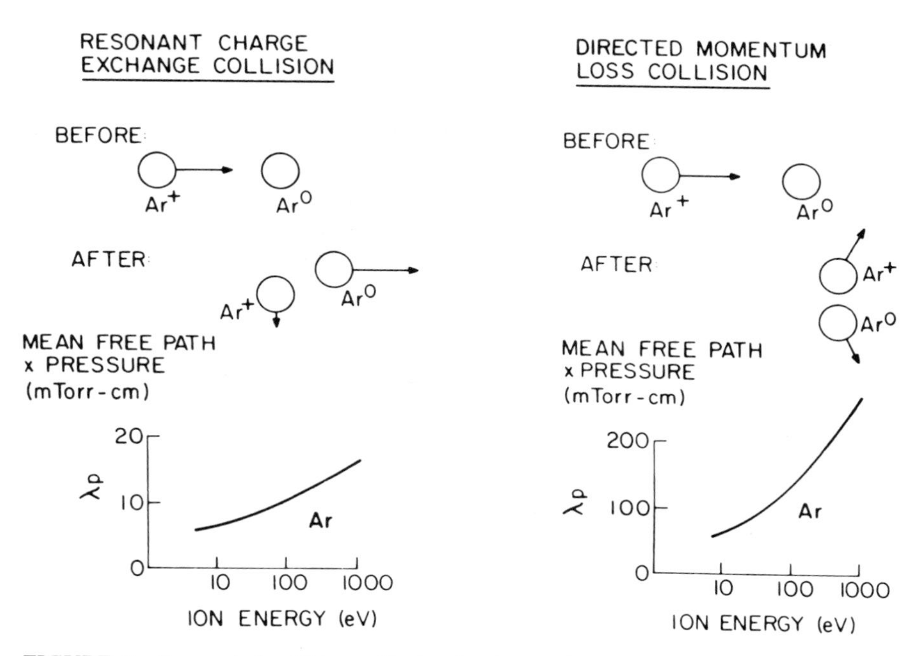

FIGURE 6. Ion-neutral collisions affecting ion beam: (a) resonant charge exchange and (b) directed momentum loss collision.

Ion energy is determined by the anode potential and is usually assumed to be within 5–10 eV of this value. The actual ion energy can be measured with a retarding grid analyzer (Fig. 5). This analyzer can separately measure the electron energy distribution by sweeping the negative retarding grid, and the positive ion energy distribution by sweeping the positive retarding grid. Alternatively, a hemispherical electrostatic analyzer may be used [9].

Two distinct types of collisions occur between beam ions and neutral background gas atoms. The first is the directed momentum loss collision (Fig. 6), which significantly alters the ion direction. These collisions may be almost neglected for most deposition conditions. The second is the resonant charge exchange collision (Fig. 6), in which the fast moving ion picks up an electron from a slow atom, and then proceeds essentially undeflected as a fast neutral, leaving behind a slow ion. This process is often significant and can lead to errors in beam current density measurements. Both interactions are described in references [6, 10]. Since the atom density is proportional to the gas pressure ($p = nkT$), a convenient parameter of the process is the product of mean-free-path and pressure. This quantity is given for argon in Fig. 6 for both collisions as a function of ion energy. As an example, for 500 eV argon ions traveling through a background pressure of 0.1 mTorr, the directed momentum loss mean free path is 20 m. This means that in a distance of less than 1 m, very few of the beam ions are scattered out of the initial direction by collisions with background gas atoms. This long mean-free-path distinguishes ion beam processing from higher pressure plasma processing, such as reactive ion etching or plasma etching, where mean-free-paths are typically centimeters or millimeters.

The cross section for resonant charge exchange is much larger than the cross section for directed momentum loss, and the mean-free-path is

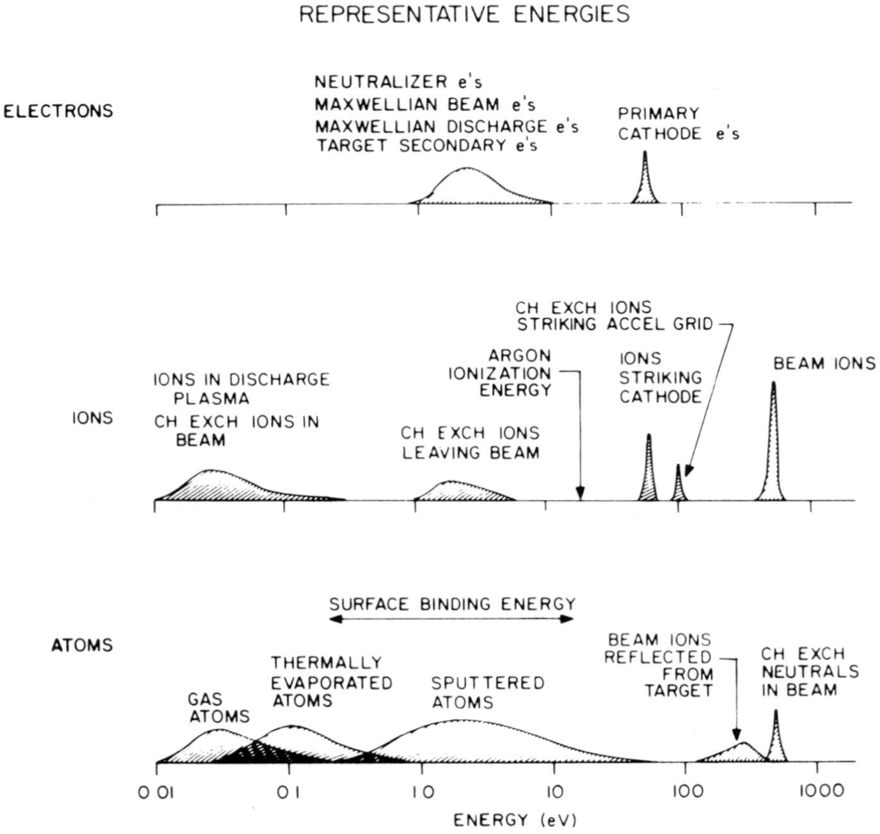

FIGURE 7. Representative energies of electrons, ions and atoms in a broad-beam ion source system. The target refers to the surface that the ion beam strikes.

correspondingly shorter. For 500 eV argon ions in 0.1 mTorr argon (Fig. 6), the mean free path is 1.2 m. The ion current traveling in the initial direction decreases exponentially with distance. For this example, 30 cm downstream only 78% of the initial beam current remains as ions, while 22% of the initial beam is converted to neutral atoms traveling with almost the full 500 eV. In order to make accurate measurements of sputtering yields, or to use previously measured values properly, it may be necessary to correct the measured ion flux upward to account for the fraction that has undergone charge exchange. The values in Fig. 6 are appropriate for argon ions moving through argon background gas, but will not apply to molecular fragment ions moving through a background of the parent molecular gas. This is because the charge exchange cross section is only large for resonant exchange collisions. These seldom occur between unlike species, with the effect that the mean-free-path for unlike species is more nearly comparable to the directed momentum loss mean free path. A summary of representative energies is given in Fig. 7 for ions, electrons and atoms in ion beam applications. Here, the target refers to any surface that the ion beam strikes.

B. MICROWAVE ION SOURCES

Microwave ion sources, also called electron cyclotron resonance (ECR) sources, have been applied to reactive ion beam etching in the past several years [11], and also to plasma deposition. They share many basic features with the Kaufman source. Ions are produced by electron bombardment in a discharge chamber (Fig. 8), accelerated through extraction grids, and directed as a beam to the surface to be etched. The major difference is in the source of electrons and means for providing electron energy in the discharge. The microwave ion source contains no thermionic filament, but ionizes the gas by microwave excitation coupled to the discharge chamber through a ceramic window. The advantage of eliminating the thermionic filament is increased lifetime of operation and reduced temperature of the discharge chamber walls. An example, described by Matsuo and Adachi [12], uses microwave power at 2.45 GHz to excite the discharge in a chamber 20 cm in diameter and 20 cm high in a resonant cavity mode. An axial magnetic field of 875 Gauss is applied to obtain the ECR condition, and the strength of this field is decreased toward the extraction grids, to provide a more uniform ion density at the grids. At a pressure of 10^{-4} to 10^{-3} Torr, an ion flux of 1 mA/cm^2 is extracted at an energy of 1 keV. To achieve reasonably high current density at lower ion energy, a shielded single grid configuration is used [12]. Here, a shield of coarse grating size separates the plasma generation region from the single grid extraction

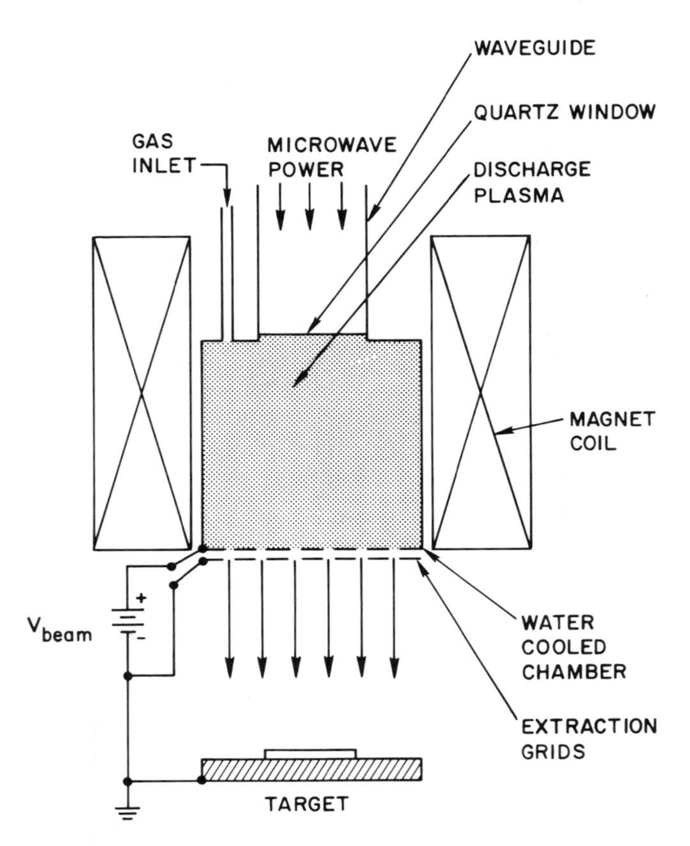

FIGURE 8. Microwave (electron cyclotron resonance) ion source (adapted from Matsuo [11]).

region. This shield acts as a microwave reflector, and also improves the region of voltage that is stable against sparkover. Additional characterization of such a microwave plasma is given by Keqiang *et al.* [13]. The considerations of beam divergence, flux measurement and neutralization are essentially the same as those described for the Kaufman source.

III. Inert Ion Beam Etching

Ion beam etching as a fabrication process became practical in the early 1960s with the invention of the Kaufman source. The ability to etch virtually any material by the sputtering process opened up a variety of applications. Limitations in selectivity, however, encouraged the development of chemically reactive dry etching processes. Plasma etching or reactive ion etching (RIE), described elsewhere in this volume, became major manufacturing processes during the 1970s. As device structures become more complicated, however, with many layers of different chemical composition, inert ion beam sputtering continues to find application. Reviews of inert ion beam etching (ion milling) have been written by Melliar-Smith and Mogab [1] and by Lee [2], and the bibliographies by Hawkins [14] list applications of this technology to many fields. Here, we discuss several basic features of the sputtering process that are central to ion beam etching, and give several recent examples.

A. SPUTTERING YIELDS AND ETCH RATES

In the ion energy range of tens of eV to several keV, the sputtering yield (number of sputtered atoms per incident ion) increases nearly linearly with ion energy. A detailed compilation of sputtering yields of different beam and target species is given by Matsunami *et al.* [15], and tabulations for common materials used in thin film processing are given by Vossen [16], Lee [2], Melliar-Smith and Mogab [1] and Kaufman [5]. Although sputtering yields of the elements are well documented, it should not be assumed that yields of the constituents of multicomponent materials will be directly related to these. In fact, large differences are observed [17].

To obtain the etch rate R in Å/min from a known value of sputtering yield s, one must know the ion flux j (mA/cm^2), the material density ρ (g/cm^3) and the atomic weight W (g/mole). The equivalence of 1 mA/cm^2 = 6.25×10^{15} ion/cm^2 sec gives the relationship:

$$R = 62.2\, sjW/\rho.$$

Inserting values for Si ($s = 0.45$ for 500 eV Ar^+ at normal incidence) gives an etch rate of 340 Å/min for an ion flux of 1 mA/cm^2.

Although a large body of data exists for normal incidence sputtering yields, these data should not be accepted unquestioned. Published yields sometimes omit the correction for the charge exchange fraction of the beam, described earlier. Also, reactive metals are particularly sensitive to small amounts of residual gas (e.g., water vapor), which will decrease the measured yield substantially. It is always wise to calibrate the etch rates of materials in a particular system. After this, the rates should be quite reproducible under repeated etch conditions.

The angular dependence of etch rates (or sputtering yields) is not commonly measured or published for many materials, yet it strongly affects the resulting shapes of etched surfaces that contain topographic structure.

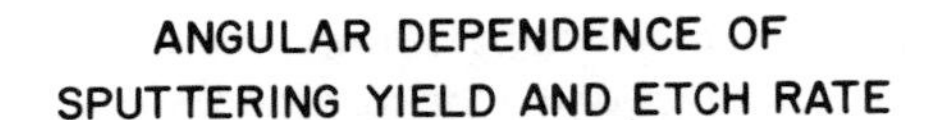

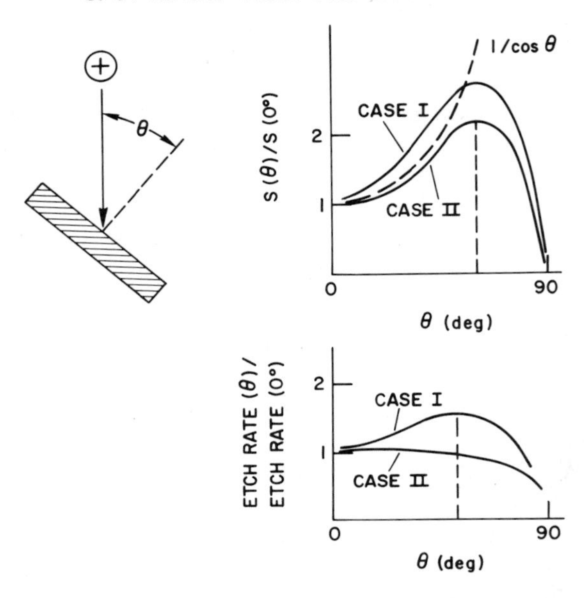

FIGURE 9. Typical angular dependence of sputtering yield and etch rate.

The typical dependence of sputtering yield on the angle of ion incidence is shown in Fig. 9. The increase of yield with tilt angle is substantial, up to an angle of typically 60–70°, at which point the yield reaches a maximum. The resulting etch rate will show a maximum as a function of angle only if the sputtering yield rises faster than $1/\cos\theta$ (Case I in Fig. 9). For some materials, the situation shown as Case II occurs, and the maximum etch rate occurs for normal incidence. In the etching of surfaces with patterns or roughness, sputtering along angles with the maximum etch rate will amplify topographic structures, leading to a cone-shaped or wavy topography over a period of time. This texture is accentuated if the surface being etched is simultaneously receiving a deposition flux, either by intentional seeding or by contamination.

B. PATTERN TRANSFER AND STEP PROFILES

The effectiveness of patterning by any process depends on the resolution achieved both in the horizontal direction and in the vertical direction. An important parameter is the aspect ratio, the depth to width ratio of an etched feature. Generally speaking, inert ion beam etching is capable of very high resolution (< 100 Å) patterning with aspect ratios less than or equal to unity. For low aspect ratios, resolution is nearly always limited by the mask material, not by the etching process itself. For aspect ratios exceeding unity, however, wall shapes become more sloped, and reactive gas ions are necessary to generate volatile etch products. These products do not redeposit on the mask and substrate, but are pumped away.

In order to accurately transfer a pattern from mask to substrate, the mask shape must be controlled. For example, if the mask edges are sloped (Fig. 10a), the edge of the etched step advances sideways during etching,

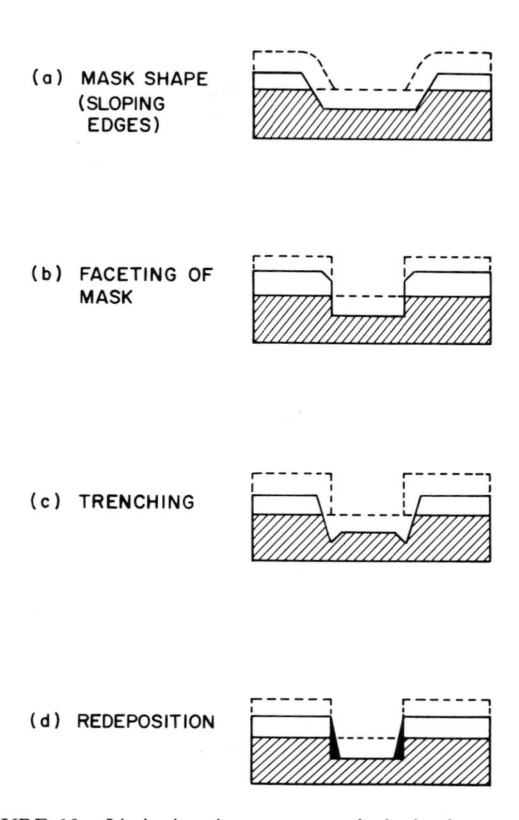

FIGURE 10. Limitations in pattern transfer by ion beam etching.

leading to sloping edges in the final substrate pattern. Another factor in mask shape is the tendency for the mask material to develop a facet at the angle of maximum etch rate (Fig. 10b). When this facet reaches the substrate surface, the mask edge moves laterally, exposing more substrate surface and giving a sloping step. If the developing step edge is sloped, then an additional artifact develops due to the reflected flux of ions striking near the bottom of the step. This reflected flux produces a trench at the bottom of the sloped step (Fig. 10c). When surfaces are etched by sputtering, the involatile sputtered atoms are deposited on the step edges themselves. This redeposition not only contributes to further sloping of the edges, but may also coat the walls of the mask (Fig. 10d). Later, when the mask material is dissolved away, these fringes of redeposited material may remain attached to the substrate.

Substantial improvements in etch profiles are achieved by tilting and rotating the substrate during etching, as shown in Fig. 11. Many commer-

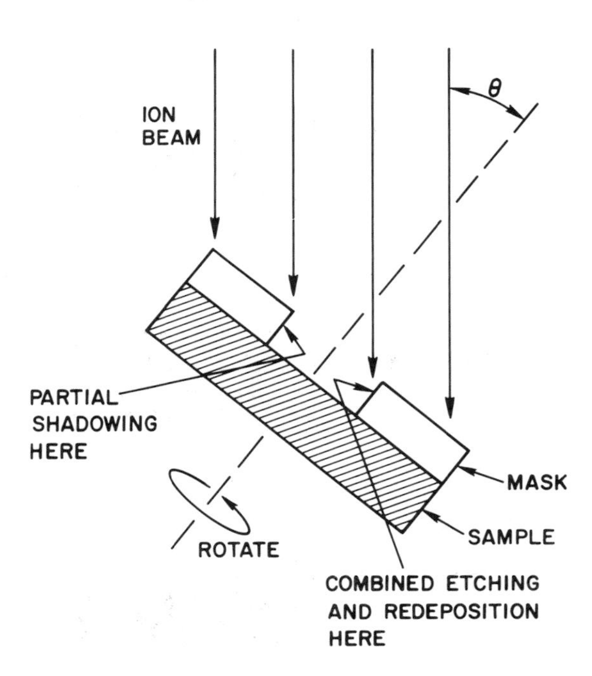

FIGURE 11. Configuration for tilting and rotating a sample during ion beam etching.

cial ion beam etching systems are equipped with tilt and rotate stages to provide this motion. The reasons for tilting are a combination of shadowing the bottom of the step (which reduces trenching), partially etching the sidewalls of the mask (which reduces redeposition), and additionally gaining more nearly vertical edges on the etched profiles in the substrate. Optimization of tilt angles for various mask/substrate combinations depends on the angular dependence of the sputtering yields (as in Fig. 9), and is thoroughly discussed by Lee [18].

C. LIMITS OF ETCHING RESOLUTION

The highest resolution structures obtained by ion beam etching have been produced by contamination resist lithography [19]. Here, a finely focused electron beam writes a pattern on a thin blanket film (several hundred Å thick), leaving a carbonaceous residue where the beam strikes the surface. Subsequent ion beam etching removes the metal film where it has not been

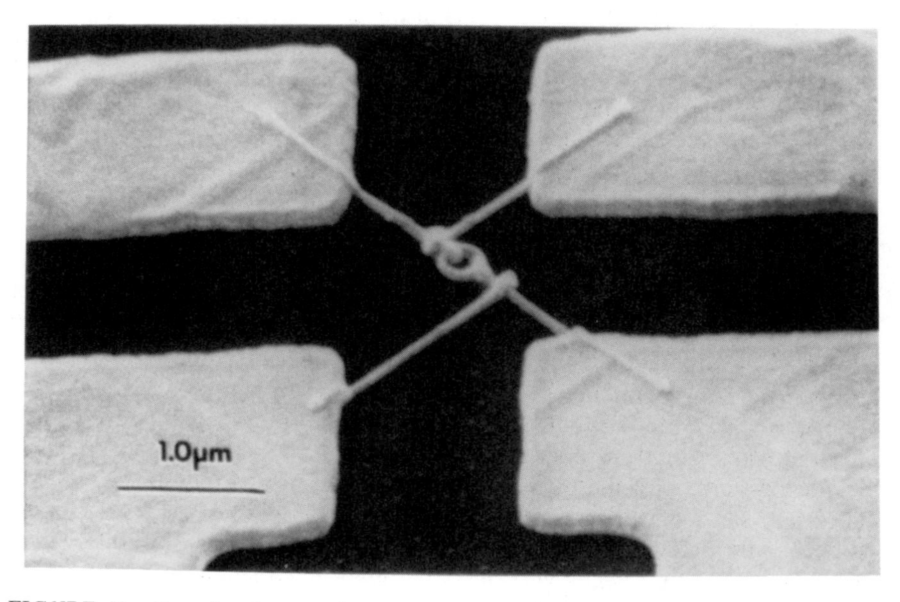

FIGURE 12. Scanning electron micrograph of Au narrow line device fabricated by contamination resist lithography and argon ion beam etching (courtesy of R. B. Laibowitz).

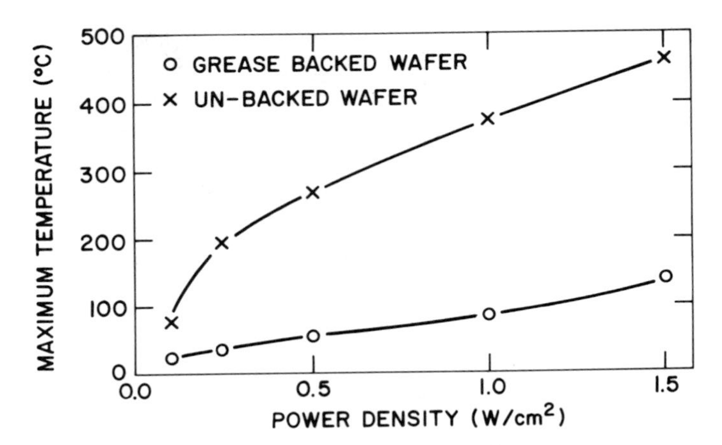

FIGURE 13. Steady-state temperature reached by 1-inch diameter Si wafer as a function of input power from argon ion beam.

protected by this residue. An example is given in Fig. 12, using a Au film 380 Å thick on a thin (500 Å) Si_3N_4 membrane. Such finely patterned conductors have revealed oscillations in the transport properties due to the fundamental wave nature of electrons [20]. Continuous metal lines of less than 100 Å width have been achieved this way. The resolution of etching is clearly established by the masking material, not by the ion etching process itself. Electron beam resists have been used to produce metal linewidths of several hundred angstroms with ion beam etching [21]. Thicker film patterns are generally limited in resolution by the aspect ratio, because of redeposition discussed previously.

Another important limitation in achieving high resolution patterning is the substrate temperature. With a beam power input easily in the range of 1 W/cm^2, the temperature of a loosely mounted substrate can reach several hundred °C. In Fig. 13, the temperature of a 1-inch diameter Si wafer is plotted as a function of the input power in an argon ion beam (power density in W/cm^2 = beam energy in keV times current density in mA/cm^2) for the wafer simply resting on a water-cooled copper block, and for the wafer heatsunk with thermally conducting grease to the block. Heatsinking is clearly needed with masking materials such as photoresist, to keep the temperature below 100°C for etching with a typical beam of 1000 eV and 1.0 mA/cm^2. Electrostatic and elastomer hold-down techniques are used to provide non-contaminating heatsinking.

IV. Reactive Ion Beam Etching

A. BEAM / TARGET COMBINATIONS WITH INVOLATILE PRODUCTS

To gain higher etch rates and additional etch rate selectivity over that offered by inert ion beam etching, chemically reactive gases are added to ion beam systems to retard the etch rate of the masking material and to enhance the etch rate of the material to be patterned.

The addition of oxygen to an argon ion beam system strongly reduces the etch rate of reactive metals, but only after a threshold level of oxygen partial pressure is added. Examples of several materials are given by Melliar-Smith and Mogab [1], and the case of Ni is illustrated in Fig. 14.

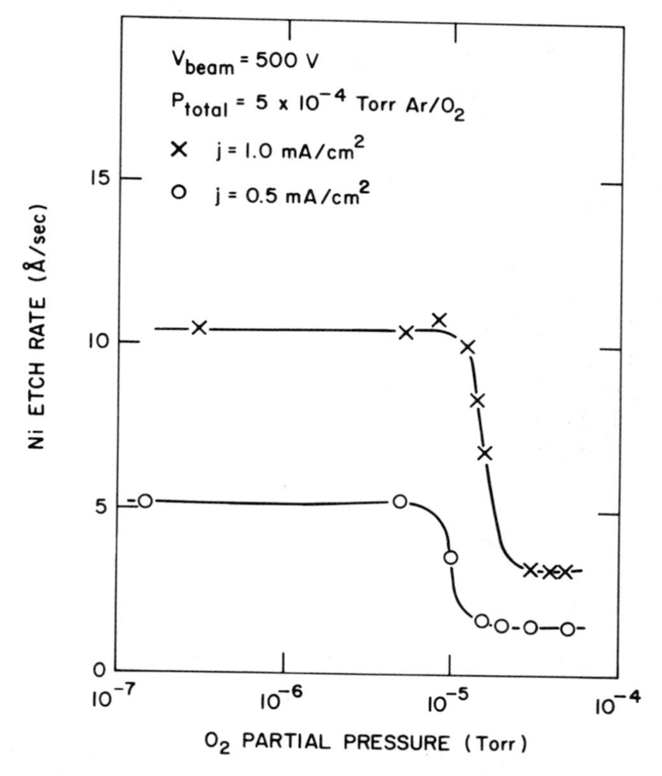

FIGURE 14. Etch rate of Ni under 500 eV ion bombardment as a function of partial pressure of O_2 in argon.

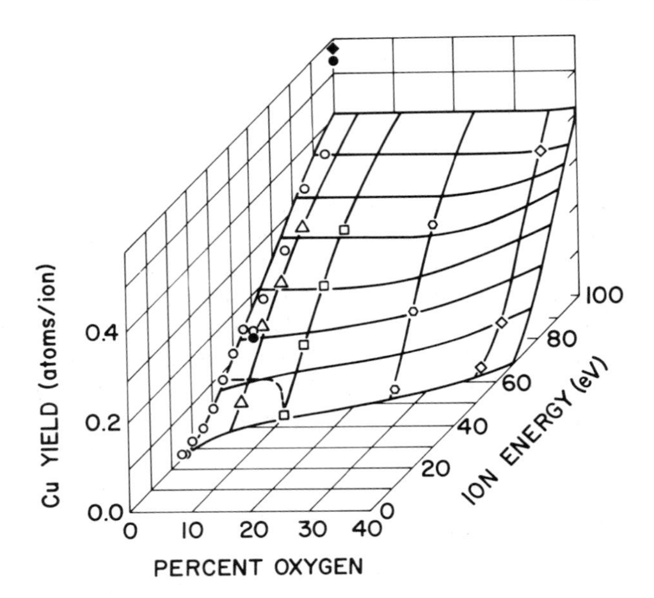

FIGURE 15. Sputtering yield of Cu as a function of ion energy and percent O_2 in Ar ion beam. The dashed line shows the suppression of etch rate to zero by adding oxygen at an ion energy of 45 eV [23].

The threshold pressure is a function of the etch rate of the metal, and increases by a factor of two when the ion flux (and hence the metal etch rate) is doubled. Typically, the etch rate decreases by a factor of 2–10 when the threshold pressure of oxygen is exceeded, giving a steady state oxidation of the surface. The threshold pressure is lower for more reactive metals, such as Ti, Cr, Nb. Castellano [22] demonstrated an inverse correlation between the threshold pressure and the heat of formation of the metallic oxide. Similar behavior occurs for nitride-forming materials in the presence of nitrogen.

If the ion energy is low enough, giving a very low yield, it is possible to decrease the etch rate to zero by added oxygen. This is demonstrated in Fig. 15 for Cu etched by an argon ion beam with various amounts of oxygen added to the ion source [23]. At an ion energy of 60 eV, the etch rate is decreased substantially, but at an energy of 40 eV, the etch rate is suppressed to zero by the addition of 15% oxygen. In fact, direct observation of the substrate mass (by quartz crystal rate monitor) demonstrates a steady-state mass increase under high oxygen partial pressure (Fig. 16), indicating the growth of oxide on the surface.

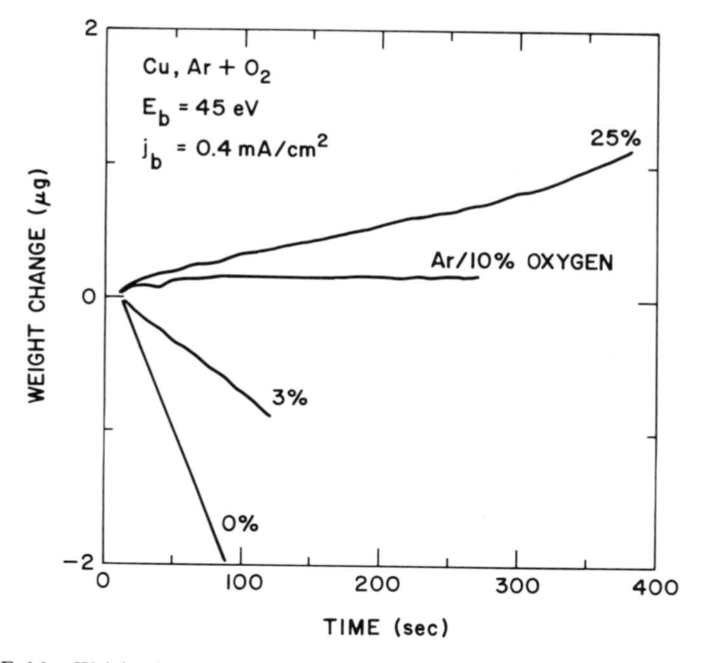

FIGURE 16. Weight change of quartz crystal rate monitor vs. time, for 45 eV ion etching of Cu, with various percentages of oxygen in argon [23].

This principle of reducing the etch rate to near zero by a combination of low ion energy and added oxygen was used by Kinoshita *et al.* [24] to gain a selectivity of 13 in etching GaAs relative to $Al_{0.5}Ga_{0.5}As$. As shown in Fig. 17, using 2% O_2 in Ar, there is little selectivity at an ion energy of 300 eV, but at 100 eV the etch rate of $Al_{0.5}Ga_{0.5}As$ is practically zero, producing a favorable etch rate ratio. This selectivity is helpful in the fabrication of solid state laser structures.

At ion energies of several hundred eV, more typically used in ion beam etching, oxygen addition is able to increase etch rate ratios substantially for some combinations of material, which may show similar etch rates in argon alone. An example of a multi-material etching problem is given in Fig. 18. Here, an alloy of Pb-Au-In was to be etched down to areas of SiO and Nb with a photoresist mask [25], to form Josephson junction tunneling devices. With pure argon, the etch rate ratio of the Pb-alloy relative to Nb was only 7 : 1, and relative to SiO only 5 : 1. However, the addition of 10% oxygen to argon decreased the Nb rate by an additional factor of 7 and that of SiO by a factor of 3, while not substantially decreasing the etch rate of the

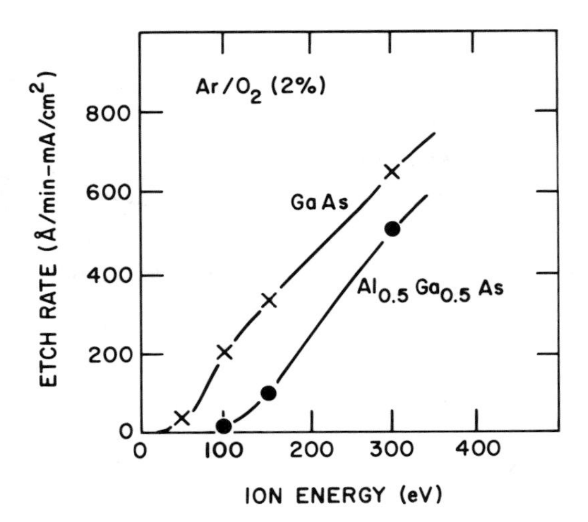

FIGURE 17. Etch rate of GaAs and $Al_{0.5}Ga_{0.5}As$ vs. ion energy with 2% oxygen in argon (adapted from Kinoshita *et al.* [24]).

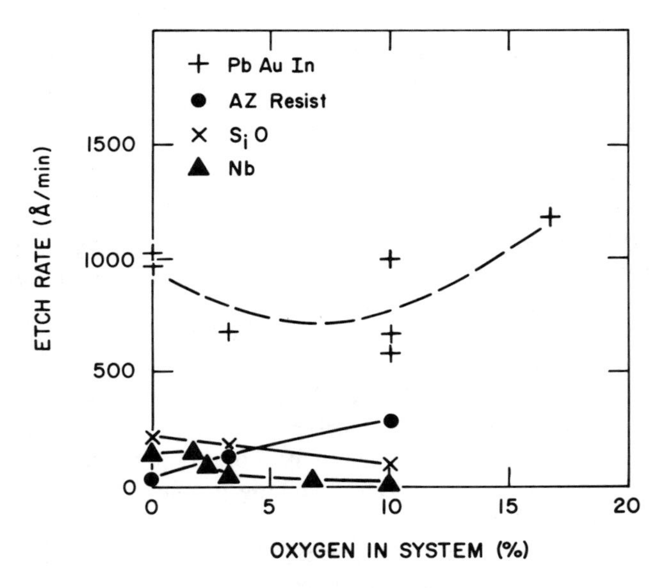

FIGURE 18. Etch rate of PbAuIn alloy, AZ photoresist, SiO and Nb as a function of percent oxygen added to 500 eV ion beam [25].

relatively non-reactive Pb-alloy. The resulting etch rate ratios obtained were about 40 : 1 relative to Nb and 15 : 1 relative to SiO. Simultaneously, the etch rate of photoresist increased by a factor of 7 with the addition of oxygen, but was still acceptable for the desired patterning process. Such a manipulation of etch rate ratios is necessary when several different materials are simultaneously exposed. In this example, the more reactive Nb and SiO were able to be oxidized preferentially relative to Pb, Au and In, which have less stable oxides.

Few examples have been published using nitrogen/argon mixtures to obtain preferential etch rates, but a similar degree of selectivity should be achieved with nitrides as with oxides. There is an equivalent range of reactivity of nitrides as with oxides, with many materials used in microelectronics having stable nitride phases (e.g., Ti, Zr, Ta, Si). Also, many nitrides are good conductors and diffusion barriers, a useful combination of properties for many applications.

B. BEAM / TARGET COMBINATIONS WITH VOLATILE PRODUCTS

An ideal application of oxygen addition is the patterning of polymers (such as polyimide). Here, a thin layer of a reactive metal (such as Ti or Cr) is deposited on the polymer, then patterned with conventional lithographic and inert ion etching techniques. Next, oxygen is added to continue etching the underlying polymer at typically several thousand Å/min while not removing the reactive metal mask. The original photoresist is also conveniently removed in this step. Polymer thicknesses of several micrometers can be patterned using metal masks of only several thousand Å thickness. Oxygen is particularly effective at etching polymer-based materials containing C, H, N, etc., which yield volatile products as in plasma etching.

The addition of gases such as oxygen and nitrogen shortens the filament lifetime in a Kaufman source, and may require a higher discharge voltage to maintain operation. Typically, argon discharges are stable at 40 V, whereas oxygen or nitrogen may require 50 V or higher, depending on the cleanliness of the anode. Prolonged oxidation of the anode surfaces will cause an increased discharge resistance, necessitating higher voltage for operation. Normal operation is easily restored by cleaning the anode.

Etching of Si and SiO_2 is accomplished by fluorocarbon and chlorocarbon gases, or chlorine, usually in a mixture with argon to minimize the effect on the filament cathode [26]. With these gases, a microwave ion source has the advantage of a filament-free discharge chamber. In a Kaufman source, operation with CF_4 deposits a polymer coating on the discharge chamber, which may extinguish the discharge [27]. However,

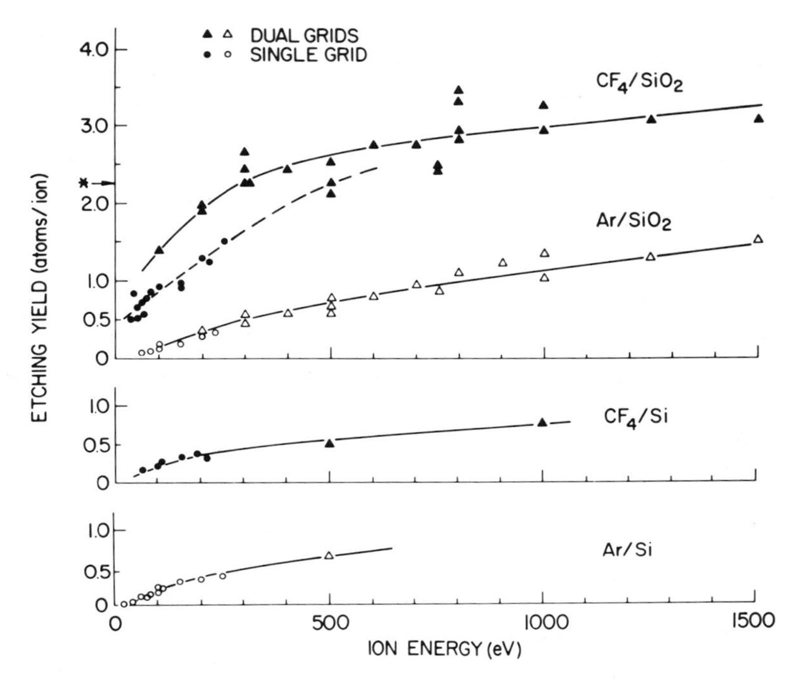

FIGURE 19. Etching yield of Si and SiO_2 as a function of ion energy, for Ar and CF_4 in ion source [27].

satisfactory performance is achieved with a graphite-lined discharge chamber [3]. The etch rates of Si and SiO_2 are shown in Fig. 19 as the etching yield (sputtered atoms per beam ion) as a function of ion energy, for pure CF_4 compared with Ar [27]. The Ar curves are the familiar sputtering yield curves, showing a threshold energy for etching of 25–50 eV. Etching Si with CF_4 provides little improvement due to the simultaneous deposition of carbon, which was identified by Coburn and Winters using other beam techniques [28]. Etching of SiO_2 by CF_4 shows a substantial etching yield, greater than 0.5 atom/ion even at the lowest energies used (about 25 eV), which required the single grid extraction method described earlier. This significant yield at low energy, and the rapid increase in yield up to 300 eV, indicates the chemical reactivity of the CF_4 beam is being activated by the energy of arrival of the beam ions.

Mass analysis of similar beams by Mayer and Barker [29] shows that the primary ion constituent is CF_3^+ with low plasma densities in the source, but that the beam composition is a strong function of the plasma conditions. The asterisk in Fig. 19 indicates the maximum etch rate achievable with

CF_3^+ ions alone. If this were the only etching process, the yield would remain constant at this value independent of ion energy. The fact that the CF_4/SiO_2 yield curve continues to rise with increasing ion energy, however, indicates a superposition of chemical etching and sputter etching above 300 eV. For this case, therefore, there is very little etching by neutral species adsorbed on the sample surface. The etch rate is limited by the arriving flux of reactive ions, in contrast with the usual case in rf reactive ion etching, for which the neutral reactive flux plays a major role. Additional information on the chemistry of reactive ion beam etching of semiconductor materials is given in the review article by Heath and Mayer [3].

C. ION-ASSISTED CHEMICAL ETCHING

The addition of a chemically reactive gas (e.g., Cl_2) to the substrate region of an argon ion beam system is generally referred to as ion-assisted chemical etching, to distinguish it from the case where the reactive gas is introduced through the ion source. The distinction is not absolute, however, since the presence of the reactive gas will produce some beam ions from species that back-diffuse into the broad-beam ion source. Only in differentially pumped systems may one create the situation in which the reactive gas ions are not a component of the ion beam. Ion assisted chemical etching is most effective when the reactive neutral species are adsorbed on the substrate surface, where they can be efficiently activated by the energy of arriving ions. In this case, the angular dependence of the sputtering yield reveals that the etch rate is limited by the chemical reaction rate, rather than the physical sputtering process. Instead of the usual peak in sputtering yield vs. angle, the yield in the chemically limited case has a maximum at normal incidence, and decreases monotonically with increasing angle [26].

The effect of added Cl_2 on the argon ion etch rate of Si is shown in Fig. 20 [30]. The sputtering yield curve is displaced upwards by increasing the flow of Cl_2, to the point where the effective etch yield rises by a factor of 3 or more. GaAs is commonly etched by Cl-containing gases, often using the directionality of the ion beam to advantage. Blazed gratings [31], angled facets on GaAs lasers [32], and other asymmetric surface structures have been fabricated using ion beam etching at non-normal incidence. Having volatile etch products allows angled structures to be etched without the problems of redeposition. Also, the directionality available with a beam technique removes the constraint of normal-incidence bombardment found in the usual plasma etching system. In fact, both the direction of the ion beam and the direction of the reactive gas flux may be controlled, producing unusually shaped etch structures [3].

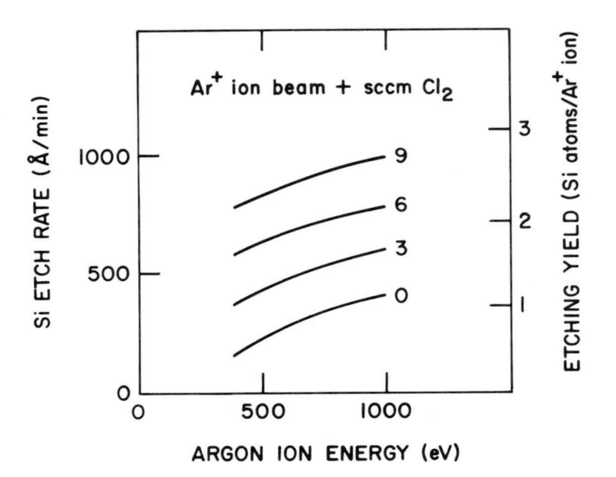

FIGURE 20. Silicon etch rate vs. argon ion energy for various levels of chlorine flow added to the ion source (adapted from Chinn *et al.* [30]).

V. Ion Etching Combined with Growth or Deposition

A. SURFACE COMPOUND LAYER FORMATION

Although this article is mainly concerned with etching applications where material removal is the objective, ion etching also induces the formation of a surface compound layer in some cases. One example is the oxidation of a metal or semiconductor surface using Ar/O_2 ion bombardment. By judicious choice of the ratio of oxygen to argon, and the ion energy, a self-limiting oxide may be grown on the bombarded surface. The oxide reaches and maintains a constant thickness even during conditions of net removal. The condition for this self-limiting thickness is that the oxide growth rate balance the removal by sputtering. Since the sputtering rate depends on the ion energy, and the oxide growth rate decreases exponentially with oxide thickness, the steady-state thickness is readily controlled by varying the ion energy. An example of this effect, shown in Fig. 21, is the oxidation of Ni at ion energies below 100 eV [33]. An ion energy of 80 eV produces a self-limiting oxide, which is indicated by the tunneling resistance, whereas 45 eV allows growth of the oxide to continue to much greater times. The conditions for optimizing such compound layer formation and its relationship to normal etching with oxygen at higher beam energies are reviewed by Kleinsasser *et al.* [34]. Clearly, any etching gas which reacts with the sample being etched to form a stable compound has the potential for self-limiting layer growth. More reactive gas/material

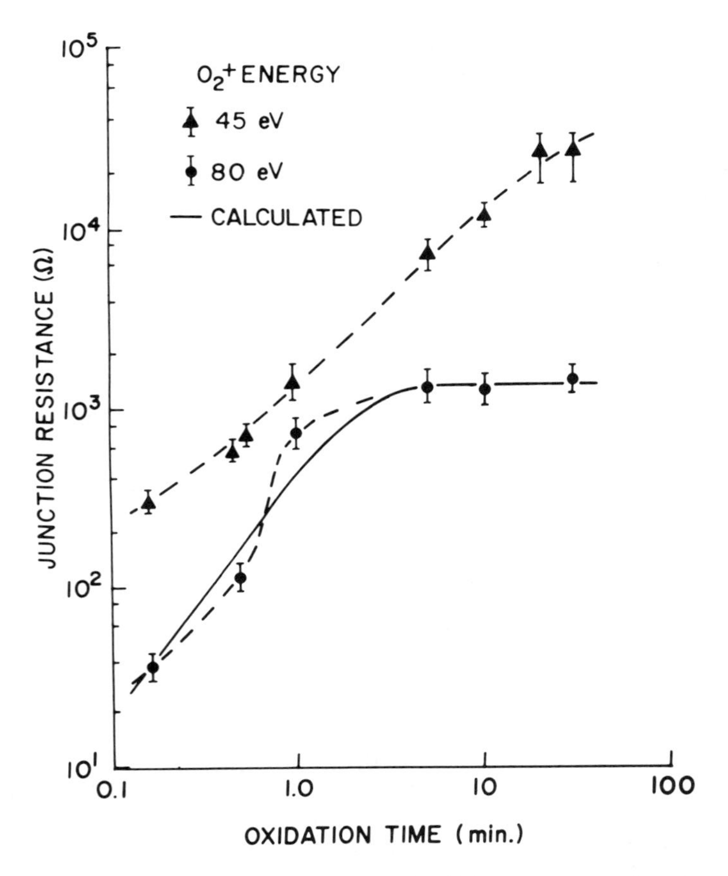

FIGURE 21. Junction resistance (indicating oxide thickness) vs. oxidation time for Ni exposed to O_2^+ ion beam at 45 eV and 80 eV. The solid line is calculated from a model predicting self-limiting oxide thickness [33].

combinations may require dilution of the gas and higher ion beam energies to achieve this balance [34]. Silicon oxidation and nitridation, performed under net etching conditions, show promising electrical properties as measured by MOS capacitor and transistor structures [35].

B. ETCHING EFFECTS DURING FILM DEPOSITION

Ion bombardment during thin film growth is a factor in all deposition techniques where a plasma interacts with the film surface. A review of

[3] B. A. Heath and T. M. Mayer, in "VLSI Electronics," Vol. 8, Microstructure Science, N. G. Einspruch and D. M. Brown, eds. Academic, New York (1984), p. 365.
[4] H. R. Kaufman, NASA Tech. Note TND-585, January 1961; H. R. Kaufman, in "Advances in Electronics and Electron Physics." L. Marton ed. Academic, New York (1974) Vol. 36, p. 265; H. R. Kaufman, *J. Vac. Sci. Technol.* **15**, 272 (1978).
[5] H. R. Kaufman, "Fundamentals of Ion Source Operation." Commonwealth Scientific Corp., Alexandria, Virginia (1985).
[6] H. R. Kaufman, J. J. Cuomo, and J. M. E. Harper, *J. Vac. Sci. Technol.* **21**, 725 (1982).
[7] J. M. E. Harper, J. J. Cuomo, and H. R. Kaufman, *J. Vac. Sci. Technol.* **21**, 737 (1982).
[8] R. S. Robinson, NASA-CR-159567 (1979).
[9] R. J. Warmack, J. A. D. Stockdale, and R. N. Compton, *Int. J. Mass Spectrometry and Ion Physics* **27**, 239 (1978).
[10] R. S. Robinson, *J. Vac. Sci. Technol.* **16**, 185 (1979).
[11] S. Matsuo, p. 75, in Applied Solid State Science, Supplement 2C (1985).
[12] S. Matsuo and Y. Adachi, *Jpn. J. Appl. Phys.* **21**, L4 (1982).
[13] C. Keqiang, A. Erli, W. Jinfa, Z. Hansheng, G. Zuoyao, and Z. Bangwei, *J. Vac. Sci. Technol.* **A4**, 828 (1986).
[14] D. T. Hawkins, *J. Vac. Sci. Technol.* **12**, 1389 (1975); D. T. Hawkins, *J. Vac. Sci. Technol.* **16**, 1051 (1979).
[15] N. Matsunami, Y. Yamamura, Y. Itikawa, N. Itoh, Y. Kazumata, S. Miyagawa, K. Morita, and R. Shimizu, IPPJ-AM-14, Institute of Plasma Physics, Nagoya University, Japan (1980).
[16] J. L. Vossen and J. J. Cuomo, in "Thin Film Processes." J. L. Vossen and W. Kern, eds. Academic, New York (1978), p. 12.
[17] J. M. E. Harper and R. J. Gambino, *J. Vac. Sci. Technol.* **16**, 1901 (1979).
[18] R. E. Lee, *J. Vac. Sci. Technol.* **16**, 164 (1979).
[19] R. B. Laibowitz, A. N. Broers, J. T. C. Yeh, and J. M. Viggiano, *Appl. Phys. Lett.* **35**, 891 (1979).
[20] R. A. Webb, S. Washburn, C. P. Umbach, and R. B. Laibowitz, *Phys. Rev. Lett.* **54**, 2696 (1985).
[21] A. N. Broers, J. M. E. Harper, and W. W. Molzen, *Appl. Phys. Lett.* **33**, 392 (1978).
[22] R. N. Castellano, *Thin Solid Films* **46**, 213 (1977).
[23] T. M. Mayer, J. M. E. Harper, and J. J. Cuomo, *J. Vac. Sci. Technol.* **A3**, 1779 (1985).
[24] H. Kinoshita, T. Ishida, and K. Kaminishi, *Appl. Phys. Lett.* **49**, 204 (1986).
[25] D. F. Moore, H. P. Dietrich, A. W. Kleinsasser, and J. M. E. Harper, *J. Vac. Sci. Technol.* **A3**, 1844 (1985).
[26] T. Mizutani, C. J. Dale, W. K. Chu, and T. M. Mayer, *Nucl. Instr. and Methods* **B7**, 825 (1985).
[27] J. M. E. Harper, J. J. Cuomo, P. A. Leary, G. M. Summa, H. R. Kaufman, and F. J. Bresnock, *J. Electrochem. Soc.* **128**, 1077 (1981).
[28] J. W. Coburn and H. F. Winters, *J. Vac. Sci. Technol.* **16**, 391 (1979).
[29] T. M. Mayer and R. A. Barker, *J. Electrochem. Soc.* **129**, 585 (1982).
[30] J. D. Chinn, W. Phillips, I. Adesida, and E. D. Wolf, *J. Electrochem. Soc.* **131**, 375 (1984).
[31] S. Matsui, T. Yamato, H. Aritome, and S. Namba, *Jpn. J. Appl. Phys.* **19**, L126 (1980).
[32] J. J. Yang, M. Sergant, M. Jansen, S. S. Ou, L. Eaton, and W. W. Simmons, *Appl. Phys. Lett.* **49**, 1138 (1986).
[33] J. M. E. Harper, M. Heiblum, J. L. Speidell, and J. J. Cuomo, *J. Appl. Phys.* **52**, 4118 (1981).
[34] A. W. Keinsasser, J. M. E. Harper, J. J. Cuomo, and M. Heiblum, *Thin Solid Films* **95**, 333 (1982).

[35] C. F. Yu, S. S. Todorov, and E. R. Fossum, *J. Vac. Sci. Technol.* **A5**, 1569 (1987); J. R. Troxell, *J. Electronic Materials* **14**, 707 (1985).
[36] J. M. E. Harper, J. J. Cuomo, R. J. Gambino, and H. R. Kaufman in "Ion Bombardment Modification of Surfaces: Fundamentals and Applications." R. Kelly and O. Auciello, eds. Elsevier, New York (1984).
[37] D. W. Skelly and L. A. Gruenke, *J. Vac. Sci. Technol.* **4**, 457 (1986).
[38] S. Berg, C. Nender, B. Gelin and M. Östling, *J. Vac. Sci. Technol.* **A4**, 448 (1986).
[39] L. S. Yu, J. M. E. Harper, J. J. Cuomo, and D. A. Smith, *Appl. Phys. Lett.* **47**, 932 (1985).
[40] J. J. Cuomo and R. J. Gambino, *J. Vac. Sci. Technol.* **12**, 79 (1975).
[41] C. R. Guarnieri, S. D. Offsey, and J. J. Cuomo, *J. Vac. Sci. Technol.* (to be published 1988).
[42] B. O. Johansson, H. T. G. Hentzell, J. M. E. Harper and J. J. Cuomo, *J. Mat. Res.* **1**, 442 (1986).
[43] J. M. E. Harper, J. J. Cuomo and H. T. G. Hentzell, *J. Appl. Phys.* **58**, 550 (1985).

7 Safety, Health, and Engineering Considerations for Plasma Processing

G. K. Herb

AT & T Bell Laboratories
Allentown, Pennsylvania

Plasma Etching:
An Introduction

ISBN 0-12-469370-9

The basic characteristic that is exploited in plasma processing is the ability of the process to achieve anisotropic pattern transfer for device designs that have aspect ratios (feature dimension/film thickness) less than three. This necessarily demands that the plasma process use feed gases that are chemically or biologically hazardous. For many of the successful plasma processes that depend on a fluorine or chlorine chemistry, there may not be available a benign source material (such as CF_4). It is generally accepted that source gases that are fluorocarbons are not hazardous; however, the addition of chlorine to produce chlorofluorocarbon or chlorine compounds (such as $CClF_3$ or $SiCl_4$) makes that material hazardous. In addition to the feed gases, a complete knowledge of the *effluent* chemistry must be available, since these effluent materials also can be hazardous and must be handled in a safe manner.

There is an increasing awareness of the conditions and the consequences that accompany the use of hazardous chemicals. All too often the use of a cavalier plasma process may result in public attention through the media, as has already occurred in several published accounts of accidents and/or chemical spills, particularly in the high density of microelectronic industry that resides in Silicon Valley, California. Safety and a total regard for the coworker and the environment *must* be the starting point of every plasma process. Safety is no accident (no pun intended)—it results from constant awareness. Many plasma installations are already operating under good state-of-the-art safety considerations. Many companies have excellent in-house safety practices and codes. For example, AT&T Bell Laboratories operates under the motto: "NO job is so important and NO service is so urgent that we cannot take time to perform our work SAFELY."

I. Legislated Safety Obligations

So that we can perform our work safely, however, we must be adequately informed about our legislated obligations. A good source for this information is often your local company safety representative, or the local state or federal OSHA (Occupational Safety and Health Agency) or EPA (Environmental Protection Agency) authority. There are laws that we *must* be informed about so that we can practice the *minimum* safety required.

A. FEDERAL LEGISLATION

In 1971 the Occupational Safety and Health Act was added to the Code of Federal Regulations (29CFR1910) that attempts to protect the worker in the workplace [1]. Its main provisions are:

a. Reduce hazards in the workplace.
b. Improve industry safety and health programs.
c. Establish employer-employee responsibilities.
d. Set mandatory job safety and health standards.
e. Provide for an effective enforcement program.
f. Encourage the states to assume responsibility for the administration of the program.
g. Provide a reporting procedure for injuries and fatalities for cases that fall under the jurisdiction of the agency.

The regulatory statements of OSHA include one that reads "No recognized hazard, whether covered specially under a standard or not, shall be permitted to exist [2]." In addition, there is no appeal to ignorance of the law, since OSHA deems that the "individual" making the mistake is responsible. Typically, in most companies in the United States, there is a vice president who carries the corporate responsibility for these issues and any penalty (financial or detention) is usually borne by that individual. This is a sweeping statement that allows the agency to render judgments that at first glance do not come under its jurisdiction. The best response to this situation is to become acquainted with the interpretations of the legislation by your company through its local safety organization.

Interpretations of the law are often necessary because of the sweeping OSHA guideline. Any new or novel case is usually dealt with on an individual basis. By the practice of the best-known and identified safety procedures, the best response to any legislated responsibilities is provided.

An even more sweeping legislation is the Environmental Protection Agency, enacted in 1970, whose responsibility is to protect the air, water, land, and public health through regulation [3]. The 1970 Clean Air Act requires development of air-quality control techniques, criteria for measuring air quality, and plans for implementing necessary programs. Some supporting legislation gives added strength to EPA:

Federal Water Pollution Control Act (1970) [4] determines that water treatment technology shall be used to its fullest extent, establishes water quality standards, and requires (when enacted in 1970) zero discharge into the country's waters by 1986.

Toxic Substances Control Act (Public Law 94-46A in 1976) [5] is a broad enactment that regulates and limits exposure to defined hazardous substances and establishes a toxic substances data base. This agency also defines regulations for processing and using hazardous chemicals and maintains records of exposure of such substances to workers.

Resource Conservation and Recovery Act (45CFR12.272-12-746 in 1976) [6, 7] tracks hazardous waste in storage, transport, and disposal.

The Department of Transportation regulates the transport of all compressed gases and the methods of containment of these gases [8].

Not all federal agencies are regulatory. There are those whose function it is to provide information and help to interested parties in order that proper regard to the legislated demands are more easily met. The National Institute for Occupational Safety and Health is a federally funded research agency responsible for identifying occupational safety and health hazards and recommmending changes in controlling regulation [9]. NIOSH supports OSHA, and also:

- periodically publishes summaries of recent evaluations for information of types of hazards, evaluation approaches and recommended control procedures,
- publishes yearly list of equipment tested and approved for respiratory devices (under the Federal Mine Safety and Health Act),
- conducts programs to determine and update PEL values for hazardous substances, and
- publishes literature on industrial hygiene.

B. STATE LEGISLATION

During the Reagan Administration, there has been an attempt to shift the responsibility of regulating the workplace and the environment from federal to state jurisdiction. The states have enacted legislation that has become known as the "sunshine laws" or the "right-to-know laws." Basic to these laws is the intent that *all* workers *must* be informed about the substances surrounding them on the job. Local fire and police officials and state and county health officials *must* be advised about materials made, stored, or emitted from plants. In general, the various state legislations are more encompassing and have tougher requirements than their federal counterparts. Hence it is even more relevant that compliance of these state laws are properly carried out. Again the best place to obtain the corporate plan of response to these state laws is through the local company safety representative.

A comparison of the New Jersey law to the United States rules (OSHA) is informative. Appendix A gives this comparison for the two governing bodies for determining those substances that are hazardous and by whom this determination is to be made. In the federal code the chemicals makers themselves decide what chemicals are to be considered as hazardous and write fact sheets that have to accompany the hazardous product to the users. The state of New Jersey determines what substances are hazardous (currently listing 2,051 chemicals) and writes the fact sheet on each substance. From this list *all* companies must notify the state of the hazardous substances in the work place and the number of people exposed to them and identify those hazards that are emitted to the environment and the amount emitted.

New Jersey requires *all* companies to update this workplace list annually and to send copies to state and county health departments and local fire and police groups. The environmental list must also be annually given to state officials, county officials, and local health, fire, and police departments. There are no corresponding federal requirements. All workers in New Jersey may see the hazardous substances lists and fact sheets, whereas only the workers at the chemical production plant have a right to see the list and fact sheets and the workers at plants using the chemicals may see the chemical fact sheet that is sent by the producer. The detailing of the responsibility of educating and training the workers on the proper safety precautions with respect to the hazardous substances is the same for state and federal agencies.

The labeling of the hazardous materials is surprisingly different in the two responsible agencies. Under the federal law companies must identify the chemical common name and an appropriate warning be placed either on the chemical container or the storage area. New Jersey requires companies to label all containers of hazardous chemicals with the precise name and index number from the state list. In addition, all pipelines, valves, vents, drains, etc., must be labeled, and ultimately (perhaps to guarantee the labeling of hazardous substances be done) all containers with *non*hazardous substances must be also labeled.

Under the federal mandate, chemical manufacturers, importers, and distributors must comply to the law. All other manufacturers must retain the fact sheet if they buy hazardous chemicals, while there is no need for nonusers of hazardous substances to comply. In New Jersey *all* manufacturers, hospitals, wholesalers, clinics, garages, public employers, museums, repair shops, electric and gas utilities, schools and transportation services (among others) must comply to the state law whether they manufacture, use, or distribute the hazardous chemicals, or not.

II. Response to Safety Legislation

A. SEMICONDUCTOR INDUSTRY STUDY—CAL/DIR–DOSH

A decade ago the state of California conducted a study to determine the compliance of the semiconductor industry to the various legislations [10]. This interest by the state was no doubt fostered by the explosive growth of this industry in the area between San Francisco and San Jose known as "Silicon Valley" and the use by this industry of chemicals on the hazard list. The study used two thrusts: a lengthy questionnaire presented to all semiconductor concerns in the state that was written by the Departments of Industrial Relations and Occupational Safety and Health and site visits to a random sample of companies to evaluate compliance to legislation.

The general conclusion of this state study was that the semiconductor industry has not ignored the consequences of hazardous chemicals and, in general, has established adequate state-of-the-art safety procedures that comply with the legislation. Some of the conclusions of this study are [10]:

- that of the facilities surveyed within the industry, health hazards appear to be generally well controlled, except for possible inorganic arsenic exposure during gallium arsenide light-emitting diode (LED) production,
- the diversity and quantities of chemical and physical agents in use present the potential for a variety of occupational health and safety hazards,
- the stringent processing controls required by the industry are reflected in the low exposure values typically found,
- there is a potential for acute exposures to hazardous chemical and physical agents, and
- nationally, the semiconductor industry has consistently achieved notable records of occupational safety and health during the past decade—ranking 12th out of the nation's 234 durable goods manufacturing industries in 1979 [10].

With respect to plasma processing, the Cal/DIR–DOSH study stated: "The potential for sporadic short-duration (acute) exposures to the range of hazardous agents and by-products during maintenance or system malfunctions is an area that warrants in-depth evaluation and control strategy prioritization" [10]. The handling of and disposal of vacuum pump oils should minimize the absorption of various fluorinated and/or chlorinated species (e.g., HF, HCl, Cl_2, and possible $COCl_2$) into the vacuum pump oils, depending on the composition of etchant source gases. The handling and cleaning of in-line filters and liquid nitrogen traps must also be properly planned since they may also contain the species.

This Cal/DIR–DOSH study presented a good review of four major areas of interest. Each of these four areas was documented with respect to the general procedure and activities that are necessary to result in four products; silicon wafers, silicon devices, GaAs wafers, and GaAs devices. The production of both silicon and III-V semiconductor substrates and devices were intensely studied. In each area a list of the common hazardous agents is given, along with a good toxicological basis for the concern of the various hazardous substances encountered in the practice of the technology by the semiconductor industry.

B. SEMI—THE EQUIPMENT MANUFACTURERS' POSITION

The equipment designers and builders have long been aware that the methods used in the semiconductor industry have seemingly incorporated some of the worst hazards from many other industries, including:

- high temperatures (up to 1200°C),
- high voltages (up to 400 kV),
- x-rays,
- laser radiation, UV radiation,
- cryogenics,
- acids (including HF),
- pyrophoric and other highly flammable and auto-explosive gases,
- lethally toxic gases, and
- combinations of the above.

In general, the equipment industry has been responsible and has encouraged built-in safety and corporate responsibility. The explosive growth of the semiconductor industry (50% increase from 1977 to 1980 alone) has resulted in less experienced operators today than several years ago, demanding equipment design to be simpler and safer. The composition of the work force in the semiconductor industry is markedly different from the norm found in manufacturing industries in general. The "typical" fabrication employee is younger, almost 39% female and about 30% of a minority ethnic background. This nontypical makeup of the semiconductor industry workforce introduces health hazard variables with respect to stability and training (age, job mobility, importance of employment, dedication, etc.), reproductive hazards (gender), and language and cultural norms.

Because the designers and builders recognized that safer apparatus must be available, equipment design must expect normal usage conditions, and there is a duty to instruct and warn about safety, the Semiconductor Equipment Manufacturers Institute began in 1973 [11]. This is an organization that recognizes that a cooperative responsibility to recognize hazards associated with the use of the systems reduces the individual responsibility and raises safety incorporation into standardized designs. In addition to regulatory considerations, the equipment manufacturer had to become legally astute. Since the maker of apparatus is legally responsible for product design defects, manufacturing defects, and operating instruction and warning label deficiencies, standard equipment represents the best response to legal liability. This area has become an increasingly active legal arena, with almost 8900 liability cases filed in 1982 (42 of which resulted in damage awards above $1 million). These penalties must be borne by the manufacturer as operating expenses that are passed on to the user. In addition, there is no statute of limitations for personal injury cases. Hence not only has the wafer benefited from properly designed equipment resulting from the SEMI influence, but the equipment may be less expensive as a result of a more informed and astute legal responsibility awareness.

C. A WORKING CORPORATE SAFETY PROGRAM

Figure 1 shows the cost to American industry for compliance to the EPA requirements for the years 1977–1983 [12]. In 1976 over $385 million was spent on personal protective devices. Compliance with the OSHA carcinogen policy caused that figure to exceed $100 billion in 1983. OSHA estimates that compliance to the new lead standard alone may cost $3.3 billion. Money alone is not the answer to the problem. Expert training and continued in-house motivation are the necessary ingredients to solve the problem for American industry.

If money is not the total solution, what is? There is a general acceptance of the concept that "If you want to get somewhere, you had better know where you are." This old saw holds excellent advice for a proper strategy for determining the best response to the requirements that are placed on us for safety and health regards for our fellow worker and the environment. "Knowing where you are..." is understanding the necessary actions de-

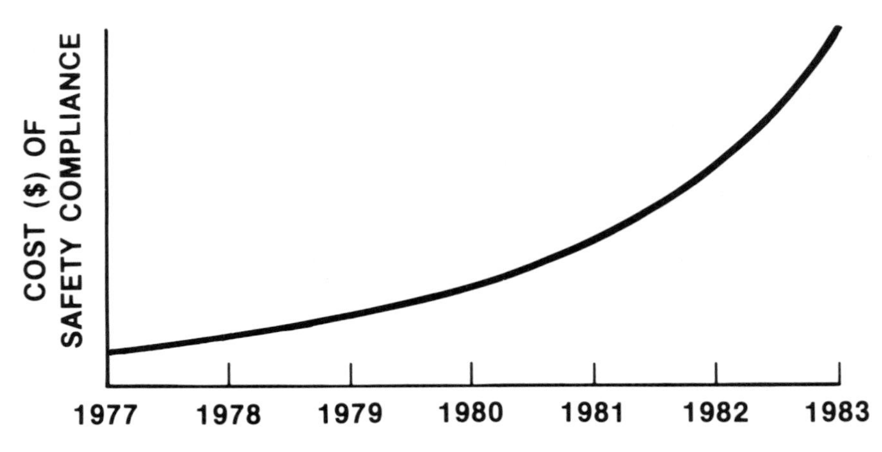

FIGURE 1. The cost to American industry for compliance to EPA regulations for the period 1978–1983 [12].

manded by the various legal agencies. A proper assessment of the *risk* that a particular condition exposes one to is an extremely important company activity. If a company ignores the continuation of an unsafe practice, legal stoppage of the operation can be demanded by the governing legislated agencies. In that case the importance of the product won't matter. The identification of the risk is one element of the equation that can be used to specify a safe operation. Since the cost of safety is considerable, the relation between risk, safety and cost must be known.

There are many paths that can be used to take a condition where the risk (or where you are) is clearly identified, according to the responsibilities that are legislated, to one of total regard for safety (where you want to be). The trip may be costly, however, and this cost must be borne by your company. The proper identification of the risk status will determine if the cost is too high. For an installation of capital-cost equipment into the semiconductor facility, AT&T Bell Laboratories has determined that an additional 20–30 percent of the cost must be added to the purchase price to provide for safety.

A strong corporate safety program is necessary to provide the mechanism that will allow the risk-cost-safety evaluation to be effectively done. The primary ingredient of such a program is the commitment of management. It is management's responsibility to provide the necessary physical resources, provide a clean, well-lighted workplace that is designed to protect workers from hazards, and to support the program by being involved, sincere, and committed. After the management has provided a proper framework for the safety program, the program itself can be fleshed out. Parts of a successful safety program are:

- training (attitude is key, training develops good attitudes; management provides attitude),
- inspections (check safety status and adherence to established procedures),
- maintenance (preventive maintenance minimizes downtime, loss of productivity, and domino effect of failure that leads to injury),
- fire prevention (frequent drills, coordination with local fire departments),
- disasters ("what if..." reaction must be thought out),
- off-the-job safety awareness should be second nature,
- accident records (learn from mistakes),
- medical and first aid systems, and
- continual review and update.

The interplay between the elements of risk-cost-safety cannot be overstated. Understanding the degree of risk and the necessary changes to guarantee an operation with total regard for safety will allow the cost to be properly defined to management for their enthusiastic support. Being informed of the necessary actions that *must* be taken in order that a particular operation be within the law provides a proper basis for the risk assessment.

Help in understanding the demanded responses to the law should be provided by the company safety representative. Federal and state help is also available to appreciate the requirements and perhaps provide interpretations for some of the legislation [13, 14].

III. Plasma Process Chemistry

The transition from installing the as-safe-as-possible machine purchased from the equipment manufacturer in an as-safe-as-possible way in the semiconductor facility for the use of the apparatus in a plasma operation is not easy. The characteristics of the feed gas(es) must be known, the type of effluent chemistry must be considered or measured, and the system appendages to the reactor (pump and gas delivery systems) must be properly designed. In this section some engineering considerations will be given for the operation of the plasma process with regard for safety.

A. PLASMA FEED GAS SOURCE

Almost without exception the plasma feed gas(es) are contained in a compressed gas cylinder. It is of utmost importance that knowledge be available about the physical state of the contents of the cylinder (gas or liquid) and the chemical consequences and hazards of the cylinder contents. The physical hazards of the tank itself are always to be kept in mind. The total force on an "A" cylinder at a gas pressure of 2500 psig is 4,000,000 pounds of force, which in the event of an accident can be converted to an extremely high thrust propelling the gas cylinder like a rocket and cause extensive injury and/or damage. The first thing that must be done when using compressed gas cylinders is to properly restrain and attach them to proper storage sites (temporary storage in a properly designed gas cabinet) so that the cylinder has no opportunity to fall. A ruptured cylinder that results from an unrestrained mounting is an extremely dangerous thing.

The Department of Transportation Office of Hazardous Materials Operations has its regulations published primarily in CFR 49 [8]. DOT controls the shipment of hazardous materials (including all compressed gases) by all common carriers. It requires proper shipping containers, acceptable shipping papers, shipper certification of contents, and placarding and labeling of the shipping container and/or transporting vehicle.

Labeling began in 1944 with the Labels and Precautionary Information Committee (LAPI) established by the Manufacturing Chemists Association (MCA, which changed its name to the Chemical Manufacturers Association in 1979) [15, 16]. Since their original guide for labeling hazardous chemicals was published in 1946, this *voluntary* standard has been used by manufacturers to label the drums, cans, bottles, jars, jugs, barrels, cylinders, and carboys containing hazardous chemicals. The sixth revision of this MCA labeling guide has been recognized and adopted by the American National Standards Institute (ANSI) in document Z129.1-1976 [17]. In the ANSI system, hazards are not classified in the same categories as in the DOT system. Most specific information relates to the following major dangers: toxics, corrosives, and flammables. In addition, many special hazards and combined hazards are recognized. These include explosives, carcinogens, teratogens, mutagens, reactive chemicals, oxidizers, pressure hazards, cryogens, radioactive materials, and biohazards. Within the classes a signal word is used to show the degree of hazard: Danger (highest degree of hazard), Warning (intermediate degree of hazard), and Caution (lowest degree of hazard). In addition, the word "POISON" and the skull and crossbones must be displayed with the word "DANGER" for highly toxic chemicals.

The DOT labeling scheme is similar to the ANSI standards and also conforms to the United Nations' eight hazardous material groupings. These groups include:

1. Explosives. Three subdivisions include those materials that are capable of detonation with a moderate spark, shock, or flame, and propagating the explosion to other packages (EXPLOSIVE A—nitroglycerine, dynamite); rapid combustibles (EXPLOSIVE B—Propellants); and those that may explode under extreme heat when densely packed (EXPLOSIVE C—flares, blasting caps).
2. Flammables. Five subdivisions include the highly flammable materials (FLAMMABLE GAS—acetylene, hydrogen, propane, butane); any liquid with a flash point less than 100°F (FLAMMABLE LIQUID—ether, acetone, gasoline); any solid that can be ignited readily and burns persistently (FLAMMABLE SOLID—charcoal, lithium, red phosphorus); any material that can ignite without externally supplied sources of ignition (COMBUSTIBLE); and any compressed gas not falling into any of the other four categories (NON FLAMMABLE GAS).
3. Dangerous When Wet. Water can cause the material to react violently (DANGEROUS—sodium, calcium carbide).
4. Oxidizers. Two classes of materials that can liberate oxygen or stimulate combustion of organic matter (OXIDIZER—calcium

hypochlorite, ammonium, and potassium nitrates); and a specific oxidizer that is a flammable organic material (ORGANIC PEROXIDE—benzyl and acetyl peroxides).

5. Poisons. Three classes: Class A poisons are life threatening with a small amount of gas or vapor (POISON GAS—cyanogen, arsine, phosphine, fluorine); Class B poisons are life threatening by contacting or ingesting the liquid or solid material (POISON—nitric acid, sodium cyanide, lead, and arsenic salts); and Class C poisons that cause severe irritation but are not life threatening (IRRITANT—chloroacetophenone).
6. Radioactive. Three classifications of materials that depend on the level of radiation outside the package: 0.5 mrem/hr at the package surface and 0.0 mrem/hr maximum at 3 feet (RADIOACTIVE I); 10 mrem/hr at the surface and 0.5 mrem/hr at 3 feet (RADIOACTIVE II); and 200 mrem/hr and 10 mrem/hr at 3 feet (RADIOACTIVE III).
7. Corrosives. Any material capable of visible destruction of tissue and/or metal (CORROSIVE—sulfuric acid, bromine, chromic acid hydrochloric acid).
8. Magnetic materials. Any material that could affect electronic guidance systems in airplanes (MAGNETIZED MATERIAL).
9. Biomedical material. Any material that may cause disease (BIOMEDICAL MATERIAL).
10. Required notice on some transport vehicles (EMPTY).
11. Special notice. Required when freight contains two or more classified materials or 5000 lbs. or more of classified materials are loaded on the vehicle (DANGEROUS).
12. Chlorine. May be used instead of the poison designation (CHLORINE).

These labels are extremely important to identify the material and the material characteristics of the contents of any package of hazardous chemical. This is especially true of the compressed gases. Unfortunately, each local supplier of compressed gases believes that its system of labeling is superior to others. This practice exists despite the requirement that all compressed gas cylinders must conform by law to the DOT labeling scheme. This required labeling is done by affixing a small DOT label to the tank, but using the supplier's own labeling scheme that tends to dominate the visual identification of the tank. There is even within the same supplier a multiplicity of labels for the same material. This unwillingness of the suppliers to adopt a single labeling scheme causes some real problems in being able to quickly identify the cylinder contents during an emergency when close scrutiny of the DOT label is impossible. The Compressed Gas Association (CGA) [18], a collection of the gas suppliers themselves, has been unable to convince the suppliers that a uniform macro-labeling system has meaning.

IV. Characteristics of Cylinder Gases

A. CYLINDER PRESSURE

The gases that are included in this group are those that have cylinder pressure greater that 500 psig at 70°F. Some of these are Ar, CF_4, He, H_2, N_2, O_2, SiH_4, and SiF_4. Typically the delivery pressure for the costly autoflow systems of many plasma machines are set at a maximum of 20 psig. If the delivery of the feed gas to the plasma system is more than of 20 psig, an incorrect flow is measured and permanent damage to the system may result. Therefore, it is necessary to reduce the delivery gas pressure to the AFC of the plasma machine.

The proper regard for the expensive AFC of the reactor and for safe operation of the gas delivery system is to reduce the gas pressure delivered from the cylinder in stages. The cylinder regulator that accomplishes this pressure reduction operates on the principle of a metallic diaphragm loading caused by the tension supplied by a spring-loaded adjustment. Higher tension allows more gas to flow into the lower pressure chamber. Double diaphragm regulators, which accomplish a reduction of pressure in two steps, are recommended for reduction of any cylinder pressure above 500 psig. The highest cylinder gas pressure of some 2500 psig should be reduced in more than two stages, perhaps four with the use of two double diaphragmed regulators in series, since diaphragms do rupture. If multiple stages of cylinder pressure reduction is used, such diaphragm ruptures will not injure the precious plasma system, and the remaining stages of safety will allow a perfectly functioning delivery system until the defective regulator can be replaced at the next cylinder change.

Another aspect of controlling the delivery gas pressure to a low value (< 20 psig) is that in many installations the gas cylinder may be a considerable distance from the use point (plasma reactor). Here the gas pressure should be kept low in the lengthy delivery line so that if a leak occurs, the escaping gas has low pressure driving it from the delivery line.

For those materials that have low (< 20 psig) cylinder pressures (liquids like BCl_3 and $SiCl_4$), no pressure regulator is necessary since the AFC operating pressure is already achieved. A suitable CGS coupling to the delivery line is used for this case since no pressure drop is needed.

B. SPECIFIC VOLUME

Since the material in the compressed gas cylinder is at high pressure, the volume of the material contained at the high pressure will be increased if the pressure is released. The specific volume of a gas is the ratio of the occupied volume at atmospheric pressure to the volume of the gas compressed in the cylinder, per unit weight of the material. The specific volumes of some of the regularly used feed gases are listed next [19].

He	(gas)	96.7 ft^3/lb
NH_3	(liquid)	22.6
SiH_4	(gas)	12.1
O_2	(gas)	12.1
PH_3	(liquid)	11.4
HCl	(liquid)	10.9
Ar	(gas)	9.7

For those materials that have high specific volumes ($> 10\ ft^3/lb$) and are hazardous (NH_3, SiH_4, PH_3, HCl), the smallest amount of material that can be stored at the gas delivery source (gas cabinet) should be used so that if an accident occurs, the contaminated volume will be minimized. However, the cylinder size of these hazardous materials should be also considered when the cylinders are changed. It is during the changing of gas cylinders that the probability for mistakes that result in the release of hazardous material is highest. Hence, a compromise between small amounts of hazardous materials to minimize the exposure volume and large cylinders to increase the time between tank changes (to decrease the possibility of accidents) must be made. The use rate of the material must be known to make the best choice.

The cylinder size is also important if a decision is made to contain the total volume of escaped gas within the gas cabinet if an accident occurs to minimize the contaminated area. The specific volume then determines what the required exhaust speed must be so that no hazardous material escapes the gas cabinet.

C. BOILING / FREEZING POINTS

For those materials that have boiling points between 0°C and 100°C, such as CCl_4 (76.5°C), $SiCl_4$ (57.6°C), BCl_3 (12.5°C) and SiH_2Cl_2 (8.4°C) [19], the placement of the tank in an outside unenclosed position is significantly different in Massachusetts than in Southern California. The AFC of the plasma system operates only with gas or vapor flowing through it. If liquid enters the AFC, violent unstable flow readings result. If the temperature extremes are large, the system operation in summer and winter are markedly different.

For liquid sources this problem of swamping the AFC is a real one. There are several installation designs that can guarantee that the AFC will not be awash with liquid. Making certain that the material flow from the cylinder to the AFC is always "up" will make gravity become an ally and force any liquid that may collect in the delivery line to flow back to the tank. Cooling the tank in the gas cabinet will ensure that the vacuum withdrawal of the material from the tank will deliver gas or vapor to the AFC. The installation of a porous plug immediately in front of the AFC will cause the material to expand and deliver only gas or vapor. Heating the AFC with tape slightly above room temperature will cause the material to vaporize in the AFC. Any of these, or better yet, a combination of them, will allow the AFC to operate as it was designed. For those installations that require long distances between the tank and the AFC, the consideration of these simple designs with the use of low boiling point materials will pay dividends in trouble-free (at least for the AFC) plasma system operation.

D. CORROSIVENESS

The main corrosive materials that are commonly used as feed source gases for plasma processes are BCl_3, $SiCl_4$ and ClF_3. The selection of a suitable gas pressure regulator with a diaphragm material that will not rapidly be digested by the corrosive is of utmost importance. The material composition of the delivery line may also be considered. The solution of the three-element equation, risk-cost-safety, must be done with all factors considered. Serious design considerations that guarantee that no corrosive material come in contact with the worker's skin and eyes are essential for good safety practice. Vapors of corrosive liquids are also hazardous and must not be allowed to enter the workplace. In addition, protection of the expensive plasma machine should be maintained.

Corrosive gases are commonly grouped according to the depth of penetration into the body, usually by inhalation [20]. Group I corrosives cause severe nose and throat irritation, but because of their high water solubility, do not penetrate deeply into the respiratory tract. Some of the Group I corrosives are ammonia, HCl, HF, formaldehyde, and sulfonyl and thionyl chlorides. Group II corrosive compounds attack the upper respiratory tract and bronchi (arsenic trichloride, bromine, chlorine, iodine, and sulfur dioxide). Group III corrosives reach the alveoli quickly without throat or nose irritation, resulting in delayed pulmonary edema, even for low exposure levels. These materials are the least soluble in water and include phosgene, nitrogen dioxide, and ozone.

E. FLAMMABILITY

The flammability of various materials that are used in plasma operations, along with the flammable limits in air, are given next [19].

SiH_4	spontaneously
PH_3	spontaneously
C_3H_8	2.2–9.5% in air
H_2	4–75% in air
CO	13–74% in air
NH_3	15–28% in air

Obviously, care must be exercised in the use of these materials by proving the properly designed gas cabinet with a water sprinkler inside the cabinet, which is connected to the sprinkler system in the immediate cabinet area. The volume of the flammable material must be as small as practical.

The listed flammable limits are with respect to air. Consideration of another material that is used for the cabinet make-up medium, such as nitrogen, eliminates the flammability characteristic entirely. However, the installation of such a system is costly.

If a fire should occur, the tank should be isolated from the delivery system with the use of a normally closed valve. Containment of the fire area by the proper fire personnel and evacuation of the affected area are the best initial responses. Except for fires with materials that also have toxic consequences (phosphine), it is probably best to allow the fire to self extinguish under contained conditions. The phosphine fire is more serious since the toxicity of the material must also be considered and dealt with.

F. CHEMICAL TOXICITY

Any chemical is classified as a toxic chemical that when ingested, inhaled, or absorbed, or when applied to, injected into, or developed with the body, in small amounts, by its chemical action may cause damage to structure or disturbance to bodily function [21, 22]. The bodily damage can be local (confined to exposed area) and/or systemic (effect spread by bloodstream) resulting from acute (short, or single contact) and/or chronic (long period) exposures to toxic chemicals.

All chemicals can be toxic to humans, depending on the mode of entry, physical condition, dose, duration, sensitivity, stress, and synergystic chemical combinations. The degree of chemical toxicity is based on an "average" human who has no chemical sensitivities. Safe exposure levels of toxic chemicals are determined for humans by direct clinical data, extrapolation of toxicological data from human surrogates (rats and guinea pigs), and assumption of some safe permissible level below which no impairment occurs no matter how often the exposure. Toxic exposure levels for chemicals need to be thoroughly understood. These are [23, 24]

TWA (time weighted average)—the sum of the products of the toxicant concentration and duration of exposure, taken over the total exposure time;

TLV (threshold limit value)—the concentration of a toxicant exposure that can be tolerated on a continuous basis by an "average" human in a normal 8-hour workday and a normal 40-hour workweek, week after week, without any adverse biological effect;

PEL (permissible exposure level)—same as TLV; and

IDLH (immediately dangerous to life and health)—the toxic concentration that the "average" human can tolerate during a 30 minutes exposure and be able to escape from the contaminated area without any help (human or equipment) and suffer no biological effect.

The hazardous concentrations are usually given in ppm (parts toxic chemical concentration per million parts nontoxic medium) and g/Kg (the weight of the toxic substance per kilogram weight of "average" human). Some commonly used chemicals in plasma processing and their PEL and IDLH values (ppm) are given next [19, 24].

Chemical	Designation	PEL	IDLH
		concentration in ppm	
AsH_3	POISON	0.05	6
PH_3	POISON	0.3	200
Cl_2	POISON	1.0	25
HCl		5.0	100
NF_3		10.0	2000
CCl_4		10.0	300

The long-term effects in man of hazardous chemicals is the subject of intense study. The tenuous cause and effect connection between exposure and biological result (because of the long time span between the two) for carcinogenic, mutagenic, and teratogenic substances make these materials particularly hazardous to man [25]. The PEL values for these carcinogens, mutagens, and/or teratogens are often based on the results of animal tests [26]. There is also a strong drive by OSHA and EPA to remove all chemicals that are even suspected of being carcinogenic, mutagenic, and/or teratogenic.

As the toxic level of a chemical that is used in a process increases, the smaller should be the amount of the chemical that should be stored at any source location for the process system. This practice is also good procedure for even the storage of such materials in a central holding, or "gas house" area. The less of the more hazardous materials that are around the less is the danger presented to the workforce. The most toxic materials such as arsine and phosphine should be used in lecture bottle sizes only and must have special precautions taken for their storage and use.

G. IMPURITIES IN GAS SOURCES

It is informative to be aware of the composition of the gas(es) that are used in plasma processes. Boron trichloride, BCl_3, is a commonly used gas in many plasma processes. It is a colorless, corrosive, nonflammable, toxic gas that has an irritating odor. It forms dense fumes in humid air and hydrolizes in moist air to form HCl. BCl_3 is irritating to the mucous membranes. It is shipped as a liquid under its own vapor pressure. Some characteristics of the material are [19]:

cylinder pressure (psig @ 70°F)	4.4,
specific volume (ft^3/lb)	3.3,
boiling point (°C)	12.5, and
freezing point (°C)	− 107.5

The common impurities contained in BCl_3 are [19]:

boron trichloride (minimum)	99.9%,
chlorine, free (maximum)	40 ppm,
silicon, as Si (maximum)	8 ppm, and
phosgene, as $COCl_2$ (maximum)	50 ppm.

The incorporation of phosgene and carbon dioxide in boron trichloride is a result of the chemistry used to make BCl_3:

$$4BO + 3C + 8Cl_2 \rightarrow 4BCl_3 + 2COCl_2 + CO_2, \text{ and}$$

$$3B_4C + 2O_2 + 20Cl_2 \rightarrow 12BCl_3 + 2COCl_2 + CO_2.$$

The impurity level of $COCl_2$ in BCl_3 from the manufacturer is high, 800–1000 ppm. After filtration and distillation by the gas supplier, if it is done at all, the phosgene level may decrease below 100 ppm. The best domestic gas supplier of BCl_3 guarantees that the phosgene level does not exceed 50 ppm, nominally 20–25 ppm. The PEL for $COCl_2$ is 0.1 ppm! The IDLH for phosgene is 2 ppm! If an accident with BCl_3 should occur, however, it is not the phosgene that is the danger, it is the rapid hydrolysis of BCl_3 to HCl that presents the most danger.

It is good engineering procedure to know what the components are that are being presented to the plasma reactor. Perhaps the low level incorporation of an unwanted component can strongly affect the process results. Also the identification of a normally present gas component can be used as the one that is used to monitor the presence of the gas. If the gas itself does not have a good specific monitor, there might be a good monitor that is specific for the gas impurity.

V. Hazardous Gas Monitoring Instruments

Being able to warn the workforce about the presence of dangerous concentrations of hazardous and toxic materials is a primary function of a working safety program. The identification of the proper instrumentation that serve this significant purpose is important. The properties that all useful gas monitors should have are that they be simple (in design) and fast (in response), stable (no drift with time), specific (they work on only one chemical, not many), properly sited (at the locations that will give the best early warning), their sensitivity be below PEL levels (to provide a safety margin) or at IDLH levles for equipment protection (which is costly and should be protected), and obeyed (*all* alarms, even false ones, must be responded to as if they were the real thing). NIOSH holds meetings under the Hazardous Materials Workshop and Exposition auspices that disseminates information on various gas monitoring instruments and their uses and applications [27].

There is an instrument that is sensitive to hazardous chemicals that all of us carry with us always. The nose can be relied on for supplying correct warning for some gases, but not for some others. Knowing the chemicals for which the nose is a good warning detector can be vital (no pun intended). Table 1 [12] shows that the nose is a good detector of ammonia, becoming irritated below PEL so that the wearer can take proper action. The nose rapidly loses its ability to be sensitive to ammonia on a continuing basis, however, so that the first warning must be heeded as the nose cannot detect continuing presence of ammonia, even at rapidly increasing concen-

Table 1 A comparison of the physical properties; TLV, IDLH, LC_{50} and nose irritation concentrations for ammonia and phosgene (after [12]).

AMMONIA		PHOSGENE
COLORLESS GAS WITH A SHARP IRRITATING ODOR. SOLUBLE IN WATER AND MANY SOLVENTS.	PHYSICAL PROPERTIES	COLORLESS GAS WHICH HYDROLYZES IN WATER TO FORM HCl AND CO_2. SOLUBLE IN SOME SOLVENTS.
NONFLAMMABLE ACCORDING TO DOT, BUT CAN BURN VIOLENTLY UNDER THE RIGHT CONDITIONS.	FLAMMABILITY	NONFLAMMABLE ACCORDING TO DOT TESTS, BUT IT WILL BURN.
25 ppm	TLV	0.1 ppm
500 ppm	IDLH	2 ppm
4837 ppm/1 hr	LC_{50}(rat)	110 ppm/0.5 hr
20 ppm	NOSE IRRITATION	10 ppm
IMMEDIATE BURNING, RESPIRATORY EDEMA.	EFFECTS	DELAYED RESPIRATORY EDEMA.

trations. On the other hand, the nose does not become irritated to the presence of phosgene until the gas concentration is 100 times the PEL value! Even at 5 times the IDLH concentration, the nose just becomes irritated to phosgene, at which time it will most undoubtedly be too late. The nose can be relied on to provide good information about the presence of some hazardous chemicals, but the possible desensitivity with time and those chemicals that cannot be detected are important data to be known and continually reviewed.

There is another effect from hazardous chemicals that must be understood so that proper attention be given to the person that inhales chlorine [27]. Since chlorine is taken directly into the lungs, the close surveillance of the patient is necessary to prevent a delayed body reaction from causing unnecessary complications, or even death. The results of chlorine inhalation are divided into four phases according to the biological response to the poison [12]. In the first stage, usually occurring from the moment of inhalation to six hours later, the patient appears to recover with perhaps a slight cough, especially after oxygen has been administered. The second stage of the normal biological reaction to the invasion of the lungs by the chlorine is the formation of edema in the lungs. It must be understood that the *normal* response of the body to almost every type of trauma is the body's attempt to reduce the effect of the trauma by flooding the area with body fluids. If pulmonary trauma should occur, edema in the lungs results from the patient's lungs flooding to reduce the injury and depending on the severity of the inhalation, the patient can drown on his/her feet. Close attention must be paid to the affected patient to prevent this edema from causing even more serious complications. This second phase can last up to eight days. The third stage of from 1–4 weeks has the patient gradually recovering, sometimes with a persistent cough. After four weeks the final phase, recovery, is reached for most "average" humans; however, those with hypersensitivity may suffer from possible kidney and/or liver damage, emphysema, or fibrosis.

VI. Checking the Workplace for Hazardous Chemicals

If there is any reason that a check be made of the work environment, perhaps because of a "strange smell," a complete chemical analysis must be made to ensure that no hazardous condition exists. Remember that the "sunshine laws" guarantee that the worker has the right to know the condition of the workplace. An excellent method to determine what chemical agents are part of the workers contact zone is to use the technique of collecting the breathing zone of the worker and concentrating the various agents so that conclusive identification can be made. The worker wears a collector appendaged to the clothing so that the worker's breathing zone is monitored. A pump with known speed pulls the breathing zone gases through the concentrating medium, often charcoal, that collects the material on its surface. After being worn for a reasonable period of time (3–7 days), the collector is put through a battery of chemical identification schemes to determine if any hazardous agents are present [28].

If a positive result occurs, the concentration of the hazardous chemical must be determined and compared to the PEL value. This investigation is quantitative in principle and relies on samples taken of the worker's environment. The known hazard from the first stage of the analysis makes this second stage a bit easier. Many different types of sample can be used: gas, liquid, or solid from many different parts of a process system or worker's environment. Care should be taken to include properly prepared blanks in order that the measuring equipment be properly calibrated for the accurate concentration determination.

NIOSH suggests that another concentration level be considered, the allowed level (AL), that is, perhaps one-tenth of the PEL level. This AL will then provide more than adequate warning potential before PEL is reached.

Appendix B lists 45 common gas and liquid sources for many plasma etch chemistries used in microelectronics technology [29]. This list represents a compilation of the material characteristics from other sources into a concise list for easy, quick information retrieval.

VII. A Hazardous Aluminum Plasma Etch Process

Almost a decade ago, a successful plasma etch process for aluminum existed in the AT&T Allentown plant. The chemistry was based on neat carbon tetrachloride, CCl_4. The quality of the aluminum etch was good and high production volumes of wafers were being etched in three etch systems, all using CCl_4. Anybody who has used carbon tetrachloride knows that an excessive polymer film is produced in a plasma application. This polymer film rapidly becomes so substantial that the reactor chamber must be regularly cleaned or the film falls on the wafers as they are being loaded and creates micromasks over the aluminum, which result in unwanted aluminum features that cause shorts in the desired aluminum pattern. The condition worsened when the humidity of the environment increased (summer) since the etch system was not load-locked to protect the reactor.

Early in the development of the CCl_4 aluminum etch process, it was observed that a strong medicinal odor was generated when the reactor was wiped with Kim Wipes dipped in warm DI water. The cleaning procedure removed all traces of the film, and an oxygen plasma was used to dry the reactor before the aluminum etch was run (without any observable effect from the water wipe of the reactor). Because of the odor, proper breathing masks and gloves quickly became mandatory pieces of apparatus for the cleaning operation. It was found that the strength of the odor could be lessened significantly if an oxygen plasma was run in the chamber before the reactor was opened for cleaning after ten etch runs.

The impetus to identify the chemical responsible for the odor was strengthened when it was observed that the building blocks—chlorine, carbon and oxygen—of phosgene are present in the reactor. The system configuration is depicted schematically in Fig. 2. Here the reactor where the film is formed is labeled A, the cold trap B, the mechanical pump with its mineral oil pumping medium C, the scrubber system that periodically received the material from the cold trap during cleaning D, and the exhausts for both the scrubber and pump E. Four independent chemical evaluations were undertaken, two within the company (AT&T) and two contracted from respected commercial companies. Samples were gathered from each section of the system and were of all types—solid, liquid and/or

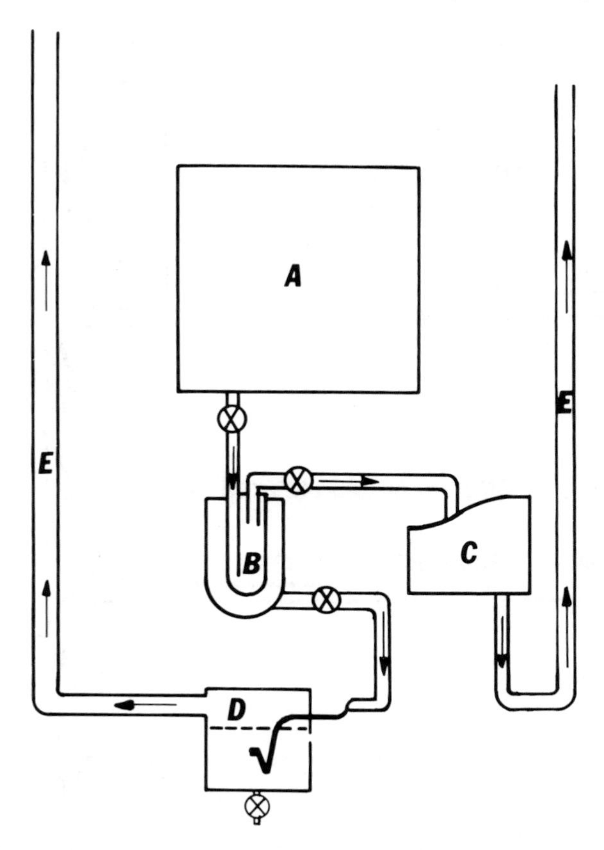

FIGURE 2. Schematic of an aluminum plasma etch system. A-etch chamber, B-cold trap, C-mechanical pump, D-scrubber system for cold trap effluents, and E exhausts for scrubber and pump.

gas. The chemical identification was done by a multiplicity of techniques, principle among them was double mass spectrometry and mass spectrometry–gas chromatography analysis. Figure 3 shows a sample spectrum of a MS-GC analysis where the intensity of the peak corresponds to the concentration and the mass number identifies the chemical.

The powders from the reactor chamber were shown to consist of major constituents (CCl_4, C_2Cl_6, C_4Cl_8, aluminum and chloride), minor components (C_2Cl_4, C_3Cl_4, C_4Cl_6, and higher weight halocarbons) and trace chemicals (C_2Cl_3 and iron), as shown in Table 2. The inorganic aluminum and iron came from the wafers and the reactor surfaces, respectively, and the chloride from the abundant chlorine etch source. All the organic

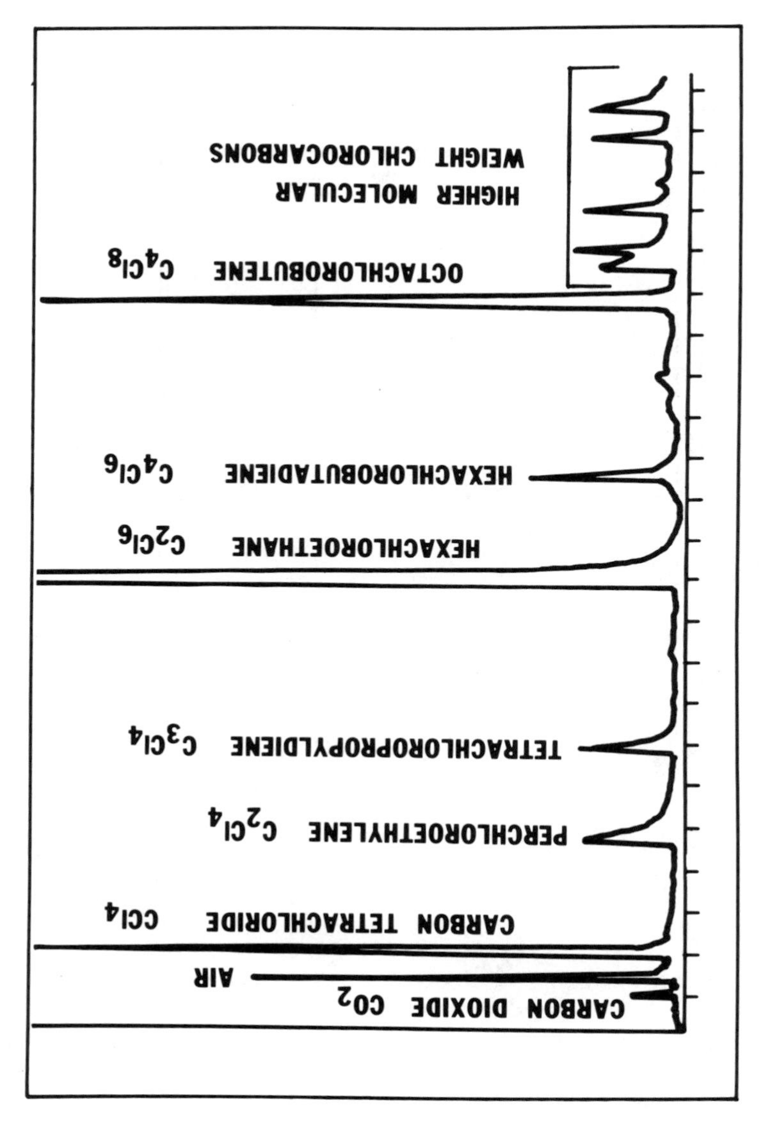

FIGURE 3. An MS-GC spectrum of the powder from the etch chamber walls of a CCl_4 aluminum etch process using the system of Fig. 2. The intensity of the mass number species is directly related to its abundance in the film.

Table 2 A list of the constituents of the powder film taken from the walls of the aluminum etch chamber. M = major component, m = minor component and T = trace constituent (just above threshold detection limit of the measuring instrument).

POWDERS FROM ETCHER

ORGANICS

1. CARBON TETRACHLORIDE (M)	CCl_4
2. TRICHLOROETHYLENE (T)	C_2Cl_3
3. PERCHLOROETHYLENE (m)	C_2Cl_4
4. TETRACHLOROPROPYLDIENE (m)	C_3Cl_4
5. HEXACHLOROETHANE (M)	C_2Cl_6
6. HEXACHLOROBUTADIENE (m)	C_4Cl_6
7. OCTACHLOROBUTENE (M)	C_4Cl_8
8. HIGHER WEIGHT HALOCARBONS (m)	

INORGANIC

1. ALUMINUM (M)
2. CHLORIDE (M)
3. IRON (T)

identifications are chlorocarbon combinations from the source gas. The results of the other four parts of the system are given in Table 3. The results of each independent investigation gave identical results. It was determined that the insecticidal odor came from hexachloroethane, C_2Cl_6, that was liberated when the system was wiped with the warm water. None of the investigations found any trace of phosgene.

The feed gas, CCl_4, is a colorless nonflammable liquid that has a characteristic odor. Oxidative decomposition by flame causes phosgene and HCl to form. Carbon Tetrachloride has been named by the National Cancer Institute as a human carcinogen [30, 31]. Perhaps an even more insidious property is its high percutaneous absorption. The TLV is 5 ppm and the IDLH is 300 ppm. Excessive exposure results in central nervous system depression and acute exposure may lead to liver and kidney damage.

The first daughter plasma product of CCl_4 is hexachloroethane, C_2Cl_6, a colorless solid with a camphorlike odor that sublimes at 189°C. It was determined by the National Cancer Institute to be a human carcinogen even before CCl_4 [32]. C_2CL_6 acts primarily as a central nervous system depressant, resulting in narcosis in high concentrations.

Table 3 A list of the identified components of the parts B, C, D and E of the aluminum etch system of Fig. 2. M, m and T have the same meaning as in Table 2.

COLD TRAP		BUBBLER WATER	
		ORGANICS	
1. CARBON TETRACHLORIDE (M)	CCl_4	1. TRICHLOROETHYLENE (m)	C_2Cl_3
2. PERCHLOROETHYLENE (M)	C_2Cl_4	2. FREON TF (m)	$C_2Cl_3F_3$
3. HEXACHLOROETHANE (m)	C_2Cl_6	INORGANICS	
4. TRICHLOROETHYLENE (m)	C_2Cl_3	1. CHLORIDE (M)	
5. FREON TF (m)	$C_2Cl_3F_3$		

PUMP OIL		AIR SAMPLES FROM EXHAUST STACK	
1. CARBON TETRACHLORIDE (M)	CCl_4	1. FREON TF (M)	$C_2Cl_3F_3$
2. PERCHLOROETHYLENE (m)	C_2Cl_4	2. METHYLENE CHLORIDE (m)	CH_3Cl
3. TRICHLOROETHYLENE (m)	C_2Cl_3	3. TRICHLOROETHANE (m)	$C_2H_3Cl_3$
4. FREON TF (T)	$C_2Cl_3F_3$	4. TRICHLOROETHYLENE (M)	C_2Cl_3
		5. PERCHLOROETHYLENE (m)	C_2Cl_4

There is a concensus of opinion among toxicologists that "multiple exposure to multiple suspected carcinogenic substances for even low doses may be much worse than a single massive exposure [33]." In addition, most of the high order chlorocarbon organic compounds are hazardous, toxic, carcinogenic, or combinations of these. Based on that opinion, it was decided that carbon tetrachloride would not be used at AT&T. During the time devoted to the identification of the hazardous agents, the aluminum etch process was shut down. An intense 4-week interval was experienced by engineering trying to respond to manufacturing's loss of shipments of product. Fortunately, an alternative chemistry was introduced to plasma etch aluminum that proved to be superior to the CCl_4 results. This chemistry used neat silicon tetrachloride, $SiCl_4$. A similar analysis was conducted for the $SiCl_4$ process without the identification of any carcinogenic chemicals. The list in Table 4 gives the results of the chemical analysis using $SiCl_4$ to plasma etch aluminum. Silicon tetrachloride is a hazardous material in its own right because of its corrosiveness and hence its safe handling cannot be minimized [34, 35, 36].

This is a good example of the solving of the risk-cost-safety equation. AT&T did not consider that any cost would allow the CCl_4 chemistry to

Table 4 A list of the components found in the same aluminum etch system as Figure 2 that uses $SiCl_4$ as the etch chemistry. The three major parts of the system show no evidence of any carcinogenic materials.

$SiCl_4$ RESIDUE ANALYSIS

CHAMBER: WHITE POWDER ANALYZED TO BE COMBINATIONS OF $AlCl_3$, $AlCl_3 \bullet 6H_2O$, Al_2O_3, SiO_2, $Al(ClO_3)_3 \bullet 6H_2O$
NO CHLOROCARBON COMPOUNDS DETECTED

COLD TRAP: HCl, $SiCl_4$, Al_2O_3, SiO_2, $AlCl_3$, $AlCl_3 \bullet 6H_2O$, $Al(ClO_3)_3 \bullet 6H_2O$

PUMP EXHAUST: $SiCl_4$, CF_4

reduce its risk and be able to be operated as a safe process. A substitute, $SiCl_4$, could be dealt with in a safe way to protect the workers and the environment. Other studies of the risk involved in the practice of plasma etch processing indicate good agreement with this study [37, 38].

VIII. Plasma Process System Configuration

Over the years a standard configurational design for the installation of plasma equipment with a high regard for safety has evolved in AT&T Bell Laboratories and Technology Systems at the Allentown, Pennsylvania, plant. This concept has spread to other AT&T facilities and has become a company standard.

The installation of a plasma process complex is viewed as the installation of three separate subsystems. Figure 4 shows these three parts. The reactor system that is purchased from the equipment supplier is placed in the clean room facility. The gas source is often remoted from the clean room as much as a few hundred feet and a different floor level and connected to the reactor by double walled delivery lines for hazardous materials. The pump complex is also remoted from the reactor center. The pumps are connected to the plasma system with welded 4-inch diameter stainless steel lines that

breathing air connected to breathing masks that totally isolate the wearer from the environment (if a toxic gas leak should occur). The pressure of the breathing air tank has a loud sounding alarm for low air pressure set high enough to offer a suitable safety margin of time to change breathing air tanks. A second SCBA system must be available. All hazardous gas cylinder maintenance is done with two people (the buddy system); one to do the work and the second to oversee the operation, to give assistance to the primary worker, and to warn away others from entering the area during the work period. It is important that the proper mask be used for this activity. The type mask that has a reserve supply of air as part of the mask is *not* the proper equipment (it should only be used to escape from a toxic contaminated area).

The cylinder is connected to the delivery line by using the proper CGA fitted pressure reduction regulator. Couplers or reducers are never used because the lowest number of couplings should be made to reduce the possibility of leaks. A normally-closed pneumatic valve is placed after the pressure reducing regulator since it isolates the high pressure from the low pressure components. This valve is operated by high pressure nitrogen that is controlled by an electrical valve outside the cabinet, interconnected to the system so that system failure will cause the pneumatic valve to be in its normal (off) state.

The use of a flow-restricting element and/or a check valve is a good practice. The flow-restricting unit will help guarantee that the pressure to the AFC does not suddenly increase and damage the unit, acting as a backup to the gas regulator. The check valve will allow flow in only one direction so that contamination from other chemicals is restricted to the proper manifold area. Many designs use an additional gas flow check valve at the gas mixing manifold to maintain the integrity of the individual gas lines. Since the flow and pressure are important parameters that should be kept below some maximum value, the inclusion of a rotameter and pressure gauge is good design. A cross-purge system should be considered for use in the system. Finally, just before the gas delivery line leaves the gas cabinet, there is a manual shut-off valve that can be used to isolate the gas cabinet from the rest of the system during major maintenance work.

The gas is delivered to the plasma reactor (with its AFC) through 1/4-inch or 3/8-inch seamless stainless steel tubing [40, 41]. No coupling is made between the gas cabinet and the reactor so that the possibility of leaks is minimal. For hazardous material delivery this tubing is surrounded by a tight (but not vacuum tight) fitted conduit that measures 8×4 inches in cross section. All lines for hazardous gases are placed in this conduit for delivery to the reactor. Inside the conduit, hazardous gas monitoring detectors, which will sense any leakage are placed at strategic locations. The

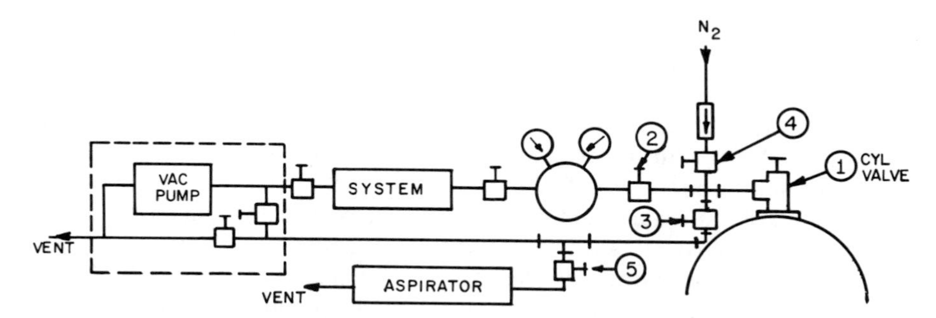

FIGURE 6. A schematic of a cross-purge system that is used to remove any hazardous material from the gas delivery line before the line is broken for cylinder replacement. The operation of this system is detailed in the text.

conduit is connected to the skins of the reactor enabling the gas cabinet exhaust to pull air from the clean room through the reactor skins along the conduit to the gas cabinet. With this flow (opposite to the hazardous gas flow in the small delivery line) any leak of hazardous material will be detected by the appropriately placed sensors, in the reactor, the delivery line, or the gas cabinet, and the location of the leak will quickly be pinpointed by the alarming sensor.

Nonhazardous gases are delivered to the reactor by similar tubing delivery system; however, they are not contained in the conduit. They are typically not contained in gas cabinets (because of the high cost of the cabinet), but have the tanks properly restrained and proper pressure reducing regulators are attached to the cylinders. Flow restrictors and/or check valves are used in these delivery lines.

The use of a cross-purge system will allow the removal of hazardous material from the delivery line in preparation for cylinder change or maintenance work. Figure 6 shows such a system and the operation of it will be illustrated with the use of numbered valves. To start the procedure the tank valve (#1) and the pneumatic isolation valve (#2) are closed. Valves #3 and #5 that are connected to an aspirator pump are opened to exhaust the line to ≈ 5 psia, then closed. Valve #4 is opened to allow the line to become over-pressured with nitrogen to ≈ 30 psia, then closed. Then the cycle is repeated, alternately under- and over-pressurizing the line. The utility of this approach is that according to the principle of dilution, the dilution factor is

$$\text{Dilution Factor} = \text{DF} = \left[\frac{\text{Lower Absolute Pressure}}{\text{Higher Absolute Pressure}}\right]^{\#\text{ of cycles}}.$$

For 8 cycles between 30 psia nitrogen and partial vacuum (5 psia by aspiration) the dilution factor is less than 1 ppm. Dilution by flowing purge gas for the equivalent time produces a dilution factor of 1000 ppm, at best. The operation of the valves can be made automatic so that a single switch can cycle the system under automatic control. This feature can easily be incorporated into the gas cabinet.

Since manifold systems are becoming popular for delivering gases to the reactors, care must be taken to prevent a contaminated gas cylinder from contaminating the entire gas delivery system [42, 43, 44]. Most of the compressed gases used in microelectronics research and production are not electronic grade chemicals (as illustrated above for BCl_3) and contain impurities. It is not yet obvious what effect the impurities have on device yield, but the effect may be considerable. Connecting a contaminated cylinder to a manifold system should lower the yield of every reactor on the system. The ability to certify the cylinder contents is a good insurance policy that protects the operating systems. The choice of the technique that will be used to identify the cylinder contents is the result of the risk-cost-safety consideration applied to this problem. Low cost approaches take the form of taking a simple spectrograph of the gas in the cylinder and comparing a standard spectrum for identification of differences. A costly, but detailed, approach analyzes the gas with the use of a MS-GC system to automatically identify the contents and concentration of the gas in the cylinder [45].

B. HAZARDOUS GAS PUMPING CONFIGURATION

The most important information that will determine the configuration of the pumping system for any gas, hazardous or not, is a thorough understanding of the chemistry of the process. The chemical characteristics of the feed gas(es) and the process effluents will determine the type of materials and pumps that can be used. With a real knowledge of the job that is asked of the pump system, the design of a compatible apparatus is easily done [46, 47]. Figure 7 shows a system schematic for a pumping package that has evolved over many years at AT&T.

To illustrate this principle, the rationale used to design pumping systems for the most hazardous, toxic, and/or corrosive chemicals will be detailed [48]. The old saw, "Do not thwart Mother Nature," has a valid message. The aluminum etch feed gas chemistry of BCl_3/Cl_2 represents the worst case of the pump's capability [49]. Understanding that the pump will be exposed to corrosives, sand and toxic agents will practically eliminate many

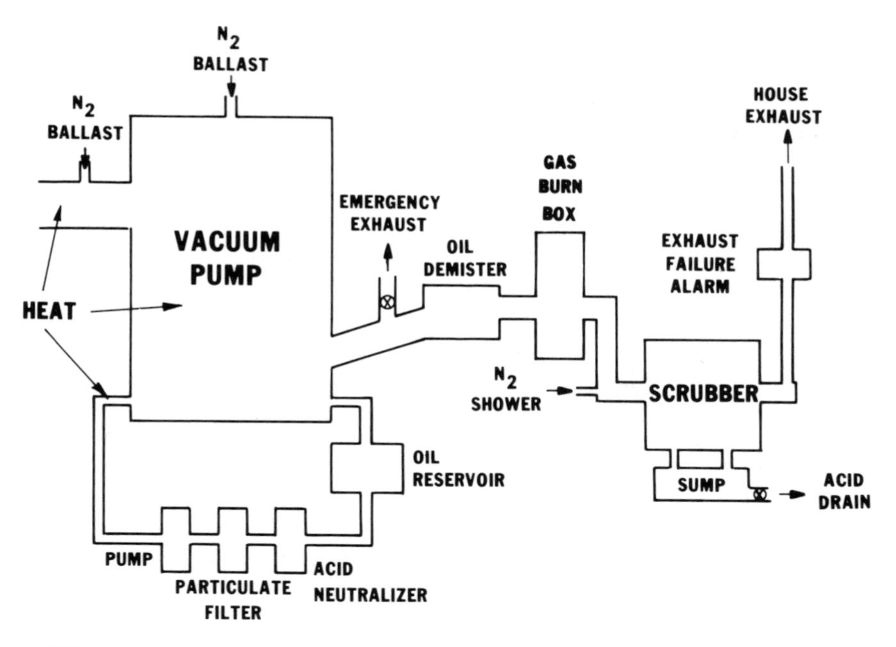

FIGURE 7. A schematic of the pump complex. Each of the components is detailed in the text.

possible choices of pump systems. The strong interaction of BCl_3 and water vapor produces HCl and B_2O_3, or acid and "sand."

There is only one type of mechanical pump that will tolerate sand dumped into its throat—the piston pump. The rotary vane pump depends on a close tolerance between the pump's moving parts. The result of putting sand, even small amounts, into the pump is the rapid abrasion of the vane, to the point of an early pump death. The piston pump, on the other hand, has large tolerances between the pump moving part (a few mils), allowing sand to be tolerated somewhat, but at least not condemning the pump to an early grave with small amounts of sand. The vacuum line connecting the pump to the reactor is leak tight to reduce the amount of moisture the BCl_3 can react with, and thus minimize the sand input to the pump. The oil medium that the pump uses has bearing on the type of pump [50]. The newer, synthetic oils are characterized by a significant decrease in their lubricity. The rotary vane pump, due to its close tolerance design, requires that the oil medium that is used be highly lubricating to the tightly fitted moving parts. The loose tolerance piston pump naturally couples better with the lower lubricity synthetic oils. The need for the newer synthetic pump oils is that the pump must be able to handle corrosive and acidic chemicals without substantial chemical reactivity [51, 52].

It has been determined that the interior parts of the pump, even in contact with corrosives and/or acids, do not react with the corrosives and/or acids if the internal parts are constantly bathed with the oil. For this main reason the pump is operated under a continuous load. The reactor gate valve is connected to a nitrogen valve that is positioned immediately in front of the pump throat. When the gate valve is closed, the nitrogen valve opens, allowing the nitrogen to flow (at a pressure of 100 mTorr) into the pump causing the pump to operate under a load so that the oil level in the pump is in constant agitation, coating the walls of the internal pump parts. When the gate valve is opened during pumpdown or device processing, the nitrogen flow ceases, but the pump continues to operate violently. There should be no period during which the pump oil is allowed to become calm. Nitrogen is also introduced into the pump head space so that a buildup of hazardous materials does not occur.

The pump is also operated as hot as possible. The cartridge heater that is contained in most sumps of piston pumps is used to heat the pump to at least 60°C. For larger pump sizes the temperature of the pump rises automatically to the needed temperature without the sump heater turned on. The combination of the violent pump operation and the high temperature of the oil promotes the gases that enter the pump throat to rapidly pass through the pump and exit the pump exhaust.

The needed pumping capacity of the system can be quickly determined in the standard way. As time goes on the pumping capacity of any system will decrease. Hence it is beneficial to initially oversize the pump to accommodate this normal attrition. In addition, when pumps are remote, the use of a Root's blower to aid the pumping capacity has been found to work well. The blower, even in spite of the fact that it is known as a close tolerance system, has no internal rapid digestion and abrasion of the working parts as in the rotary vane pump. A standard pump configuration that uses a 210 cfm blower mounted with a 97 cfm piston mechanical pump is being used in all installations at AT&T, for plasma and CVD applications (see Fig. 7).

The oil reservoir of the pump should be large, preferable 3–5 times the volume of the pump sump. The oil is circulated from this reservoir into a filtration stage(s) that removes the sand or other particulate matter from the oil. The pressure drop across this filter determines the frequency of filter element replacement. The spent oil filter cartridges are stored in a closed container where the oil drains and the filter is disposed of as hazardous waste. The use of an acid neutralizing medium can lower the acidity of the pump oil, particularly since for aluminum plasma etching a few Lewis

Acids (BCl_3, Al_2Cl_3) are naturally-occurring chemicals. It is our experience, however, that the acidity level rises at the startup of a pump system and levels off, not rising further, independent of whether the acid neutralizer is used or not.

Since the oil is being violently agitated, the release of considerable oil mist from the pump exhaust results. The loss of this amount of expensive oil cannot be tolerated. With the use of a commercial demister almost all the oil can be returned to the pump. The only oil that should be lost from the system is the oil taken along with the spent filter cartridge. There are pumping systems at the AT&T Allentown plant that have conserved the same oil charge for two and one-half years. A good case can be made for the cost savings that can be accrued with the use of the much more expensive-synthetic pump oil [53]. Given a cost factor of 10 between the mineral oil and synthetic oil, but a factor of 100 or more in the life of the two oils, the economic advantage can be startling.

The exhaust products are introduced to a burn box where a flame or glowing wire causes the completion of any reaction of the effluent gases. The flow continues into a scrubber system where the gases are brought into contact with water. Typically four pumps are connected to a single scrubber, while in turn the exhausts from 4–6 scrubbers are introduced into a second stage of scrubbing. The quality of the exhaust of the second scrubber stage and the sumps of every scrubber are evaluated often so that assurance is gained that only good quality air and water are sent to the environment (scrubber exhaust) or the house recycling water system (water from scrubber sumps) [54]. The exhausts of the scrubbers are armed with exhaust failure alarms that are interlocked to the main process center.

C. REACTOR CHAMBER / SYSTEM

The equipment that is purchased from the vendor is placed in the clean room facility (for production) or the research area. A schematic of the system is shown in Fig. 8. Little alteration of the system is attempted, but extensive installation must be done to connect the gas delivery (two-walled for hazardous chemicals) and pump systems to the reactor itself. The vendor does offer pumping packages that accompany the reactor, but we use our own standard design as stated above. The inclusion of hazardous gas detectors in the AFC portion of the reactor is necessary and easily done since most AFC subsystems from the vendor have enough space (usually all delivery lines in the reactor are contained within an exhausted enclosure) to accommodate the detector.

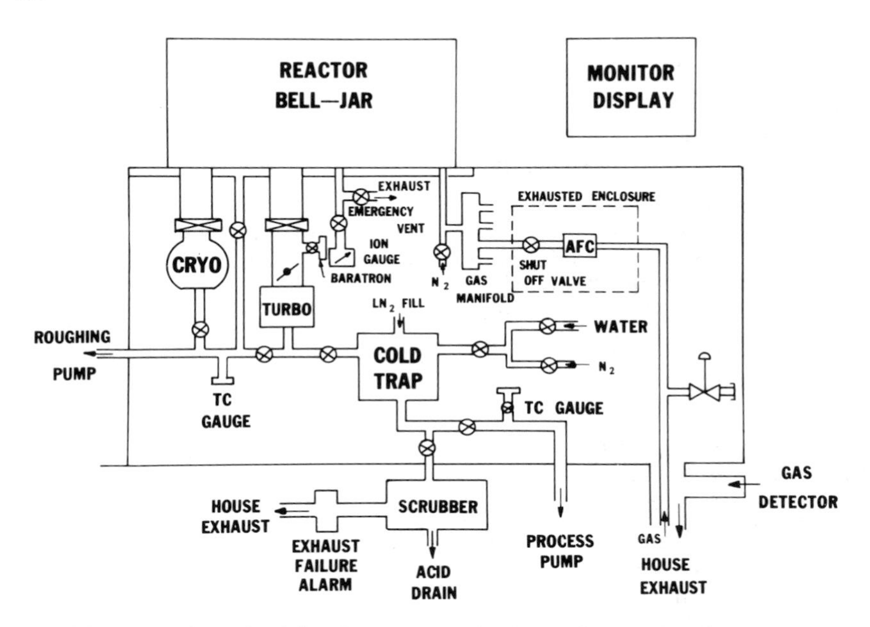

FIGURE 8. A schematic of the plasma reactor that is usually placed in the semiconductor clean room. Most of the components shown here are part of the system that is delivered from the equipment vendor. The scrubber, connections to the pump and gas source and monitor display are explained in the text.

There are a few engineering principles that have been established over the years at AT&T about the design of purchased reactors that determine the operation of the total system. With respect to the use of hazardous chemicals one of two fundamental choices can be made about handling them. One can use trapping methods (cold traps) to attempt to remove most, if not all, of the volatile hazardous materials from the reactor exhaust. The rationale stated for this approach is that protection of the pumping system is maintained. Periodically, the highly concentrated hazardous material must be removed from the traps, however, creating an extremely dangerous condition. Cold traps are notorious for unscheduled regurgitations, causing emergencies. Since the above designed pump complex accommodates the hazardous chemicals on a continuous, low level basis, the better choice is to eliminate entirely the use of traps from the system and not allow any concentration of hazardous chemicals to occur.

The aversion to traps is continued even to the elimination of cyropumps from the system. The cyrosystem does nothing more than concentrate material. The appeal that the cryopump is necessary to remove water vapor from the reactor chamber walls for better process control is mitigated by the (long) time necessary to pump on a chamber to remove a monolayer of water [55]. In addition, if water is the problem, running the reactor chamber hot so that minimal water can absorb on walls during open chamber periods when wafers are exchanged is a more acceptable way to conduct the process cycle than to lower process temperatures to gain reduced resist digestion rates. There is a controversy about what value the "blank-off" pressure needs to be for most vacuum processes. Common wisdom wants the lowest pressure possible to ensure the process itself is consistent and not influenced by impurities from walls. If the chamber is run hot, most, if not all, reaction products are not condensed on walls and leave the chamber through the pumping system that is designed to handle them. Heating of the pump line may be necessary to keep chamber effluents volatized. In addition, the feed gases themselves are not electronic grade and all the care taken before admitting them to the chamber by continued pumping to have a clean reactor may go for naught. A reasonable rule-of-thumb for the pre-process chamber pressure is that it need be no lower than one order of magnitude below the operating process pressure [53]. If this procedure does not allow good results, a thorough review of the process itself needs to be done.

If the process needs a pressure below 25–30 mTorr (even for blankoff), the attachement of a mechanical turbomolecular pump directly below the chamber baseplate and gate valve is recommended. Now the mechanical pump complex that is remote (blower and piston pump) becomes a foreline pump for the turbo. For the case of any plasma chemistry that uses BCl_3, two blower-piston pump packages are used for a single reactor. The two pump packages are labeled the "roughing pump" and the "foreline pump." In an attempt to not have moisture and BCl_3 mix and produce excessive sand (even though the pump is designed to accept it), the roughing pump does not see BCl_3 ever, while the foreline pump is protected from moisture. During the process cycle the two pumps are connected to, and disconnected from, the chamber automatically to insure that each pump "sees" only what it is supposed to. For those systems that use no BCl_3, only one blower-piston pump (with a turbo if required) is necessary.

All indications of the state of activity of interlocks, hazardous gas detectors, and normal process monitors are made at the process reactor, since that is the area most likely to have a worker present. The alarm shuts down the process and isolates the gas cylinder, but audible alarms designed to warn workers in the intermediate vicinity of the reactor for predetermined response (possible evacuation) are present. Alarms are also positioned at the remoted gas cabinets and pump rooms for the same response by anyone in those areas. A well-thought-out plan for handling such emergencies must be in place. Whoever goes into the alarmed area to evaluate the condition must wear the proper protective equipment. The "all-clear" condition must be properly determined so that workers are not put at jeopardy.

IX. Recommendations

It is important that a clear understanding of the required regulations that must be obeyed and the complete chemical makeup of the processes being operated is necessary to be able to assess risk and safety conditions. The continual encouragement of respect, rather than fear and confusion, by the work force will promote acceptance and commitment for safety programs instead of a psychosomatic rejection. A well-designed working safety program is to be promoted by corporate management, and continual training of the operators to update and reinforce good safety procedures so that emergencies are not catastrophic.

The design and installation of equipment and sub-systems should be done with safety foremost of the design criteria. The designs should be on a no-fail basis, but redundancy for systems handling toxic materials for emergencies is necessary. The selection of the materials used in the installation are often dictated by the chemical properties of the feed gases and/or liquids that will be part of the chemistry of the process. The correct volume of hazardous material (cylinder sizing) should be selected according to the frequency of cylinder change and the toxicity of the chemical. The use of the proper hazardous gas detectors is vital.

Consultation with local safety representatives for proper interpretation of the legal safety requirements is recommended. The California-DIR/DOSH 1981 Semiconductor Industry Report is a good source of information. The Semiconductor Safety Association is a group of concerned engineers that collect and disseminate information on serious accidents and incidents with high injury potential, and to promote an exchange of non-proprietary safety information within the electronics components industry. Publications from OSHA, EPA, NIOSH, and state and municipal agencies contain helpful information.

The job should never be considered finished until the safety considerations have been completed. Safety can and should come first. Remember, the life or health you protect may not be only your co-worker's, it is more than likely your own.

Appendix A: Comparison of Federal and State Occupational Safety and Health Requirements.

New Jersey Law	U.S. Rules (OSHA)
	Hazards
State determines which substances are hazardous. 2051 chemicals on current list.	Chemical makers decide which substances are hazardous based on federal criteria. 2300 chemicals listed as potentially dangerous.
All companies list hazards in their work places and the number of people exposed.	Only chemical makers list work place hazards.
All companies list hazards emitted to environment and amounts.	No requirement.
State writes fact sheet on each substance, giving safety data.	Chemical maker writes fact sheet and sends it with each shipment of hazardous product. All manufacturers that buy the product must retain the fact sheet.
	Disclosure
Companies send work place hazard list annually to state and county health departments and local fire and police. Public may see lists.	No requirement.
Companies send environment hazard list annually to state officials, county health departments, and "pertinent" data to local fire and police. Public may see lists.	No requirement.
Workers may see hazardous substances list and fact sheet.	Workers at chemical production plant may see list and fact sheets. Workers at plants buying chemical may see fact sheets supplied by producer.

Worker Training

Companies must educate and train workers on substances and safety precautions.	Same.

Labeling

Companies must label all containers of hazardous substances with the precise name and index.	Companies must identify chemical common name, pictures or symbols and an appropriate warning; storage areas may be labeled instead of each container. precise name and index number not required.
Pipelines, valves, vents, drains, etc. must be labeled.	No requirement.
Label all containers with non-hazardous substances within two years (1986).	No requirement.

Who Must Comply

All manufacturers, hospitals, wholesalers, clinics, garages, public employers, museums, gas and electric utilities, repair shops, schools, transportation services.	Chemical manufacturers, importers and distributors. All other manufacturers must retain substance fact sheet if they buy hazardous chemical. Non-manufacturing sector need not comply.

Trade Secrets

Companies may seek to protect disclosure of data on trade secrets, but state may reject claim; courts then decide. In emergency, state has right to demand disclosure of trade secrets.	Companies may withhold data in substances they consider trade secrets if they report the general properties and effects of the substances. Doctors can require disclosure in emergencies.
No trade secret protection for 835 special substances that may cause cancer, birth defects, or are explosive, caustic, flammable, etc.	No requirement.

Appendix B: Characteristics of Common Plasma Etch Feeds

Gas	Flammable Limits in Air (%)	Toxic	Corrosive	State of Cylinder Contents	Cylinder Pressure (PSIG@70°F)	Specific Volume (ft^3/lb)	Boiling Point (°C)	Freezing Point (°C)	Typical Impurities (ppm max)	PEL (ppm)	IDLH (ppm)	Comments
Acetylene-C_2H_2	2.5-81			dissolved gas	250	14.5	−74.8					
Ammonia-NH_3	15-28	yes	yes	liquified gas	114	22.6	−33.4	−77.7	H_2O-8 O_2-2 THC-1	50	500	pungent odor
Argon-Ar				gas	2000-2500	9.7	−185					
Arsine-AsH_3	pyrophoric	high		diluted gas	205	5.0	−62.5	−117		0.05	6	garlic-like odor
Boron Trichloride-BCl_3		yes*	yes	liquid	4.4	3.3	12.5	−107.5	Cl_2-40 Si-8 $COCl_2$-50	5	100	hydrolizes to form HCl
Carbon Dioxide-CO_2		low		liquified gas	225-830	8.76	subl	−78	N_2-20 O_2-4 CO-1 H_2-4 THC-5	5000	50,000	slightly acidic
Carbon Monoxide-CO	12.5-74	yes		gas	225-1500	13.8	−191		N_2-100 O_2-20 Ar-20 CO_2-10 H_2-10 THC-2	50	1500	
Carbon Tetrachloride-CCl_4		yes		liquid		76.5				10	300	human carcinogen, percutaneous absorption
Chlorine-Cl_2		yes	yes	liquified gas	85	5.4	−34.1	−150	CO_2-50 N_2-8 O_2-2 H_2O-3 THC-6	1	25	greenish-yellow gas; heavier than air; strong irritating effect on mucous membranes
Chlorodifluoromethane-$CHClF_2$ (Freon-22)		low		liquified gas	123	4.4	−40.5	−157		1000		
Chloroform-$CHCl_3$				liquid			61.7			50	1000	suspected human carcinogen
Chlorotrifluoroethylene-C_2ClF_3 (Genetron-1113)	8.4-39	yes		liquified gas	62	3.3	−28.2	−158		20		very reactive and must be stabilized with an inhibitor
Chlorotrifluoromethane-$CClF_3$ (Freon-13)		low		liquified gas	458	3.5	−81.2	−181		1000		

Gas	Flammable Limits in Air (%)	Toxic	Corrosive	State of Cylinder Contents	Cylinder Pressure (PSIG@70°F)	Specific Volume (ft^3/lb)	Boiling Point (°C)	Freezing Point (°C)	Typical Impurities (ppm max)	PEL (ppm)	IDLH (ppm)	Comments
Dichlorodifluoromethane-CCl_2F_2 (Freon-12)				liquified gas	70.2	3.13	−29.6	−158				
Dichlorofluoromethane-$CHCl_2F$ Freon-21)				liquified gas	8.4	3.5	9.1	−135				
1,2-Dichlorotetrafluoroethane-$C_2Cl_2F_4$ (Freon-114)				liquified gas	13	2.3	3.9	−93.8				
Dichlorosilane-SiH_2Cl_2	4.1-98.8	yes*	yes	liquified gas	9	3.83	8.4	−122	Fe-50 ppbw C-6.5 ppbw	5	100	suffocating odor, hydrolizes to form HCl and polymeric siloxanes
Ethane-C_2H_6	3-12.5			liquified gas	543	12.8	−88.5					
Ethylene-C_2H_4	3.1-32	low		gas	1200	13.8	−103.6			5000		sweet odor and taste
Fluoroform-CHF_3 (Freon-23)				liquified gas	635	5.5	−81.9	−155				
Helium-He				gas	225-1650	96.7	−270		Ar-0.1 N_2-0.1 O_2-0.1 Ne-0.1 H_2-0.1 CO_2-0.1 THC-0.1			
Hexafluoroethane-C_2F_6 (Freon-116)				gas	435	2.8	−78.2	−100.4	H_2O-10 HCl-10			
Hydrogen-H_2	4-75			gas	225-1650	191.7	−253		N_2-0.3 O_2-0.1 Ar-0.1 CO_2-0.1 CO-0.1 THC-0.1			lighter than air
Hydrogen Chlorode-HCl		yes	yes	liquified gas	613	10.6	−85		N_2-50 CO_2-15 CH_4-5	5	100	suffocating odor; hydrolizes in moist air
Hydrogen Fluoride-HF		yes	yes	liquid	0.9	19.3	19.7			3	20	
Hydrogen Sulfide-H_2S	4.3-46	yes		liquified gas	252	11.23	−60.2			20	300	odor of rotten eggs
Methane-CH_4	5-15.4			gas	1585	23.7	−161.3					sweet, oil-like odor
Nitric Oxide-NO		yes		gas	500-2000	13	−151			25	100	

Gas	Flammable Limits in Air (%)	Toxic	Corrosive	State of Cylinder Contents	Cylinder Pressure (PSIG@70°F)	Specific Volume (ft^3/lb)	Boiling Point (°C)	Freezing Point (°C)	Typical Impurities (ppm max)	PEL (ppm)	IDLH (ppm)	Comments
Nitrogen-N_2				gas	225-1650	13.8	−196		O_2-1 H_2-1 Ar-1 He-1 CO_2-0.5 THC-0.5			
Nitrogen Dioxide-NO_2				liquified gas	745	8.7	−88.3		N_2-8 O_2-2 H_2O-3 CO_2-2 THC-1			oxidizer at elevated temperature and pressure
Nitrogen Trifluoride-NF_3		yes		gas	1000	5.4	−128.9			10	2000	moldy odor
Oxygen-O_2				gas	225-1650	12.1	−183		N_2-20 Ar-20 THC-20 CO_2-10			supports combustion
Phosphine-PH_3	pyrophoric	high		liquified gas	594	11.4	−88		As-2	0.3	200	odor of decaying fish
Propane-C_3H_8	2.2-9.5			liquified gas	109	8.5	−42			1000	20,000	natural gas odor
Silane-SiH_4	pyrophoric; detonatable	low		gas	150-1700	12.1	−111.4	−185				often shipped as mixture with H_2, Ar, He or N_2; repulsive odor
Silicon Tetrachloride-$SiCl_4$		yes*	yes	liquid			57.6	−70	B-0.3 ppbw P-0.3 ppbw As-0.5 ppbw	5	100	hydrolizes rapidly releasing HCl
Silicon Tetrafluoride-SiF_4				gas	500-1500	3.7	−65	−187	Air-0.35% SO_2-0.05% As-0.5 ppbw			hydrolizes rapidly to form fluosilicic acid and HF
Sulfur Hexafluoride-SF_6				liquified gas	320	2.5	−50.6					
Tetrafluoromethane-CF_4 (Freon-14)				gas	500-2000	4.4	−127.9	−186.6	Air-1.5% H_2O-15			
Trichlorofluoromethane-CCl_3F (Freon-11)				liquid	13.4	2.9	24	−111		1000	10,000	water-white liquid
1,1,2-Trichlorotrifluoroethane-$C_2Cl_3F_3$ (Freon-113)				liquid			47.7	−34.8		1000	4500	water-white liquid; also known as TTE

* PEL and IDLH values are based on HCl, the hydrolized product

References

[1] Occupational Safety and Health Administration, "General Industry Standards, 29 CFR 1910." Superintendant of Documents, U.S. Government Printing Office, Washington, District of Columbia.

[2] D. Peterson, *The OSHA Compliance Manual*. McGraw-Hill Book Company, New York, New York.

[3] *U.S. Environmental Protection Agency Guidebook*. Government Institutes, (1983).

[4] *Environmental Engineer's Handbook*. Vol. 1, "Water Pollution," Chilton (1974).

[5] G. Dominquez, ed., *Guidebook: Toxic Substances Control Act*. CRC Press, Inc., Cleveland, Ohio.

[6] P. Powers, *How to Dispose of Toxic Substances and Industrial Waste*. Noyes Data Corporation, Park Ridge, New Jersey.

[7] *Hazardous Waste and Consolidation Permit Regulations*. Environmental Protection Agency, Cincinnati, Ohio.

[8] U.S. Department of Transportation, "Hazardous Materials Regulations, 49 CFR, Parts 170–179." Superintendant of Documents, U.S. Government Printing Office, Washington, District of Columbia.

[9] National Institute for Occupational Safety and Health, *Occupational Health Guidelines for Chemical Hazards*. NIOSH, Government Printing Office (1981).

[10] Task Force on the Electronics Industry, *Semiconductor Industrial Study—1981*. State of California, Department of Industrial Relations, Division of Occupational Safety and Health, Sacramento, California.

[11] Semiconductor Equipment and Materials Institute, *SEMI Handbook*, SEMI, Mountain View, California (1983).

[12] Training Manual, *Hazardous Chemical Safety*. J. T. Baker Chemical Co., Phillipsburg, New Jersey.

[13] U.S. Government, *Job Safety and Health*. U.S. Government Printing Office, Washington, District of Columbia.

[14] J. Olishifski, *Fundamentals of Industrial Hygiene*. National Safety Council, Chicago, Illinois.

[15] Manufacturing Chemists Association, *Guide for Safety in the Chemical Laboratory*. Van Nostrand Reinhold Co., New York.

[16] Chemical Manufacturers' Association, *Chemical Safety Data Sheets*. CMA, Washington, District of Columbia.

[17] *American National Standard for the Precaution of Hazardous Industrial Chemicals*. ANSI Z129.1–1976, American National Standards Institute, New York.

[18] *Handbook of Compressed Gases*. Compressed Gas Association, Inc., New York.

[19] *Matheson Gas Data Book*. Matheson Gas Products, East Rutherford, New Jersey.

[20] International Technical Information Institute, *Toxic and Hazardous Industrial Chemicals Safety Manual*. Lab Safety Supply, Janesville, Wisconsin.

[21] N. Sax, *Dangerous Properties of Industrial Materials*. Van Nostrand Reinhold Company, New York.

[22] American Society of Safety Engineers, *Professional Safety*. ASSE, Park Ridge, Illinois.

[23] *Effects of Exposure to Toxic Gases*. Matheson Gas Products, East Rutherford, New Jersey.

[24] American Conference of Government Hygienists, *Threshold Limit Values for Chemical Substances and Physical Agents in the Work Environment*. ACGIH, Cincinnati, Ohio.

[25] D. Walters, ed., *Safe Handling of Chemical Carcinogens, Mutagens, Teratogens and Highly Toxic Substances*. Ann Arbor Science Publishers, Inc., Ann Arbor, Michigan.

[26] N. Sax, *Cancer Causing Chemicals*. Van Nostrand Reinhold Company, New York.

[27] J. H. Meidl, *Hazardous Materials Handbook*, Glencoe, (1972).

[28] Drager Detector Handbook, *Air Investigations and Technical Gas Analysis with Drager Tubes*. 3rd ed., (1976) (with 1981 certification update).

[29] G. K. Herb, et al., "Plasma Processing: Some Safety, Health and Engineering Considerations." *Solid State Technology* **26**, 8, 185 (1983).

[30] U.N. International Agency for Research on Cancer (IARC), *IARC Monograph—Carbon Tetrachloride*. WHO Publications Center USA, Albany, New York.

[31] National Institute of Occupational Safety and Health, *Carcinogens—Regulations and Control*. NIOSH, Division of Technical Services, Cincinnati, Ohio.

[32] National Cancer Institute, *Bioessay of Hexachloroethane for Possible Carcinogenicity*. Technical Report Series No. 68, Bethesda, Maryland.

[33] T. Loomis, "Essentials of Toxicology," Lea & Febiger, Philadelphia, Pennsylvania.

[34] M. Steere and E. B. Sansone, *Handbook of Laboratory Safety*. CRC Press, Cleveland, Ohio.

[35] *Best's Safety Directory*. A. M. Best Company, Inc., Oldwick, New Jersey.

[36] Gosselin, *Clinical Toxicology of Commercial Products*. Williams and Wilkins, Baltimore, Maryland.

[37] David G. Baldwin, *Chemical Exposure from Carbon Tetrachloride Plasma Aluminum Etchers*. Abstract No. 296, Electrochemical Society Fall Meeting, Las Vegas, Nevada (1985).

[38] Jean T. Ohlsen, "Dry Etch Chemical Safety," TEGAL Plasma Seminar, Palo Alto, California (1985).

[39] *Safety Precautions and Emergency Procedures for Speciality Gases*. A publication of the Linde Division, Union Carbide Corporation, National Specialty Gases Office, Somerset, New Jersey.

[40] L. Bretherick, *Handbook of Reactive Chemical Hazards*. Butterworth, Inc., Woburn, Massachusetts.

[41] Robert Zawierucha, "High Purity Materials for Gas Distribution Systems." *Microelectronic Manufacturing & Testing*, **16** (1986).

[42] Phil Danielson, "Gas Analysis and Process Technology." *Microelectronic Manufacturing & Testing*, **61** (1986).

[43] David Lowies, "How to Keep Process Gas Clean By Reducing Contamination." *Microelectronic Manufacturing & Testing*, **49** (1986).

[44] George B. Bunyard, R. E. Pecsar, and Kuo-chin Lin, "Improved Product Yield Through Gas Purity Assurance." *Semiconductor International* **101** (1983).

[45] Robert E. Dunkel, "Gas Risk Management—A Safer Approach to Monitoring for Hazardous Gases." *Microelectronic Manufacturing & Testing* **24** (1986).

[46] D. Van der Linden, "Vacuum Systems: A Review of the Basic Components," *Industrial Research and Development* **131** (1983).

[47] D. B. Fraser and J. F. O'Hanlon, "Pumping Hazardous Gases", an AVS Short Course, c/o American Institute of Physics, Woodbury, New York.

[48] Peter Connock, "Corrosion Resistance in Mechanical Pumps for Semiconductor Processes," *Industrial Research and Development* **131** (1981).

[49] Christine B. Whitman, "Vacuum Pump Fluids: Choosing the Right Ones," *Industrial Research and Development* **157** (1982).

[50] J. F. O'Hanlon, "Mechanical Pump Fluids for Plasma Deposition and Etching Systems," Research Report No. 8910, IBM Research Division, Yorktown Heights, New York.

[51] R. A. DuBoisson and S. F. Sellers, "The Resistance of Perfluoropolyether Fluids to Lewis Acid Attack." *Microelectronic Manufacturing & Testing* **85** (1986).

[52] R. A. Boisson and S. F. Sellers, "Some Subtle But Significant Differences Among Perfluoropolyether Vacuum Pump Fluids," *Microelectronic Manufacturing & Testing*, **19** (1986).

[53] Pieter S. Burggraaf, "Success With Handling Hazardous Gases." *Semiconductor International* **55** (1982).

[54] Peter H. Singer, "Services for Monitoring Hazardous Conditions." *Semiconductor International* **83** (1982).

[55] V. J. McClurg, "Use Per Cent Nigrogen to Monitor, Diagnose Systems," *Industrial Research and Development* **147** (1980).

Index

A

B

C

D

M

N

O

P

Q

R

V

X, Z

Plasma–Materials Interactions

Dennis M. Manos and Daniel L. Flamm, *Plasma Etching: An Introduction*

Orlando Auciello and Daniel L. Flamm, *Plasma Diagnostics: Volume 1, Discharge Parameters and Chemistry; Volume 2, Surface Analysis and Interactions*

D'Agostino, R., *Plasma Deposition: Treatment and Etching of Polymers*